B

ISNM 91:
International Series of Numerical Mathematics
Internationale Schriftenreihe zur Numerischen Mathematik
Série internationale d'Analyse numérique
Vol. 91

Edited by
K.-H. Hoffmann, Augsburg; H. D. Mittelmann, Tempe;
J. Todd, Pasadena

Birkhäuser Verlag
Basel · Boston · Berlin

Control and Estimation of Distributed Parameter Systems

4th International Conference on Control
of Distributed Parameter Systems,
Vorau, July 10–16, 1988

Edited by

F. Kappel
K. Kunisch
W. Schappacher

1989

Birkhäuser Verlag
Basel · Boston · Berlin

Editors

F. Kappel und
W. Schappacher
Institut für Mathematik
Karl-Franzens-Universität Graz
Elisabethenstrasse 16
A–8010 Graz

K. Kunisch
Institut für Mathematik
Technische Universität Graz
Kopernikusgasse 24
A–8010 Graz

CIP-Titelaufnahme der Deutschen Bibliothek

Control and estimation of distributed parameter systems / 4.
Internat. Conference on Control of Distributed Parameter
Systems, Vorau, July 10–16, 1988. Ed. by F. Kappel ... – Basel ;
Boston ; Berlin : Birkhäuser, 1989
    (International series of numerical mathematics ; Vol. 91)
    ISBN 3-7643-2345-0 (Basel ...) Gb.
    ISBN 0-8176-2345-0 (Boston) Gb.
NE: Kappel, Franz [Hrsg.]; International Conference on Control of
    Distributed Parameter Systems < 04, 1988, Vorau>; GT

© 1989 Birkhäuser Verlag Basel
Printed in Germany on acid-free paper
ISBN 3-7643-2345-0
ISBN 0-8176-2345-0

Preface

About 50 participants from 14 different countries attended the "4th International Conference on Control of Distributed Parameter Systems" held at the Chorherrenstift Vorau (Styria), July 10 - 16, 1988. The main purpose of this conference was to provide up-to-date information on important directions of research in the field of control theory for infinite dimensional systems. The contributions contained in this volume cover a variety of areas which are presently subject to intensive research efforts. Since most of the papers are concerned with more than one aspect of control theory, it is by no means an easy task to give a one-to-one correspondence between these papers and a catologue of different areas in control theory. The following attempt certainly reflects this difficulty:

Wellposedness and Representation. In his contribution G. WEISS gives conditions which guarantee that an abstract linear system (defined axiomatically, guided among other considerations by finite dimensional linear systems theory) can be represented by a controlled linear abstract differential equation and an algebraic equation describing the output. Related questions are studied in the paper by R.F. CURTAIN and G. WEISS using transfer function techniques. These contributions are partially in the spirit of earlier work by D. Salamon and aim at a unified theory for abstract linear systems in infinite dimensional spaces allowing for unbounded input and output operators. R.H. FABIANO and K. ITO study integro-partial differential equations with strongly and weakly singular kernels arising in the theory of linear viscoelasticity. They derive well-posedness results in the sense of generation of a contraction semigroup through the flow of the differential equation. If the kernels are weakly singular and the delay is finite the semigroup is differentiable.

Aspects of Controllability. J.A. BURNS and G.H. PEICHL consider the preservation of (approximate) controllability of the functional state of a controlled delay differential equation under approximation of the state by Galerkin-type approximations. Controllability and stabilizability measures are introduced and are calculated for specific examples. Controllability of viscoelastic Kirchhoff plate models is studied by I. LASIECKA using a "direct" method which is in an appropriate sense a generalization of the use of the controllability Grammian for finite dimensional systems to infinite dimensional systems. An alternative for the study of controllability to the direct method is the Hilbert uniqueness method (HUM) due to J.L. Lions. It was used by G. LEUGERING in his work on reachability of the states of a mathematical model for the slewing maneuver of a viscoelastic beam.

Optimal Control. H.O. FATTORINI in his contribution considers Pontriagin's maximum principle for constrained optimal control problems governed by a semilinear state equation the linear part of which is characterized by an analytic semigroup. Among the problems addressed are the following: existence of the Hamiltonian, constancy of the Hamiltonian along controls (and corresponding trajectories) which satisfy the maximum principle. An infinite horizon optimal control problem for semilinear systems is studied by F. GOZZI. He shows that the associated Hamilton Jacobi equation has a unique strict solution, provided that the nonlinearity is sufficiently small. A class of moment problems that arises from minimum norm control problems is treated in the contribution by W. KRABS. D. and M. TIBA derive uniform estimates for the violation of pointwise constraints in the approximation of constrained optimal control problems using a variational inequality method. In numerical examples this method is compared to the classical method of incorporating the constraints by means of penalty terms. The final two papers in this category are devoted to the linear quadratic regulator problem. K. ITO and H.T. TRAN develop an abstract approximation framework for the numerical treatement of the Riccati operators for a class of linear infinite-dimensional systems with unbounded in- and output operators. The results are applied to functional differential equations. I.G. ROSEN studies Galerkin type approximations of the algebraic Riccati operator equations in the Hilbert-Schmidt norm, for a class of linear quadratic regulator problems which are described by state equations of parabolic type.

Stabilization. Extending earlier work of Datko, Lagnese and Polis, W. DESCH and R.L. WHEELER derive an abstract framework for the description of the destabilization effect due to time delays in the feedback loop of elastic systems. In specific cases these delays may in fact leed to loss of well-posedness of the system. K.B. HANNSGEN studies stabilization of a viscoelastic Timoshenko beam model via feedback through the boundary. Both these contributions are based on transfer function techniques. In his paper, J.E. LAGNESE derives energy estimates for Kirchhoff plate models with viscoelastic damping. Again, feedback damping is activated through the boundary of the physical domain. T. NAMBU studies the structure of the Ljapunov equation arising in stabilization of self-adjoint parabolic systems in terms of the spectrum of the governing elliptic differential operator. R. TRIGGIANI obtains uniform exponential decay estimates (and wellposedness) for the Euler Bernoulli beam for various boundary feedback configurations.

For linear infinite dimensional systems the asymptotic behavior is determined by the location of the spectrum if only certain conditions are met by the governing equation. This is also the case if the input-output behavior of the infinite dimensional system is known. Y. YAMAMOTO provides a class of inpulse response matrices for which internal stability agrees with external L^2-input/output stability and for which internal exponential stability is determined by the location of the spectrum. H. ZWART finally compares the concepts of open and closed loop stabilizability.

Sensitivity and Shape design. In his contribution K. MALANOWSKI establishes weak directional differentiability of the solution to a class of convex constrained optimization problems in Hilbert space. The directional derivative is shown to be the solution of a quadratic optimization problem, and the general results are applied to contrained linear quadratic optimal control problems. J. SIMON characterizes the second variation of the state and of the cost functional with respect to the domain of the governing partial differential equation in optimal design problems. The remaining two papers in this group also address aspects of optimal shape design. R. MÄKINEN describes a numerical approximation procedure for state constrained optimal shape design problems. The constraint is incorporated by means of penalization. P. NEITAANMÄKI and T.I. SEIDMAN consider a free boundary control problem arising in modelling of crystal growth. Optimality conditions and regularity of the minimizer are obtained, and numerical examples are given.

Identification. H.T. BANKS, C.K. LO, S. REICH and I.G. ROSEN give a semigroup theoretic approximation framework for nonlinear parameter estimation problems. Parameter estimation problems in single and multiphase flow problems are described in the paper by R.E. EWING and T. LIN. For the single phase flow problems a space-marching scheme employing hyperbolic regularization is developed to estimate the unknown diffusion coefficient from overspecified boundary data. In a companion paper, T. LIN and R.E. EWING study the identifiability of the source term in one dimensional parabolic equations, and for the constant coefficient case uniqueness and aspects of illposedness of the inverse problem are investigated by means of a closed form solution. A. FRIEDMAN and M. VOGELIUS in their contribution derive a-priori estimates on the (parameterized) unknown conductivity coefficient from boundary measurements of the static voltage potential satisfying an elliptic equation. Problems of this kind arise for instance in nondestructive testing.

Z. BARTOSIEWICZ investigates the power of ringed spaces as state spaces for linear infinite dimensional systems. In the work of P.C. SABATIER a Green's function for a chain of Helmholtz equations is constructed and used to write a chain of Schrödinger equations as integral equations.

The contributions to this volume distinctively reflect certain trends in that part of control theory which was the theme of the conference. For instance we see considerable efforts to develop abstract state space methods for various problems in control theory for infinite dimensional systems, with special emphasis on unbounded input and/or output operators. In this context one should also draw attention to the increasing importance of transfer function techniques. A further development in recent years is towards more realistic models in continuum mechanics. There are more and more papers on control problems for beams and plates with various damping mechanisms as well as on control problems involving visco-elastic materials. Of course, numerical methods play a central role when it comes to applications. Problems in control theory for infinite dimensional systems require algorithms which preserve certain qualitive properties of the original problem (as for instance controllability) for the approximating problems. One can observe an increasing number of publications addressing these questions.

The success of the conference depends first of all on the participants, and here our thanks go to all of them for their contributions, but equally important is a stimulating environment. This certainly was provided by the Bildungshaus Chorherrenstift Vorau. We thank the staff and especially Direktor P. Riegler for all their efforts which made the stay at Vorau so pleasant.

Organization of the conference was made possible by grants from the Amt der Steiermärkischen Landesregierung, Graz, Bundesministerium für Wissenschaft und Forschung, Vienna and the "US Air Force EOARD" and the "US Army European Research Office", London. We greatly appreciate the financial support rendered by those agencies.

Finally, special thanks and appreciation to Mrs. G. Krois for her help in all administrational matters and for her excellent and careful preparation of the manuscript for these proceedings.

Graz, August 1989

F. Kappel, K. Kunisch, W. Schappacher

List of Participants

H.T. BANKS	(Brown University)
Z. BARTOSIEWICZ	(Politechnika Białostocka)
M. BROKATE	(Universität Kaiserslautern)
J.A. BURNS	(Virginia Polytechnic Institute and State University)
F. COLONIUS	(Universität Augsburg)
R.F. CURTAIN	(University of Groningen)
W. DESCH	(Universität Graz)
R.E. EWING	(University of Wyoming)
H.O. FATTORINI	(University of California)
J. GOLDSTEIN	(Tulane University)
F. GOZZI	(Scuola Normale Superiore Pisa)
K.B. HANNSGEN	(Virginia Polytechnic Institute and State University)
K. ITO	(Brown University)
M. JAKSCHE	(Technische Universität Graz)
F. KAPPEL	(Universität Graz)
A. KHAPALOV	(IIASA)
R. KLUGE	(Akademie der Wissenschaften der DDR)
W. KRABS	(Technische Hochschule Darmstadt)
M. KROLLER	(Universität Graz)
K. KUNISCH	(Technische Universität Graz)
A. KURZHANSKI	(IIASA)
J.E. LAGNESE	(Georgetown University)
I. LASIECKA	(University of Virginia)
G. LEUGERING	(Technische Hochschule Darmstadt)
T. LIN	(University of Wyoming)
K. MALANOWSKI	(Polish Academy of Sciences)
R. MÄKINEN	(University of Jyväskylä)
M. MÜLLER	(Technische Universität Graz)
G.H. PEICHL	(Universität Graz)

W. PRAGER   (Universität Graz)

G. PROPST   (Universität Graz)

J. PRÜSS   (Universität Paderborn)

V.G. ROMANOV   (Sibirean Branch of the Academy)

I.G. ROSEN   (University of Southern California)

P.C. SABATIER   (U. S. T. L.)

E. SACHS   (Universität Trier)

W. SCHAPPACHER   (Universität Graz)

T.I. SEIDMAN   (University of Maryland Baltimore County)

J. SIMON   (Université Blaise Pascal)

D. TIBA   (INCREST)

H.T. TRAN   (Brown University)

R. TRIGGIANI   (University of Virginia)

M. VOGELIUS   (University of Maryland)

G. WEISS   (The Weizmann Institute)

R.L. WHEELER   (Virginia Polytechnic Institute and State University)

A. WYLER   (ETH Zürich)

Y. YAMAMOTO   (Kyoto University)

H. ZWART   (University of Twente)

Contents

International Series of
Numerical Mathematics, Vol. 91
© 1989 Birkhäuser Verlag Basel

Numerical studies of identification in nonlinear distributed parameter systems[1]

H.T. Banks [2,5], C.K. Lo [3], Simeon Reich [4], and I.G. Rosen[3,5]

Center for Control Sciences
Division of Applied Mathematics
Brown University

Department of Mathematics,
University of Southern California

Department of Mathematics
University of Southern California
and
Department of Mathematics
The Technion-Israel Institute of Technology

Department of Mathematics
University of Southern California

Abstract. An abstract approximation framework and convergence theory for the identification of first and second order nonlinear distributed parameter systems developed previously by the authors and reported on in detail elsewhere are summarized and discussed. The theory is based upon results for systems whose dynamics can be described by monotone operators in Hilbert space and an abstract approximation theorem for the resulting nonlinear evolution systems. The application of the theory together with numerical evidence demonstrating the feasibility of the general approach are discussed in the context of the identification of a first order quasi-linear parabolic model for one dimensional heat conduction/mass transport and the identification of a nonlinear dissipation mechanism (i.e. damping) in a second order one dimensional wave equation. Computational and implementational considerations, in particular with regard to supercomputing, are addressed.

Keywords. Parameter identification, nonlinear evolution equation, Galerkin approximation, nonlinear heat conduction, nonlinear damping.

1980 *Mathematics subject classifications*: 65J15, 65N30, 93B30

1 A portion of this research was carried out with computational resources made available through a grant from the San Diego Supercomputer Center operated for the National Science Foundation by General Atomics, San Diego, Ca.

2 This research was supported in part under grants NSF MCS-8504316, NASA NAG-1-517, AFOSR-F49620-86-C-0111.

3 This research was supported in part under grant AFOSR-87-0356.

4 This research was supported in part by the Fund for the Promotion of Research at the Technion and by the Technion VPR Fund.

5 Part of this research was carried out while these authors were visiting scientists at the Institute for Computer Applications in Science and Engineering (ICASE) at the NASA Langley Research Center, Hampton, VA, which is operated under NASA contract NAS1-18107.

1. Introduction

In this paper we report on the results of our efforts in the area of approximation for the identification of nonlinear distributed parameter systems. The central focus of the present paper is our numerical studies. However, we shall also provide a brief outline, summary, and discussion of the essential features of the underlying theory. The theoretical basis (i.e. convergence analysis) for the computational results to be presented below has been treated in detail in two of our earlier papers, [3] and [5]. In [3] and [5] we developed nonlinear analogs of the general abstract approximation framework for the identification of linear systems given by Banks and Ito in [2]. Inverse problems for first order nonlinear evolution equations are handled in [3], while [5] is concerned with second order systems. We note that in another earlier paper, [4], we have developed an approximation framework for the estimation of parameters in nonautonomous nonlinear distributed systems. Although the general approach taken in [4] differs somewhat from the ideas in [3] and [5], and from those to be discussed here, the results presented there are certainly related, and at present, remain our only means of dealing with either linear or nonlinear nonautonomous systems.

In the next section we briefly review some abstract functional analytic existence, uniqueness, regularity, and approximation results for nonlinear evolution equations in Banach spaces. In section 3 we consider first order systems, define the class of inverse problems and evolution systems with which we shall be dealing and sketch the relevant approximation and convergence theory. We then consider an example involving a quasi-linear model for heat conduction and present the results (including those aspects related to supercomputing) of our computational study and numerical investigations. In the fourth section we treat inverse problems for second order systems. In particular we consider the estimation or identification of nonlinear damping or dissipation mechanisms in distributed parameter models for mechanical systems. A brief fifth section contains some concluding remarks.

2. Nonlinear evolution equations in Banach spaces-existence, uniqueness, and approximation results

We consider quasi-autonomous, in general nonlinear, initial value problems of the form

$$(2.1) \qquad \dot{x}(t) + Ax(t) \ni f(t), \quad 0 < t \le T,$$

$$(2.2) \qquad x(0) = x^0$$

set in a Banach space X with norm $|\cdot|_X$. In (2.1), (2.2) above we assume that $T > 0, x^0 \in \overline{Dom(A)} = \overline{\{x \in X : Ax \ne \phi\}}, f \in L_1(0, T; X)$, and that for some $\omega \in \mathbf{R}$ the operator $A + \omega I : X \to 2^X$ is m-accretive. In other words, that for some $\omega \in \mathbf{R}$ we have (i) $|x_1 - x_2|_X \le |(1 + \lambda\omega)(x_1 - x_2) + \lambda(y_1 - y_2)|_X$ for every $x_1, x_2 \in Dom(A), y_1 \in Ax_1, y_2 \in Ax_2$, and $\lambda > 0$, and (ii) $\mathcal{R}(I + \lambda(A + \omega I)) \equiv \bigcup_{x \in Dom(A)} (I + \lambda(A + \omega I))x = X$ for some $\lambda > 0$. We note that $A + \omega I$ m-accretive implies that for each $\lambda > 0$ the resolvent of $A + \omega I$ at λ, $J(\lambda; A + \omega I) : X \to X$, a single valued, everywhere defined, nonexpansive, nonlinear operator on X, can be defined by $J(\lambda; A + \omega I) = (I + \lambda(A + \omega I))^{-1}$.

A nonlinear evolution system on a subset $\Omega \subset X$ is a two parameter family of nonlinear operators, $\{U(t, s) : 0 \le s \le t \le T\}$, on Ω satisfying $U(t, s)\varphi \in \Omega, U(s, s)\varphi = \varphi$ and

$U(t,s)U(s,r)\varphi = U(t,r)\varphi$ for every $\varphi \in \Omega$ and $0 \leq r \leq s \leq t \leq T$ with the mapping $(s,t) \rightarrow U(t,s)\varphi$ continuous from the triangle $\Delta = \{(s,t) : 0 \leq s \leq t \leq T\} \subset \mathbf{R}^2$ into X for each $\varphi \in \Omega$. A strongly continuous function $x : [0,T] \rightarrow X$ is said to be a strong solution to the initial value problem (2.1), (2.2) if $x(0) = x^0$, it is absolutely continuous on compact subintervals of $(0,T)$, differentiable almost everywhere, and satisfies $f(t) - \dot{x}(t) \in Ax(t)$ for almost every $t \in (0,T)$.

It can be shown (see [7], [9], and [3]) that under the assumptions on A, f, and x^0 made above, a unique nonlinear evolution system $\{U(t,s) : 0 \leq s \leq t \leq T\}$, on $\overline{Dom(A)}$ with the following properties can be constructed:

(i) $|U(t,s)\varphi - U(t,s)\psi|_X \leq e^{\omega(t-s)}|\varphi - \psi|_X$, for all $\varphi, \psi \in \overline{Dom(A)}$ and $0 \leq s \leq t \leq T$;

(ii) $|U(s+t,s)\varphi - U(r+t,r)\varphi|_X \leq 2\int_0^t e^{\omega(t-\tau)}|f(\tau+s) - f(\tau+r)|_X \, d\tau$, for all $\varphi \in \overline{Dom(A)}$ and all $t > 0$ such that $s + t, r + t \leq T$;

(iii) If the initial value problem (2.1), (2.3) has a strong solution x, then $x(t) = U(t,s)x(s)$, for $0 \leq s \leq t \leq T$.

The strongly continuous function x given by $x(t) = U(t,0)x^0, t \in [0,T]$, is referred to as the unique mild, generalized, or integral (see [6]) solution to (2.1), (2.2). It is immediately clear from (iii) above that when the initial value problem (2.1), (2.2) admits a strong solution, this strong solution and the mild solution coincide.

The abstract approximation result which will play a fundamental role in the discussions to follow, is given in Theorem 2.1 below. The theorem we state here is similar in spirit to other related approximation results for nonlinear evolution equations - in particular, those that can be found in [8] and [10]. The proof of Theorem 2.1 is given in [3]. In the statement of Theorem 2.1 and elsewhere, for sets $S_n, n = 0, 1, 2, \ldots$ we use the notation $\lim_{n\to\infty} S_n \supset S_0$ to mean that given any $s_0 \in S_0$, there exists $s_n \in S_n, n = 1, 2, \ldots$ for which $\lim_{n\to\infty} s_n = s_0$.

Theorem 2.1. *For each* $n \in \mathbf{Z}^+ = \{1, 2, 3, \ldots\}$ *let* X_n *be a closed linear subspace of X, let* $f_n \in L_1(0,T;X_n)$, *and let* $A_n : X_n \rightarrow 2^{X_n}$ *be an operator on* X_n *with* $A_n + \omega I$ *m-accretive. Suppose that there exists a function* $g \in L_1(0,T)$ *for which* $|f_n(t)|_X \leq g(t)$, *and that* $\lim_{n\to\infty} \overline{Dom(A_n)} \supset \overline{Dom(A)}$. *Suppose further that*

(i) $\lim_{n\to\infty} f_n(t) = f(t), \quad a.e. \ t \in (0,T),$

and that

(ii) $\lim_{n\to\infty} J(\lambda; A_n + \omega I)\varphi_n = J(\lambda; A + \omega I)\varphi$ *for each* $\varphi \in X$ *whenever* $\varphi_n \in X_n$ *with* $\lim_{n\to\infty} \varphi_n = \varphi$, *for some* $\lambda > 0$.

Then if $\{U_n(t,s) : 0 \leq s \leq t \leq T\}$ *is the evolution system on* $\overline{Dom(A_n)}$ *generated by* A_n *and* f_n, *we have*

$$\lim_{n\to\infty} U_n(t,s)\varphi_n = U(t,s)\varphi$$

uniformly in s and t for $(s,t) \in \Delta$, *for each* $\varphi \in \overline{Dom(A)}$ *and* $\varphi_n \in \overline{Dom(A_n)}$ *with* $\lim_{n\to\infty} \varphi_n = \varphi$.

3. The Identification of first order systems with dynamics governed by monotone operators on Hilbert space

Let H be a real Hilbert space with inner product denoted by $\langle\cdot,\cdot\rangle$ and corresponding norm $|\cdot|$, and let V be a real reflexive Banach space with norm $\|\cdot\|$. We shall assume that V is densely and continuously embedded in H. It follows therefore that $V \hookrightarrow H \hookrightarrow V^*$ and that there exists a constant $\mu > 0$ for which $\|\varphi\|_* \leq \mu|\varphi|$, for all $\varphi \in H$ and $|\varphi| \leq \mu\|\varphi\|$, for all $\varphi \in V$ where $\|\cdot\|_*$ denotes the usual uniform operator norm on V^*. We shall also use $\langle\cdot,\cdot\rangle$ to denote the natural extension of the H inner product to the duality pairing between V and V^*. Let Q and Z be metric spaces and let Q be a fixed, nonempty, compact subset of $\mathcal{Q}$.

For each $q \in Q$ let $\mathcal{A}(q) : V \to V^*$ be a single valued, everywhere defined, semicontinuous (see [6]), in general nonlinear, operator from V into V^* satisfying the following conditions:

(A) (Continuity): For each $\varphi \in V$, the map $q \to \mathcal{A}(q)\varphi$ is continuous from $Q \subset \mathcal{Q}$ into V^*.

(B) (Equi-V-montonicity): There exist constants $\omega \in \mathbf{R}$ and $\alpha > 0$, both independent of $q \in Q$, for which

$$\langle \mathcal{A}(q)\varphi - \mathcal{A}(q)\psi, \varphi - \psi \rangle + \omega|\varphi - \psi|^2 \geq \alpha\|\varphi - \psi\|^2,$$

for every $\varphi, \psi \in V$.

(C) (Equi-boundedness): There exists a constant $\beta > 0$, independent of $q \in Q$, for which

$$\|\mathcal{A}(q)\varphi\|_* \leq \beta(\|\varphi\| + 1),$$

for every $\varphi \in V$.

For each $q \in Q$ define the operator $A(q) : Dom(A(q)) \subset H \to H$ to be the restriction of the operator $\mathcal{A}(q)$ to the set $Dom(A(q)) = \{\varphi \in V : \mathcal{A}(q)\varphi \in H\}$. It can be shown (see [3]) that $A(q)$ is densely defined (i.e. that $\overline{Dom(A(q))} = H$) and that $A(q) + \omega I$ is m-accretive. Let $T > 0$ and for each $q \in Q$ let $f(\cdot\ ;q) \in L_1(0,T;H)$ and let $u^0(q) \in H$. We assume that the mapping $q \to u^0(q)$ is continuous from $Q \subset \mathcal{Q}$ into H and that the mapping $q \to f(t;q)$ is continuous from $Q \subset \mathcal{Q}$ into H for almost every $t \in (0,T)$. Also, for every $z \in Z$, let $u \to \tilde{\Phi}(u;z)$ be a continuous map from $C(0,T;H)$ into $\mathbf{R}^+$. We consider the abstract parameter identification problem given by:

(ID)   Given observations $z \in Z$, determine parameters $\bar{q} \in Q$ which minimize the functional

$$\Phi(q) = \tilde{\Phi}(u(q); z)$$

where for each $q \in Q$ $u(q) = u(\cdot\ ;q)$ is the mild solution to the initial value problem in H given by

(3.1) $\dot{u}(t) + A(q)u(t) = f(t;q),\ 0 < t \leq T$

(3.2) $u(0) = u^0(q).$

Recalling the discussion in section 2 and our remarks above, it is clear that for each $q \in Q, A(q)$ and $f(\cdot \ ; q)$ generate a nonlinear evolution system $\{U(t, s; q) : 0 \leq s \leq t \leq T\}$ on H with the mild solution to the initial value problem (3.1), (3.2) given by

$$u(t; q) = U(t, 0; q)u^0(q), \ 0 \leq t \leq T.$$

We develop an abstract Galerkin based approximation theory for problem (ID). For each $n = 1, 2, \ldots$ let H_n be a finite dimensional subspace of H with $H_n \subset V$ for all n. We let $P_n : H \rightarrow H_n$ denote the orthogonal projection of H onto H_n with respect to the inner product $\langle \cdot, \cdot \rangle$ and we make the standing assumption

(D) $\lim_{n \to \infty} \|P_n \varphi - \varphi\| = 0,$ for each $\varphi \in V.$

It is clear that assumption (D) together with the dense and continuous embedding of V in H yield that the P_n tend strongly to the identity on H as well, as $n \rightarrow \infty$.

For each $q \in Q$ we define the Galerkin approximation $A_n(q), n = 1, 2, \ldots$, to $A(q)$ in the usual manner. That is, for $\varphi_n \in H_n$ we set $A_n(q)\varphi_n = \psi_n$ where ψ_n is the unique element in H_n (guaranteed to exist by the Riesz Representation Theorem applied to H_n) satisfying $\langle A(q)\varphi_n, \chi_n \rangle = \langle \psi_n, \chi_n \rangle, \chi_n \in H_n$. We set $f_n(\cdot \ ; q) = P_n f(\cdot \ ; q) \in L_1(0, T; H_n)$ and $u_n^0(q) = P_n u^0(q) \in H_n$. With these definitions, it is not difficult to argue that for each $n = 1, 2, \ldots$ and each $q \in Q$, $A_n(q)$ and $f_n(\cdot \ ; q)$ generate a nonlinear evolution system, $\{U_n(t, s; q) : 0 \leq s \leq t \leq T\}$ on H_n. Thus we consider the sequence of approximating identification problems given by:

(ID_n) Determine parameters $\bar{q}_n \in Q$ which minimize

$$\Phi_n(q) = \tilde{\Phi}(u_n(q); z)$$

where for each $q \in Q, u_n(q) = u_n(\cdot \ ; q)$ is the mild solution to the initial value problem in H_n given by

(3.3) $\dot{u}_n(t) + A_n(q)u_n(t) = f(t; q), \ 0 < t \leq T$

(3.4) $u_n(0) = u_n^0(q).$

The mild solution, $u_n(q)$, to (3.3), (3.4) is given by $u_n(t; q) = U_n(t, 0; q)u_n^0(q), \ 0 \leq t \leq T.$

We may summarize the existence and convergence theory for solutions to problem (ID_n) given in [3] as follows. Let $\{q_n\}_{n=1}^\infty$ be a sequence in Q with $\lim_{n \to \infty} q_n = q_0 \in Q$. Assumption (D) and the continuity of the map $q \rightarrow u^0(q)$ imply $\lim_{n \to \infty} u_n^0(q_n) = u^0(q_0)$ in H. Similarly we have $\lim_{n \to \infty} f_n(t; q_n) = f(t; q_0)$ in H for almost every $t \in (0, T)$ with $|f_n(\cdot \ ; q_n)|$ dominated by an L_1 function which is independent of n. Assumption (D) also implies that $\lim_{n \to \infty} \overline{Dom(A_n(q_n))} = \lim_{n \to \infty} H_n \supset H = \overline{Dom(A(q_0))}$, and that for each

$\lambda > 0$, $\lim\limits_{n \to \infty} J(\lambda; A_n(q_n) + \omega I)\varphi_n = J(\lambda; A(q_0) + \omega I)\varphi$ in H whenever $\varphi_n \in H_n$ with $\lim\limits_{n \to \infty} \varphi_n = \varphi \in H$. Thus, Theorem 2.1 yields

$$(3.5) \qquad \lim_{n \to \infty} u_n(q_n) = \lim_{n \to \infty} U_n(\cdot, 0; q_n) u_n^0(q_n) = U(\cdot, 0; q_0) u^0(q_0) = u(q_0),$$

in $C(0, T; H)$. Similar arguments can be used to demonstrate that if $\{q_m\}_{m=1}^{\infty} \subset Q$ with $\lim\limits_{m \to \infty} q_m = q_0$, then

$$(3.6) \qquad \lim_{m \to \infty} u_n(q_m) = \lim_{m \to \infty} U_n(\cdot, 0; q_m) u_n^0(q_m) = U_n(\cdot, 0; q_0) u_n^0(q_0) = u_n(q_0)$$

in $C(0, T; H)$ for each $n = 1, 2, \ldots$

The compactness of Q, the continuity assumption on $\tilde{\Phi}$, and (3.6) are sufficient to conclude that for each $n = 1, 2, \ldots$, problem (ID_n) admits a solution $\bar{q}_n \in Q$. Since $\{\bar{q}_n\}_{n=1}^{\infty} \subset Q$ and Q compact, a convergent subsequence, $\{\bar{q}_{n_j}\}_{j=1}^{\infty}$, may be extracted from $\{\bar{q}_n\}_{n=1}^{\infty}$. If $\bar{q} = \lim\limits_{j \to \infty} \bar{q}_{n_j}$, then for each $q \in Q$ (3.5) implies

$$
\begin{aligned}
\Phi(\bar{q}) = \tilde{\Phi}(u(\bar{q}); z) &= \tilde{\Phi}\big(\lim_{j \to \infty} u_{n_j}(\bar{q}_{n_j}); z\big) \\
&= \lim_{j \to \infty} \tilde{\Phi}(u_{n_j}(\bar{q}_{n_j}); z) = \lim_{j \to \infty} \Phi_{n_j}(\bar{q}_{n_j}) \\
&\leq \lim_{j \to \infty} \Phi_{n_j}(q) = \lim_{j \to \infty} \tilde{\Phi}(u_{n_j}(q); z) \\
&= \tilde{\Phi}\big(\lim_{j \to \infty} u_{n_j}(q); z\big) = \tilde{\Phi}(u(q); z) \\
&= \Phi(q)
\end{aligned}
$$

(3.7)

and consequently that $\bar{q}$ is a solution to problem (ID).

When the admissible parameter set Q is also infinite dimensional (when, for example, as is frequently the case, the unknown parameters to be identified are elements in a function space) it must be discretized as well. When this is in fact the case, the theory presented above requires the following modification. For each $m = 1, 2, \ldots$ let $I^m : Q \subset Q \to Q$ be a continuous map with range $Q^m = I^m(Q)$ in a finite dimensional space and with the property that $\lim\limits_{m \to \infty} I^m(q) = q$, uniformly on Q. We consider the doubly indexed sequence of approximating identification problems (ID_n^m) where for each n and m (ID_n^m) is defined to be problem (ID_n) with Q replaced by Q^m. It can be shown that each of these problems admits a solution $\bar{q}_n^m \in Q^m$, and that the sequence $\{\bar{q}_n^m\}$ will have a Q-convergent subsequence, $\{\bar{q}_{n_k}^{m_j}\}$, with limit in Q. If $\lim\limits_{j, k \to \infty} \bar{q}_{n_k}^{m_j} = \bar{q} \in Q$, then $\bar{q}$ can be shown to be a solution to problem (ID). The problems (ID_n^m) involve the minimization of functionals over compact subsets of Euclidean space subject to finite dimensional state space constraints. Consequently they may be solved using standard computational techniques.

We illustrate the application of our theory with an example involving the identification of a quasi-linear model for one dimensional heat flow (see [13], [14]). We consider the quasi-linear parabolic partial differential equation

$$\frac{\partial u}{\partial t}(t, x) - \frac{\partial}{\partial x}\left\{ q\big(\frac{\partial u}{\partial x}(t, x)\big)\frac{\partial u}{\partial x}(t, x) \right\} = f(t, x), \qquad t > 0, 0 < x < 1,$$

together with the Dirichlet boundary conditions

$$u(t,0) = 0, \ u(t,1) = 0, \ t > 0$$

and initial data

$$u(0,x) = u^0(x), \ \ 0 < x < 1.$$

We assume that $u^0 \in L_2(0,1), f \in L_1(0,T;L_2(0,1))$, and that $q \in C_B(\mathbf{R})$, the space consisting of all bounded continuous functions defined on the entire real line and endowed with the usual supremum metric which we shall denote by $d_\infty(\cdot,\cdot)$. We assume further that q satisfies

$$(3.8) \qquad (q(\theta)\theta - q(\eta)\eta)(\theta - \eta) \geq \alpha|\theta - \eta|^2, \ \theta, \eta \in \mathbf{R}$$

for some $\alpha > 0$. (We note that if q is differentiable on $\mathbf{R}$, then the Mean Value Theorem implies that the condition $q'(\theta)\theta + q(\theta) \geq \alpha > 0, \ \theta \in \mathbf{R}$, is sufficient to conclude that condition (3.8) holds.) To apply our framework we set $H = L_2(0,1)$ endowed with the standard inner product and norm, and set $V = H_0^1(0,1)$ with norm $\|\cdot\|$ given by $\|\phi\| = \left(\int_0^1 |D\phi(x)|^2\,dx\right)^{\frac{1}{2}}$. In this case we have $V^* = H^{-1}(0,1)$ and $V \hookrightarrow H \hookrightarrow V^*$ with the embeddings dense and continuous ($\mu = 1$). We take $Q = C_B(\mathbf{R}), Z = C(0,T;L_2(0,1))$, and for given fixed values of $\alpha_0, \rho_0, \sigma_0, \theta_0 > 0$ we take Q to be the Q-closure of the set

$$\{q \in C_B(\mathbf{R}) : q(\theta) = q(-\theta), \ |q(\theta)| \leq \rho_0, \ |q'(\theta)| \leq \sigma_0, \ q'(\theta)\theta + q(\theta) \geq \alpha_0,$$
$$\text{for } \theta \in \mathbf{R}, \ q(\theta) = \text{constant for } |\theta| \geq \theta_q$$
$$\text{for some numbers } \theta_q \text{ satisfying } 0 \leq \theta_q \leq \theta_0\}.$$

A straightforward application of the Arzelá-Ascoli Theorem reveals that Q is a compact subset of $C_B(\mathbf{R})$. If for each $q \in Q$ we define the operator $\mathcal{A}(q) : V \to V^*$ by

$$\langle \mathcal{A}(q)\varphi, \psi \rangle = \int_0^1 q(D\varphi(x))D\varphi(x)D\psi(x)dx, \ \ \varphi, \psi \in V,$$

then it is not difficult to show that assumptions (A), (B), and (C) are satisfied. Let $\{t_i\}_{i=1}^\nu$ with $0 \leq t_1 < t_2 \cdots < t_\nu \leq T$ be given, and for each $z \in Z$ define the least squares performance index $\tilde{\Phi} : C(0,T;L_2(0,1)) \to \mathbf{R}^+$ by

$$(3.9) \qquad \tilde{\Phi}(v;z) = \sum_{i=1}^{\nu} \int_0^1 |v(t_i,x) - z(t_i,x)|^2\,dx.$$

We consider the parameter estimation problem (ID) with $Q, \tilde{\Phi}, \mathcal{A}(q), f$, and u^0 as defined above.

For each $n = 1, 2, \ldots$ let $H_n = \text{span}\{\phi_n^j\}_{j=1}^{n-1}$ where ϕ_n^j is the j-th linear B-spline on the interval [0,1] defined with respect to the uniform mesh $\{0, 1/n, 2/n, \ldots, 1\}$. That is,

$$(3.10) \qquad \phi_n^j(x) = \begin{cases} 0 & 0 \leq x \leq \frac{i-1}{n} \\ nx - j + 1 & \frac{i-1}{n} \leq x \leq \frac{i}{n} \\ j + 1 - nx & \frac{i}{n} \leq x \leq \frac{i+1}{n} \\ 0 & \frac{i+1}{n} \leq x \leq 1, \end{cases}$$

$j = 1, 2, \ldots, n - 1$. Clearly $H_n \subset V = H_0^1(0,1), n = 1, 2, \ldots$ Let $P_n : H \to H_n$ denote the orthogonal projection of $L_2(0,1)$ onto H_n with respect to the usual L_2-inner product. Standard approximation results for interpolatory splines (see [15]) can be used to argue that assumption (D) is satisfied.

We discretize the admissible parameter set as follows. For $m \in \mathbf{Z}^+$ and $q \in Q$ set

$$(I^m q)(\theta) = \sum_{j=0}^{m} q(j\theta_q/m)\psi_j^m(|\theta|; \theta_q), \quad \theta \in \mathbf{R}$$

where the $\psi_j^m(\cdot\; ; \theta_q), j = 0, 1, 2, \ldots, m$ are the standard linear B-splines on the interval $[0, \theta_q]$ defined with respect to the uniform mesh $\{0, \theta_q/m, 2\theta_q/m, \ldots, \theta_q\}$ and then extended to a continuous function on the entire positive real line via $\psi_j^m(\theta; \theta_q) = \psi_j^m(\theta_q; \theta_q), \theta \geq \theta_q$. Using the Peano Kernel Theorem (see [12]) it can be argued that

$$d_\infty(I^m q, q) = \sup |I^m q - q| \leq \frac{1}{2}\left(\frac{\theta_0}{m}\right)^{1/2} \sigma_0, \quad q \in Q,$$

and consequently that $\lim_{m \to \infty} I^m q = q$ in $C_B(\mathbf{R})$, uniformly in q for $q \in Q$.

For $q^m \in Q^m = I^m(Q)$ the finite dimensional initial value problem (3.3), (3.4) takes the form

$$(3.11) \qquad M_n \dot{w}_n(t) + K_n(w_n(t); q^m) = F_n(t), \quad 0 < t \leq T$$

$$(3.12) \qquad M_n w_n(0) = w_n^0$$

where $w_n(t) \in \mathbf{R}^{n-1}$, M_n is the $(n-1) \times (n-1)$ - Gram matrix whose (i,j)-th entry is given by $M_n^{i,j} = \langle \varphi_n^i, \varphi_n^j \rangle$, $F_n(t)$ and w_n^0 are the $(n-1)$-vectors whose i-th elements are given by $F_n^i(t) = \langle f_n(t, \cdot), \varphi_n \rangle$ and $w_n^{0i} = \langle u^0, \varphi_n^i \rangle$, respectively, and $K_n(\cdot\; ; q^m) : \mathbf{R}^{n-1} \to \mathbf{R}^{n-1}$ is given by

$$K_n^i(v; q^m) = \begin{cases} nq^m(nv^1)v^1 - nq^m(n\{v^2 - v^1\})\{v^2 - v^1\}, & i = 1 \\ nq^m(n\{v^i - v^{i-1}\})\{v^i - v^{i-1}\} & \\ \quad -nq^m(n\{v^{i+1} - v^i\})\{v^{i+1} - v^i\}, & i = 2, 3, \ldots, n - 2 \\ nq^m(n\{v^{n-1} - v^{n-2}\})\{v^{n-1} - v^{n-2}\} & \\ \quad +nq^m(-nv^{n-1})v^{n-1}, & i = n - 1 \end{cases}$$

for $v \in \mathbf{R}^{n-1}$.

If $q^m \in Q^m$ is given by $q^m(\theta) = \sum\limits_{j=0}^{m} q_j^m \psi_j^m(|\theta|; \theta_{q^m})$, for $\theta \in \mathbf{R}$, solving the identification problem (ID_n^m) involves the choosing of parameters $(q_0^m, q_1^m, \ldots, q_m^m, \theta_{q^m})^T$ from a compact subset of $\mathbf{R}^{m+2}$ so as to minimize the functional $\Phi(q^m) = \tilde{\Phi}(u_n(q^m); z)$ where $u_n(q^m)$ is given by $u_n(t; q^m) = \sum\limits_{j=1}^{n-1} w_n^j(t; q^m) \varphi_n^j$, $t \in [0, T]$ with $w_n(\cdot; q^m)$ the solution to the initial value problem (3.11), (3.12) in $\mathbf{R}^{n-1}$ corresponding to q^m.

In order to actually test our scheme we let $q^*(\theta) = a(1 - .5e^{-b\theta^2})$ for $\theta \in \mathbf{R}$ with $a = .9$ and $b = .5$ (note that $\theta D q^*(\theta) + q^*(\theta) \geq .45$ for $\theta \in \mathbf{R}$ and consequently that condition (3.8) is satisfied by q^*) and set $z(t, x) = 5e^{-t}(x - x^3)$, $x \in [0, 1], t > 0$. Then setting

$$f(t, x) = \frac{\partial z}{\partial t}(t, x) - \frac{\partial}{\partial x}\left\{ q^* \left(\frac{\partial z}{\partial x}(t, x) \right) \frac{\partial z}{\partial x}(t, x) \right\},$$

$x \in [0, 1], t > 0$, and $u^0(x) = 5(x - x^3)$, $x \in [0, 1]$, we used our scheme to attempt to estimate q^* based upon the observations $\{z(.5j, \cdot)\}_{j=1}^{12}$.

All integrals that had to be computed numerically (i.e. some of the L_2 inner products and the integral appearing in the definition of the least-squares performance index (3.9)) were computed using a composite two point Gauss-Legendre quadrature rule on $[0,1]$. For each n and m the IMSL implementation (routine ZXSSQ) of the Levenberg-Marquardt algorithm, an iterative steepest descent/Newton's method hybrid, was used to solve the finite dimensional nonlinear least-squares minimization problem (ID_n^m). For a given choice of the parameters q^m, the initial value problem (3.11), (3.12) was solved in each iteration using the IMSL routine DGEAR with the stiff option operative. As an initial guess for q^* we took $q^0(\theta) = 1$, for $\theta \in \mathbf{R}$ with $\theta_{q^0} = 4$.

All computations were carried out on a Cray X-MP/48 at the San Diego Supercomputer Center. Standard coding techniques which permit optimal vectorization were used whenever possible. These included the nesting of loops with the largest ranges the deepest, and the separation of vectorizable and non-vectorizable code into different loops. In general, we observed that in the absence of vectorization, the Cray was able to run our codes in approximately 1% of the time that it took an IBM 3081. Vectorization on the Cray then yielded an additional speed-up factor of 17. Representative results that we obtained for various values of n and m are shown in Figures 3.1 - 3.3 below.

The CPU times on the Cray for these runs ranged from about 3 seconds for $n = 8, m = 1$ to about 180 seconds for $n = 20, m = 4$. The value of the performance index was reduced from $\Phi_n(q^0) \approx 10^{-2}$ to $\Phi_n(\bar{q}_n^m) \approx 10^{-4}$. We solved the problem (ID_n^m) unconstrained. That is we did not enforce the constraints in the definition of Q which render it and $I^m(Q) = Q^m$ compact. Thus it was not surprising that, as we have seen before in the case of linear system identification, for each n, the inherent ill-posedness of the problem of identifying functional parameters began to cause difficulties as m was increased (see [1]). We were, to a certain degree, able to mitigate these instability effects with the introduction of Tikhonov regularization (see [11]). However, at least from a qualitative point of view, this is probably unnecessary since we seem to obtain reasonably good estimates with m relatively small. It is worth noting that we have also tested our approach on the much simpler problem of identifying constant parameters (for example, the estimation of the parameters a and b in the definition of q^*). In these tests it performed superbly with convergence to the true values of the parameters as $n \to \infty$ immediately apparent. Finally, we also tested our scheme using discrete or sampled rather than distributed observations in the spatial variable although strictly speaking these examples cannot be treated with our theory. With measurements taken at only one spatial point, $x = .58$ (i.e. with the observations $\{z(.5j, .58)\}_{j=1}^{12}$), the scheme's performance remained essentially unchanged from that observed with distributed observations. We note that the existing theory for the case of linear dynamics (see [2]) can handle spatially discrete measurements. An extension of these results to the nonlinear case is currently being investigated but at present remains an open question.

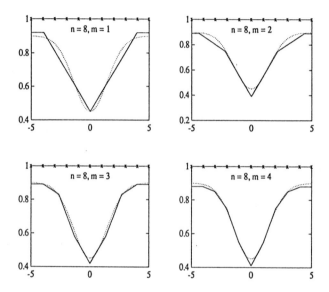

Figure 3.1 Legend: $q^* \cdots\cdots$, $\bar{q}_n^m$ $------$, $q^0 - \times - \times - \times$.

4. The identification of nonlinear damping in second order systems

In this section we consider the identification of nonlinear dissipation mechanisms in abstract infinite dimensional second order elastic systems. In our treatment below we assume that the stiffness operator is linear. However we note that a similar approach can be used to identify a nonlinear stiffness operator in the presence of linear damping. We are currently looking into the extension of our theory to systems which involve both nonlinear stiffness and damping. A more detailed presentation along with proofs for the theoretical results we summarize below can be found in [5].

Let the spaces H, V, V^*, Q, and Z, and the set Q be as they were defined for abstract first order systems at the beginning of section 3. For each $q \in Q$ let the operator $\mathcal{A}(q) \in$

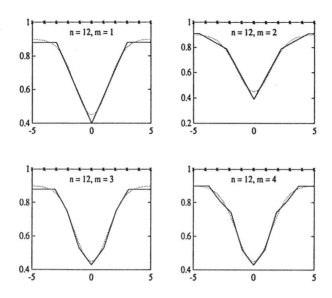

Figure 3.2 Legend: $q^* \cdots\cdots$, $\bar{q}_n^m$ $-----$, $q^0 - \times - \times - \times$.

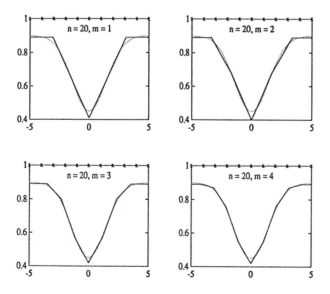

Figure 3.3 Legend: $q^* \cdots\cdots$, $\bar{q}_n^m$ $-----$, $q^0 - \times - \times - \times$.

$\mathcal{L}(V, V^*)$ satisfy the following conditions:

(A1) (Symmetry) For all $\varphi, \psi \in V$ $\langle \mathcal{A}(q)\varphi, \psi \rangle = \langle \varphi, \mathcal{A}(q)\psi \rangle$;

(A2) (Continuity) For each $\varphi \in V$ the map $q \to \mathcal{A}(q)\varphi$ is continuous from $Q \subset \mathcal{Q}$ into V^*;

(A3) (Equi-V-Coercivity) There exist constants $\omega \in \mathbf{R}$ and $\alpha > 0$, both independent of $q \in Q$ for which $\langle \mathcal{A}(q)\varphi, \varphi \rangle + \omega|\varphi|^2 \geq \alpha\|\varphi\|^2$, for all $\varphi \in V$ and $q \in Q$;

(A4) (Equi-Boundedness) The operators $\mathcal{A}(q)$ are uniformly bounded in q for $q \in Q$. That is, there exists a constant $\beta > 0$, independent of $q \in Q$, for which $\|\mathcal{A}(q)\|_* \leq \beta\|\varphi\|$, for all $\varphi \in V$.

Also, for each $q \in Q$ let the operator $\mathcal{B}(q) : Dom(\mathcal{B}(q)) \subset V \to 2^{V^*}$ satisfy the following conditions:

(B1) (Domain Uniformity) $Dom(\mathcal{B}(q)) = Dom(\mathcal{B})$ is independent of q for $q \in Q$, and $0 \in Dom(\mathcal{B})$;

(B2) (Continuity) For each $\varphi \in Dom(\mathcal{B})$ the map $q \to \mathcal{B}(q)\varphi$ is lower semi-continuous from $Q \subset \mathcal{Q}$ into 2^{V^*};

(B3) (Maximal Monotonicity) For all $(\varphi_1, \psi_1), (\varphi_2, \psi_2) \in B_q \equiv \{(\varphi, \psi) \in V \times V^* : \varphi \in Dom(\mathcal{B}), \psi \in \mathcal{B}(q)\varphi\}$ we have $\langle \psi_1 - \psi_2, \varphi_1 - \varphi_2 \rangle \geq 0$ with B_q not properly contained in any other subset of $V \times V^*$ for which this monotonicity condition holds;

(B4) (Equi-Boundedness) The operators $\mathcal{B}(q)$ map V-bounded subsets of $Dom(\mathcal{B})$ into subsets of V^* which are uniformly bounded in q for $q \in Q$. That is, if S is a V-bounded subset of $Dom(\mathcal{B})$, the set $\mathcal{B}(q)S$ is V^*-bounded, uniformly in q for $q \in Q$.

Let $T > 0$ and for each $q \in Q$ let $u^0(q) \in V$, $u^1(q) \in H$, and $f(\cdot\ ;q) \in L_1(0, T; H)$. We assume that the mappings $q \to u^0(q), q \to u^1(q)$, and $q \to f(t;q)$ are continuous from $Q \subset \mathcal{Q}$ into V, H, and H respectively, for almost every $t \in (0, T)$. For every $z \in Z$ let $(u, v) \to \tilde{\Phi}(u, v; z)$ be a continuous mapping from $C(0, T; V \times H)$ into $\mathbf{R}^+$. The identification problem, which we shall again denote by (ID), takes the form:

(ID)   Given observations $z \in Z$, determine parameters $\bar{q} \in Q$ which minimize the functional

$$\Phi(q) = \tilde{\Phi}(u(q), \dot{u}(q); z)$$

where for each $q \in Q, u(q) = u(\cdot\ ;q)$ is the mild solution to the initial value problem

(4.1) $$\ddot{u}(t) + \mathcal{B}(q)\dot{u}(t) + \mathcal{A}(q)u(t) \ni f(t;q),\ 0 < t \leq T$$

(4.2) $$u(0) = u^0(q),\ \dot{u}(0) = u^1(q).$$

To make the notion of a mild solution to a second order initial value problem of the form (4.1), (4.2) precise, we rely on a reformulation as an equivalent first order system in a product space and then apply the abstract theory outlined in section 2. For each $q \in Q$ define the Hilbert space $X_q = V \times H$ with inner product $(\cdot, \cdot)_q$ given by

(4.3) $$((\varphi_1, \psi_1), (\varphi_2, \psi_2))_q = \langle \mathcal{A}(q)\varphi_1, \varphi_2 \rangle + \omega\langle \varphi_1, \varphi_2 \rangle + \langle \psi_1, \psi_2 \rangle.$$

We denote the corresponding induced norm on X_q by $|\cdot|_q$. We note that our assumptions on the operators $\mathcal{A}(q)$ guarantee that (4.3) indeed defines an inner product on $V \times H$

and that the Banach spaces $\{X_q, |\cdot|_q\}$ are norm equivalent, uniformly in q for $q \in Q$, to the Banach space $X = V \times H$ endowed with the standard product topology induced by the norm $|(\varphi, \psi)|_X = (\|\varphi\|^2 + |\psi|^2)^{\frac{1}{2}}$. For each $q \in Q$ define the operator $A(q)$: $Dom(A(q)) \subset X_q \to 2^{X_q}$ by

$$A(q)(\varphi, \psi) = (-\psi, \{\mathcal{A}(q)\varphi + \mathcal{B}(q)\psi\} \cap H),$$

for $(\varphi, \psi) \in Dom(A(q)) = \{(\varphi, \psi) \in V \times H : \psi \in V, \{\mathcal{A}(q)\varphi + \mathcal{B}(q)\psi\} \cap H \neq \phi\}$. It can be shown (see [5]) that there exists a $\gamma > 0$, independent of $q \in Q$, for which the operator $A(q) + \gamma I$ is m-accretive on $Dom(A(q)) \subset X_q$ for each $q \in Q$. Also, for each $q \in Q$ define $F(\cdot \; ; q) \in L_1(0, T; X_q)$ by $F(t; q) = (0, f(t; q))$, for almost every $t \in (0, T)$ and set $x^0(q) = (u^0(q), u^1(q)) \in X_q$. It follows that for every $q \in Q$, $A(q)$ and $F(\cdot \; ; q)$ generate a nonlinear evolution system $\{U(t, s; q) : 0 \leq s \leq t \leq T\}$ on $\overline{Dom(A(q))} \subset X_q$ satisfying conditions (i) - (iii) given in section 2. Henceforth we shall assume that $x^0(q) \in \overline{Dom(A(q))}$ for each $q \in Q$, and by a mild solution to the second order initial value problem (4.1), (4.2) we shall mean the V-continuous function $u(\cdot \; ; q)$ given by the first component of the X_q (or X)-continuous function $x(\cdot \; ; q) = U(\cdot, 0; q)x^0(q)$. We shall take $\dot{u}(\cdot \; ; q)$ to be the H-continuous second component of $x(\cdot \; ; q)$.

We note that if assumption (B4) is strengthened to the condition that the operators $\mathcal{B}(q)$ map H-bounded subsets into V^*-bounded subsets, uniformly in q for $q \in Q$, it can be argued (see [5]) that $\overline{Dom(A(q))} = X_q = X$. In addition, since conditions (A3) and (A4) imply that $Dom (\mathcal{A}(q)) = \{\varphi \in V : \mathcal{A}(q)\varphi \in H\}$ is dense in V (see [16]) it is clear that the operators $A(q)$ will also be densely defined when the set $\{\varphi \in V : \mathcal{B}(q)\varphi \in H\}$ is dense in H. In particular this will in fact be the case for all of the standard linear dissipation mechanisms - for example, air $(\mathcal{B} \sim I)$, so called structural $(\mathcal{B} \sim \mathcal{A}^{\frac{1}{2}})$, and Kelvin-Voigt viscoelastic $(\mathcal{B} \sim \mathcal{A})$ damping.

With the existence and uniqueness of mild solutions on X_q now demonstrated for each $q \in Q$, the q-uniform norm equivalence of X_q and X will allow us to subsequently ignore the q-dependence of the state spaces and to develop our approximation theory and convergence results on the q-independent space X.

Once again, as was the case with first order systems, our approximation theory is of Galerkin type. For each $n = 1, 2, \ldots$ let H_n be a finite dimensional subspace of H with $H_n \subset V$. Let $P_n; H \to H_n$ denote the orthogonal projection of H onto H_n with respect to the standard inner product on $H, \langle \cdot, \cdot \rangle$, and we assume that $P_n(Dom(\mathcal{B})) \subset Dom(\mathcal{B})$, for all n. We also again assume that condition (D) given in section 3 is satisfied. For each $n = 1, 2, \ldots$ and each $q \in Q$ let $\mathcal{A}_n(q) \in \mathcal{L}(H_n)$ and $\mathcal{B}_n(q) : Dom(\mathcal{B}_n) \subset H_n \to 2^{H_n}$ denote the Galerkin approximations to $\mathcal{A}(q)$ and $\mathcal{B}(q)$ respectively. That is for $\varphi_n \in H_n, \mathcal{A}_n(q)\varphi_n = \psi_n$ where ψ_n is the unique element in H_n which satisfies $\langle \mathcal{A}(q)\varphi_n, \chi_n \rangle = \langle \psi_n, \chi_n \rangle, \chi_n \in H_n$, and for $\varphi_n \in Dom(\mathcal{B}_n) = Dom(\mathcal{B}) \cap H_n, \mathcal{B}_n(q)\varphi_n = \{\psi_n \in H_n : \langle \psi, \chi_n \rangle = \langle \psi_n, \chi_n \rangle, \chi_n \in H_n$ for some $\psi \in \mathcal{B}(q)\varphi_n\}$. We set $f_n(\cdot \; ; q) = P_n f(\cdot \; ; q) \in L_1(0, T; H_n), u_n^0(q) = P_n u^0(q)$, and $u_n^1(q) = P_n u^1(q)$, and consider the sequence of approximating parameter identification problems given by:

(ID_n) Determine parameters $\bar{q}_n \in Q$ which minimize the functional

$$\Phi_n(q) = \tilde{\Phi}(u_n(q), \dot{u}_n(q); z)$$

where for each $q \in Q$, $u_n(q) = u_n(\cdot\ ; q)$ is the mild solution to the initial value problem in H_n given by

$$(4.4) \qquad \ddot{u}_n(t) + \mathcal{B}_n(q)\dot{u}_n(t) + \mathcal{A}_n(q)u_n(t) \ni f_n(t; q), \quad 0 < t \leq T$$

$$(4.5) \qquad u_n(0) = u_n^0(q), \quad \dot{u}_n(0) = u_n^1(q).$$

We again use the theory in section 2 to define what is meant by a mild solution to the second order initial value problem (4.4), (4.5) in H_n. For each $n = 1, 2, \ldots$ let $X_n = H_n \times H_n$, and for each $q \in Q$ define the operator $A_n(q) : Dom(A_n) \subset X_n \to 2^{X_n}$ by

$$A_n(q)(\varphi_n, \psi_n) = (-\psi_n, \mathcal{A}_n(q)\varphi_n + \mathcal{B}_n(q)\psi_n)$$

for $(\varphi_n, \psi_n) \in Dom(A_n) = H_n \times Dom(\mathcal{B}_n)$. Set $F_n(\cdot\ ; q) = (0, f_n(\cdot\ ; q)) \in L_1(0, T; X_n)$ and $x_n^0(q) = (u_n^0(q), u_n^1(q))$. We assume $u_n^1(q) \in \overline{Dom(\mathcal{B}_n)}$ so that $x_n^0(q) \in \overline{Dom(A_n)}$. We define the mild solution to (4.4), (4.5) to be the first component of the function $x_n(\cdot\ ; q) = U_n(\cdot, 0; q)x_n^0(q) \in C(0, T; X_n)$ where $\{U_n(t, s; q) : 0 \leq s \leq t \leq T\}$ is the nonlinear evolution system on $\overline{Dom(A_n)}$ generated by $A_n(q)$ and $F_n(\cdot\ ; q)$. That such an evolution system in fact exists can be argued as it was for the corresponding infinite dimensional second order system using the definitions of the operators $A_n(q)$ and $\mathcal{B}_n(q)$, and the function $f_n(\cdot\ ; q)$ (see [5]). The function $\dot{u}_n(\cdot\ ; q)$ is obtained from the second component of $x_n(\cdot\ ; q)$.

By using condition (D) to argue resolvent convergence, i.e., that for each $\lambda > 0$ sufficiently large, $\lim_{n \to \infty} J(\lambda; A_n(q_n) + \tilde{\omega}I)(\phi_n, \psi_n) = J(\lambda; A(q_0) + \tilde{\omega}I)(\phi, \psi)$ in X for some $\tilde{\omega} \in \mathbf{R}$ whenever $(\varphi, \psi) \in X$ and $(\varphi_n, \psi_n) \in X_n$ with $\lim_{n \to \infty} (\varphi_n, \psi_n) = (\varphi, \psi)$ and $q_n, q_0 \in Q$ with $\lim_{n \to \infty} q_n = q_0$, we are able to apply Theorem 2.1 to obtain that $\lim_{n \to \infty} u_n(q_n) = u(q_0)$ in $C(0, T; V)$ and $\lim_{n \to \infty} \dot{u}_n(q_n) = \dot{u}(q_0)$ in $C(0, T; H)$ whenever $q_n, q_0 \in Q$ with $\lim_{n \to \infty} q_n = q_0$. A continuous dependence result analogous to (3.6) can also be obtained in this fashion. Then using estimates in the spirit of those given in (3.7) we find that solutions $\bar{q}_n$ to the problem (ID_n) exist and that the sequence $\{\bar{q}_n\}_{n=1}^{\infty}$ admits a Q-convergent subsequence, $\{\bar{q}_{n_j}\}_{j=1}^{\infty}$, with $\lim_{n \to \infty} \bar{q}_{n_j} = \bar{q}$ and $\bar{q}$ a solution to problem (ID). The discretization of the admissible parameter set Q can be carried out, and a subsequent convergence theory established exactly as they were in section 3 for first order systems. A complete and detailed discussion of the results for second order inverse problems which we have summarized above can be found in [5].

To illustrate the application of our approach we consider an inverse problem involving the identification of a nonlinear damping functional in a one dimensional wave equation. Let $\mathcal{Q} = C_B(\mathbf{R})$ endowed with the usual supremum norm, and let Q be the $\mathcal{Q}$-closure of the set

$$\{q \in \mathcal{Q} : q(\theta) = -q(-\theta),\ \theta q(\theta) \geq 0, \quad \text{for } \theta \in \mathbf{R}, |q(\theta)| = q(\theta_0),$$
$$\text{for } |\theta| \geq \theta_0, q \in H^1(-\theta_0, \theta_0),\ |q|_{H^1(-\theta_0,\theta_0)} \leq K_0,$$
$$q'(\theta) \geq 0,\ \text{for a.e. } \theta \in (-\theta_0, \theta_0)\}$$

where θ_0 and K_0 are given positive constants. It is not difficult to show that $\bar{Q}$ is a compact subset of Q.

For each $q \in Q$ we consider the one dimensional wave equation with nonlinear damping given by

$$\frac{\partial^2 u}{\partial t^2}(t,x) + q(\frac{\partial u}{\partial t}(t,x)) - \frac{\partial}{\partial x}\left(E(x)\frac{\partial u}{\partial x}(t,x)\right) = f(t,x), \quad t > 0, \quad 0 < x < 1$$

with boundary and initial conditions

$$u(t,0) = 0, \quad u(t,1) = 0, \quad t > 0,$$

$$u(0,x) = u^0(x), \quad \tfrac{\partial u}{\partial t}(0,x) = u^1(x), \quad 0 < x < 1$$

where $E \in L_\infty(0,1)$ with $E(x) \geq E_0 > 0$, a.e. $x \in (0,1), f \in L_2((0,T) \times (0,1)), u^0 \in H_0^1(0,1)$, and $u^1 \in L_2(0,1)$ are given. We set $H = L_2(0,1), V = H_0^1(0,1)$ and $V^* = H^{-1}(0,1)$. The operator $\mathcal{A} \in \mathcal{L}(V,V^*)$ given by

$$\langle \mathcal{A}\varphi, \psi \rangle = \int_0^1 E(x)D\varphi(x)D\psi(x)dx, \quad \varphi, \psi \in H_0^1(0,1),$$

is easily shown to satisfy conditions (A1) - (A4). For each $q \in Q$ we define the operator $\mathcal{B}(q) : V \to V^*$ via

$$\langle \mathcal{B}(q)\varphi, \psi \rangle = \int_0^1 q(\varphi(\theta))\psi(\theta)d\theta, \quad \varphi, \psi \in H_0^1(0,1).$$

(Note that in this case we in fact have $\mathcal{R}(\mathcal{B}(q)) \subset H$.) With the set Q as it has been defined above, it is clear that conditions (B1)-(B4) are satisfied and moreover that $\mathcal{B}(q)$ maps H-bounded subsets of V into V^*-bounded subsets, uniformly in q for $q \in Q$.

We take the observation space Z to be $\times_{i=1}^\nu \{\mathbf{R}^\ell \times L_2(0,1)\}$ and a weighted least-squares performance index, $\tilde{\Phi}$, of the form

$$(4.6) \qquad \tilde{\Phi}(\varphi, \psi; z) = \sum_{i=1}^\nu \left\{ \rho_i \sum_{j=1}^\ell |\varphi(t_i, x_j) - z_{i,j}^1|^2 + \sigma_i \int_0^1 |\psi(t_i, x) - z_i^2(x)|^2 dx \right\}$$

for $\varphi \in C(0,T;V), \psi \in C(0,T;H)$; and $z = ((z_1^1, z_1^2), (z_2^1, z_2^2), \ldots, (z_\nu^1, z_\nu^2)) \in Z$ with $\rho_i, \sigma_i \geq 0, i = 1, 2, \ldots, \nu, 0 < t_1 < t_2 < \cdots < t_\nu \leq T$, and $0 < x_1 < x_2 < \cdots < x_\ell < 1$.

As in our first order example, we employ linear spline based state approximation. For each $n = 1, 2, \ldots$ let $H_n = \text{span}\{\varphi_n^j\}_{j=1}^{n-1}$ where the φ_n^j are the standard linear B-splines on the interval $[0,1]$ defined with respect to the uniform mesh $\{0, 1/n, 2/n, \ldots, 1\}$ as given by (3.10). Let $P_n : H \to H_n$ denote the orthogonal projection of $L_2(0,1)$ onto H_n with respect to the standard L_2-inner product. We again use linear interpolating splines to discretize the admissible parameter set Q. For $m \in \mathbf{Z}^+$ and $q \in Q$ set

$$(4.7) \qquad (I^m q)(\theta) = \sum_{j=1}^{m} q(j\theta_q/m)\psi_j^m(|\theta|;\theta_q)sgn(\theta)$$

for $\theta \in \mathbf{R}$, where the $\psi_j^m(\cdot;\theta_q)$ are as they were defined in section 3, and θ_q is that number in $(0,\theta_0)$ for which $|q(\theta)| = q(\theta_q)$, $|\theta| \geq \theta_q$. (Note that in this case the lower limit of the sum in (4.7) is 1 rather than 0 since $q \in Q$ implies q(0) = 0.) We again have that condition (D) is satisfied and that $\lim_{m \to \infty} I^m q = q$, uniformly in q for $q \in Q$. We set $Q^m = I^m(Q)$.

For H_n and Q^m as defined above, the finite dimensional initial value problem (4.4), (4.5) takes the form

$$(4.8) \qquad M_n \ddot{w}_n(t) + C_n(\dot{w}_n(t);q^m) + K_n w_n(t) = F_n(t), \; 0 < t \leq T,$$

$$(4.9) \qquad M_n w_n(0) = w_n^0, \; M_n \dot{w}_n(0) = w_n^1,$$

where the $(n-1) \times (n-1)$ matrix M_n and the $(n-1)$-vectors $f_n(t)$ and w_n^0 are as they were defined in section 3, w_n^1 is the $(n-1)$-vector whose i-th component is given by $w_n^{1i} = \langle u^1, \varphi_n^i \rangle$, K_n is the $(n-1) \times (n-1)$ matrix whose (i,j)-th entry is given by $K_n^{i,j} = \langle E\varphi_n^i, \varphi_n^j \rangle$, and $C_n(\cdot;q^m): \mathbf{R}^{n-1} \to \mathbf{R}^{n-1}$ is given by

$$C_n^i(v;q^m) = \int_{\frac{i-1}{n}}^{\frac{i}{n}} \{nx - i + 1\}\, q^m\left(\{nx - i\}\{v^i - v^{i-1}\} + v^i\right) dx$$

$$+ \int_{\frac{i}{n}}^{\frac{i+1}{n}} \{i + 1 - nx\}\, q^m\left(\{nx - i\}\{v^{i+1} - v^i\} + v^i\right) dx,$$

$i = 1, 2, \ldots, n-1$, for $v \in \mathbf{R}^{n-1}$ with $v^0, v^n \equiv 0$. If $w_n(\cdot;q^m)$ is the solution to the second order initial value problem (4.8), (4.9) corresponding to $q^m \in Q^m$, then $u_n(t;q^m) = \sum_{j=1}^{n-1} w_n^j(t;q^m)\varphi_n^j$ and $\dot{u}_n(t;q^m) = \sum_{j=1}^{n-1} \dot{w}_n^j(t;q^m)\varphi_n^j$, for $t \in [0,T]$. If $q^m \in Q^m$ is given by $q^m(\theta) = \sum_{j=1}^{m} q_j^m \psi_j^m(|\theta|;\theta_{q^m})\, sgn(\theta)$, $\theta \in \mathbf{R}$, the identification problem (ID_n^m) becomes one of determining parameters $(\bar{q}_1^m, \ldots, \bar{q}_m^m, \bar{\theta}_{q^m})$ in some compact subset of $\mathbf{R}^{m+1}$ which minimize $\Phi_n(q^m) = \tilde{\Phi}_n(u_n(q^m), \dot{u}_n(q^m); z)$.

In order to actually test our scheme, we set

$$q^*(\theta) = \begin{cases} \beta|\theta|^\alpha sgn(\theta) & -\theta_{q^*} \leq \theta \leq \theta_{q^*} \\ \beta|\theta_{q^*}|^\alpha sgn(\theta) & |\theta| \geq \theta_{q^*} \end{cases}$$

with $\beta = .15, \alpha = 2$, and $\theta_{q^*} = 2.5$. With

$$y(t, x) = \{3 \cos(\frac{1}{3} \pi t) + 2 \sin(\frac{1}{2} \pi t)\} \sin \pi x$$

for $t > 0$ and $x \in [0, 1]$, we set

$$f(t, x) = \frac{\partial^2 y}{\partial t^2}(t, x) + q^*(\frac{\partial y}{\partial t}(t, x)) - E \frac{\partial^2 y}{\partial x^2}(t, x),$$

with $E = 1$, $u^0(x) = y(0, x) = 3 \sin \pi x$, and $u^1(x) = \frac{\partial y}{\partial t}(0, x) = \pi \sin \pi x$, for $t > 0$ and $x \in [0, 1]$. For observations upon which to base our fit, we took $z = \{(z_{i,1}^1, z_i^2)\}_{i=1}^{10}$ with $z_{i,1}^1 = y(.5i, .12)$ and $z_i^2(x) = \frac{\partial y}{\partial t}(.5i, x)$, $x \in [0, 1]$, $i = 1, 2, \ldots, 10$. As an initial guess we set

$$q^0(\theta) = \begin{cases} .6\theta & -1.2 \le \theta \le 1.2 \\ .6 sgn(\theta) & |\theta| \ge 1.2. \end{cases}$$

The weights $\{\rho_i\}_{i=1}^{10}$ and $\{\sigma_i\}_{i=1}^{10}$ in the performance index (4.6) were all set equal to 1.

Using the same computational techniques and resources (both hardware and software) that we used for the first order example described in the previous section, we obtained the results plotted in Figures 4.1 - 4.3 below. The CPU times on the Cray X-MP/48 for these runs ranged from 84.96 seconds for $n = 8$, $m = 3$ to 1032.78 seconds for $n = 20$, $m = 3$. When $n = 20$, the value of the performance index Φ_n was reduced from $\Phi_n(q^0) \approx 6.0 \times 10^{-2}$ to $\Phi_n(\bar{q}_n^m) \approx 2.0 \times 10^{-3}$. For other values of n, the reduction in Φ_n was less pronounced. In this particular example we found (and it is apparent from the figures) that truly satisfactory results could not be obtained until n was chosen sufficiently large. However, as is clear from Figure 4.3 the scheme performed extremely well when $n = 20$. Once again, as in the case of a first order system, we found that although our theory does not apply, similar results could be obtained using a performance index involving spatially discrete measurements of velocity. As expected, since the problems (ID_n^m) were solved unconstrained (i.e. the compactness assumption on Q, and therefore Q^m for each m, was not enforced) the presence of instabilities became apparent for each n with m sufficiently large.

5. Concluding remarks

In this paper we have summarized the theoretical framework for the identification of non-linear distributed parameter systems which we have developed elsewhere ([3] and [5]), and, more importantly, have for the first time provided numerical evidence that our approach is indeed feasible and in fact performs well. In the case of second order systems, while

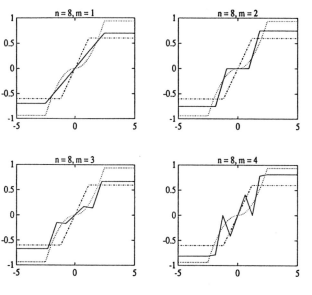

Figure 4.1 Legend: q^* ······ , $\bar{q}_n^m$ — — — — , q^0 — · — · — · .

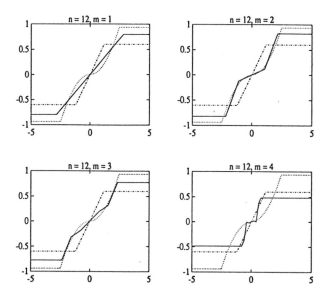

Figure 4.2 Legend: q^* ······ , $\bar{q}_n^m$ — — — — , q^0 — · — · — · .

our focus here has been on the identification of nonlinear damping in systems with linear stiffness, our theory is easily modified to handle the estimation of a nonlinear stiffness operator in the presence of linear damping. We are currently studying the extension of

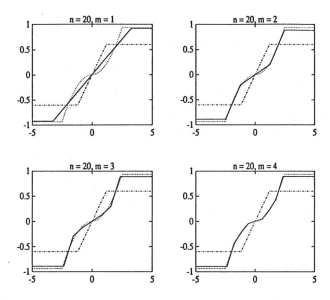

Figure 4.3 Legend: $q^* \cdots\cdots$, $\bar{q}_n^m - - - - -$, $q^0 - \cdot - \cdot - \cdot$.

our results to second order systems which simultaneously involve both nonlinear stiffness and damping. Further numerical studies involving supercomputing are also presently underway. In addition to continuing our efforts using simulation data, we intend to test our schemes using experimental data in the near future.

References

[1] H.T. Banks and D. Iles, *On compactness of admissible parameter sets: convergence and stability in inverse problems for distributed parameter systems*, in Control Problems for Systems Described by Partial Differential Equations and Applications, Proceedings of the IFIP-WG7.2 Working Conference, Gainsville, Florida, February 3-6, 1986, (I. Lasiecka and R. Triggiani, eds.), Lecture Notes in Control and Information Sciences 97, Springer-Verlag, New York (1987), 130–142.

[2] H.T. Banks and K. Ito, *A unified framework for approximation and inverse problems for distributed parameter systems*, Control-Theory and Advanced Technology **4** (1988), 73-90.

[3] H.T. Banks, S. Reich, and I.G. Rosen, *An approximation theory for the identification of nonlinear distributed parameter systems*, ICASE Report No. 88-26, Institute for Computer Applications in Science and Engineering, NASA Langley Research Center, Hampton, VA, April, 1988; SIAM J. Control and Opt., to appear.

[4] H.T. Banks, S. Reich, and I.G. Rosen, *Galerkin approximation for inverse problems for nonautonomous nonlinear distributed systems,*, ICASE Report No. 88-38 , Institute for Computer Application in Science and Engineering, NASA Langley Research Center, Hampton, VA, June, 1988; Applied Mathematics and Optimization, submitted.

[5] H.T. Banks, S. Reich, and I.G. Rosen, *Estimation of nonlinear damping in second order distributed parameter systems*, in preparation.

[6] V. Barbu, "Nonlinear Semigroups and Differential Equations in Banach Spaces," Noordhoff International, Leyden, The Netherlands, 1976.

[7] M.G. Crandall and L.C. Evans, *On the relation of the operator $\partial/\partial s + \partial/\partial t$ to evolution governed by accretive operators*, Israel J. Math. **21** (1975), 261-278.

[8] M.G. Crandall and A. Pazy, *Nonlinear evolution equations in Banach space*, Israel J. Math. **11** (1972), 57-94.

[9] L.C. Evans, *Nonlinear evolution equations in an arbitrary Banach space*, Israel J. Math **26** (1977), 1-42.

[10] J.A. Goldstein, *Approximation of nonlinear semigroups and evolution equations*, J. Math. Soc. Japan **24** (1972), 558-573.

[11] C. Kravaris and J.H. Seinfeld, *Identification of parameters in distributed systems by regularization*, SIAM J. Control and Opt. **23** (1985), 217-241.

[12] M.H. Schultz, "Spline Analysis," Prentice-Hall, Englewood Cliffs, N.J., 1973.

[13] J.C. Slattery, *Quasi-linear heat and mass transfer I. The constitutive equations*, Appl. Sci. Res. **12** Sec **A** (1963), 51-56.

[14] J.C. Slattery, *Quasi-linear heat and mass transfer II. Analyses of experiments*, Appl. Sci. Res. **12** Sec **A** (1963), 57-65.

[15] B.K. Swartz and R.S. Varga, *Error bounds for spline and L-spline interpolation*, J. Approx. Theory **6** (1972), 6-49.

[16] H. Tanabe, "Equations of Evolution," Pitman, London, 1979.

H.T. Banks
Center for Control Sciences
Division of Applied Mathematics
Brown University
Providence, RI 02912
U.S.A.

C.K. Lo
Department of Mathematics,
University of Southern California
Los Angeles, CA 90089
U.S.A.

Simeon Reich
Department of Mathematics
University of Southern California
Los Angeles, CA 90089
U.S.A.
and
Department of Mathematics
The Technion-Israel Institute of Technology
Haifa 32000
Israel

I.G. Rosen
Department of Mathematics
University of Southern California
Los Angeles, CA 90089
U.S.A.

International Series of
Numerical Mathematics, Vol. 91
© 1989 Birkhäuser Verlag Basel

Nonlinear systems on algebraic spaces

Z. Bartosiewicz

Politechnika Białostocka

Keywords. Ringed space, vector field, nonlinear system, observability, observation algebra.

0. Introduction

The aim of this paper is to advocate for a new concept of state space in infinite dimensional systems theory. It is known [CP] that Banach and Hilbert spaces form a nice class of objects which can serve as state spaces in the *linear* theory. However, in the nonlinear case this choice does not seem to be correct. Simple system-theoretic operations lead to systems whose state space are no longer linear spaces (e.g. when passing from unobservable systems to observable ones). Moreover, one can obtain a space which has singular points, so it is pointless to substitute Banach spaces by Banach manifolds as one could be tempted to. We propose ringed spaces and complete ringed spaces for state spaces of infinite dimensional systems. Our concept of system is almost purely algebraic. Actually we want to investigate how far we can go into systems theory using mainly algebraic tools. The first obstacle on this way is existence and uniqueness of solutions of ordinary differential equations defined on ringed spaces. We do not solve this problem here, so we have to assume these properties. However we show that if solutions do exist and are unique then one can develop a theory similarly as in the linear case. We study a specific problem of observability and construct a natural passing from an unobservable system to an observable one.

1. Ringed spaces and systems on them

To motivate introducing our concepts let us study a simple example.

1.1. Example. Let us consider the following delay system

$$\dot{x}(t) = x(t - h)$$
$$y(t) = x^2(t)$$

where $x(t) \in \mathbf{R}$. One may take $X = C([-h, 0]; \mathbf{R})$ as a state space [H], but it is clear that the system is not observable. For example two initial conditions φ and $-\varphi$ give the same output $y(t)$, $t \geq 0$, i.e. φ and $-\varphi$ are *indistinguishable*. We want to glue up all indistinguishable states in order to obtain an observable system. However the space obtained in this procedure is no longer linear. Moreover, it has a singularity at 0. Observe that when $h = 0$, i.e. the state space is finite dimensional, the situation does not improve. In this case the reduced state space is just the halfline $[0, +\infty)$.

1.2. Definition [P]. *A ringed space (over $\mathbf{R}$) is a pair (X, A), where X is a set and A is an algebra of real valued functions on X such that if $\varphi(x_1) = \varphi(x_2)$ for all $\varphi \in A$ then $x_1 = x_2$. Let e_x denote the homomorphism $e_x : A \to \mathbf{R} : \varphi \mapsto \varphi(x)$. We say that a ringed space (X, A) is complete if every homomorphism of algebras $A \to \mathbf{R}$ is of the form e_x for some $x \in X$. The (algebraic) dimension of a ringed space (X, A) is $\dim X = tr \deg A$.*

1.3. Examples of ringed spaces.

1.3.1. Let X be a linear space over $\mathbf{R}$, X' – its algebraic dual space, i.e. the space of linear functionals $X \to \mathbf{R}$. Put A to be the algebra generated by functionals in X'. If $\dim X$ is finite then the ringed space (X, A) is complete.

1.3.2. Let X be a Banach space over $\mathbf{R}$ and A be the algebra generated by linear *continuous* functionals $X \to \mathbf{R}$.

1.3.3. Let X be a completely regular (i.e. $T_{3^{1/2}}$) topological space and A be the algebra of real continuous functions on X [P].

1.3.4. Let $X = \mathbf{R}^n$ and $A = \mathbf{R}[x_1, \dots, x_n]$ – the algebra of polynomial functions on $\mathbf{R}^n$. It is a complete ringed space. In this case $\dim X = n$.

1.4. Definition. *Let (X, A) be a ringed space. A vector field f defined on $D \subseteq X$ is a derivation $A \to A_{|D}$, i.e. f is linear and $f(\varphi_1 \cdot \varphi_2) = f(\varphi_1) \cdot \varphi_2 + \varphi_1 \cdot f(\varphi_2)$.*

1.5. Remark. If X is a C^∞ real manifold, A is the algebra of C^∞ real functions on X and $D = X$ then the above definition is a standard definition in differential geometry of a C^∞ vector field on X. We introduce a domain D in order to study also distributed parameter systems where vector fields will not be everywhere defined.

1.6. Example. Let X be a linear space and $A = \mathbf{R}[X']$ be the algebra of functions generated by X'. Take a linear operator $F : X \to X$. It induces a vector field $f : A \to A$ (i.e. everywhere defined) as follows: for $\varphi \in X'$ $(f\varphi)(x) := \varphi(Fx)$ and $f(\varphi_1 \cdot \varphi_2) := (f\varphi_1) \cdot \varphi_2 + \varphi_1 \cdot (f\varphi_2)$ for $\varphi_1, \varphi_2 \in X'$. The same may be done for a linear operator $F : D \to X$, where D is a subspace of a Banach space X. Then D is the domain of f.

1.7. Definition. *A trajectory of a vector field f defined on D through $x_0 \in D$ is a map $x : [a, b] \to D$ such that $x(c) = x_0$ for some $c \in [a, b]$, and for all $\varphi \in A$ and $t \in [a, b]$*

$$\frac{d}{dt}(\varphi(x(t))) = (f\varphi)(x(t)).$$

1.8. Remark. Existence and uniqueness of trajectories seems to be a hard problem. We know how to solve it in many specific ringed spaces like Banach spaces or smooth manifolds. Trajectories also exist on certain spaces with singularities like algebraic varieties [B].

1.9. Example. Let X be a completely regular topological space with the algebra A of real continuous functions on X. Then there exists only the trivial vector field on X, i.e. $f\varphi = 0$ for all $\varphi \in A$ [P]. It is easy to see that trajectories exist and are unique; they are constant.

1.10. Definition. *By a (input/state/output) system we mean a 5-tuple $\Sigma = (U, X, Y, \mathcal{F}, h)$, where U is a set of control (input) values, X – a (ringed) state space (with and algebra A), Y – a (ringed) output space (with an algebra B), $\mathcal{F} = \{f_\alpha : \alpha \in U\}$ is a family of vector fields on domains $D_\alpha \subseteq X$, $\alpha \in U$, and represents dynamics of Σ, and $h : X \to Y$ is an output map. We assume that h is a ringed space morphism [P], i.e. for $\varphi \in B$ $\varphi \circ h \in A$. Let $\mathcal{U}$ be the set of all piecewise constant functions $u : [0, T] \to U$, where $T \geq O$ depends on u. We call it the set of controls (inputs). We also assume that all f_α are integrable, i.e. they have unique trajectories through all points $x_0 \in D_\alpha$.*

1.11. Remark. A system Σ may be represented by the following pair of equations

$$\begin{cases} \dot{x}(t) & = f_{u(t)}(x(t)) \\ y(t) & = h(x(t)) \end{cases}$$

where $u(t) \in U$, $x(t) \in X$ and $y(t) \in Y$. However, one should be aware that the first equation has only a symbolic meaning because $\dot{x}(t)$ was not defined. If we specify a control $u : [0,T] \to U$, $u(t) = \alpha_i$ for $t \in [t_{i-1}, t_i]$, $t_0 = 0$, $t_1 + \cdots + t_k = T$, then a *solution* of this equation is a map $x : [0,T] \to X$ such that $x_{|[t_{i-1}, t_i]}$ is a trajectory of the vector field f_{α_i} and $\varphi(x(\cdot))$ is continuous for all $\varphi \in A$. If we fix an initial state x_0 then one may define the *input-output map* $u \mapsto y$ corresponding to x_0. This map may be not defined for all $u \in \mathcal{U}$.

2. Observability

2.1 Definition. *Let Σ be a system. Two states x_1 and x_2 are said to be indistinguishable if the input-output maps corresponding to x_1 and x_2 are the same. Σ is observable if there are no indistinguishable states. This means the same as initial observability for linear systems in [CP].*

2.2. Remark. The definition of indistinguishability is clear when all f_α are defined on the whole space X and they are *complete*, i.e. their trajectories are defined for all times $t \in \mathbf{R}$. In other cases some comment is needed. Namely, in the noncomplete case we should compare only such inputs $u : [0,T] \to U$ for which the outputs corresponding to x_1 and x_2 are defined on the whole interval $[0,T]$. Moreover, if $u(t) = \alpha_1$ on $[0, t_1]$ then x_1 and x_2 should be the domain D_{α_1} of f_{α_1}.

2.3. Definition. *Let $\Sigma = (U, X, Y, \mathcal{F}, h)$. We assume for simplicity that $Y = \mathbf{R}$. Consider $x_0 \in X$, a control u, $u(t) = \alpha_i$ for $t \in [t_{i-1}, t_i]$, $i = 1, \ldots, k$, and the map $g : (t_1, \ldots, t_k) \mapsto y(t_1 + \cdots + t_k) \in \mathbf{R}$, where $y(\cdot)$ is the output corresponding to x_0 and u. If for every k and every sequence $(\alpha_1, \ldots, \alpha_k)$ the map g has an analytic extension onto a neighbourhood of $0 \in \mathbf{R}^k$ then the point x_0 is said to be analytic.*

2.4. Proposition. *If x_0 is analytic and $u(t) = \alpha$ is a constant control then $y^{(k)}(0) = (f_\alpha^k h)(x_0)$ (f_α^k means composition of the vector field f_α k-times).*

Proof: We have $\frac{d}{dt}y(t) = \frac{d}{dt}h(x(t)) = (f_\alpha h)(x(t))$ and so on. ∎

2.5. Proposition. *Let x_1 and x_2 be analytic. Then x_1 and x_2 are indistinguishable $\Leftrightarrow$ $h(x_1) = h(x_2)$ and $(f_{\alpha_1}, \ldots, f_{\alpha_k})(x_1) = (f_{\alpha_1}, \ldots, f_{\alpha_k} h)(x_2)$ for every $\alpha_1, \ldots, \alpha_k \in U$ and $k \in \mathbf{N}$.*

Proof: Let $y_1(\cdot)$ and $y_2(\cdot)$ denote the outputs corresponding to a control u and the initial states x_1 and x_2 respectively. The outputs are equal iff the maps g_1 and g_2 defined as in 2.3 are equal iff all their derivatives at 0 are equal. It is easy to see that $\frac{\partial}{\partial t_1}, \cdots, \frac{\partial}{\partial t_k} g_j(0, \cdots, 0) = (f_{\alpha_1}, \cdots, f_{\alpha_k} h)(x_j)$ (similarly as in 2.4). This ends the proof. ∎

Let X_α denotes the set of analytic points and let $\mathcal{H}(\Sigma)$ be the algebra of real functions on X_α generated by h and all $f_{\alpha_1}, \ldots, f_{\alpha_k} h$. We call $\mathcal{H}(\Sigma)$ the *observation algebra* of the system Σ.

2.6. Corollary. *If $X_\alpha = X$ then Σ is observable iff the observation algebra distinguishes points of X (i.e. if $\varphi(x_1) = \varphi(x_2)$ for all $\varphi \in \mathcal{H}(\Sigma)$ then $x_1 = x_2$).* ∎

2.7. Corollary. *If Σ is observable then $\mathcal{H}(\Sigma)$ distinguishes analytic points of Σ.* ∎

2.8. Example. Let us consider the following linear system

$$\begin{cases} \dot{x} &= Fx + Gu \\ y &= Hx \end{cases}$$

where $x(t) \in X$, $u(t) \in U$, X and U are Banach spaces, $y(t) \in Y = \mathbf{R}$ and $F : D \to X$ generates an analytic semigroup. This induces a system $\Sigma = (U, X, Y, \mathcal{F}, h)$, where $\mathcal{F} = \{f_\alpha : \alpha \in U\}$, f_α is the vector field induced by the operators $x \mapsto Fx + G\alpha$ and $h(x) = Hx$. It is easy to see that $\mathcal{H}(\Sigma)$ is generated by the functions $Hx, HF^2x, \ldots,$ where $x \in \cap_{i=1}^\infty D(F^i)$ (notice that the control part is not important). By 2.7 we obtain that if Σ is observable then

$$\ker \begin{bmatrix} H \\ HF \\ HF^2 \\ \vdots \end{bmatrix} = 0.$$

3. Reduction of the state space

We describe here passing from an unobservable system to an observable one. To simplify the procedure we assume that *the system is analytic*, i.e. all points of X are analytic. Then indistinguishability is an equivalent relation denoted by $\sim$. Actually, by 2.5, $x_1 \sim x_2$ iff $\varphi(x_1) = \varphi(x_2)$ for all $\varphi \in \mathcal{H}(\Sigma)$. Let us put $X' = X/_\sim$ – the set of equivalent classes. Denote by $\pi : X \to X'$ the canonical projection. Then $\pi^*\varphi' := \varphi' \circ \pi$ defines the pullback of real functions $\varphi' : X' \to R$. Observe that π^* is 1:1. Put $A' = (\pi^*)^{-1}(\mathcal{H}(\Sigma))$. It is clear that π^* is an isomorphism between A' and $\mathcal{H}(\Sigma)$. Hence, the ringed space (X', A') is our new state space. Let us construct a new dynamics $\mathcal{F}' = \{f'_\alpha : \alpha \in U\}$. For $\varphi' \in A'$ we put $(f'_\alpha\varphi')(\pi(x)) := (f'_\alpha(\pi^*\varphi'))(x)$. The definition does not depend on the choice of x. At last, put $h' = (\pi^*)^{-1}(h)$. Our new system is $\Sigma' = (U, X', Y, \mathcal{F}', h')$.

3.1 Proposition. *If $x(\cdot)$ is a trajectory of f_α through x_0 then $x' = \pi \circ x$ is a trajectory of f'_α through $x'_0 = \pi(x_0)$.*

Proof: Observe that $x'(c) = \pi(x(c)) = \pi(x_0) = x'_0$ and

$$\frac{d}{dt}\varphi'(x'(t)) = \frac{d}{dt}\varphi'(\pi(x'(t))) = \frac{d}{dt}(\pi^*\varphi')(x(t))$$
$$= (f_\alpha(\pi^*\varphi'))(x) = (f'_\alpha\varphi')(x'(t)). \quad ∎$$

3.2. Remark. If may be proved that if $x'(\cdot)$ is a trajectory of f'_α then $x' = \pi \circ x$, where $x(\cdot)$ is a trajectory of f_α. It implies uniqueness of trajectories of f'_α and this together with 3.1 gives integrability of f'_α.

3.3. Proposition. *For every point* $x \in X$ *we have*

$$(f_{\alpha_1}, \ldots, f_{\alpha_k} h)(x) = (f'_{\alpha_1}, \ldots, f'_{\alpha_k} h')(\pi(x)).$$

Proof: Similarly as in 3.1. ∎

3.4. Corollary. *Input-output maps corresponding to* $x \in X$ *and* $\pi(x) \in X'$ *are the same.* ∎

REFERENCES

[B] Bartosiewicz, Z., *Ordinary differential equations on real affine varieties*, Bull. Polish Academy of Sciences **35** (1987).

[CP] Curtain, R., and A. Pritchard, "Infinite Dimensional System Theory," Springer-Verlag, New York, 1978.

[H] Hale, J., "Theory of Functional Differential Equations," Springer-Verlag, New York, 1977.

[P] Palais, R., "Real Algebraic Differential Topology," Publish or Perish, Inc., 1981.

Z. Bartosiewicz
Politechnika Białostocka
Wiejska 45
Białystok
Poland

International Series of
Numerical Mathematics, Vol. 91
© 1989 Birkhäuser Verlag Basel

A note on the asymptotic behavior of controllability radii for a scalar hereditary system

John A. Burns*
Gunther H. Peichl**

Interdisciplinary Center for Applied Mathematics
Department of Mathematics
Virginia Polytechnic Institute and State University

Institut für Mathematik
Karl-Franzens-Universität Graz

Abstract. In this note we discuss the preservation and robustness of controllability of finite dimensional approximations to an (M_2-approximate) controllable delay system. We present a complete answer for a particular example, the general problem itself, however, is still open.

Keywords. Functional differential equations, controllability, approximation, robustness.

1980 *Mathematics subject classifications*: 34K35, 93B35, 93B05, 41A15

1. Introduction

During the past fifteen years several approximation schemes have been developed specifically for control problems governed by delay differential equations (see [2,14,15,16]). Most of this effort is devoted to the linear quadratic control problem and much of the literature is concerned with the convergence of approximating gain operators. This work lead to the discovery that not only should stabilizability and detectability be preserved under approximation, but the preservation of stabilizability needs to be uniform with respect to the approximation (see [3,14,16]). In recent years we have considered the problem of preserving other system properties under approximation. In particular, we have investigated several questions concerning the preservation of controllability under approximation for delay systems and special heat equations (see [5,6,7]).

We mention two additional issues that are closely related to the problem of preserving a specific system property under approximation. The first issue is concerned with determining what, if any, system properties are preserved by "parameter perturbations". The second issue involves the condition number of the numerical problem one must solve after the "approximate control problem" is constructed. For example, it is noted in [1,17] that if a finite dimensional linear quadratic regulator problem is "nearly unstabilizable"

*The work of this author was supported in part by the Air Force Office of Scientific Research under grant AFOSR-89-0001, the Defense Advanced Research Projects Agency under grant F49620-87-C-0116 and SDIO under contract F49620-87-C-0088. Parts of the research were carried out while the author was a visitor at the Institut für Mathematik at Karl-Franzens-Universität Graz, Austria, supported by the Fonds zur Förderung der wissenschaftlichen Forschung, Austria, under project S3206.
**This work was supported in part by DARPA under contract F49620-87-C-0116, while the author was a visiting professor at the Interdisciplinary Center for Applied Mathematics, VPI & SU, Blacksburg, Virginia 24061, USA

then the algebraic Riccati equation will be numerically ill-conditioned. The following simple example illustrates that the same issue occurs if one considers the finite dimensional pole-placement problem and "nearly uncontrollable" systems. Consider

$$(1.1) \qquad \frac{d}{dt}\begin{pmatrix} x_1(t) \\ x_2(t) \end{pmatrix} = \begin{pmatrix} 0 & 0 \\ 0 & -1 \end{pmatrix}\begin{pmatrix} x_1(t) \\ x_2(t) \end{pmatrix} + \begin{pmatrix} \delta \\ \epsilon \end{pmatrix}u(t).$$

Observe that the system is controllable if and only if $\epsilon\delta \neq 0$. Consider the problem of finding a feedback matrix $K = [k_1, k_2]$ such that the poles of the closed-loop system

$$(1.2) \qquad \frac{d}{dt}\begin{pmatrix} x_1(t) \\ x_2(t) \end{pmatrix} = \{\begin{pmatrix} 0 & 0 \\ 0 & -1 \end{pmatrix} + \begin{pmatrix} \delta \\ \epsilon \end{pmatrix}[k_1, k_2]\}\begin{pmatrix} x_1(t) \\ x_2(t) \end{pmatrix}$$

are located at $-\frac{1}{2} \pm \frac{\sqrt{3}}{2}i$. If $\epsilon\delta \neq 0$, then the unique solution to the problem is given by

$$(1.3) \qquad K = [-\frac{1}{\delta}, \frac{1}{\epsilon}].$$

Observe that $\|K\| \to \infty$ if $\epsilon\delta \to 0$ (i.e. if (1.1) becomes "nearly uncontrollable").

We review some basic definitions needed in the paper. Consider the infinite dimensional control system

$$(1.4) \qquad \dot{z}(t) = \mathcal{A}z(t) + \mathcal{B}u(t), \quad z(0) = z_0.$$

We assume that $\mathcal{A}$ generates a C_0-semigroup $S(t)$ on a Hilbert space Z and $\mathcal{B} : \mathbf{R}^m \to Z$ is a bounded linear operator. The control function u is assumed to be locally square-integrable. The mild solution of (1.4) is given by

$$(1.5) \qquad z(t) = S(t)z_0 + \int_0^t S(t-s)\mathcal{B}u(s)ds.$$

For $t > 0$, the reachable set at time t is given by

$$(1.6) \qquad \mathcal{R}(t) = \{\int_0^t S(t-s)\mathcal{B}u(s)ds \big| u \in L^2(0, t; \mathbf{R}^m)\}$$

and (1.4) is said to be exactly controllable in time t if $\mathcal{R}(t) = Z$, exactly controllable if $\cup_{t\geq0}\mathcal{R}(t) = Z$, approximately controllable in time t if $\overline{\mathcal{R}(t)} = Z$ and approximately controllable if $\overline{\cup_{t\geq0}\mathcal{R}(t)} = Z$. It is known that if $\mathcal{A}$ generates an analytic semigroup or if $\mathcal{A}$ is the generator of the solution semigroup for a retarded delay equation, then (1.4) is approximately controllable if and only if there is a finite time $\hat{t}$ such that $\overline{\mathcal{R}(\hat{t})} = Z$ (see [13,19]). System (1.4) is said to be exponentially stabilizable if there is a bounded linear operator $\mathcal{F} : Z \to \mathbf{R}^m$ such that the closed loop operator $\mathcal{A}_c = \mathcal{A} + \mathcal{B}\mathcal{F}$ generates a C_0-semigroup $S_c(t)$ satisfying

$$(1.7) \qquad \|S_c(t)\| \leq Me^{-\beta t}, \quad t \geq 0$$

for some $M \geq 1$, $\beta > 0$. One can add an output equation to (1.4) and also consider the properties of observability and detectability. Duality arguments can be used to reduce the questions considered in this paper to questions involving controllability and stabilizability.

In order to make precise the terms "nearly unstabilizable" and "nearly uncontrollable", we use the system radii measures found in [11,12,17]. As to a stability radius we mention [23] in which the norm of the smallest destabilizing perturbation for several distributed parameter systems is calculated. We remind the reader that we shall ultimately be interested in determining these radii for finite dimensional systems that arise as approximations to infinite dimensional control systems governed by delay equations. Although we shall concentrate on the single system property of controllability, we shall close the paper with some comments on stabilizability.

Given a $p \times q$ complex matrix M, the spectral norm is given by

$$\|M\| = \sup_{\|x\|_2 = 1} \|Mx\|_2$$

where $\| \cdot \|_2$ denotes the standard Euclidean norm on $\mathbf{C}^p$ (and $\mathbf{C}^q$). We shall identify the n-dimensional complex control system

(1.8) $$\dot{x}(t) = Ax(t) + Bu(t)$$

with the $n \times (n + m)$ matrix

(1.9) $$\Sigma = [A, B] \in \mathbf{C}^{n \times (n+m)},$$

where A and B are $n \times n$ and $n \times m$ (complex) matrices, respectively. In particular, let

(1.10) $$\Gamma_{n,m} = \{\Sigma = [A, B] | A \in \mathbf{C}^{n \times n}, B \in \mathbf{C}^{n \times m}\}$$

and let $\Omega_{n,m}$ denote the subspace of real systems

(1.11) $$\Omega_{n,m} = \{\Sigma = [A, B] | A \in \mathbf{R}^{n \times n}, B \in \mathbf{R}^{n \times m}\}.$$

The distance between the two systems

$$\Sigma_1 = [A_1, B_1] \in \Gamma_{n,m} \quad \text{and} \quad \Sigma_2 = [A_2, B_2] \in \Gamma_{n,m}$$

is defined by

(1.12) $$\delta(\Sigma_1, \Sigma_2) = \|[A_1 - A_2, B_1 - B_2]\|_2.$$

Given $\Sigma \in \Gamma_{n,m}$ and a subset $\mathcal{S} \subseteq \Gamma_{n,m}$ the distance between Σ and $\mathcal{S}$ is defined to be

(1.13) $$d(\Sigma; \mathcal{S}) = \inf\{\delta(\Sigma, \Sigma_\alpha) | \Sigma_\alpha \in \mathcal{S}\}.$$

Let $\tilde{C}_{n,m}$ be the subset of $\Gamma_{n,m}$ consisting of all systems that are *not* controllable and $\tilde{S}_{n,m}$ be the subset of $\Gamma_{n,m}$ consisting of all systems that are *not* stabilizable, i.e.

(1.14) $$\tilde{C}_{n,m} = \{[A, B] \in \Gamma_{n,m} | (A, B) \quad \text{is not controllable}\}$$

and

(1.15) $\tilde{S}_{n,m} = \{[A, B] \in \Gamma_{n,m} | (A, B) \text{ is not stabilizable}\}.$

Given $\Sigma \in \Gamma_{n,m}$ the controllability radius $\gamma_c(\Sigma)$ and the stabilizability radius $\gamma_s(\Sigma)$ are defined by

(1.16) $\gamma_c(\Sigma) = d(\Sigma; \tilde{C}_{n,m})$

and

(1.17) $\gamma_s(\Sigma) = d(\Sigma; \tilde{S}_{n,m}),$

respectively. These definitions provide a measure of the degree of controllability and sta-bilizability of *finite dimensional systems*. We note that $\gamma_c(\Sigma) = 0$ if and only if Σ is uncontrollable and $\gamma_s(\Sigma) = 0$ if and only if Σ is unstabilizable (see [17]).

Consider the example described by (1.1). Observe that if Σ becomes "nearly uncontrol-lable", i.e. if $\gamma_c(\Sigma) \to 0$, then either $\epsilon \to 0$ or $\delta \to 0$. Therefore, $\gamma_c(\Sigma) \to 0$ implies that $\|K\| \to \infty$ and the pole-placement problem becomes ill-posed. This is a special case of a more general result that relates the condition number of the pole-placement problem to the inverse of $\gamma_c(\Sigma)$ (see [8,9,10]). Thus, the controllability radius provides a measure of the numerical conditioning of the pole-placement problem. It is shown in [1,17] that the condition number of the algebraic Riccati equation associated with the linear quadratic regulator problem is proportional to the inverse of $\gamma_s(\Sigma)$. Thus these system radii provide information about the nature of the numerical problems one faces when solving various control problems described by Σ. For further results concerning the conditioning of con-trollability problems we refer to [22].

2. Delay Systems

In this section we will illustrate some of the ideas discussed above by means of a simple single delay equation

(2.1)
$$\dot{x}(t) = A_0 x(t) + A_1 x(t - h) + Bu(t),$$
$$x(0) = \eta, \quad x(s) = \varphi(s) \quad \text{a.e. on} \quad [-h, 0),$$

where A_0, A_1 are $n \times n$ matrices, B is an $n \times m$ matrix, $\eta \in \mathbf{R}^n$, $\varphi \in L^2(-h, 0; \mathbf{R}^n)$ and $u \in L^2_{\text{loc}}(0, \infty; \mathbf{R}^m)$. It is well known [2,4], that (2.1) is well posed on the state space $Z = \mathbf{R}^n \times L^2(-h, 0; \mathbf{R}^n)$. In particular, define $\mathcal{A}$ on the domain

$$\mathcal{D}(\mathcal{A}) = \{(\eta, \varphi) | \varphi \in H^1(-h, 0; \mathbf{R}^n), \quad \varphi(0) = \eta\}$$

by

$$\mathcal{A}(\eta, \varphi) = (A_0 \eta + A_1 \varphi(-h), \dot{\varphi})$$

and let $\mathcal{B} : \mathbf{R}^m \to Z$

$$\mathcal{B}u = (Bu, 0),$$

then $\mathcal{A}$ generates a C_0-semigroup on Z and (2.1) is equivalent to an abstract Cauchy problem (1.4). In view of the smoothing property of the semigroup the best one can hope for is approximate controllability. An algebraic criterion for approximate controllability of (2.1) is given in [18,19]. It is quite natural to ask whether this property is robust under perturbations in the system matrices A_0, A_1, B and the delay h. Recently L. Pandolfi ([21]) has shown that exact controllability for general abstract control systems (1.4) is preserved under sufficiently small bounded perturbations of $\mathcal{A}$ and $\mathcal{B}$. Although (2.1) is only approximately controllable and perturbations in A_1 and h amount to unbounded perturbations of the generator $\mathcal{A}$, we have a similar result for (2.1). More specifically, we consider the perturbed system

$$(2.2) \qquad \dot{x}(t) = (A_0 + \delta A_0)x(t) + (A_1 + \delta A_1)x(t - h + \delta h) + (B + \delta B)u,$$

where δA_0, δA_1, δB are **real** matrices of the appropriate format and $\delta h \in \mathbf{R}$. The following result has been shown in [6].

Theorem 2.1. *Let (2.1) be approximately controllable and assume $m = 1$. Then there exists $\rho > 0$ such that (2.2) is approximately controllable provided*

$$\|\delta A_0\|^2 + \|\delta A_1\|^2 + \|\delta B\|^2 + |\delta h|^2 < \rho.$$

It is still an open problem to calculate the optimal value of ρ such that Theorem 2.1 remains true.

Example 2.2 (L. Pandolfi, [21]):
Consider

$$(2.3) \qquad \dot{x}(t) = A_0 x(t) + \sum_{i=1}^{3} A_i x(t - h_i) + B u(t),$$

where

$$A_0 = 0, \quad A_1 = \begin{pmatrix} 0 & -1 \\ 0 & 0 \end{pmatrix}, \quad A_2 = \begin{pmatrix} 0 & \alpha \\ 0 & 0 \end{pmatrix}, \quad A_3 = \begin{pmatrix} 1 & -2 \\ 0 & -1 \end{pmatrix},$$

$$B = \begin{pmatrix} 1 \\ 1 \end{pmatrix}.$$

If we choose $0 < h_1 < h_2 < h_3 = 2$ then the nominal system which corresponds to the parameter $\alpha = 0$ is approximately controllable. However, there exist $\alpha \in \mathbf{C}\backslash\mathbf{R}$ which may be chosen arbitrarily small such that (2.3) is *not* spectrally controllable. Since A_3 has full rank this implies that (2.3) is not approximately controllable.

This example does not contradict Theorem 2.1 since it involves more than one delay and since the applied perturbation is necessarily complex. Hence it is not clear whether Theorem 2.1 extends to more general delay differential equations.

We conclude this section with a few remarks concerning finite dimensional approximations to (2.1). From the numerous schemes developed we just single out the scheme of averaging projections (AVE) [2] and the spline based scheme [15] for which we will discuss some numerical experiments in section 3. All of these schemes lead to a system of ordinary differential equations

(2.4) $$\dot{x}^N(t) = A^N x^N(t) + B^N u^N(t).$$

For the AVE-scheme — which will be dealt with subsequently — A^N is an $n(N + 1) \times n(N + 1)$ matrix, B^N an $n(N + 1) \times m$ matrix. They are given by

$$(\Sigma_{\mathrm{AVE}}^N) \qquad A^N = \begin{pmatrix} A_0 & 0 & \cdots\cdots\cdots & 0 & A_1 \\ \frac{N}{h}I & -\frac{N}{h}I & & & \\ & \ddots & \ddots & & 0 \\ & & \ddots & \ddots & \\ 0 & & & \ddots & \\ & & & \frac{N}{h}I & -\frac{N}{h}I \end{pmatrix}, \quad B^N = \begin{pmatrix} B \\ 0 \\ \vdots \\ 0 \end{pmatrix}.$$

Since one reason for constructing finite dimensional approximations to infinite dimensional systems such as those which were discussed above is their use in control design a prominent question is whether (approximate) controllability carries over to the approximating systems. This question is still unresolved in general. However for a special case of (2.1) there is a positive result. The proof of the following theorem may be found in [6].

Theorem 2.3. *Let (2.1) be approximately controllable and assume* $rkB = 1$. *Then* Σ_{AVE}^N *is controllable for all N sufficiently large.*

As noted in section 1 the fact that Σ_{AVE}^N is controllable (this is almost impossible to check by numerical means alone [20]) is not sufficient to imply that this model is good for control design. The more important issue is the distance of Σ^N to the nearest uncontrollable system. Using the results in [7,11] we calculated $\gamma_c^N = \gamma_c(\Sigma_{\mathrm{AVE}}^N)$ for the following example

(2.5) $$\dot{x}(t) = x(t) + x(t-1) + u(t).$$

The system (2.5) is evidently approximately controllable, hence by Theorem 2.3 its finite dimensional approximations based on averaging projections are controllable for N sufficiently large (a straightforward calculation shows that indeed Σ_{AVE}^N is controllable for all N). In the following table $k(N)$ denotes the dimension of Σ_{AVE}^N which is $N + 1$. We also give an upper bound for γ_c^N which is shown in remark 2 below. This bound is not useful for $N = 2, 4, 6$.

$k(N)$	$[\gamma_c^N]^2$	upper bound
3	0.20123	–
4	0.15781	0.2260
5	0.16379	–
6	0.14370	0.2021
7	0.14147	–
8	0.12875	0.1820
9	0.12510	0.2756
10	0.11610	0.1665

Table 2.1

The numerical results do not give sufficient evidence as to the asymptotic behavior of $[\gamma_c^N]^2$. The following theorem provides an answer to this question for the particular example (2.5). It requires a technical lemma which we state below. Since its proof is tedious, though elementary, we defer it to the end of the paper.

Lemma 2.4. Let $\mathcal{J}^N = \{z = \sigma + i\tau \,|\, (\sigma, \tau) \in [-1 + \frac{1}{N}, -1 + \frac{1}{\sqrt{N}}] \times [0, \frac{\pi}{N}] \subset \mathbf{C}$. Then for N sufficiently large $\mathcal{J}^N$ contains at least one root y_N of (2.8) and

$$(2.6) \qquad (1 - \frac{1}{\sqrt{N}})^2 \le |y_N|^2 \le (1 - \frac{1}{N})^2 + \frac{\pi^2}{N^2}.$$

Theorem 2.5. Let the finite dimensional approximations of (2.5) based on averaging projections be denoted by Σ_{AVE}^N. Then

$$\lim_{N \to \infty} [\gamma_c^N]^2 = 0,$$

where $\gamma_c^N = \gamma_c(\Sigma_{AVE}^N)$.

Proof: The proof is based on the observation that for any normalized left eigenvector ν of A^N, $\|\nu^* B^N\|_2^2$ provides an upper bound for $[\gamma_c^N]^2$ ([7]). It is easy to see that any left eigenvector $\nu = \mathrm{col}(\overline{\xi}_0, \overline{\xi}_1, \ldots, \overline{\xi}_N)$, $\xi_i \in \mathbf{C}$, $i = 0, \ldots, N$ of A^N associated with an eigenvalue λ is characterized by the following set of equations

$$(2.7) \qquad \begin{aligned} \xi_0 + N\xi_1 &= \lambda \xi_0 \\ -N\xi_i + N\xi_{i+1} &= \lambda \xi_i, \quad i = 1, \ldots, N-1, \\ \xi_0 - N\xi_N &= \lambda \xi_N. \end{aligned}$$

Since $\lambda \notin \{1, -N\}$ the solution of (2.7) is given by

$$\xi_i(\lambda) = N^{-1}(1 + \frac{\lambda}{N})^{i-1}(\lambda - 1)\xi_0, \quad i = 1, \ldots, N,$$

where λ is a root of the characteristic equation

$$1 = (1 + \frac{\lambda}{N})^N (\lambda - 1).$$

Since ξ_0 is left arbitrary we set $\xi_0 = 1$. Performing the transformation $\lambda = N(y - 1)$ we finally obtain

$$\xi_0 = 1$$

$$\xi_i(y) = (y - 1 - \frac{1}{N})y^{i-1}, \quad i = 1, \ldots, N,$$

where y is a root of

(2.8) $$Ny^{N+1} - (N+1)y^N - 1 = 0.$$

Consequently,

(2.9) $$\|\nu(y)\|^2 = 1 + |y - 1 - \frac{1}{N}|^2 \frac{1 - |y|^{2N}}{1 - |y|^2}.$$

As a consequence of lemma 2.4 we have that for N sufficiently large

(2.10) $$|y_N|^{2N} \leq (1 - \frac{1}{N})^{2N}(1 + \frac{4\pi^2}{N^2})^N < 1.$$

Inserting (2.10) into (2.9) and using (2.6) one obtains the following lower bound

$$\|\nu(y_N)\|^2 \geq 1 + (2 - \frac{1}{\sqrt{N}} + \frac{1}{N})^2 \frac{1 - (1 - \frac{1}{N})^{2N}(1 + \frac{4\pi^2}{N^2})^N}{1 - (1 - \frac{1}{\sqrt{N}})^2}$$

which is easily seen to diverge. Observing that $\|\nu^* B^N\|^2 = \|\nu(y_N)\|_2^{-2}$ completes the proof of the theorem.

Remarks:

(1) It is not unexpected that γ_c^N converges to 0. The reason lies in the fact that the definition of γ_c^N allows for arbitrary perturbations of A^N and B^N. In particular, the structural zeros of A^N and B^N may be destroyed. As a consequence there will be no relationship between a perturbation of (A^N, B^N) and the finite dimensional approximation to a perturbation of the original delay equation. Furthermore it is known that for general control canonical systems the controllability radii tend to 0 as their dimension increases (see [17]). Since however perturbations are likely to occur only in a few distinct entries of A^N and B^N this example also shows the need for a measure of the controllability radius based on structured (real) perturbations of A^N and B^N only.

(2) Obviously the above proof also provides the following upper bound for $[\gamma_c^N]^2$

$$[\gamma_c^N]^2 \leq [1 + (2 - \frac{1}{\sqrt{N}} + \frac{1}{N})^2 \frac{1 - [(1 - \frac{1}{N})^2 + \kappa \frac{\pi^2}{N^2}]^N}{1 - (1 - \frac{1}{\sqrt{N}})^2}]^{-1}$$

where $\kappa = 0$ if N is odd and $\kappa = 1$ if N is even ($N \geq 8$). Table 3.1 shows some values for this upper bound. As a consequence we also infer that $[\gamma_c^N]^2 = O(\frac{1}{\sqrt{N}})$ as $N \to \infty$.

(3) Although we conjecture that $\gamma_c(\Sigma^N)$ converges to zero for any of the spline schemes in [15,16] applied to (2.1), we do not have a rigorous proof of this fact.

3. Closing remarks

It is interesting to note that the proof that $\gamma_c(\Sigma_{\text{AVE}}^N) \to 0$ even for the simple delay system (2.5) is not straightforward. In fact, the proof does not (easily) extend to more complex systems. The following table (see also [6]) shows that the different schemes produce significantly different radii. Again we consider (2.5) and the averaging scheme AVE and the Ito-Kappel spline scheme SPL. Table 3.1 compares the values of γ_c for these two schemes.

$k(N)$	$\gamma_c(\Sigma_{\text{AVE}}^N)^2$	$\gamma_c(\Sigma_{\text{SPL}}^N)^2$	$\gamma_s(\Sigma_{\text{AVE}}^N)^2$	$\gamma_s(\Sigma_{\text{SPL}}^N)^2$
3	0.20123	0.30390	0.8821	0.8361
4	0.15781	0.27015	0.9088	0.8898
5	0.16379	0.25545	0.9259	0.9112
6	0.14370	0.24865	0.9377	0.9277
7	0.14147	0.22123	0.9462	0.9390
8	0.12875	0.20929	0.9527	0.9473
9	0.12510	0.20016	0.9578	0.9607
10	0.11610	0.18412	0.9619	0.9585

Table 3.1

In Table 3.1 we also present the stabilizability radii $\gamma_s(\Sigma_{\text{AVE}}^N)$ and $\gamma_s(\Sigma_{\text{SPL}}^N)$. Apparently they seem to converge to a non-zero value. It is not known if this property is related to the uniform preservation of stabilizability under approximation.

Proof of Lemma 2.4: Obviously y is a zero of (2.8) if and only if it is also a zero of

$$(3.1) \qquad w(y) = y^N[y - (1 + \frac{1}{N})] - \frac{1}{N}.$$

First we show that for N odd there is a zero of w in the interval $(-1 + \frac{1}{N}, -1 + \frac{1}{\sqrt{N}})$. This is evident because

$$w(-1 + \frac{1}{N}) = 2(1 - \frac{1}{N})^N - \frac{1}{N} > 0,$$

$$w(-1 + \frac{1}{\sqrt{N}}) = \frac{1}{N}[(-1 + \frac{1}{\sqrt{N}})^N N(-2 + \frac{1}{\sqrt{N}} - \frac{1}{N}) - 1] < 0,$$

the inequalities being true for N sufficiently large. In case N is even we apply the argument principle to $\mathcal{J}^N$. Let $\partial \mathcal{J}^N$, the boundary of $\mathcal{J}^N$, be decomposed as $\partial \mathcal{J}^N = \mathcal{J}_1^N \cup \mathcal{J}_2^N \cup \mathcal{J}_3^N \cup \mathcal{J}_4^N$ with

$$(3.2)$$
$$\mathcal{J}_1^N = [-1 + \frac{1}{N}, -1 + \frac{1}{\sqrt{N}}],$$

$$\mathcal{J}_2^N = \{y = -1 + \frac{1}{\sqrt{N}} + i\pi\frac{\tau}{N} | \tau \in [0,1]\},$$

$$\mathcal{J}_3^N = \{y = -1 + \frac{1}{\sqrt{N}} + \xi(\frac{1}{N} - \frac{1}{\sqrt{N}}) + i\frac{\pi}{N} | \xi \in [0,1]\},$$

$$\mathcal{J}_4^N = \{y = -1 + \frac{1}{N} + i\pi\frac{\tau}{N} | \tau \in [0,1]\}.$$

It is straightforward to show that there is no change of the argument of w on $\mathcal{J}_1^N$. Next we consider $y \in \mathcal{J}_2^N$. Expanding the arctan we find after a short calculation

$$(3.3) \qquad \arg\left(y^N(y - 1 - \frac{1}{N})\right) = \pi - \pi\tau\left(1 + \frac{1}{\sqrt{N}} + \frac{3}{2N}\right) + o\left(\frac{\tau\pi}{N}\right).$$

and similarly

$$|y^N| = \exp\left(-\sqrt{N} - \frac{1}{2}\right)\left(1 + O\left(\frac{1}{\sqrt{N}}\right)\right),$$

$$|y - 1 - \frac{1}{N}| = 2 + O\left(\frac{1}{\sqrt{N}}\right),$$

so that

$$(3.4) \qquad |y|^N|y - 1 - \frac{1}{N}| = 2\exp\left(-\sqrt{N} - \frac{1}{2}\right)\left(1 + O\left(\frac{1}{\sqrt{N}}\right)\right).$$

Equations (3.3) and (3.4) imply that

$$\operatorname{Im} w(y) = \operatorname{Im}\left(y^N(y - 1 - \frac{1}{N})\right)$$
$$= 2\exp\left(-\sqrt{N} - \frac{1}{2}\right)\left(1 + O\left(\frac{1}{\sqrt{N}}\right)\right)\left[\sin\left(\pi\tau(1 + \frac{1}{\sqrt{N}} + \frac{3}{2N})\right) + o\left(\frac{\tau\pi}{N}\right)\right]$$

and

$$\operatorname{Re}\left(y^N(y - 1 - \frac{1}{N})\right) = 2\exp\left(-\sqrt{N} - \frac{1}{2}\right)\left(1 + O\left(\frac{1}{\sqrt{N}}\right)\right)\times$$
$$\times\left[\cos\left(\pi\tau(1 + \frac{1}{\sqrt{N}} + \frac{3}{2N})\right) + o\left(\frac{\tau\pi}{N}\right)\right].$$

This shows that $y^N(y - 1 - \frac{1}{N})$ follows approximately a segment of a circle with radius $2e^{-\sqrt{N}-1/2}$ and opening $\pi(1 + \frac{1}{\sqrt{N}} + \frac{3}{2N})$ as y traverses $\mathcal{J}_2^N$. Since (up to higher order terms)

$$\operatorname{Re}\left(y^N(y - 1 - \frac{1}{N})\right) < 2e^{-\sqrt{N}-\frac{1}{2}}\left(1 + O\left(\frac{1}{N}\right)\right) < \frac{1}{N},$$

we conclude that

$$\operatorname{Re} w(y) < 0 \quad \text{for all} \quad \tau \in [0, 1].$$

In accordance with the above remark we also have

$$\operatorname{Im} w(y) \begin{cases} \geq 0 & \text{for } \tau \in [0, (1 + \frac{1}{\sqrt{N}} + \frac{3}{2N})^{-1}], \\ < 0 & \text{for } \tau \in ((1 + \frac{1}{\sqrt{N}} + \frac{3}{2N})^{-1}, 1]. \end{cases}$$

Next we choose $y \in \mathcal{J}_3^N$ which leads to

$$(3.5) \qquad \arg y^N(y - 1 - \frac{1}{N}) = -\pi\left[\frac{1}{\sqrt{N}} + \frac{3}{2N} - \xi\left[\frac{1}{\sqrt{N}} + \frac{1}{N} - \frac{\xi}{N}\right]\right] + o\left(\frac{\pi}{N}\right)$$

and this coincides (up to higher order terms) for $\xi = 0$ with the estimate obtained along $\mathcal{J}_2^N$. We note that

$$|y|^N|y - 1 - \frac{1}{N}| = [(-1 + \frac{1}{\sqrt{N}} + \xi(\frac{1}{N} - \frac{1}{\sqrt{N}}))^2 + \frac{\pi^2}{N^2}]^{N/2} \times$$

$$\times [(-2 + (\frac{1}{\sqrt{N}} - \frac{1}{N})(1 - \xi))^2 + \frac{\pi^2}{N^2}]^{1/2}$$

increases because each factor is monotonically increasing as ξ increases. This together with the estimate

$$\sin[\arg(y^N(y - 1 - \frac{1}{N}))] = -\sin\pi[\frac{1}{\sqrt{N}} + \frac{3}{2N} - \xi[\frac{1}{\sqrt{N}} + \frac{1}{N} - \frac{\xi}{N}]] + o(\frac{\pi}{N})$$

$$\leq -\sin\frac{3\pi}{2N}, \qquad \xi \in [0, 1],$$

shows that

$$\operatorname{Im} w(z) \leq -|y^N(y - 1 - \frac{1}{N})|_{\xi=0} \sin\frac{3\pi}{2N} < 0 \quad \text{for all} \qquad \xi \in [0, 1].$$

Analogously one may show that

$$\operatorname{Re}(y^N(y - 1 - \frac{1}{N})) = |y^N(y - 1 - \frac{1}{N})|[\cos\pi(\frac{1}{\sqrt{N}} + \frac{3}{2N} - \xi[\frac{1}{\sqrt{N}} + \frac{1}{N} - \frac{\xi}{N}])] + o(\frac{\pi}{N})$$

strictly increases as $\xi \to 1$ and attains its maximum at the left endpoint of $\mathcal{J}_3^N$. Consequently

$$\operatorname{Re}(y^N(y - 1 - \frac{1}{N})|_{\xi=1}) = ((1 - \frac{1}{N})^2 + \frac{\pi^2}{N^2})^{N/2}(4 + \frac{\pi^2}{N^2})^{1/2} \times$$

$$\times [\cos\frac{3\pi}{2N} + o(\frac{\pi}{N})]$$

$$= 2(1 - \frac{1}{N})^N (1 + O(\frac{1}{N}))[\cos\frac{3\pi}{2N} + o(\frac{\pi}{N})] > \frac{1}{N},$$

the last inequality being valid for N sufficiently large. Thus the above considerations show that $w(y)$ crosses the imaginary axis in the open lower halfplane as y traverses $\mathcal{J}_3^N$. Finally we consider $y \in \mathcal{J}_4^N$ which implies that

(3.6) $$\arg(y^N(y - 1 - \frac{1}{N})) = \pi - \pi\tau(1 + \frac{3}{2N}) + o(\frac{\pi\tau}{N}).$$

Again this estimate matches continuously the one obtained along $\mathcal{J}_3^N$. It is easily seen that

(3.7) $$|y^N||y - 1 - \frac{1}{N}| \geq 2(1 - \frac{1}{N})^N$$

and that (within the indicated order of accuracy)

$$\operatorname{Im} w(z) \begin{cases} < 0 & \text{for} \quad \tau \in ((1 + \frac{3}{2N})^{-1}, 1], \\ \geq 0 & \text{for} \quad \tau \in [0, (1 + \frac{3}{2N})^{-1}]. \end{cases}$$

Since $\tau \to \cos \pi \tau (1 + \frac{3}{2N})$ strictly increases for $\tau \in ((1 + \frac{3}{2N})^{-1}, 1]$ and decreases for $\tau \in (0, (1 + \frac{3}{2N})^{-1}]$ we conclude (for N sufficiently large) that

$$\operatorname{Re}(y^N(y - 1 - \frac{1}{N})) \geq 2(1 - \frac{1}{N})^N > \frac{1}{N} \quad \text{for} \quad \tau \in ((1 + \frac{3}{2N})^{-1}, 1]$$

and

$$\operatorname{Re}(y^N(y - 1 - \frac{1}{N})) < 0 \qquad \text{for} \quad \tau \in [0, \frac{1}{2}(1 + \frac{3}{2N})^{-1}).$$

This shows that $w(z)$ crosses the real axis in the right half plane, remains thereafter in the (open) upper half plane and intersects the negative real axis again at $-2(1 - \frac{1}{N})^N - \frac{1}{N}$. Invoking the argument principle this shows that $\mathcal{J}^N$ contains (exactly) one zero of $w(y)$ for N even.

REFERENCES

[1] Arnold, W.F. and Laub, A.J., *Generalized algorithms and software for algebraic Riccati equations*, Proc. IEEE **12** (1984), 1746 - 1754.

[2] Banks, H.T. and Burns, J.A., *Hereditary control problems: numerical methods based on averaging approximations*, SIAM J. Control Opt. **16** (1978), 169-208.

[3] Banks, H.T. and Kunisch, K., *The linear regulator problem for parabolic systems*, SIAM J. Control Opt. **22** (1984), 684-698.

[4] Bernier, C. and Manitius, A., *On semigroups in $\mathbf{R}^n \times L^p$ corresponding to differential equations with delay*, Can. J. Math. **30** (1978), 897-914.

[5] Burns, J.A. and Peichl, G., *On robustness of controllability for finite dimensional approximations of distributed parameter systems*, in "Proceedings of the IMACS/IFAC International Symposium on Modelling and Simulation of Distributed Parameter Systems," . Sunahara, Y., Tzafestas, S.G., Futagami, T., Eds., Hiroshima Institute of Technology, Hiroshima, 1987, pp. 491-496.

[6] Burns, J.A. and Peichl, G., *Preservation of controllability under approximation and controllability radii for hereditary systems*, Differential and Integral Equations (to appear).

[7] Burns, J.A. and Peichl, G., *Open questions concerning approximation and control of infinite dimensional systems*, submitted.

[8] Demmel, J.W., *On condition numbers and the distance to the nearest ill posed problem*, Numer. Math. **51** (1987), 251-289.

[9] Demmel, J.W. and Kagstrom, B., *Accurate solutions of illposed problems in control theory*, SIAM J. Matrix Anal. Appl. **9** (1988), 126-145.

[10] Demmel, J.W., *The probability that a numerical analysis problem is difficult*, Math. of Computations **50** (1988), 449-480.

[11] Eising, R., *The distance between a system and the set of uncontrollable systems*, in "Lecture Notes in Control and Information Sciences," Proc. MTNS-83, Beer Sheva, Springer, New York, 1984, pp. 303-314.

[12] Eising, R., *Between controllable and uncontrollable*, Systems Control Lett. **4** (1984), 263-264.

[13] Fattorini, H., *On complete controllability of linear systems*, JDE **3** (1967), 391-462.

[14] Gibson, J.S., *Linear quadratic optimal control of hereditary differential systems: Infinite dimensional Riccati equations and numerical approximations*, SIAM J. Control Opt. **21** (1983), 95-139.

[15] Ito, K. and Kappel, F., *A uniformly differentiable approximation scheme for delay systems using splines*, J. Appl. Math. Opt. (submitted).

[16] Kappel, F. and Salamon, D., *On the stability properties of spline approximations for retarded systems*, SIAM J. Control Opt. **27** (1989), 407-431.

[17] Kenney, C. and Laub, A.J., *Controllability and stability radii for companion form systems*, Math. Control Signals Systems (to appear).

[18] Manitius, A., *Necessary and sufficient conditions of approximate controllability for general linear retarded systems*, SIAM J. Control Opt. **19** (1981), 516-532.

[19] Manitius, A. and Triggiani, R., *Function space controllability of linear retarded systems: a derivation from abstract operator conditions*, SIAM J. Control Opt. **16** (1978), 599-645.

[20] Paige, Chr., *Properties of Numerical Algorithms Related to Computing Controllability*, IEEE Trans. Autom. Control **AC-26** (1981), 130-138.

[21] Pandolfi, L., *Controllability properties of perturbed distributed parameter systems*, Technical Report, Politechnio di Torino (1988).

[21] Pijnacker R., Pritchard, A.J. and Townley, S., *Conditioning of controllability problems*, report No. 146, Control theory Center Warwick (1987).

[23] Pritchard, A.J. and Townley, S., *A stability radius for infinite dimensional systems*, in "Distributed parameter systems," (F.Kappel, K.Kunisch, W.Schappacher, eds.), Lecture Notes in Control and Information Sciences 102, 1986, pp. 272-291.

John A. Burns
Interdisciplinary Center for Applied Mathematics
Department of Mathematics
Virginia Polytechnic Institute and State University
Blacksburg, VA 24061
USA

Gunther H. Peichl
Institut für Mathematik
Karl-Franzens-Universität Graz
Elisabethstraße 16, A-8010 Graz
Austria

International Series of
Numerical Mathematics, Vol. 91
© 1989 Birkhäuser Verlag Basel

Well posedness of triples of operators
(in the sense of linear systems theory)

Ruth F. Curtain and George Weiss

Department of Mathematics
University of Groningen

Department of Theoretical Mathematics
The Weizmann Institute

Abstract. A triple of operators (A, B, C) is called well posed if there exists an abstract linear system having A as the generator of its semigroup, B as its control operator and C as its observation operator. The main result of this paper is a set of necessary and sufficient conditions for (A, B, C) to be well posed. Essential use is made of the concept of transfer function and the theory is illustrated by two examples. In the second example we solve the delicate problem of modelling the one dimensional heat equation with Dirichlet boundary control and point observation with an abstract linear system.

Keywords. Unbounded control and observation operators, admissibility, well posedness.

1980 *Mathematics subject classifications*: 93C25

1. Introduction

This article is a sequel to the article *"The Representation of Regular Linear Systems on Hilbert Spaces"*, by George Weiss, appearing in this volume. We will use freely the terminology and the results from this paper using the prefix A; e.g., Definition A1.1 refers to Definition 1.1 from the latter paper.

Here we address the following problem: Given a triple of operators, (A, B, C), is there some abstract linear system Σ such that A is the generator of the semigroup of Σ, B is the control operator of Σ and C is the observation operator of Σ ? (For definitions of the terminology used see Sections A1 and A3.) If the answer is yes, we call the triple (A, B, C) *well posed* (a formal definition of well posedness as well as its interpretation in terms of well posedness of differential equations will be given in Section 2).

Our main result, Theorem 5.1, establishes necessary and sufficient conditions for the well posedness of a triple. These conditions are partly based on properties of the transfer function of an abstract linear system, which is the reason for investigating the notion of transfer function more carefully in Section 4. We illustrate our results with two examples in Section 6.

2. Admissibility and well posedness

Our first definition follows Weiss [12, Section 4].

Definition 2.1. Suppose that U and X are Hilbert spaces, $\mathbb{T}$ is a strongly continuous semigroup of operators on X and the space X_{-1} is defined as in Section A3. An operator

$B \in \mathcal{L}(U, X_{-1})$ is an *admissible control operator* for $\mathbb{T}$ if for some (and hence any) $\tau > 0$, the operator $\mathbf{\Phi}_\tau : L^2([0, \infty), U) \to X_{-1}$ defined by

$$(2.1) \qquad\qquad \mathbf{\Phi}_\tau u = \int_0^\tau \mathbb{T}_{\tau-\sigma} B \, u(\sigma) \, d\sigma$$

has its range in X.

Remark 2.2. It is easy to verify (using the closed graph theorem) that if B is an admissible control operator for $\mathbb{T}$ and the family $\mathbf{\Phi} = (\mathbf{\Phi}_\tau)_{\tau \geq 0}$ is defined by (2.1), then $\mathbf{\Phi}_\tau$ is bounded from $L^2([0, \infty), U)$ to X and $(\mathbb{T}, \mathbf{\Phi})$ is an abstract linear control system (see the text after Definition A1.1). Conversely, if $(\mathbb{T}, \mathbf{\Phi})$ is an abstract linear control system, then there is a unique admissible control operator for $\mathbb{T}$, denoted by B, such that the representation (2.1) holds (see A(3.1)). B is called the *control operator of the system* $(\mathbb{T}, \mathbf{\Phi})$. (If $\Sigma = (\mathbb{T}, \mathbf{\Phi}, \mathbb{L}, \mathbb{F})$ is an abstract linear system, then obviously the control operator of Σ, as defined in Section A3, is the same as the control operator of $(\mathbb{T}, \mathbf{\Phi})$.) The control operator B can be obtained from $\mathbf{\Phi}$ by the formula

$$(2.2) \qquad\qquad B \mathrm{v} = \lim_{\tau \to 0} \frac{1}{\tau} \mathbf{\Phi}_\tau \chi_\mathrm{v} \,, \qquad \forall \, \mathrm{v} \in U \,,$$

where for any $\mathrm{v} \in U$, χ_v is the constant function on $[0, \infty)$ equal to v everywhere ($\mathbf{\Phi}_\tau$ can be applied to χ_v, see the beginning of Section A2). For the proof of (2.2) see Weiss [12, formula (3.6)].

Remark 2.3. Various necessary conditions as well as sufficient conditions for the admissibility of an operator $B \in \mathcal{L}(U, X_{-1})$ are known, see Weiss [12] and the references therein.

The next definition follows Weiss [14, Section 6].

Definition 2.4. Suppose that X and Y are Hilbert spaces, $\mathbb{T}$ is a strongly continuous semigroup of operators on X and the space X_1 is defined as in Section A3. An operator $C \in \mathcal{L}(X_1, Y)$ is an *admissible observation operator* for $\mathbb{T}$ if for some (and hence any) $\tau > 0$, the operator $\mathbb{L}_\tau : X_1 \to L^2([0, \infty), Y)$ defined by

$$(2.3) \qquad\qquad (\mathbb{L}_\tau x)(t) = C \mathbb{T}_t x \,, \qquad \text{for} \qquad t \in [0, \tau) \,,$$

and $(\mathbb{L}_\tau x)(t) = 0$, for $t \geq \tau$, has a continuous extension to X.

Remark 2.5. It is easy to verify that if C is an admissible observation operator for $\mathbb{T}$ and the family of operators $\mathbb{L} = (\mathbb{L}_\tau)_{\tau \geq 0}$ is defined by (2.3) and continuous extension, then $(\mathbb{L}, \mathbb{T})$ is an abstract linear observation system (see the text after Definition A1.1). Conversely, if $(\mathbb{L}, \mathbb{T})$ is an abstract linear observation system, then there is a unique admissible observation operator for $\mathbb{T}$, denoted C, such that the representation (2.3) holds (see A(3.4)). C is called the *observation operator of the system* $(\mathbb{L}, \mathbb{T})$. (If $\Sigma = (\mathbb{T}, \mathbf{\Phi}, \mathbb{L}, \mathbb{F})$ is an abstract linear system, then obviously the observation operator of Σ, as defined in Section A3, is the same as the observation operator of $(\mathbb{L}, \mathbb{T})$.) The observation operator C can be obtained from $\mathbb{L}$ by the formula

$$(2.4) \qquad\qquad C x = (\mathbb{L}_\infty x)(0) , \qquad \forall x \in X_1 ,$$

where $\mathbb{L}_\infty$ is defined as in Section A2. For $x \in X_1$, the function $\mathbb{L}_\infty x$ is continuous, so the point evaluation appearing in (2.4) makes sense. For the proof of (2.4) see Weiss [14, Remark 3.4].

Remark 2.6. Let $\mathbb{T}^*$ be the adjoint semigroup of $\mathbb{T}$, then $C \in \mathcal{L}(X_1, Y)$ is an admissible observation operator for $\mathbb{T}$ if and only if C^*, which belongs to $\mathcal{L}(Y, X_1^*)$, is an admissible control operator for $\mathbb{T}^*$. We have $X_1^* = (X^*)_{-1}$, which is the analogue of X_{-1} for the semigroup $\mathbb{T}^*$. The proof is not difficult. For more information on this topic see Weiss [14, Section 6].

Definition 2.7. Suppose that $\Sigma = (\mathbb{T}, \Phi, \mathbb{L}, \mathbb{F})$ is an abstract linear system. If A is the generator of $\mathbb{T}$, B is the control operator of Σ and C is the observation operator of Σ, then we say that (A, B, C) is the triple *associated* with Σ. A triple of operators (A, B, C) will be called *well posed* if there is an abstract linear system Σ such that (A, B, C) is the triple associated with Σ.

In the following two remarks we try to clarify what well posedness of a triple means in terms of differential equations. (These remarks will not be needed later.)

Remark 2.8. Suppose that U, X and Y are Hilbert spaces, A is the generator of a semigroup on X, $B \in \mathcal{L}(U, X_{-1})$ and $C \in \mathcal{L}(X_1, Y)$. If C_L is the Lebesgue extension of C, and if the operator $C_L(\beta I - A)^{-1} B$ is well defined for some (and hence any) $\beta \in \rho(A)$, then (A, B, C) is well posed if and only if the system of equations

$$(2.5) \qquad\qquad \begin{cases} \dot{x}(t) & = A x(t) + B u(t) , \\ y(t) & = C_L x(t) , \end{cases}$$

is well posed in a certain natural sense, as explained in Theorem A4.6. This follows from Proposition A4.3 and Theorem A4.6. If the triple (A, B, C) is well posed, but $C_L(\beta I - A)^{-1} B$ does not exist, then (2.5) is no longer well posed.

Remark 2.9. Let U, X, Y, A, B, C and C_L be as in the previous remark, but we do not assume that $C_L(\beta I - A)^{-1} B$ makes sense. Then (A, B, C) is well posed if and only if the following (more complicated) system of equations is well posed:

$$\begin{cases} \dot{x}(t) & = A x(t) + B u(t) , \\ y(t) & = C_L [x(t) - (\beta I - A)^{-1} B u(t)] , \end{cases}$$

in the same sense as (2.5). This follows from what is written in Section A3, after formulae A(3.1) and A(3.9).

3. Shifted systems and Laplace transforms

This section comprises some technical preliminaries needed in the following sections.

For any $\lambda \in \mathbb{C}$, e_λ will denote the function defined on $[0, \infty)$ by $e_\lambda(t) = e^{\lambda t}$. If W is a Hilbert space and λ is as above, then the symbol e_λ will be used also to denote the operator of pointwise multiplication by $e^{\lambda t}$ on $L^2_{loc}([0, \infty), W)$:

$$(e_\lambda f)(t) = e^{\lambda t} f(t), \qquad \forall\, t \geq 0.$$

Definition 3.1. Let $\Sigma = (\mathbb{T}, \mathbf{\Phi}, \mathbb{L}, \mathbb{F})$ be an abstract linear system and $\lambda \in \mathbb{C}$. Then Σ *shifted* by λ, denoted by $\mathbf{S}_\lambda\Sigma$, is a quadruple of families of operators $\mathbf{S}_\lambda\Sigma = (\mathbb{T}', \mathbf{\Phi}', \mathbb{L}', \mathbb{F}')$, defined by

$$\mathbb{T}'_\tau = e^{\lambda \tau} \mathbb{T}_\tau, \qquad\qquad \mathbf{\Phi}'_\tau = e^{\lambda \tau} \mathbf{\Phi}_\tau e_{-\lambda},$$
$$\mathbb{L}'_\tau = e_\lambda \mathbb{L}_\tau, \qquad\qquad \mathbb{F}'_\tau = e_\lambda \mathbb{F}_\tau e_{-\lambda}.$$

Proposition 3.2. *For Σ, λ and $\mathbf{S}_\lambda\Sigma$ as above, $\mathbf{S}_\lambda\Sigma$ is an abstract linear system. The triple (A', B', C') associated with $\mathbf{S}_\lambda\Sigma$ is given by*

$$A' = A + \lambda I, \qquad B' = B, \qquad C' = C,$$

where (A, B, C) is the triple associated with Σ.

Proof. Using the notation of Definition 3.1, we check the functional equation A(1.4) for $\mathbf{\Phi}'$. We have

$$\mathbf{\Phi}'_{t+\tau}(u \underset{\tau}{\diamondsuit} v) = e^{\lambda(t+\tau)} \mathbf{\Phi}_{t+\tau} \left(e_{-\lambda} u \underset{\tau}{\diamondsuit} e^{-\lambda \tau} e_{-\lambda} v \right)$$

$$= e^{\lambda(t+\tau)} \mathbb{T}_t \mathbf{\Phi}_\tau\, e_{-\lambda} u + e^{\lambda(t+\tau)} \mathbf{\Phi}_t\, e^{-\lambda \tau} e_{-\lambda} v$$

$$= \mathbb{T}'_t \mathbf{\Phi}'_\tau u + \mathbf{\Phi}'_t v.$$

The functional equations A(1.5) and A(1.6) can be checked similarly. It is clear that the generator of $\mathbb{T}'$ is $A + \lambda I$. The fact that $B' = B$ follows from (2.2) and A(1.7), using that

$$\lim_{\tau \to 0} \frac{1}{\tau} \left\| \mathrm{P}_\tau \left(e^{\lambda \tau} e_{-\lambda} \chi_v - \chi_v \right) \right\|_{L^2} = 0$$

for any $v \in U$ (U is the input space of Σ), where P_τ is the projection defined in A(1.2). The fact that $C' = C$ follows from (2.4). ∎

We shall use the following terminology and notation. The *growth bound* of a semigroup $\mathbb{T}$ is the smallest element $\omega_0 \in [-\infty, \infty)$ such that for any $\omega > \omega_0$ there is an $M_\omega \geq 1$ for which

$$\| \mathbb{T}_t \| \leq M_\omega e^{\omega t}, \qquad \forall\, t \geq 0,$$

see e.g. Nagel *et al* [7]. For any $\alpha \in [-\infty, \infty)$, $\mathbb{C}_\alpha$ will denote the right open half-plane delimited by α:

$$\mathbb{C}_\alpha = \{ s \in \mathbb{C} \mid \operatorname{Re} s > \alpha \} .$$

If W is a Hilbert space, $p \in [0, \infty)$ and $w \in L^1_{loc}([0, \infty), W)$, the *abscissa of L^p-convergence of the Laplace integral* of w, which we denote $\alpha_p(w)$, is the smallest element in $[-\infty, \infty)$ such that for any $\alpha > \alpha_p(w)$

$$\int_0^\infty e^{-p\alpha t} \|w(t)\|^p \, dt < \infty .$$

If the above integral diverges for any $\alpha \in \mathbb{R}$, then we put $\alpha_p(w) = \infty$. In this paper we are interested only in the values $p = 1$ and $p = 2$. It is easy to see that

$$(3.1) \qquad\qquad\qquad\qquad \alpha_1(w) \leq \alpha_2(w) .$$

If $\alpha_1(w) < \infty$ then the *Laplace transform* of w is the function $\hat{w} : \mathbb{C}_{\alpha_1(w)} \to W$ defined by

$$\hat{w}(s) = \int_0^\infty e^{-st} w(t) \, dt .$$

For properties of the Laplace transformation we refer to Doetsch [3].

Among our aims in the next section is to show that if u is the input function of an abstract linear system and y is the corresponding output function for zero initial state, then under certain conditions $\hat{y} = \mathbf{H} \cdot \hat{u}$, where $\mathbf{H}$ is the transfer function of the system. For that we have to give sufficient conditions for the existence of $\hat{y}$, or equivalently for $\alpha_1(y) < \infty$. It would be nice if the conditions $u \in L^2_{loc}$ and $\alpha_1(u) < \infty$ would imply $\alpha_1(y) < \infty$, but we do not know if this is true (and our guess is that it is not). However, we can prove that $\alpha_2(u) < \infty$ implies $\alpha_2(y) < \infty$, which in turn implies $\alpha_1(y) < \infty$, by (3.1). More precisely, we have the following refinement of Remark A2.3.

Proposition 3.3. *Let $\Sigma = (\mathbb{T}, \boldsymbol{\Phi}, \mathbb{L}, \mathbb{F})$ be an abstract linear system with input space U and output space Y, and let ω_0 be the growth bound of $\mathbb{T}$. If $u \in L^2_{loc}([0, \infty), U)$ and $y = \mathbb{F}_\infty u$, then*

$$\alpha_2(y) \leq \max\{\omega_0, \alpha_2(u)\} .$$

Proof. If $\alpha_2(u) = \infty$, there is nothing to prove. Suppose $\alpha_2(u) < \infty$. We have to show that for any $\alpha > \max\{\omega_0, \alpha_2(u)\}$,

$$(3.2) \qquad\qquad\qquad\qquad \int_0^\infty e^{-2\alpha t} \|y(t)\|^2 \, dt < \infty .$$

Let $\Sigma' = S_{-\alpha}\Sigma = (\mathbb{T}', \boldsymbol{\Phi}', \mathbb{L}', \mathbb{F}')$, then $\mathbb{T}'$ is exponentially stable. We have $e_{-\alpha} y = \mathbb{F}'_\infty e_{-\alpha} u$. Since $e_{-\alpha} u \in L^2([0, \infty), U)$, it follows by Remark A2.2 that $e_{-\alpha} y \in L^2([0, \infty), Y)$, i.e. (3.2) holds. $\blacksquare$

4. Transfer functions

Let $\Sigma = (\mathbb{T}, \Phi, \mathbb{L}, \mathbb{F})$ be an abstract linear system with input space U, state space X and output space Y, and let (A, B, C) be the triple associated with Σ. We have seen in Section 2 that the triple (A, B, C) determines the triple $(\mathbb{T}, \Phi, \mathbb{L})$ in a straightforward way. The representation of $\mathbb{F}$ is more complicated. We recall from Section A3 (see formula A(3.9) and the text after it) that for any $u \in W_{0,loc}^{1,2}([0, \infty), U)$ and any $t \geq 0$

$$(4.1) \qquad (\mathbb{F}_\infty u)(t) = C \left[\int_0^t \mathbb{T}_{t-\sigma} B u(\sigma) d\sigma - (\beta I - A)^{-1} B u(t) \right] + \mathbf{H}(\beta) u(t) .$$

Here β is any element of $\rho(A)$ and $\mathbf{H}(\beta) \in \mathcal{L}(U, Y)$. This formula makes sense, since denoting by $z(t)$ the expression in the paranthesis in (4.1):

$$z(t) = \int_0^t \mathbb{T}_{t-\sigma} B u(\sigma) d\sigma - (\beta I - A)^{-1} B u(t) ,$$

by a simple computation we have that

$$z(t) = (\beta I - A)^{-1} \int_0^t \mathbb{T}_{t-\sigma} [\beta u(\sigma) - u'(\sigma)] d\sigma .$$

This shows that z is a (continuous) function from $[0, \infty)$ to X_1, and so C can be applied to $z(t)$.

Formula (4.1) determines $\mathbb{F}_\infty$, because $W_{0,loc}^{1,2}([0, \infty), U)$ is dense in $L_{loc}^2([0, \infty), U)$ (or equivalently, for any $\tau \geq 0$, $W_0^{1,2}([0, \tau], U)$ is dense in $L^2([0, \tau], U)$). The representation formula (4.1) shows that (A, B, C) determines $\mathbb{F}_\infty$ only up to an additive constant, namely $\mathbf{H}(\beta)$.

The function $\mathbf{H} : \rho(A) \to \mathcal{L}(U, Y)$ appearing in (4.1) is called the *transfer function* of Σ. This definition of the transfer function (already mentioned in Section A3) is quite unusual, but we shall see in Theorem 4.2 below that $\mathbf{H}$ has the properties which one expects the transfer function to have.

Remark 4.1. As already mentioned, (4.1) is due to Salamon [10]. In Weiss [16] it is proved that (4.1) remains true for any $u \in L_{loc}^2([0, \infty), U)$, if we replace C by its Lebesgue extension C_L. For regular systems this latter version of (4.1) becomes much simpler and the transfer function has a simple expression, see Propositions A4.5 and A4.7.

The following theorem summarizes some simple but important properties of the transfer function.

Theorem 4.2. Let $\Sigma = (\mathbb{T}, \Phi, \mathbb{L}, \mathbb{F})$ be an abstract linear system with input space U and output space Y. If (A, B, C) is the triple associated with Σ, $\mathbf{H}$ is its transfer function and if ω_0 is the growth bound of $\mathbb{T}$, then we have the following:

    ($\imath$) If $u \in L_{loc}^2([0, \infty), U)$ with $\alpha_2(u) < \infty$, then denoting $y = \mathbb{F}_\infty u$ and $\alpha_0 = \max\{\omega_0, \alpha_2(u)\}$, we have

$$(4.2) \qquad\qquad \hat{y}(s) = \mathbf{H}(s) \cdot \hat{u}(s) , \qquad \forall s \in \mathbb{C}_{\alpha_0} .$$

(*ii*) **H** *satisfies the equation*

(4.3)
$$\frac{\mathbf{H}(s) - \mathbf{H}(\beta)}{s - \beta} = -C(sI - A)^{-1}(\beta I - A)^{-1} B,$$

for any $s, \beta \in \rho(A)$ *with* $s \neq \beta$.

(*iii*) **H** *is analytic and for any* $\alpha > \omega_0$ *it is bounded on* $\mathbb{C}_\alpha$.

(*iv*) *For any* $\lambda \in \mathbb{C}$, *the transfer function* $\mathbf{H}'$ *of* $\Sigma' = \mathbf{S}_\lambda \Sigma$ *is given by*

(4.4)
$$\mathbf{H}'(s) = \mathbf{H}(s - \lambda), \qquad \forall s \in \rho(A) + \lambda.$$

Proof. First we prove (*ii*). If we rewrite (4.1) with s instead of β and then subtract the two equalities side by side, we get

$$[\mathbf{H}(s) - \mathbf{H}(\beta)] u(t) = C \left[(sI - A)^{-1} - (\beta I - A)^{-1} \right] Bu(t),$$

which is true for any $u \in W^{1,2}_{0,loc}([0,\infty), U)$ and any $t \geq 0$, and hence true for any $v \in U$ instead of $u(t)$. Applying the resolvent identity, we get (4.3).

Next we prove (*i*). We know from Proposition 3.3 and inequality (3.1) that for Re $s > \alpha_0$ the Laplace integral of y is convergent. If we assume additionally that $u \in W^{1,2}_{0,loc}([0,\infty), U)$ then, applying the Laplace transformation to both sides of (4.1), we get that for Re $s > \alpha_0$

$$\hat{y}(s) = C \left[(sI - A)^{-1} B\hat{u}(s) - (\beta I - A)^{-1} B\hat{u}(s) \right] + \mathbf{H}(\beta)\hat{u}(s).$$

By the resolvent identity and (4.3) we get that for our special u (4.2) holds. Using the continuity of the Laplace transformation with respect to suitable weighted L^2-norms, we get that (4.2) holds in general.

We prove (*iv*). Equation (4.3) applied once for **H** and once for **H**' implies that for $s, \beta \in \rho(A + \lambda I) = \rho(A) + \lambda$

$$\mathbf{H}'(s) - \mathbf{H}'(\beta) = \mathbf{H}(s - \lambda) - \mathbf{H}(\beta - \lambda).$$

Hence it is enough to show that (4.4) holds for some nonempty subset of $\rho(A) + \lambda$. Let $u \in L^2([0,\infty), U)$ and let $y = \mathbb{F}_\infty u$. By the definition of Σ', $e_\lambda y = \mathbb{F}'_\infty e_\lambda u$, where $\mathbb{F}'_\infty$ is the extended input/output map of Σ'. By (4.2) applied to $\mathbb{F}'_\infty$, $\hat{y}(s - \lambda) = \mathbf{H}'(s)\hat{u}(s - \lambda)$, for $s \in \mathbb{C}$ with large real part. Comparing this with (4.2) applied to $\mathbb{F}_\infty$ we get that for $s \in \mathbb{C}$ with large real part $\mathbf{H}'(s) = \mathbf{H}(s - \lambda)$. As explained before, this implies (4.4).

Finally, let us prove (*iii*). The equation (4.3) shows (by letting $s \to \beta$) that **H** is differentiable, and hence analytic. It remains to prove that for $\alpha > \omega_0$, **H** is bounded on $\mathbb{C}_\alpha$. We need the following notation: if W is a Hilbert space then $\langle \cdot, \cdot \rangle$ stands for the bilinear pairing between W and its dual W^*. The dual of $L^2([0,\infty), W)$ will be identified with $L^2([0,\infty), W^*)$ in the natural way. Then for any $z \in L^2([0,\infty), W)$, $s \in \mathbb{C}_0$ and $w \in W^*$ we have

(4.5)
$$\langle \hat{z}(s),\, \mathrm{w}\rangle \;=\; \langle z, e_{-s}\mathrm{w}\rangle\,,$$

where $\hat{z}$ denotes, as usual, the Laplace transform of z.

Let $\Sigma'' = S_{-\alpha}\Sigma = (\mathbb{T}'', \Phi'', \mathbb{L}'', \mathbb{F}'')$, then $\mathbb{T}''$ is exponentially stable. By Remark A2.2, $\mathbb{F}''_{\infty}$ is a bounded linear operator from $L^2([0,\infty), U)$ to $L^2([0,\infty), Y)$. Let $\mathbf{H}''$ be the transfer function of Σ''. For any $u \in L^2([0,\infty), U)$ and any $\mathrm{w} \in Y^*$, (4.5) together with (4.2) applied to Σ'' implies that for any $s \in \mathbb{C}_0$

$$\begin{aligned}
\langle \mathbb{F}''_{\infty} u, e_{-s}\mathrm{w}\rangle &= \langle \mathbf{H}''(s)\hat{u}(s), \mathrm{w}\rangle \\
&= \langle \hat{u}(s), \mathbf{H}''^*(s)\mathrm{w}\rangle \\
&= \langle u, e_{-s}\mathbf{H}''^*(s)\mathrm{w}\rangle\,,
\end{aligned}$$

whence

$$\mathbb{F}''^*_{\infty} e_{-s}\mathrm{w} \;=\; e_{-s}\mathbf{H}''^*(s)\mathrm{w}\,.$$

This implies that for $\mathrm{w} \in Y^*$ with $\mathrm{w} \neq 0$

$$\frac{\|\mathbf{H}''^*(s)\mathrm{w}\|}{\|\mathrm{w}\|} \;=\; \frac{\|e_{-s}\mathbf{H}''^*(s)\mathrm{w}\|_{L^2}}{\|e_{-s}\mathrm{w}\|_{L^2}} \;=\; \frac{\|\mathbb{F}''^*_{\infty} e_{-s}\mathrm{w}\|_{L^2}}{\|e_{-s}\mathrm{w}\|_{L^2}} \;\leq\; \|\mathbb{F}''^*_{\infty}\|\,,$$

so $\|\mathbf{H}''^*(s)\| \leq \|\mathbb{F}''^*_{\infty}\|$, or equivalently

$$\|\mathbf{H}''(s)\| \;\leq\; \|\mathbb{F}''_{\infty}\|\,.$$

Since the above formula holds for any $s \in \mathbb{C}_0$, by (4.4) and the definition of Σ'',

$$\|\mathbf{H}(s)\| \;\leq\; \|\mathbb{F}''_{\infty}\|\,, \qquad \forall\, s \in \mathbb{C}_{\alpha}\,. \qquad \blacksquare$$

Remark 4.3. Equation (4.3) determines $\mathbf{H}$ up to an additive constant element of $\mathcal{L}(U,Y)$. The restriction of $\mathbf{H}$ to any half-plane $\mathbb{C}_{\alpha}$ with $\alpha > \omega_0$ determines $\mathbb{F}_{\infty}$ via (4.2).

Remark 4.4. Statement (*iii*) of Theorem 4.2 can be obtained also by other methods, see e.g. Fourés and Segal [4] or Weiss [15].

5. The main result

The following theorem gives a list of necessary and sufficient conditions for a triple of operators to be well posed. The first version of this theorem appeared in Curtain [1, p. 12], but the proof was incomplete.

Theorem 5.1. *If U, X and Y are Hilbert spaces and (A, B, C) is a triple of operators such that:*

(*i*) *A is the generator of a strongly continuous semigroup $\mathbb{T}$ on X (so $\mathbb{T}$ determines two new Hilbert spaces, X_1 and X_{-1}, see Section A3),*

(*ii*) *$B \in \mathcal{L}(U, X_{-1})$ is an admissible control operator for $\mathbb{T}$ (see Definition 2.1),*

(*iii*) *$C \in \mathcal{L}(X_1, Y)$ is an admissible observation operator for $\mathbb{T}$ (see Definition 2.4),*

(*iv*) *there is an $\alpha \in \mathbf{R}$ such that some (and hence any) solution $\mathbf{H} : \rho(A) \to \mathcal{L}(U,Y)$ of the equation (4.3) is bounded on $\mathbb{C}_{\alpha}$,* *then (A, B, C) is well posed. In particular, $\mathbb{H}$ is bounded on any half-plane $\mathbb{C}_\alpha$ with $\alpha > \omega_0$, where ω_0 is the growth bound of $\mathbb{T}$.*

Conversely, if (A, B, C) is well posed, and hence associated with a system Σ, and if U, X and Y are the input, state and output spaces of Σ respectively, then conditions $(\imath)$, $(\imath\imath)$, $(\imath\imath\imath)$ and $(\imath v)$ above are satisfied.

Proof. First we prove that if $(\imath)$, $(\imath\imath)$, $(\imath\imath\imath)$ and $(\imath v)$ are satisfied, then (A, B, C) is well posed, i.e. there is a system Σ such that (A, B, C) is the triple associated with Σ. We have to find the four components $\mathbb{T}$, Φ, $\mathbb{L}$ and $\mathbb{F}$ of Σ and to show that they satisfy the conditions listed in Definition A1.1. $\mathbb{T}$ is already defined in $(\imath)$. Φ is defined by (2.1) and satisfies the functional equation A(1.4) (see Remark 2.2). $\mathbb{L}$ is defined by (2.3) and satisfies the functional equation A(1.5) (see Remark 2.5).

The crux of the matter is to define $\mathbb{F}$ and verify the functional equation A(1.6). For this it is enough to define a continuous operator $\mathbb{F}_\infty : L^2_{loc}([0, \infty), U) \to L^2_{loc}([0, \infty), Y)$ (see Section A2 for the topologies on these spaces) and to show that $\mathbb{F}_\infty$ satisfies the functional equation A(2.3). The operators $\mathbb{F}_\tau$ for $\tau \geq 0$ are then defined by A(2.1) and they satisfy A(1.6).

Let $\mathbb{H}$ be a solution of (4.3). For $u \in W^{1,2}_{0,loc}([0, \infty), U)$ we define $\mathbb{F}_\infty u$ by (4.1). As explained in the text after formula (4.1), this definition makes sense and $\mathbb{F}_\infty u$ is a continuous function (from $[0, \infty)$ to Y), in particular $\mathbb{F}_\infty u \in L^2_{loc}([0, \infty), Y)$. We have to show that the operator $\mathbb{F}_\infty : W^{1,2}_{0,loc}([0, \infty), U) \to L^2_{loc}([0, \infty), Y)$ thus defined has a continuous extension to $L^2_{loc}([0, \infty), U)$. Equivalently, we have to show that for any $T \geq 0$ there is a $K_T \geq 0$ such that for any $u \in W^{1,2}_{0,loc}([0, \infty), U)$

$$(5.1) \qquad \int_0^T \|(\mathbb{F}_\infty u)(t)\|^2 \, dt \ \leq \ K_T^2 \int_0^T \|u(t)\|^2 \, dt \,.$$

Let $W^{1,2}_C$ denote the subspace of $W^{1,2}_{0,loc}([0, \infty), U)$ consisting of functions having compact support. Let $u \in W^{1,2}_C$, then there is a $\tau \geq 0$ such that $u(t) = 0$ for $t \geq \tau$. We denote $y = \mathbb{F}_\infty u$. It is clear from (4.1) that

$$y(t) \ = \ C \mathbb{T}_{t-\tau} \Phi_\tau u \,, \qquad \text{for } t \geq \tau \,.$$

From the explanations given after formula (4.1) it follows that $\Phi_\tau u \in X_1$. Since $\mathbb{T}$ is a semigroup on X_1 with the same growth bound ω_0 as on X (see the beginning of Section A3), it follows that $\alpha_2(y) \leq \omega_0$ (see Section 3 for the definition of α_2). This implies that we can apply the Laplace transformation to both sides of (4.1). By some simple manipulations (see the proof of point $(\imath)$ of Theorem 4.2) we get

$$\hat{y}(s) \ = \ \mathbb{H}(s) \cdot \hat{u}(s) \,, \qquad \forall \, s \in \mathbb{C}_{\omega_0} \,.$$

Let $\alpha^+ = \max\{0, \alpha\}$ and let $w = e_{-\alpha^+} y$ (see Section 3 for notation), so

$$\hat{w}(s) \ = \ \mathbb{H}(s + \alpha^+) \cdot \hat{u}(s + \alpha^+) \,.$$

By the Plancherel theorem

$$\sup_{\sigma>0} \int_{-\infty}^{\infty} \|\hat{u}(\sigma+i\gamma)\|^2 d\gamma \;=\; 2\pi\|u\|_{L^2}^2\;,$$

whence

$$\sup_{\sigma>0} \int_{-\infty}^{\infty} \|\hat{w}(\sigma+i\gamma)\|^2 d\gamma \;\leq\; 2\pi \sup_{s\in\mathbb{C}_\alpha} \|\mathbf{H}(s)\|^2 \cdot \|u\|_{L^2}^2\;.$$

We apply for $\hat{w}$ the Paley-Wiener theorem, namely the version for the half-plane and Hilbert space-valued functions (see e.g. Rosenblum and Rovnyak [8, p. 91]), and we get that $w \in L^2([0,\infty),Y)$ and

$$\|w\|_{L^2} \;\leq\; \sup_{s\in\mathbb{C}_\alpha} \|\mathbf{H}(s)\| \cdot \|u\|_{L^2}\;.$$

This implies that for any $T \geq 0$

$$\int_0^T \|y(t)\|^2 dt \;\leq\; e^{2\alpha^+ T} \int_0^T e^{-2\alpha^+ t} \|y(t)\|^2 dt$$

$$\leq\; e^{2\alpha^+ T} \int_0^{\infty} \|w(t)\|^2 dt$$

$$\leq\; e^{2\alpha^+ T} \sup_{s\in\mathbb{C}_\alpha} \|\mathbf{H}(s)\|^2 \cdot \|u\|_{L^2}^2\;.$$

Denoting

$$K_T \;=\; e^{\alpha^+ T} \sup_{s\in\mathbb{C}_\alpha} \|\mathbf{H}(s)\|$$

we get

(5.2)
$$\int_0^T \|(\mathbb{F}_\infty u)(t)\|^2 dt \;\leq\; K_T^2 \int_0^{\infty} \|u(t)\|^2 dt\;,$$

valid for any $T \geq 0$ and any $u \in W_C^{1,2}$. Let P_T be the projection operator defined in A(1.2). For any $u \in W_C^{1,2}$, any $T \geq 0$ and any $\epsilon > 0$ there is a $v \in W_C^{1,2}$ such that $P_T u = P_T v$ and $\|(I-P_T)v\|_{L^2} \leq \epsilon$. Since the left-hand side of (5.2) depends only on $P_T u$, by letting $\epsilon \to 0$ we get that (5.1) is valid for any $u \in W_C^{1,2}$. But since both sides of (5.1) depend only on $P_T u$, it follows that in fact (5.1) is valid for any $u \in W_{0,loc}^{1,2}([0,\infty),U)$. Thus we have proved that $\mathbb{F}_\infty$ can be extended continuously to $L_{loc}^2([0,\infty),U)$.

It remains to prove that $\mathbb{F}_\infty$ satisfies A(2.3). Let $\tau \geq 0$ be fixed. The space of pairs (u,v) with $u, v \in W_{0,loc}^{1,2}([0,\infty),U)$ and $u(\tau) = 0$ is dense in $L_{loc}^2([0,\infty),U) \times L_{loc}^2([0,\infty),U)$. For such pairs we have $(u \underset{\tau}{\diamondsuit} v) \in W_{0,loc}^{1,2}([0,\infty),U)$ and a simple computation using (4.1) shows that A(2.3) holds. Since both sides of A(2.3) depend continuously on the pair (u,v), it follows that A(2.3) holds in general. Thus we have proved that (A, B, C) is well posed. The fact that $\mathbf{H}$ is bounded on any half-plane $\mathbb{C}_\alpha$ with $\alpha > \omega_0$ follows from part (iii) of Theorem (4.2).

Conversely, suppose (A, B, C) is well posed. Then conditions $(\imath)$, $(\imath\imath)$ and $(\imath\imath\imath)$ are obviously satisfied and condition $(\imath\upsilon)$ follows from part $(\imath\imath\imath)$ of Theorem (4.2). ∎

Remark 5.2. Conditions $(\imath)$, $(\imath\imath)$ and $(\imath\imath\imath)$ in Theorem 5.1 are not sufficient for (A, B, C) to be well posed (i.e., they do not imply condition $(\imath\upsilon)$), as we show in Example 6.1 to follow.

Remark 5.3. In Salamon [11, Section 2] a list of five conditions is given as the definition of well posedness of a certain system of equations. It is not difficult to show that Salamon's definition is equivalent with well posedness of the triple (A, B, C) involved, as defined here. Conditions $(\imath)$, $(\imath\imath)$ and $(\imath\imath\imath)$ in Theorem 5.1 appear also in Salamon's list, while condition $(\imath\upsilon)$ replaces two other conditions of which one is more difficult to verify directly.

Remark 5.4. A triple of operators (A, B, C) is well posed if and only if the dual triple (A^*, C^*, B^*) is well posed. This follows from Theorem 5.1 and Remark 2.6.

6. Two examples

Example 6.1. Let $X = l^2$ and $U = Y = \mathbb{C}$. We will show that the triple (A, B, C) given by

$$
A = \begin{bmatrix} -1 & & & \\ & -2 & & \\ & & -3 & \\ & & & \ddots \end{bmatrix}, \qquad B = \begin{bmatrix} 1 \\ 1 \\ 1 \\ \vdots \end{bmatrix},
$$

$$
C = \begin{bmatrix} 1 & 1 & 1 & \cdots \end{bmatrix}
$$

is not well posed, but if we replace C by

$$
C^a = \begin{bmatrix} 1 & -1 & 1 & -1 & \cdots \end{bmatrix}
$$

(a stands for alternating), then (A, B, C^a) is well posed.

Let us make the above statement more precise. Let the semigroup $\mathbb{T}$ on l^2 be given by

$$
(\mathbb{T}_t x)_k = e^{-kt} x_k, \qquad \forall k \in \mathbb{N},
$$

for any $x = (x_k) \in l^2$ and any $t \geq 0$. The generator A of $\mathbb{T}$ is given by $(Ax)_k = -kx_k$, $\forall k \in \mathbb{N}$, and its domain $D(A)$ consists of those $(x_k) \in l^2$ for which $(kx_k) \in l^2$. X_{-1} consists of those sequences (x_k) for which $\left(\frac{x_k}{k}\right) \in l^2$. The definition of the norms $\|\cdot\|_1$ and $\|\cdot\|_{-1}$ is obvious. The operators $B \in \mathcal{L}(U, X_{-1})$ and $C, C^a \in \mathcal{L}(X_1, Y)$ are given by $(Bu)_k = u$, $\forall k \in \mathbb{N}$, $Cx = \sum_{k \in \mathbb{N}} x_k$ and $C^a x = \sum_{k \in \mathbb{N}} (-1)^{k+1} x_k$. If we represent sequences as infinite column matrices, then we get the matrix representations with which we started this example.

Using the Carleson measure criterion (see Ho and Russell [5] and Weiss [13]) and Remark 2.6, we get that B is an admissible control operator for $\mathbb{T}$ and both C and C^a are admissible observation operators for $\mathbb{T}$. Conditions $(\imath)$, $(\imath\imath)$ and $(\imath\imath\imath)$ of Theorem 5.1 are therefore satisfied for both triples (A, B, C) and (A, B, C^a).

Let us now check condition $(\imath v)$ for (A, B, C). Equation (4.3) with $\beta = 0$ implies

$$\mathbf{H}(s) = \mathbf{H}(0) - \sum_{k \in \mathbb{N}} \frac{s}{(s+k)k} \ .$$

We see that even for real s, $\lim_{s \to \infty} |\mathbf{H}(s)| = \infty$, so $(\imath v)$ is not satisfied and (A, B, C) is not well posed.

We check $(\imath v)$ for (A, B, C^a). Equation (4.3) with $\beta = 0$ implies

$$\mathbf{H}(s) = \mathbf{H}(0) - \sum_{k \in \mathbb{N}} (-1)^{k+1} \left(\frac{1}{k} - \frac{1}{s+k} \right)$$

$$= \mathbf{H}(0) - \sum_{k \in \mathbb{N}} \frac{1}{(2n-1)2n} + \sum_{k \in \mathbb{N}} \frac{1}{(s+2n-1)(s+2n)} \ ,$$

which shows that $\mathbf{H}$ is bounded on $\mathbb{C}_0$, so (A, B, C^a) is well posed.

We mention that the systems corresponding to (A, B, C^a) are regular. Indeed, denoting by C^a_L the Lebesgue extension of C^a, the series corresponding to $C^a_L(\beta I - A)^{-1} B$ for $\beta = 0$ by formally multiplying the associated matrices is

$$S = \sum_{k \in \mathbb{N}} \frac{(-1)^{k+1}}{k} \ ,$$

which is (conditionally) convergent. By a result in Weiss [14, Proposition 7.2] it follows that $A^{-1}B \in D(C^a_L)$ and $S = -C^a_L A^{-1} B$. By Proposition A4.3, any system corresponding to (A, B, C^a) is regular.

While this last example may seem a little contrived, the following one is well motivated physically.

Example 6.2. We consider the one dimensional heat equation with Dirichlet boundary control and point observation:

(6.1)

$$\begin{cases} \frac{\partial}{\partial t} z(\xi, t) = \frac{\partial^2}{\partial \xi^2} z(\xi, t) , & 0 \le \xi \le 1 , \\ z(0, t) = 0 , & z(1, t) = u(t) , \\ z(\xi, 0) = x_0(\xi) , & \\ y(t) = z(\xi_0, t) , & \end{cases}$$

where $\xi_0 \in (0, 1)$ is fixed. Here u is the input function, x_0 is the initial state and y is the output function.

We want to find an abstract linear system $\Sigma = (\mathbb{T}, \Phi, \mathbb{L}, \mathbb{F})$ which *models* (6.1). By that we mean the following: the input and output space of Σ is $\mathbb{C}$ and denoting by X the state space of Σ, there is a dense subspace $\mathcal{D} \subset X \times L^2[0, \infty)$ such that

(a) $\mathcal{D} \subset C^2[0, 1] \times C^1[0, \infty)$,
(b) if $(x_0, u) \in \mathcal{D}$ and $\tau \geq 0$, then $(\mathbf{x}(\tau), u_\tau) \in \mathcal{D}$, where $\mathbf{x}(\tau) = \mathbb{T}_\tau x_0 + \Phi_\tau u$ and $u_\tau(t) = u(\tau + t)$, $\forall t \geq 0$,
(c) if $(x_0, u) \in \mathcal{D}$ then, denoting for $\xi \in [0, 1]$ and $t \geq 0$

$$
(6.2) \quad
\begin{cases}
z(\xi, t) &= (\mathbb{T}_t x_0)(\xi) + (\Phi_t u)(\xi)\,, \\
y(t) &= (\mathbb{L}_\infty x_0)(t) + (\mathbb{F}_\infty u)(t)\,,
\end{cases}
$$

the functions u, z and y satisfy (6.1) (in the classical sense).

At first sight it is far from clear if this is possible and, as far as we know, there is no general recipe for finding a linear system Σ which models a given system of equations like (6.1). Instead of simply giving our particular solution, we will describe how we arrived to it. Accordingly, the subsequent text is divided into two parts, titled "The guesswork" and "The solution".

The guesswork. First we are looking for possible state spaces X and semigroups $\mathbb{T}$ on X which are compatible with (6.1) with $u = 0$. We will define a scale of Hilbert spaces W_α, $\alpha \in \mathbf{R}$, which will be candidates to be our state space X. We will then see that for one (and only one) α, the construction works.

Taking $W_0 = L^2[0, 1]$, it is well known that the following operator

$$
A = \frac{d}{d\xi^2}\,, \qquad D(A) = \{\, h \in W_0 \mid h', h'' \in W_0, \ h(0) = h(1) = 0 \,\}\,,
$$

generates an analytic semigroup on W_0, which we denote by $\mathbb{T}$. A has the eigenvalues $\lambda_n = -n^2\pi^2$ and the corresponding normalized eigenvectors $e_n(\xi) = \sqrt{2}\sin n\pi\xi$, where $n \in \mathbf{N}$. These eigenvectors form a basis for W_0, and denoting $x_n = \langle x, e_n \rangle$, for any $x \in W_0$ and any $n \in \mathbf{N}$, it is most convenient to express $\mathbb{T}$ and A in terms of the sequence (x_n):

$$
(6.3) \qquad (\mathbb{T}_t x)_n = e^{\lambda_n t} x_n\,, \qquad \forall\, n \in \mathbf{N}\,,
$$

and for $x \in D(A)$

$$
(6.4) \qquad (Ax)_n = \lambda_n x_n\,, \qquad \forall\, n \in \mathbf{N}\,.
$$

For $\alpha \geq 0$ we put $W_\alpha = D((-A)^\alpha)$, with $\|x\|_\alpha = \|(-A)^\alpha x\|_{L^2}$, and for $\alpha < 0$, W_α will denote the completion of W_0 with respect to the norm $\|x\|_\alpha = \|(-A)^\alpha x\|_{L^2}$. In particular, the definition of W_1 and W_{-1} agrees with that given in Section A3. Since A is selfadjoint, we can identify for any $\alpha \in \mathbf{R}$ the dual of W_α with $W_{-\alpha}$. For any $\alpha \in \mathbf{R}$ and any $x \in W_\alpha$, the corresponding sequence (x_n) can be defined as for $x \in W_0$, by extending the scalar product of W_0 to pairs (z, v) with $z \in W_\alpha$ and $v \in W_{-\alpha}$ (we have $e_n \in W_\alpha$ for any

$\alpha \in \mathbf{R}$). It is easy to verify that for any $\alpha \in \mathbf{R}$, a sequence (x_n) represents an element of W_α if and only if $((-\lambda_n)^\alpha x_n) \in l^2$. The norm of x in W_α is the l^2-norm of $((-\lambda_n)^\alpha x_n)$.

For any $\alpha > 0$, $\mathbb{T}$ has a restriction to W_α which is a semigroup on W_α, and for any $\alpha < 0$, $\mathbb{T}$ has a unique continuous extension to W_α which is a semigroup on W_α. We will denote all these restrictions (extensions) of $\mathbb{T}$ by the same symbol $\mathbb{T}$ and their generator, which is a restriction (extension) of the original A, by A (we have $A \in \mathcal{L}(W_{\alpha+1}, W_\alpha)$ for any real α). All these semigroups and their generators can be expressed in terms of sequences by (6.3) and (6.4), and they are all compatible with (6.1) in the sense that $\mathcal{D}_0 = D(A^\infty)$ is dense in W_α, and for $x_0 \in \mathcal{D}_0$ and $u = 0$, $z(\xi, t) = (\mathbb{T}_t x_0)(\xi)$ satisfies (6.1).

Operators in $\mathcal{L}(\mathbb{C}, W_\alpha)$ will be identified with elements of W_α, and operators in $\mathcal{L}(W_\alpha, \mathbb{C})$ with elements of $W_{-\alpha}$. A sequence (b_n) represents an admissible control operator for $\mathbb{T}$ acting on the state space W_α if and only if

$$(6.5) \qquad \sum_{k=1}^{n} k^{4\alpha} |b_k|^2 \leq Mn^2 ,$$

for some fixed $M \geq 0$ and any $n \in \mathbf{N}$. Indeed, the sequence $((-\lambda_n)^\alpha b_n)$ has to represent an admissible control operator for $\mathbb{T}$ acting on W_0, and we can apply the Carleson measure criterion as presented in Weiss [13].

By a similar argument we get that (c_n) represents an admissible observation operator for $\mathbb{T}$ acting on W_α if and only if

$$(6.6) \qquad \sum_{k=1}^{n} k^{-4\alpha} |c_k|^2 \leq M_1 n^2 ,$$

for some $M_1 \geq 0$ and any $n \in \mathbf{N}$.

The next step is to identify the operators B and C which are compatible with (6.1). The way the equations (6.1) look suggests that we should first reformulate this system of equations as an abstract boundary control system, as defined in Salamon [11, Section 2]. We try to do this with the state space $H = W_0$ (this is an obvious choice for a first try, but we shall see later that it is not the best choice). We put

$$Z = \{h \in H \mid h', h'' \in H, \ h(0) = 0\}$$

and define

$$\Delta \in \mathcal{L}(Z, H), \qquad \Gamma \in \mathcal{L}(Z, \mathbb{C}), \qquad K \in \mathcal{L}(Z, \mathbb{C}),$$

by

$$\Delta z = \frac{d^2 z}{d\xi^2}, \qquad \Gamma z = z(1), \qquad Kz = z(\xi_0),$$

so the restriction of Δ to Ker Γ is the generator A of $\mathbb{T}$ on H. The abstract boundary control system

$$
(6.7) \qquad \begin{cases} \dot{\mathbf{x}}(t) \;=\; \Delta \mathbf{x}(t)\,, \\ \Gamma \mathbf{x}(t) \;=\; u(t)\,, \\ \mathbf{x}(0) \;=\; x_0\,, \\ y(t) \;=\; K\mathbf{x}(t)\,, \end{cases}
$$

satisfies assumptions (B0) and (B1) in [11, Section 2]. By [11, Proposition 2.8], the first two equations in (6.7) are equivalent with the equation $\dot{\mathbf{x}}(t) = A\mathbf{x}(t) + Bu(t)$, where B is determined by the fact that $G = -A^{-1}B$ is the solution of the (trivial) "abstract elliptic problem" $\Delta G = 0$, $\Gamma G = 1$, i.e.

$$
(6.8) \qquad \frac{d^2 G}{d\xi^2} \;=\; 0\,, \qquad G(0) \;=\; 0\,, \qquad G(1) \;=\; 1\,.
$$

Clearly $G(\xi) = \xi$, which implies (by an elementary computation) that the sequence (b_n) corresponding to B is given by

$$
(6.9) \qquad b_n \;=\; \sqrt{2}\pi(-1)^{n+1}n
$$

(in particular, $B \in W_{-\frac{3}{4}-\epsilon}$ for any $\epsilon > 0$). The above result can be obtained also by the method proposed in Curtain and Pritchard [2, Chapter 8], which yields that B can, in some sense, be identified with δ_1' (δ_1 being the "delta function" supported at $\xi = 1$). By (6.5), B is admissible for $\mathbb{T}$ acting on W_α if and only if $\alpha \leq -\frac{1}{4}$. (We mention that this implies that the system (6.7) with $H = W_0$ does not satisfy assumption (B2) in [11, Section 2].)

It is easy to see that for $u = 0$ and $x_0 \in \operatorname{Ker} \Gamma = D(A)$, (6.7) is equivalent with $\dot{\mathbf{x}}(t) = A\mathbf{x}(t)$, $y(t) = C\mathbf{x}(t)$, where C is the restriction of K to $D(A)$. An easy computation gives that C is represented by the sequence (c_n), where

$$
(6.10) \qquad c_n \;=\; \sqrt{2}\sin n\pi\xi_0
$$

(in particular, $C \in W_{-\frac{1}{4}-\epsilon}$ for any $\epsilon > 0$). By (6.6), C is admissible for $\mathbb{T}$ on W_α if and only if $\alpha \geq -\frac{1}{4}$. So $\alpha = -\frac{1}{4}$ is the only value for which both B and C are admissible for $\mathbb{T}$ on W_α, which leads us to the state space $X = W_{-\frac{1}{4}}$. A characterization of $W_{-\frac{1}{4}}$ in terms of Sobolev spaces is given in the Appendix.

A last piece of guesswork: by applying the Laplace transformation to equations (6.1) we easily get that for $s \in \mathbb{C}_0$, the transfer function $\mathbf{H}$ of our system Σ should be given by

$$
(6.11) \qquad \mathbf{H}(s) \;=\; \frac{\sinh\sqrt{s}\,\xi_0}{\sinh\sqrt{s}}\,.
$$

(The same result can be obtained from Salamon's formula $\mathbf{H}(s) = K(sI - A)^{-1}B$, given in [11, p. 390].)

The solution. The spaces W_α, $\alpha \in \mathbb{R}$, are defined as after (6.4). Let the operators $A \in \mathcal{L}(W_{\frac{3}{4}}, W_{-\frac{1}{4}})$, $B \in \mathcal{L}(\mathbb{C}, W_{-\frac{5}{4}}) = W_{-\frac{5}{4}}$ and $C \in \mathcal{L}(W_{\frac{3}{4}}, \mathbb{C}) = W_{-\frac{3}{4}}$ be defined by (6.4), (6.9) and (6.10), respectively. We claim that the triple (A, B, C) is well posed (in the sense of Definition 2.7). To prove this claim, we use Theorem 5.1. Conditions $(\imath)$, $(\imath\imath)$ and $(\imath\imath\imath)$ in Theorem 5.1 are immediately verified: A is the generator of the semigroup $\mathbb{T}$ defined by (6.3) on $X = W_{-\frac{1}{4}}$ (so $X_1 = W_{\frac{3}{4}}$, $X_{-1} = W_{-\frac{5}{4}}$) and B and C are admissible for $\mathbb{T}$ by (6.5) and (6.6). It remains to check condition $(\imath v)$. Equation (4.3) with $\beta = 0$ becomes

$$(6.12) \qquad \frac{\mathbf{H}(s) - \mathbf{H}(0)}{s} = 2 \sum_{n \in \mathbb{N}} \frac{(-1)^n \sin n\pi\xi_0}{(s + n^2\pi^2)n\pi} ,$$

for any $s \in \rho(A)$, in particular for $s \in \mathbb{C}_0$. By an elementary (but tiresome) computation we can check that $\mathbf{H}$ given by (6.11) satisfies (6.12). Indeed, if for $s \in \mathbb{C}_0$ fixed we regard $\frac{1}{s}[\mathbf{H}(s) - \mathbf{H}(0)]$ as a function of $\xi_0 \in (0, 1)$ and compute its Fourier expansion with respect to the basis (e_n) of $L^2[0, 1]$ (recall that $e_n(\xi) = \sqrt{2} \sin n\pi\xi$), we get exactly (6.12). The series in (6.12) being uniformly convergent, the equality (6.12) holds for every $\xi_0 \in (0, 1)$. It is clear that $\mathbf{H}$ given by (6.11) has a continuous extension to the closed half-plane $\overline{\mathbb{C}_0}$. For $s \in \overline{\mathbb{C}_0}$ with large $|s|$, $\mathbf{H}(s)$ is practically equal to $e^{\sqrt{s}(\xi_0 - 1)}$, so $\lim_{|s| \to \infty} \mathbf{H}(s) = 0$. This implies that $\mathbf{H}$ is bounded on $\mathbb{C}_0$, so condition $(\imath v)$ is satisfied. Thus we have proved that (A, B, C) is well posed.

There is a unique abstract linear system $\Sigma = (\mathbb{T}, \Phi, \mathbb{L}, \mathbb{F})$ such that the triple associated with Σ is (A, B, C) and for some (and hence any) $s \in \mathbb{C}_0$, the transfer function of Σ is given by (6.11) (see Remarks 2.2 and 2.5 and the beginning of Section 4). We will show that Σ models (6.1), as defined at the beginning of this example.

First we show that Σ is regular. As in Example 6.1, we look at the formal series expansion of $-C_L A^{-1} B$:

$$S = -\frac{2}{\pi} \sum_{n \in \mathbb{N}} \frac{(-1)^n \sin n\pi\xi_0}{n} .$$

This series is conditionally convergent, so by the argument used at the end of Example 6.1 we conclude that Σ is regular. By A(4.15) the feedthrough operator of Σ is $D = 0$.

We introduce a semigroup $(\mathcal{G}_\tau)_{\tau \geq 0}$ and investigate some of its properties. Let $(\mathcal{L}_\tau)_{\tau \geq 0}$ denote the left shift semigroup on $L^2[0, \infty)$, i.e., for any $u \in L^2[0, \infty)$,

$$(\mathcal{L}_\tau u)(t) = u(\tau + t), \qquad \forall \, \tau, t \geq 0 .$$

We define the Hilbert space $\mathcal{X}$ and the semigroup $(\mathcal{G}_\tau)_{\tau \geq 0}$ on $\mathcal{X}$ by

$$\mathcal{X} = \begin{matrix} X \\ \times \\ L^2[0, \infty) \end{matrix}, \qquad \mathcal{G}_\tau = \begin{bmatrix} \mathbb{T}_\tau & \Phi_\tau \\ 0 & \mathcal{L}_\tau \end{bmatrix}$$

(recall that $X = W_{-\frac{1}{4}}$). We will write elements of $\mathcal{X}$ sometimes as pairs, sometimes as column vectors with two entries. Using some facts reviewed at the beginning of Section A3, it is easy to check that $(\mathcal{G}_\tau)_{\tau \geq 0}$ is indeed a strongly continuous semigroup on $\mathcal{X}$. A simple computation shows that the generator $\mathcal{A}$ of $(\mathcal{G}_\tau)_{\tau \geq 0}$ is given by

(6.13)

$$\mathcal{A} = \begin{bmatrix} A & B\delta \\ 0 & \frac{d}{dt} \end{bmatrix}, \quad D(\mathcal{A}) = \left\{ \begin{bmatrix} x_0 \\ u \end{bmatrix} \in \mathcal{X} \mid u \in H^1[0,\infty), \; Ax_0 + Bu(0) \in X \right\},$$

where $\delta u = u(0)$, and

$$(6.14) \quad D(\mathcal{A}^2) = \left\{ \begin{bmatrix} x_0 \\ u \end{bmatrix} \in \mathcal{X} \mid u \in H^2[0,\infty), \; A^2 x_0 + ABu(0) + Bu'(0) \in X \right\}.$$

As explained after (6.4), if $x \in W_\alpha$, $\alpha \in \mathbf{R}$, then $x = \sum_{n \in \mathbb{N}} x_n e_n$, where $((-\lambda_n)^\alpha x_n) \in l^2$. It is easy to check that for $\alpha > \frac{1}{4}$, the series converges uniformly on $[0,1]$. Since $e_n(0) = e_n(1) = 0$, it follows that

$$(6.15) \qquad\qquad W_\alpha \subset C_0^0[0,1], \qquad \forall\, \alpha > \frac{1}{4},$$

where $C_0^0[0,1] = \{x \in C^0[0,1] \mid x(0) = x(1) = 0\}$. We denote by $\mathcal{P}$ the projection of $\mathcal{X}$ onto its first component X, and by $C_z^n[0,1]$, $n \geq 0$, the space of all $x \in C^n[0,1]$ with $x(0) = 0$. We have

$$(6.16) \qquad\qquad \mathcal{P}D(\mathcal{A}) \subset C_z^0[0,1], \qquad \mathcal{P}D(\mathcal{A}^2) \subset C_z^2[0,1].$$

Indeed, if $(x_0, u) \in D(\mathcal{A})$ then by (6.13) $x_0 = \varphi - A^{-1}Bu(0)$, where $\varphi \in X_1 = W_{\frac{3}{4}}$. By (6.15), $\varphi \in C_0^0[0,1]$. The term $-A^{-1}Bu(0)$ belongs to $C_z^2[0,1]$ (see (6.8)), so $x_0 \in C_z^0[0,1]$. By a similar argument, using (6.14) and $X_2 = D(A^2) \subset C_z^2[0,1]$, one can check the second inclusion in (6.16).

We claim that the space $\mathcal{D}$ appearing in the definition of modelling (at the beginning of this example) can be taken to be $D(\mathcal{A}^2)$ (this choice is, of course, not unique, for example $D(\mathcal{A}^n)$ with $n > 2$ or even $D(\mathcal{A}^\infty)$ would also work). Let us check that Σ and $\mathcal{D} = D(\mathcal{A}^2)$ satisfy conditions (a), (b) and (c) imposed at the beginning of this example.

Condition (a) follows from (6.16) and (6.14), using that $H^2[0,\infty) \subset C^1[0,\infty)$.

Condition (b) reduces to the fact that $D(\mathcal{A}^2)$ is an invariant subspace for $\mathcal{G}_\tau$.

Let us check condition (c). Let $(x_0, u) \in \mathcal{D}$ be fixed and let $z : [0,1] \times [0,\infty) \to \mathbb{C}$ and $y : [0,\infty) \to \mathbb{C}$ be defined by (6.2). We have to show that u, z and y satisfy (6.1). Since the function $\tau \to \mathcal{G}_\tau(x_0, u)$ belongs to $C^0([0,\infty), D(\mathcal{A}^2)) \cap C^1([0,\infty), D(\mathcal{A}))$, by (6.16) the function $\tau \to \mathbf{x}(\tau) = \mathbb{T}_\tau x_0 + \Phi_\tau u$ satisfies

$$(6.17) \qquad\qquad \mathbf{x}(\cdot) \in C^0([0,\infty), C_z^2[0,1]) \cap C^1([0,\infty), C_z^0[0,1])$$

and by (6.13), $\dot{\mathbf{x}}(t) = A\mathbf{x}(t) + Bu(t)$. Recall the notations H, Z, Δ and Γ, introduced before (6.7). Since $C_z^0[0,1] \subset H$ and $C_z^2[0,1] \subset Z$, by a result in Salamon [11, Proposition

2.8] we conclude that $\mathbf{x}(\cdot)$ satisfies the first three equations in (6.7), $\dot{\mathbf{x}}(t)$ being computed with respect to the norm of H. But by (6.17) $\dot{\mathbf{x}}(t)$ exists also with respect to the norm of $C_z^0[0,1]$ (the sup-norm), which implies that it exists pointwise and

$$\dot{\mathbf{x}}(t)(\xi) \;=\; \frac{\partial}{\partial t} z(\xi,t)\,, \qquad 0 \le \xi \le 1\,.$$

Thus we have proved that z satisfies the first three equations in (6.1).

By Theorem A(4.6), y is given by $y(t) = C_L \mathbf{x}(t)$ for a.e. $t \ge 0$ (because, as we have seen when we discussed regularity, $D = 0$). For $x \in X_1 = W_{\frac{3}{4}}$, $Cx = x(\xi_0)$, because $c_n = e_n(\xi_0)$. Any $x \in C_z^0[0,1]$ belongs to $D(C_L)$ and $C_L x = x(\xi_0)$. This can be shown in various ways, for example as in Weiss [14, Section 4, Example 1.1]. With (6.17) it follows that $y(t) = \mathbf{x}(t)(\xi_0) = z(\xi_0,t)$, so the last equation in (6.1) is also satisfied.

Note that the pairs $(x_0,u) \in \mathcal{D}$ satisfy the compatibility conditions

$$x_0(1) \;=\; u(0)\,, \qquad\qquad x_0''(1) \;=\; u'(0)\,,$$

as one can expect from (6.1) (for the proof use (6.13) and (6.14)).

Appendix. We show that the state space $X = W_{-\frac{1}{4}}$ of the system discussed in Example 6.2 has the following equivalent definition:

$$(6.18) \qquad\qquad X \;=\; \left(H_{00}^{\frac{1}{2}}[0,1] \right)^* ,$$

where $H_{00}^{\frac{1}{2}}[0,1]$ is defined as in Lions and Magenes [6, p. 66], namely

$$(6.19) \qquad\qquad H_{00}^{\frac{1}{2}}[0,1] \;=\; \left[L^2[0,1],\, H_0^1[0,1] \right]_{\frac{1}{2}} .$$

In what follows we omit to write $[0,1]$, e.g. H^2 means $H^2[0,1]$. The proof of (6.18) goes as follows: $W_{\frac{1}{2}} = [W_0, W_1]_{\frac{1}{2}}$, where $W_0 = L^2$ and $W_1 = H^2 \cap H_0^1$, so $H_0^2 \subset W_1 \subset H^2$. Therefore

$$\left[L^2, H_0^2 \right]_{\frac{1}{2}} \subset W_{\frac{1}{2}} \subset \left[L^2, H^2 \right]_{\frac{1}{2}} .$$

Using interpolation theorems in [6, p. 43 and 64], the above inclusions become

$$H_0^1 \subset W_{\frac{1}{2}} \subset H^1 .$$

By (6.15) $W_{\frac{1}{2}} \subset C_0^0$, so we can replace on the right-hand side of the above formula H^1 by H_0^1, getting

$$(6.20) \qquad\qquad W_{\frac{1}{2}} \;=\; H_0^1 .$$

By the reiteration property [6, p. 28] we have $W_{\frac{1}{4}} = \left[W_0, W_{\frac{1}{2}} \right]_{\frac{1}{2}}$. By (6.20) and (6.19) we get $W_{\frac{1}{4}} = H_{00}^{\frac{1}{2}}$, and since $W_{-\frac{1}{4}} = \left(W_{\frac{1}{4}} \right)^*$, this is equivalent with (6.18). For a more concrete characterization of the space $\left(H_{00}^{\frac{1}{2}} \right)^*$ see Lions and Magenes [6, p. 71].

Note. While work on this paper was in progress, we received the research note of Russell [9], in which some related problems are discussed.

References

[1] R.F. Curtain, *Well-posedness of Infinite Dimensional Linear Systems in Time and Frequency Domain*, Research report, University of Groningen, August 1988.

[2] R.F. Curtain, A.J. Pritchard, *Infinite Dimensional Linear Systems Theory*, in "Lecture Notes in Information Sciences," vol. 8, Springer-Verlag, Berlin, 1978.

[3] G. Doetsch, "Handbuch der Laplace-Transformation," Birkhäuser Verlag, Basel, 1950.

[4] Y. Fourés, I.E. Segal, *Causality and Analyticity*, Transactions of the A.M.S. **78** (1955), 385–405.

[5] L.F. Ho, D.L. Russell, *Admissible Input Elements in Hilbert Space and a Carleson Measure Criterion*, SIAM J. Control and Optim. **21** (1983), 614–640.

[6] J.L. Lions, E. Magenes, *Non Homogeneous Boundary Value Problems and Applications, vol.I*, in "Die Grundlehren der mathematischen Wissenschaften in Einzeldarstellungen," **181**, Springer-Verlag, Berlin, 1972.

[7] R. Nagel (editor), *One-parameter Semigroups of Positive Operators*, in "Lecture Notes in Mathematics," vol. 1184, Springer-Verlag, New York, 1986.

[8] M. Rosenblum, J. Rovnyak, "Hardy Classes and Operator Theory," Oxford University Press, New York, 1985.

[9] D.L. Russell, *Some Remarks on Transfer Function Methods for Infinite Dimensional Linear Systems*, Research report, Virginia Polytechnic Institute and State University, October 1988.

[10] D. Salamon, *Realization Theory in Hilbert Space*, University of Wisconsin-Madison, Technical Summary Report **2835** (1985). Revised version submitted in 1988.

[11] D. Salamon, *Infinite Dimensional Systems with Unbounded Control and Observation: A Functional Analytic Approach*, Transactions of the A.M.S. **300** (1987), 383–431.

[12] G. Weiss, *Admissibility of Unbounded Control Operators*, SIAM J. Control & Optim. **27** (1989), 527–545.

[13] G. Weiss, *Admissibility of Input Elements for Diagonal Semigroups on l^2*, Systems and Control Letters **10** (1988), 79–82.

[14] G. Weiss, *Admissible Observation Operators for Linear Semigroups*, Israel J. Math. **65** (1989), 17–43.

[15] G. Weiss, *Representation of Shift Invariant Operators on L^2 by H^∞ Transfer Functions: an Elementary Proof, a Generalization to L^p and a Counterexample for L^∞*, submitted.

[16] G. Weiss, *The Representation of Linear Systems on Banach Spaces*, in preparation.

Ruth F. Curtain
Department of Mathematics
University of Groningen
9700 AV Groningen
The Netherlands

George Weiss
Department of Theoretical Mathematics
The Weizmann Institute
Rehovot 76100
Israel

International Series of
Numerical Mathematics, Vol. 91
© 1989 Birkhäuser Verlag Basel

61

Destabilization due to delay in one dimensional feedback

Wolfgang Desch and Robert L. Wheeler

Institut für Mathematik
Karl-Franzens-Universität Graz

Department of Mathematics
Virginia Polytechnic Institute and State University

Abstract. Delay in the feedback of a boundary stabilized elastic system causes destabilization and may even destroy the wellposedness of the system. We prove the destabilizing effect of delays in an abstract framework, and characterize the situations where wellposedness is lost in terms of the transfer function of the open loop problem.

Keywords. Boundary control, destabilization by delay, Euler Bernoulli beam.

1980 *Mathematics subject classifications:* 35B35, 35B37, 93D15

1. Introduction

We develop an abstract framework to study the destabilizing effects of feedback time delays in one-dimensional boundary stabilization schemes for infinite dimensional systems. Our results are obtained by an analysis of the behavior of a complex-valued transfer function for the stabilized input-output system.

This work is motivated by a recent paper by Datko, Lagnese and Polis [8] where it is shown that a velocity feedback scheme that uniformly exponentially stabilizes a damped wave equation can cause exponential instabilities when arbitrarily small feedback delays are present. Subsequent work has shown that this phenomemon occurs for other undamped systems [6], [7], as well as for systems with viscoelastic damping mechanisms [11], [12]. In fact, fixed delays may even lead to solutions with arbitrarily large exponential growth rates (a loss of wellposedness) [9].

This paper is organized in the following manner. In Section 2 we develop our abstract framework. It is general enough to treat one-dimensional boundary stabilization schemes for elastic systems such as the wave equation on an interval and the Euler-Bernoulli beam equation. We show that in this setting it is always true that arbitrarily small feedback delays can cause exponential instabilities. In Section 3 we turn to our main results, namely, criteria that are necessary and sufficient for arbitrarily small feedback delays to lead to a loss of wellposedness. These criteria are given in terms of the transfer function for the stabilized open loop system. In Section 4 we show that this loss of wellposedness due to

The work of W. D. was partially supported by the Air Force Office of Scientific Research under grant AFOSR–86–0085 and by Fonds zur Förderung der Wissenschaftlichen Forschung within the US-Austrian Cooperative Research Program, P-5691.
The work of R.L.W. was partially supported by the Air Force Office of Scientific Research under grant AFOSR–86–0085, the National Science Foundation under grant DMS–8500947, and by Fonds zur Förderung der Wissenschaftlichen Forschung, Austria, under Project S3206 during a visit to Universität Graz.

feedback delay is equivalent to the corresponding unstabilized open loop system being illposed in the sense of local $\mathbf{L}^2$ controls and observations. The study of input-output wellposedness for infinite dimensional systems with unbounded controls and observations is an area of much current activity [5], [13]. Finally, in Section 5 we apply the criteria developed in Sections 3 and 4 to some model mechanical examples.

2. Abstract framework

In this section we develop an abstract framework and show that boundary stabilization as it fits into this setting is not robust with respect to time delays in the feedback.

Throughout the paper let X and $\tilde{X}$ be two infinite dimensional complex Hilbert spaces with scalar products $\langle \cdot, \cdot \rangle_X$ and $\langle \cdot, \cdot \rangle_{\tilde{X}}$, respectively. (The subscripts will be omitted if no confusion is to be expected.) We consider the problem

$$u'(t) = \tilde{D}\tilde{u}(t), \ \tilde{u}'(t) = Du(t),$$
$$u(0) = u_0, \ \tilde{u}(0) = \tilde{u}_0.$$

(2.1)

(Here prime $'$ denotes derivative with respect to t.)

u and $\tilde{u}$ take values in X and $\tilde{X}$, respectively. $D : X \supseteq \operatorname{dom} D \to \tilde{X}$ and $\tilde{D} : \tilde{X} \supseteq \operatorname{dom} \tilde{D} \to X$ are — generally unbounded — linear operators.

Hypothesis D.

(a) *There exist nonzero linear operators* $P : \operatorname{dom} D \to \mathbf{C}$ *and* $\tilde{P} : \operatorname{dom} \tilde{D} \to \mathbf{C}$ *such that for all* $x \in \operatorname{dom} D$ *and* $\tilde{x} \in \operatorname{dom} \tilde{D}$

$$\langle Dx, \tilde{x} \rangle + \langle x, \tilde{D}\tilde{x} \rangle + Px\overline{\tilde{P}\tilde{x}} = 0.$$

(2.2)

(b) *The system* $D, \tilde{D}, P, \tilde{P}$ *is maximal with respect to (2.2), i.e. if* $D_1, \tilde{D}_1, P_1, \tilde{P}_1$ *are —* *possibly multivalued — extensions of* $D, \tilde{D}, P, \tilde{P}$ *such that for all* $x \in \operatorname{dom} D_1$ *and* $\tilde{x} \in \operatorname{dom} \tilde{D}_1, \langle D_1 x, \tilde{x} \rangle + \langle x, \tilde{D}_1 \tilde{x} \rangle + P_1 x \overline{\tilde{P}_1 \tilde{x}} = 0$, *then* $D = D_1, \tilde{D} = \tilde{D}_1, P = P_1, \tilde{P} = \tilde{P}_1$.

(c) *The domains of* D *and* $\tilde{D}$ *with the graph norms* $\|x\|_D = \|x\| + \|Dx\|$ *and* $\|\tilde{x}\|_{\tilde{D}} = \|\tilde{x}\| + \|\tilde{D}\tilde{x}\|$ *are compactly embedded in* X *and* $\tilde{X}$, *respectively.*

(d) *If* $x \in \operatorname{dom} D$, $\tilde{x} \in \operatorname{dom} \tilde{D}$, $s \in \mathbf{C}$ *are such that* $sx = \tilde{D}\tilde{x}$, $s\tilde{x} = Dx$, *and* $Px = \tilde{P}\tilde{x} = 0$, *then* $x = 0$ *and* $\tilde{x} = 0$.

We think of X and $\tilde{X}$ as $\mathbf{L}^2$-spaces over a compact interval J, and of D and $\tilde{D}$ as formally skew-adjoint partial differential operators, so that (2.1) is a hyperbolic PDE. By formal skew-adjointness, a Green's identity holds. We suppose that the domains of D and $\tilde{D}$ are restricted by some boundary conditions so that the Green's identity can be expressed in terms of two scalar trace operators P and $\tilde{P}$ in the abstract form (2.2). Compactness of the embedding will follow from the boundedness of the interval J. Part (d) states that the problem with boundary condition $Pu = 0$ is detectable by the observation $y = \tilde{P}\tilde{u}$.

We will consider various types of boundary conditions for (2.1):

Neutrally stable boundary condition

(NS) $$Pu(t) = 0.$$

Unstabilized open loop problem

(UO)
$$Pu(t) = f(t),$$
$$y(t) = \tilde{P}\tilde{u}(t).$$

Stabilizing feedback boundary condition

(SF) $$Pu(t) = \tilde{P}\tilde{u}(t).$$

Delayed feedback boundary condition

(DF) $$Pu(t) = \tilde{P}\tilde{u}(t - \epsilon).$$

Stabilized open loop problem

(SO)
$$\frac{1}{2}(Pu(t) - \tilde{P}\tilde{u}(t)) = f(t),$$
$$y(t) = \frac{1}{2}(Pu(t) + \tilde{P}\tilde{u}(t)).$$

(NS) yields a skew-adjoint, energy conserving Cauchy problem. (UO) is a corresponding open loop problem with boundary control and observation. We assume that closing the loop to obtain (SF) stabilizes the system exponentially. If the feedback is subject to some time delay we obtain (DF). The instability of (DF) is the main concern of this paper. The open loop problem (SO) is considered only for technical reasons. It is more tractable than (UO) since it is wellposed in a stronger sense than the latter problem.

Before we discuss stabilization we settle some simple questions of wellposedness. As this is not a central concern of our paper and none of the results is surprising, we do it in a somewhat sketchy manner. We show first that (2.1) with the neutrally stable boundary condition is wellposed for initial data $(u_0, \tilde{u}_0)$ in the space $\mathbf{X} = X \times \tilde{X}$.

Proposition 2.1. *Let $\mathbf{A}$ be defined on $\mathrm{dom}\mathbf{A} = \{(x, \tilde{x}) \in \mathrm{dom}\, D \times \mathrm{dom}\, \tilde{D} : Px = 0\}$ by $\mathbf{A}(x, \tilde{x}) = (\tilde{D}\tilde{x}, Dx)$. Then $\mathbf{A}$ is skew-adjoint. In particular, it generates a unitary group $\mathbf{S}(t)$ of operators in $\mathbf{X}$.*

Proof: Using the abstract Green's identity (2.2) and straightforward computation one sees that $\mathbf{A}$ is skew-symmetric. To obtain that it is skew-adjoint, let $(v, \tilde{v})$ be in the domain of the adjoint $\mathbf{A}^*$ with $\mathbf{A}^*(v, \tilde{v}) = -(w, \tilde{w})$. Thus for all $x \in \operatorname{dom} D$ with $Px = 0$ and all $\tilde{x} \in \operatorname{dom} \tilde{D}$ we have

$$\langle x, w \rangle + \langle \tilde{x}, \tilde{w} \rangle + \langle \tilde{D}\tilde{x}, v \rangle + \langle Dx, \tilde{v} \rangle = 0.$$

Pick an arbitrary $z \in \operatorname{dom} D$ with $Pz = 1$. We obtain for any $x \in \operatorname{dom} D$ and all $\tilde{x} \in \operatorname{dom} \tilde{D}$,

$$\langle x, w \rangle - \langle Px \cdot z, w \rangle + \langle \tilde{x}, \tilde{w} \rangle + \langle \tilde{D}\tilde{x}, v \rangle + \langle Dx, \tilde{v} \rangle - \langle Px \cdot Dz, \tilde{v} \rangle = 0.$$

From Hypothesis D, (b) we infer that $v \in \operatorname{dom} D$, $\tilde{v} \in \operatorname{dom} \tilde{D}$, with

$$Dv = \tilde{w}, \quad \tilde{D}\tilde{v} = w,$$
$$\tilde{P}\tilde{v} = -\langle z, w \rangle - \langle Dz, \tilde{v} \rangle, \quad \text{and} \quad Pv = 0.$$

Hence $(w, \tilde{w}) = \mathbf{A}(v, \tilde{v})$. ∎

For later use we note that it is easy to verify that $i\omega$ is an eigenvalue of $\mathbf{A}$ with eigenvector $(x, \tilde{x})$ if and only if $-i\omega$ and $(x, -\tilde{x})$ is an eigenvalue-eigenvector pair. Also, we will assume that $\mathbf{A}$ is unbounded so $\mathbf{A}$ has a sequence of eigenvalues $\lambda_n = i\omega_n$ with $\omega_n \to \infty$.

The following propositions deal with (SO). We start out proving wellposedness and proceed to discuss properties of the transfer function T, which will be a main tool in our discussion.

Proposition 2.2. *Problem* (SO) *is wellposed in the following sense: Given initial data* $u_0 \in \operatorname{dom} D$ *and* $\tilde{u}_0 \in \operatorname{dom} \tilde{D}$ *and a forcing function* $f \in \mathbf{W}^{1,2}([0, \infty))$ *such that* $\frac{1}{2}(Pu_0 - \tilde{P}\tilde{u}_0) = f(0)$, *there exist unique* $u \in \mathbf{W}^{1,2}([0, \infty), X)$ *and* $\tilde{u} \in \mathbf{W}^{1,2}([0, \infty), \tilde{X})$ *such that for all* $t > 0$, $u(t) \in \operatorname{dom} D$ *and* $\tilde{u}(t) \in \operatorname{dom} \tilde{D}$, (2.1) *is satisfied with the boundary condition* (SO), *and the output* $y = \frac{1}{2}(Pu(t) + \tilde{P}\tilde{u}(t))$ *is in* $\mathbf{W}^{1,2}([0, \infty))$. *Moreover,* $u(t) \in X$, $\tilde{u}(t) \in \tilde{X}$, *and* $y \in \mathbf{L}^2([0, \infty))$ *depend continuously on* $u_0 \in X$, $\tilde{u}_0 \in \tilde{X}$, *and* $f \in \mathbf{L}^2([0, \infty))$, *so that generalized solutions can be defined for all* $u_0 \in X$, $\tilde{u}_0 \in \tilde{X}$, *and* $f \in \mathbf{L}^2([0, \infty))$.

proof: To construct a semigroup in a state space including the control and output functions, we consider the state

$$(u(t), \tilde{u}(t), f(t + \cdot), y(t - \cdot)) \in \mathcal{X} = X \times \tilde{X} \times \mathbf{L}^2([0, \infty)) \times \mathbf{L}^2([0, \infty)),$$

normed by

$$\|(x, \tilde{x}, f, y)\|^2 = \|x\|^2 + \|\tilde{x}\|^2 + 2\int_0^\infty |f(s)|^2 ds + 2\int_0^\infty |y(s)|^2 ds.$$

Using the methods of the proposition above, one proves that the operator $\mathcal{A}$ defined on

$$\operatorname{dom} \mathcal{A} = \{(x, \tilde{x}, f, y) \in \operatorname{dom} D \times \operatorname{dom} \tilde{D} \times \mathbf{W}^{1,2} \times \mathbf{W}^{1,2} :$$
$$f(0) = \frac{1}{2}(Px - \tilde{P}\tilde{x}), y(0) = \frac{1}{2}(Px + \tilde{P}\tilde{x})\}$$

by

$$\mathcal{A}(x, \tilde{x}, f, y) = (\tilde{D}\tilde{x}, Dx, f', -y')$$

is skew-adjoint and therefore generates a semigroup $\mathcal{S}(t)$ in $\mathcal{X}$. Suppose now that $(u_0, \tilde{u}_0, f(\cdot), y_0(\cdot)) \in \operatorname{dom} \mathcal{A}$ and

$$\mathcal{S}(t)(u_0, \tilde{u}_0, f, y_0) = (u(t), \tilde{u}(t), g(t, \cdot), z(t, \cdot)).$$

From the definition of $\mathcal{A}$ it is clear that u and $\tilde{u}$ satisfy (2.1), and that

$$\frac{1}{2}(Pu(t) - \tilde{P}\tilde{u}(t)) = g(t, 0),$$

$$\frac{1}{2}(Pu(t) + \tilde{P}\tilde{u}(t)) = z(t, 0).$$

Moreover we have the differential equation

$$\frac{\partial g(t, s)}{\partial t} = \frac{\partial g(t, s)}{\partial s}$$

implying

$$g(t, s) = f(t + s),$$

thus

$$\frac{1}{2}(Pu(t) - \tilde{P}\tilde{u}(t)) = f(t).$$

The partial differential equation

$$\frac{\partial z(t, s)}{\partial t} = -\frac{\partial z(t, s)}{\partial s}$$

implies

$$z(t, s) = z(t - s, 0) = y(t - s)$$

for $0 < s < t$, where

$$y(t) = \frac{1}{2}(Pu(t) + \tilde{P}\tilde{u}(t))$$

if $t \geq 0$. We have therefore obtained a solution to (2.1) with (SO), and from the fact that $\mathcal{S}$ is a semigroup of contractions we infer easily the continuous dependence of solution and observation on initial values and forcing function. ∎

Corollary 2.3. *The operator* $\mathbf{A}_f$ *defined on* $\operatorname{dom} \mathbf{A}_f = \{(x, \tilde{x}) \in \operatorname{dom} D \times \operatorname{dom} \tilde{D} : Px - \tilde{P}\tilde{x} = 0\}$ *by* $\mathbf{A}_f(x, \tilde{x}) = (\tilde{D}\tilde{x}, Dx)$ *generates a semigroup of contractions in* $\mathbf{X}$. *(Thus* (SF) *yields a wellposed problem.)*

We make a strong stability hypothesis on (SF):

Hypothesis S. *There exists some $\omega < 0$ such that $\|(s - \mathbf{A}_f)^{-1}\|$ is uniformly bounded in each half-plane $s : \Re s \geq \rho$ with $\rho > \omega$.*

We are now going to introduce the transfer function T of (2.1) with (SO).

Lemma 2.4. *For some complex s let $p, q \in \mathbf{C}$ and $v, w \in \operatorname{dom} D$, $\tilde{v}, \tilde{w} \in \operatorname{dom} \tilde{D}$ satisfy*

$$sv - \tilde{D}\tilde{v} = w, \quad s\tilde{v} - Dv = \tilde{w},$$

$$\frac{1}{2}(Pv - \tilde{P}\tilde{v}) = p, \quad \frac{1}{2}(Pv + \tilde{P}\tilde{v}) = q.$$

Then

$$\Re(s)(\|v\|^2 + \|\tilde{v}\|^2) + |q|^2 - |p|^2 = \Re(\langle v, w \rangle + \langle \tilde{v}, \tilde{w} \rangle).$$

Proof:

$$\Re(s)(\|v\|^2 + \|\tilde{v}\|^2) + |q|^2 - |p|^2 = \Re(\langle v, sv \rangle + \langle s\tilde{v}, \tilde{v} \rangle) + |q|^2 - |p|^2 =$$
$$\Re(\langle v, w + \tilde{D}\tilde{v} \rangle + \langle \tilde{w} + Dv, \tilde{v} \rangle) + |q|^2 - |p|^2 =$$
$$\Re(\langle v, w \rangle + \langle \tilde{v}, \tilde{w} \rangle) - \Re(Pv\overline{\tilde{P}\tilde{v}}) + |q|^2 - |p|^2 =$$
$$\Re(\langle v, w \rangle + \langle \tilde{v}, \tilde{w} \rangle). \quad\blacksquare$$

Lemma 2.5. *For each s in the resolvent set of $\mathbf{A}_f$ there exist unique $x(s) \in \operatorname{dom} D$, $\tilde{x}(s) \in \operatorname{dom} \tilde{D}$, such that*

(2.3)
$$sx(s) = \tilde{D}\tilde{x}(s), \quad s\tilde{x}(s) = Dx(s),$$
$$\frac{1}{2}(Px(s) - \tilde{P}\tilde{x}(s)) = 1.$$

Moreover, $x(s)$, $\tilde{x}(s)$, and the function

(2.4)
$$T(s) = \frac{1}{2}(Px(s) + \tilde{P}\tilde{x}(s))$$

satisfy the following assertions:

(a) *$x(s)$, $\tilde{x}(s)$, and $T(s)$ and their derivatives depend analytically on s. Moreover, they are uniformly bounded on any half-plane $\{s : \Re(s) \geq \omega\}$ where $(s - \mathbf{A}_f)^{-1}$ is uniformly bounded.*

(b)

(2.5)
$$sx'(s) - \tilde{D}\tilde{x}'(s) = -x(s), \quad s\tilde{x}'(s) - Dx'(s) = -\tilde{x}(s),$$
$$Px'(s) - \tilde{P}\tilde{x}'(s) = 0, \quad T'(s) = \frac{1}{2}(Px'(s) + \tilde{P}\tilde{x}'(s)),$$

(where prime denotes derivative with respect to s—no confusion with the t-derivative is to be expected.)

(c) *T is the transfer function of (2.1) with (SO), i.e., given zero initial data $u(0)$ and $\tilde{u}(0)$, the Laplace transforms of a control $f \in \mathbf{L}^2$ and the corresponding observation y satisfy $\hat{y}(s) = T(s)\hat{f}(s)$.*

Proof: To show existence of x and $\tilde{x}$ we pick a pair $(x_0, \tilde{x}_0) \in \operatorname{dom} D \times \operatorname{dom} \tilde{D}$ such that $\frac{1}{2}(Px - \tilde{P}\tilde{x}) = 1$. With $z(s) = x(s) - x_0$ and $\tilde{z}(s) = \tilde{x}(s) - \tilde{x}_0$ (2.3) can be rewritten as

$$sz(s) - \tilde{D}\tilde{z}(s) = \tilde{D}\tilde{x}_0 - sx_0, \quad s\tilde{z}(s) - Dz(s) = Dx_0 - s\tilde{x}_0,$$
$$Pz(s) - \tilde{P}\tilde{z}(s) = 0.$$

This is solved by

(2.6) $$(z(s), \tilde{z}(s)) = (s - \mathbf{A}_f)^{-1}(\tilde{D}\tilde{x}_0 - sx_0, Dx_0 - s\tilde{x}_0).$$

Thus we infer existence, uniqueness and analyticity of x and $\tilde{x}$. It is easily seen from Hypothesis D (b) that P and $\tilde{P}$ are relatively bounded with respect to D and $\tilde{D}$, thus the operator $(P, \tilde{P})$ is relatively bounded with respect to $\mathbf{A}_f$. This implies analyticity of T by analyticity of $(x, \tilde{x})$ as a vector in $\operatorname{dom} \mathbf{A}_f$.

From Lemma 2.4 one obtains uniform boundedness of x, $\tilde{x}$, and T on all half-planes $\{s : \Re(s) \geq \rho\}$, where $\rho > 0$. To extend it to nonpositive real parts, we put $s = \sigma + i\tau$ and fix some $\rho > 0$. Let v, $\tilde{v}$ solve

$$(\rho + i\tau)v - \tilde{D}\tilde{v} = 0, \quad (\rho + i\tau)\tilde{v} - Dv = 0,$$
$$\frac{1}{2}(Pv - \tilde{P}\tilde{v}) = 1.$$

Then by (2.3) $(x - v, \tilde{x} - \tilde{v}) = (\sigma - \rho)(s - \mathbf{A}_f)^{-1}(v, \tilde{v})$. Thus uniform boundedness of v and $\tilde{v}$ implies uniform boundedness of x and $\tilde{x}$. We rewrite (2.3) once again as

$$(\rho + i\tau)x - \tilde{D}\tilde{x} = (\rho - \sigma)x, \quad (\rho + i\tau)\tilde{x} - Dx = (\rho - \sigma)\tilde{x},$$
$$\frac{1}{2}(Px - \tilde{P}\tilde{x}) = 1,$$

and obtain uniform boundedness of T by Lemma 2.4 and the uniform boundedness of x and $\tilde{x}$.

Differentiating (2.6) we obtain

$$\frac{d}{ds}(x(s), \tilde{x}(s)) = \frac{d}{ds}(z(s), \tilde{z}(s)) =$$
$$- (s - \mathbf{A}_f)^{-2}(\tilde{D}\tilde{x}_0 - sx_0, Dx_0 - s\tilde{x}_0) - (s - \mathbf{A}_f)^{-1}(x_0, \tilde{x}_0) =$$
$$- (s - \mathbf{A}_f)^{-1}(x(s), \tilde{x}(s)).$$

This implies formula (2.5) as well as uniform boundedness of x' and $\tilde{x}'$. Uniform boundedness of T' is checked similarly as above. Higher derivatives can be treated inductively.

To prove (c) we may of course assume that the control is in $\mathbf{W}^{1,2}$ with $f(0) = 0$, so that we have strict solutions rather than generalized ones. As the solutions are obtained by a semigroup, they are exponentially bounded and admit Laplace transforms, satisfying

$$s\hat{u}(s) = \tilde{D}\hat{\tilde{u}}(s), \quad s\hat{\tilde{u}}(s) = D\hat{u}(s),$$
$$\frac{1}{2}(P\hat{u}(s) - \tilde{P}\hat{\tilde{u}}(s)) = \hat{f}(s), \quad \frac{1}{2}(P\hat{u}(s) + \tilde{P}\hat{\tilde{u}}(s)) = \hat{y}(s).$$

Dividing the whole system by $\hat{f}(s)$ and comparing it to (2.3), (2.4) we obtain that $\hat{y}(s)/\hat{f}(s) = T(s)$. ∎

Lemma 2.6. *Let s be in the resolvent set of $\mathbf{A}_f$ and $x(s)$, $\tilde{x}(s)$, and $T(s)$ be as in Lemma 2.5. Then the following formulae hold:*

$$(2.7) \qquad |T(s)|^2 = 1 - \Re(s)(\|x(s)\|^2 + \|\tilde{x}(s)\|^2),$$

$$(2.8) \qquad T'(s)\overline{T(s)} = -\frac{1}{2}(\|x(s)\|^2 + \|\tilde{x}(s)\|^2) - \Re(s)(\langle x', x \rangle + \langle \tilde{x}', \tilde{x} \rangle),$$

$$(2.9) \qquad |T'(s)|^2 = -\Re(s)(\|x'(s)\|^2 + \|\tilde{x}'(s)\|^2) - \Re(\langle x', x \rangle + \langle \tilde{x}', \tilde{x} \rangle).$$

Moreover

$$(2.10) \qquad |T(s)| < 1 \text{ if } \Re(s) > 0, \ |T(s)| = 1 \text{ if } \Re(s) = 0, \ |T(s)| > 1 \text{ if } \Re(s) < 0.$$

Proof: Proof of (2.7): (For simplicity we will omit the argument (s).)

$$|T|^2 = \frac{1}{4}(Px + \tilde{P}\tilde{x})\overline{(Px + \tilde{P}\tilde{x})} =$$

$$\frac{1}{4}(Px - \tilde{P}\tilde{x})\overline{(Px - \tilde{P}\tilde{x})} + \frac{1}{2}(\tilde{P}\tilde{x})\overline{(Px)} + \frac{1}{2}(Px)\overline{(\tilde{P}\tilde{x})} =$$

$$1 - \Re(\langle x, \tilde{D}\tilde{x} \rangle + \langle Dx, \tilde{x} \rangle) = 1 - \Re(\langle x, sx \rangle + \langle s\tilde{x}, \tilde{x} \rangle) =$$

$$1 - \Re(s)(\|x\|^2 + \|\tilde{x}\|^2).$$

Proof of (2.8):

$$T'\overline{T} = \frac{1}{4}(Px' + \tilde{P}\tilde{x}')\overline{(Px + \tilde{P}\tilde{x})} =$$

$$\frac{1}{4}(Px' - \tilde{P}\tilde{x}')\overline{(Px - \tilde{P}\tilde{x})} + \frac{1}{2}(\tilde{P}\tilde{x}')\overline{(Px)} + \frac{1}{2}(Px')\overline{(\tilde{P}\tilde{x})} =$$

$$-\frac{1}{2}(\langle \tilde{x}', D\tilde{x} \rangle + \langle \tilde{D}\tilde{x}', x \rangle + \langle Dx', \tilde{x} \rangle + \langle x', \tilde{D}\tilde{x} \rangle) =$$

$$-\frac{1}{2}(\langle \tilde{x}', s\tilde{x} \rangle + \langle sx' + x, x \rangle + \langle s\tilde{x}' + \tilde{x}, \tilde{x} \rangle + \langle x', sx \rangle) =$$

$$-\Re(s)(\langle \tilde{x}', \tilde{x} \rangle + \langle x', x \rangle) - \frac{1}{2}(\|\tilde{x}\|^2 + \|x\|^2).$$

Proof of (2.9): From (2.5) and Lemma 2.4 we infer

$$\Re(s)(\|x'\|^2 + \|\tilde{x}'\|^2) + |T'|^2 - 0 = -\Re(\langle x', x \rangle + \langle \tilde{x}', \tilde{x} \rangle).$$

(2.10) is an obvious consequence of (2.7). ∎

The detectability assumption in Hypothesis D (d) implies that the spectral properties of the various abstract boundary value problems listed above are reflected by the transfer function T.

Lemma 2.7. *Let s be in the resolvent set of A_f and $\epsilon > 0$.*

(a) *s is an eigenvalue of A iff $T(s) = -1$.*

(b) *There exist $v \in \operatorname{dom} D$, $\tilde{v} \in \operatorname{dom} \tilde{D}$, not both zero, such that*

$$u(t) = e^{st}v, \quad \tilde{u}(t) = e^{st}\tilde{v}$$

solves (2.1) with the delayed boundary feedback condition (DF) iff s satisfies the following "characteristic equation":

(CE)
$$F(s) := T(s) + \frac{1 + e^{-s\epsilon}}{1 - e^{-s\epsilon}} = 0$$

Proof:

Proof of (a):

Let $A(v, \tilde{v}) = s(v, \tilde{v})$ for a nonzero pair $(v, \tilde{v})$. In particular we have $Pv = 0$, hence by Hypothesis D (d) $\tilde{P}\tilde{v}$ cannot vanish. We may rescale $(v, \tilde{v})$ so that $\frac{1}{2}(Pv - \tilde{P}\tilde{v}) = -\frac{1}{2}\tilde{P}\tilde{v} = 1$ and obtain $x(s) = v$, $\tilde{x}(s) = \tilde{v}$, and $T(s) = \frac{1}{2}(Pv + \tilde{P}\tilde{v}) = -1$.

Conversely assume that $T(s) = -1$. Then $sx(s) = \tilde{D}\tilde{x}(s)$, $s\tilde{x}(s) = Dx(s)$, and $Px(s) = \frac{1}{2}(Px(s) + \tilde{P}\tilde{x}(s)) + \frac{1}{2}(Px(s) - \tilde{P}\tilde{x}(s)) = \frac{1}{2}(T(s) + 1) = 0$. Thus $(x(s), \tilde{x}(s))$ is an eigenvector of A to the eigenvalue s.

Proof of (b):

Let $u(t) = e^{st}v$ and $\tilde{u}(t) = e^{st}\tilde{v}$ solve (2.1) with (DF). Direct computation yields $sv = \tilde{D}\tilde{v}$, $s\tilde{v} = Dv$, and $Pv = e^{-s\epsilon}\tilde{P}\tilde{v}$. The latter implies

$$(Pv - \tilde{P}\tilde{v}) = (e^{-s\epsilon} - 1)\tilde{P}\tilde{v}, (Pv + \tilde{P}\tilde{v}) = (e^{-s\epsilon} + 1)\tilde{P}\tilde{v}.$$

As s is in the resolvent set of A_f, $Pv - \tilde{P}\tilde{v}$ cannot vanish, otherwise $(v, \tilde{v})$ would be an eigenvector of A_f. We may therefore rescale $(v, \tilde{v})$ so that $(Pv - \tilde{P}\tilde{v}) = 1$ and obtain by (2.4) that

$$T(s) = \frac{(e^{-s\epsilon} + 1)}{(e^{-s\epsilon} - 1)}.$$

Conversely let $T(s)$ satisfy (CE). We put $u(t) = e^{st}x(s)$ and $\tilde{u}(t) = e^{st}\tilde{x}(s)$. We obtain (2.1) by direct computation, and

$$Pu(t) - \tilde{P}\tilde{u}(t - \epsilon) = e^{st}Px(s) - e^{s(t-\epsilon)}\tilde{P}\tilde{x}(s) =$$
$$e^{st}(\frac{1}{2}(T(s) + 1) - \frac{1}{2}e^{-s\epsilon}(T(s) - 1)) = 0. \quad \blacksquare$$

Our results concerning periodic and exponentially growing solutions of problem (2.1) with the delayed feedback condition (DF) are easy consequences of the behavior of $T(s)$ derived above.

Theorem 2.8. *There exist arbitrarily small positive ϵ such that problem (2.1) with (DF) admits a nontrivial periodic solution.*

Proof: By (2.10) we know that $|T(i\tau)| = 1$ for real τ, and it follows from (2.8) that $\arg T(i\tau)$ decreases as τ increases. Since $T(\lambda) = -1$ whenever λ is an eigenvalue of $\mathbf{A}$, we can find a sequence $\tau_n \to \infty$ such that $T(i\tau_n) = i$. Let $\epsilon_n = \pi/2\tau_n$. Then $\epsilon_n \to 0$, and for each ϵ_n, $i\tau_n$ satisfies the characteristic equation $F(i\tau_n) = 0$. Thus, by Lemma 2.7 there exists a nontrivial solution of (2.1) and (DF) with delay ϵ_n having period $2\pi/\tau_n$. ∎

We comment that Theorem 2.8 can be extended to certain situations where the control and observations are multiple dimensional or even infinite dimensional, although the complex-valued transfer function T cannot be used to analyze these situations. A similar result for distributed feedback has been obtained by Datko [6, Theorem 3.1]. Since our primary goal is to investigate the loss of wellposedness of problem (2.1) with (DF) (see Section 3), and since our study of loss of wellposedness relies on the mapping properties of the analytic transfer function T, we will not consider extensions of Theorem 2.8 to higher dimensional controls and observations in this paper.

Theorem 2.9. *There exist arbitrarily small positive ϵ such that problem (2.1) with (DF) admits a nontrivial exponentially growing solution.*

Proof: Let τ_n and $\epsilon_n = \pi/2\tau_n$ be as in the proof of Theorem 2.8. For fixed integer n, set

$$g(s) = \frac{1}{s} \log \frac{T(s) - 1}{T(s) + 1}$$

for s near $i\tau_n$, where the branch of the logarithm is chosen so that $\log i = \pi i/2$. Then $g(i\tau_n) = \epsilon_n$. Moreover, calculating g' and evaluating it at $s = i\tau_n$ gives

$$g'(i\tau_n) = \tau_n^{-1} i T'(i\tau_n) + \tau_n^{-2} \pi i/2$$
$$= -\tau_n^{-1} \overline{T(i\tau_n)} T'(i\tau_n) + \tau_n^{-2} \pi i/2.$$

By (2.8), $\Re g'(i\tau_n) > 0$, hence, the open mapping theorem guarantees that g is a one-to-one map from a neighborhood of $i\tau_n$ onto a neighborhood of ϵ_n. In addition, since the real and imaginary parts of $g'(i\tau_n)$ are both positive, the inverse image under g of an interval of the form $(\epsilon_n, \epsilon_n + \delta_n)$, for some $\delta_n > 0$, lies in the open right half-plane. Thus, for any $\epsilon \in (\epsilon_n, \epsilon_n + \delta_n)$, the pair (ϵ, s_ϵ), with $s_\epsilon = g^{-1}(\epsilon)$ in this neighborhood, satisfies the characteristic equation (CE), and Theorem 2.9 follows from Lemma 2.7. ∎

3. Loss of wellposedness due to feedback delay

In Section 2 we saw that there always exist arbitrarily small positive ϵ such that problem (2.1) with (DF) has exponentially growing solutions. In certain cases, for any fixed $\epsilon > 0$, problem (2.1) with (DF) has solutions with arbitrarily large exponential growth rates. We have the following characterization of this phenomenon.

Theorem 3.1. *The following assertions are equivalent:*

(a) *For some $\epsilon > 0$, problem (2.1) with (DF) admits a sequence of solutions $u_n(t) = \exp(s_n t)v_n, \tilde{u}_n(t) = \exp(s_n t)\tilde{v}_n$, with $v_n \in \operatorname{dom} D, \tilde{v}_n \in \operatorname{dom} \tilde{D}$ not both zero, and $\Re s_n \to \infty$.*

(b) *For any $\epsilon > 0$, there exist solutions as in (a).*

(c) *There exists a sequence $\tau_n \to \infty$ such that $T(i\tau_n) = -1$ and $T'(i\tau_n) \to 0$.*

(d) *There exists a sequence $\tau_n \to \infty$ such that $T(i\tau_n) \to -1$ and $T'(i\tau_n) \to 0$.*

(e) *For some $\sigma > 0$ there exists a sequence $\tau_n \to \infty$ such that $T(\sigma + i\tau_n) \to -1$.*

(f) *There exists a sequence $\tau_n \to \infty$ such that $T(\sigma + i\tau_n) \to -1$ uniformly for $\sigma \geq 0$ in compact intervals.*

If any of (a) - (f) holds, we say that feedback delay causes a <u>loss of wellposedness</u> in problem (2.1) with (DF). The proof of Theorem 3.1 depends on several additional lemmas concerning the mapping properties of the transfer function $T(s)$. As a consequence of the proof, we also have a necessary condition on the location of the eigenvalues of the operator **A**, defined in Proposition 2.1, that must be satisfied before such a loss of wellposedness due to delay can occur.

Corollary 3.2. *Let $\lambda_n = \pm i w_n, 0 < w_1 < w_2 < \ldots,$ be the nonzero eigenvalues of **A**. The assertions (a) - (f) of Theorem 3.1 cannot hold if $\limsup_{n\to\infty}(w_n - w_{n-1}) < \infty$.*

In particular, from Corollary 3.1 it follows that loss of wellposedness due to feedback delay does not occur when (2.1) describes a wave equation as in the model for torsional vibrations in an elastic rod (see Section 5).

Before proceeding to the proofs of Theorem 3.1 and its Corollary, we pause to interpret conditions (a) - (f) in that Theorem.

First, recall (cf Lemma 2.7(a)) that $T(i\omega) = -1$ specifies the resonance frequencies $\omega/2\pi$ of the neutrally stable problem (2.1) with (NS). The value $T'(i\omega)$ can formally be regarded as the time for a signal on a carrier frequency $\omega/2\pi$ to pass through the control-observation system (2.1) with (SO). To see this, consider the input

$$f(t) = e^{i\mu t}e^{i\nu t}$$

in the stabilized open loop system (2.1), (SO), regarded as a low frequency signal $e^{i\mu t}$ modulated on a high frequency carrier $e^{i\nu t}$. For very small μ, the output is approximately

(3.1) $$y(t) \approx T(i\nu)e^{i\nu t} \exp\{i\mu(t + T'(i\nu)/T(i\nu))\}.$$

To verify (3.1) write

$$\begin{aligned}
y(t) &= T(i\mu + i\nu)e^{i(\mu+\nu)t} \\
&\approx [T(i\nu) + i\mu T'(i\nu)]e^{i(\mu+\nu)t} \\
&= [1 + i\mu T'(i\nu)/T(i\nu)]T(i\nu)e^{i(\mu+\nu)t} \\
&\approx T(i\nu)\exp\{i\mu T'(i\nu)/T(i\nu)\}e^{i(\mu+\nu)t}.
\end{aligned}$$

Thus, the number $-T'(i\nu)/T(i\nu)$ can be regarded as the transition time for a signal on carrier frequency $\nu/2\pi$.

We caution the reader that in general one must be careful when attempting to relate the group velocity in systems with absorbing boundary conditions to the velocity with which the energy contained within a travelling group of waves is propagated through the system. (See, e.g., [1].)

We show in Section 4 that at resonance frequencies $2\pi/\omega_n$ of problem (2.1) with (NS) (i.e., $T(i\omega_n) = -1$), $1/T'(i\omega_n)$ is proportional to $|\tilde{P}\tilde{\phi}_n|^2$ where $\Phi_n = (\phi_n, \tilde{\phi}_n)$ is a normalized eigenvector corresponding to the eigenvalue $i\omega_n$ of the operator $\mathbf{A}$. We also show that $1/T'(i\omega_n)$ is proportional to the rate at which the unstabilized open loop problem (2.1) with (UO) picks up energy with input $f(t) = \exp(i\omega_n t)$ at resonance.

In Section 4 we also use the transfer function V of the unstabilized open loop problem (2.1) with (UO) to show that condition (e) or (f) of Theorem 3.1 is equivalent to (2.1) with (UO) being ill-posed in a sense of weighted $\mathbf{L}^2(0, \infty)$ controls and observations, and hence in a sense of $\mathbf{L}^2_{\mathrm{loc}}$ controls and observations.

We now turn to the additional estimates needed to prove Theorem 3.1.

Lemma 3.3. *For* $s = \sigma + i\tau$ *with* $\sigma > 0$,

$$\sigma|T'(s)T(s)| \leq \frac{3}{2}\sigma(\|x(s)\|^2 + \|\tilde{x}(s)\|^2) = \frac{3}{2}(1 - |T(s)|^2).$$

Proof: The equation part is precisely (2.7).

To prove the inequality part put $\rho = \sqrt{\sigma}$. By (2.9)

$$\|\rho\tilde{x}' + \frac{1}{2\rho}\tilde{x}\|^2 + \|\rho x' + \frac{1}{2\rho}x\|^2$$

$$= \sigma(\|x'\|^2 + \|\tilde{x}'\|^2) + \Re(\langle x', x\rangle + \langle \tilde{x}', \tilde{x}\rangle) + \frac{1}{4\sigma}(\|x\|^2 + \|\tilde{x}\|^2)$$

$$\leq \frac{1}{4\sigma}(\|x\|^2 + \|\tilde{x}\|^2).$$

Since in any Hilbert space the inequality $\|a + b\| \leq \|b\|$ implies $\|a\| \leq 2\|b\|$, we infer, with $a = \rho(x', \tilde{x}') \in \mathbf{X}$ and $b = \frac{1}{2\rho}(x, \tilde{x})$, that $\|x'\|^2 + \|\tilde{x}'\|^2 \leq \sigma^{-2}(\|x\|^2 + \|\tilde{x}\|^2)$. Thus, by (2.8) and the Schwarz inequality we have

$$|T'(s)T(s)| \leq \frac{1}{2}(\|x\|^2 + \|\tilde{x}\|^2) + \sigma(\|x\|^2 + \|\tilde{x}\|^2)^{1/2}(\|x'\|^2 + \|\tilde{x}'\|^2)^{1/2}$$

$$\leq \frac{3}{2}(\|x\|^2 + \|\tilde{x}\|^2). \quad \blacksquare$$

Lemma 3.4. *Let* $s_0 = \sigma_0 + i\tau_0$, $\sigma_0 > 0$, *and* $1 - |T(s_0)|^2 \leq \eta < \frac{1}{2}$.

  (a) *If* $|s - s_0| < \frac{\sigma_0}{12}$, *then* $|T(s) - T(s_0)| < \frac{1}{2}\eta(1 - 2\eta)^{-1/2}$.

  (b) *If* $0 < \sigma < \sigma_0$, *then* $1 - |T(\sigma + i\tau_0)|^2 \leq \eta(\sigma_0/\sigma)^3$. *Moreover, if* $\eta(\sigma_0/\sigma)^3 < 1/2$, *then*

$$|T(\sigma + i\tau_0) - T(\sigma_0 + i\tau_0)| \leq \eta\sigma_0^3/\sqrt{2}\sigma^3.$$

Proof: (a) Fix s satisfying $|s - s_0| < \sigma_0/12$. Write $z = s_0 + t\nu$ where $\nu = (s - s_0)/|s - s_0|$, and set $h(t) = 1 - |T(z)|^2$ for $0 \le t \le |s - s_0|$. By Lemma 3.3,

$$|h'(t)| \le 3h(t)(\Re z)^{-1} \le 4h(t)\sigma_0^{-1};$$

hence,

$$h(t) \le \exp\left(4t\sigma_0^{-1}\right)h(0) \le e^{1/3}h(0) \le 2\eta$$

for $0 \le t \le |s - s_0|$. By Lemma 3.3 again,

$$|T'(z)| \le 3h(t)(2\Re z|T(z)|)^{-1} \le 6\eta\sigma_0^{-1}(1 - 2\eta)^{-1/2},$$

so by the mean value inequality

$$|T(s) - T(s_0)| \le 6\eta|s - s_0|\sigma_0^{-1}(1 - 2\eta)^{-1/2} \le \frac{1}{2}\eta(1 - 2\eta)^{-1/2}.$$

(b) Set $g(\sigma) = 1 - |T(\sigma + i\tau_0)|^2$. By Lemma 3.3, $\sigma g'(\sigma) \ge -3g(\sigma)$; hence,

$$g(\sigma) \le g(\sigma_0)(\sigma_0/\sigma)^3 \le \eta(\sigma_0/\sigma)^3,$$

and the first inequality in (b) is proved. To prove the second inequality, use Lemma 3.3 again to write

$$|T'(\sigma + i\tau_0)| \le \frac{3g(\sigma)}{2\sigma|T(\sigma + i\tau_0)|} \le \frac{3\eta(\sigma_0/\sigma)^3}{2\sigma[1 - \eta(\sigma_0/\sigma)^3]^{1/2}} \le \frac{3\eta\sigma_0^3}{2^{1/2}\sigma^4}$$

when $2\eta(\sigma_0/\sigma)^3 < 1$. The second inequality in (b) now follows by the Fundamental Theorem of Calculus. ∎

Proof of Theorem 3.1: (b) $\Rightarrow$ (a), (c) $\Rightarrow$ (d) and (f) $\Rightarrow$ (e) are each trivial.

(d) $\Rightarrow$ (c): Let $i\tau_n$ be the roots of $T(i\tau) = -1$ with $\tau > 0$, and assume that (c) fails to hold. Then there is $\eta > 0$ such that for all sufficienlty large n, $|T'(i\tau_n)| > \eta$. By boundedness of T'' on the imaginary axis, there exists $\delta > 0$ such that $|T'(i\tau)| > \eta/2$ if $|\tau - \tau_n| < \delta$. Recall that by (2.7) and (2.8), $T(i\tau)$ rotates around the unit circle in the clockwise direction as τ increases. This combined with the last inequality yields that $|\arg T(i\tau) - \arg T(i\tau_n)| \ge \delta\eta/2$ whenever $|\tau - \tau_n| \ge \delta$ for the τ_n closest to τ. Hence, if $T(i\tau)$ is so close to -1 that $|argT(i\tau) - (2k+1)\pi| < \delta\eta/2$ for some integer k, then $|T'(i\tau)| > \eta/2$ and (d) does not hold.

(c) $\Rightarrow$ (f): Let τ_n be as in (c). From (2.9) we see that

$$Re(\langle x(\sigma + i\tau_n), x'(\sigma + i\tau_n)\rangle + \langle \tilde{x}(\sigma + i\tau_n), \tilde{x}'(\sigma + i\tau_n)\rangle) \le 0;$$

hence $\|x(\sigma + i\tau_n)\|^2 + \|\tilde{x}(\sigma + i\tau_n)\|^2$ is nonincreasing in σ. It follows from Lemma 3.3 and (2.8) that

$$|T'(\sigma + i\tau_n)T(\sigma + i\tau_n)| \le 3|T'(i\tau_n)|.$$

For fixed $\sigma_0 > 0$ and $\eta \in (0, 1/2)$ choose n such that $|T'(i\tau_n)| < \eta/12\sigma_0$. Suppose that for some $\sigma \in [0, \sigma_0)$, $|T(\sigma + i\tau_n) + 1| \geq \eta$, and let σ_i be the infimum of all such σ. Since for $\sigma \in [0, \sigma_i]$,

$$|T'(\sigma + i\tau_n)| \leq 3|T'(i\tau_n)|/|T(\sigma + i\tau_n)| \leq \eta/2\sigma_0,$$

we obtain

$$|T(\sigma_i + i\tau_n) + 1| \leq \int_0^{\sigma_i} |T'(\sigma + i\tau_n)| d\sigma \leq \eta/2,$$

in contradiction to our choice of σ_i. Thus (f) holds.

(e) $\Rightarrow$ (d): Let $\sigma_0 > 0$ be so that $T(\sigma_0 + i\tau_n) \to -1$. By repeated use of Lemma 3.4(a), we find intervals $J_n = [\tau_n^-, \tau_n^+]$ such that $\tau_n \in J_n$, $\tau_n^+ - \tau_n^- \to \infty$, and $T(\sigma_0 + i\tau) \to -1$ as $\tau \to \infty$ in $\bigcup J_n$. Pick $\eta > 0$ arbitrarily small. By uniform boundedness of derivatives of T in the right half-plane, we may choose $\sigma_1 > 0$ so that $\sigma_1 |T^{(k)}(z)| \leq \eta/2$ $(k = 1, 2)$ for $\Re z \geq 0$. By Lemma 3.4(b), $T(\sigma_1 + i\tau) \to -1$ as $\tau \to \infty$ in $\bigcup J_n$, and we now infer from Lemma 3.3 that $T'(\sigma_1 + i\tau) \to 0$ as $\tau \to \infty$ in $\bigcup J_n$. Choosing n so large that $|T(\sigma_1 + i\tau) + 1| \leq \eta/2$ and $|T'(\sigma_1 + i\tau)| \leq \eta/2$ for $\tau \in J_n$, we obtain (by our choice of σ_1) $|T(i\tau) + 1| \leq \eta$ and $|T'(i\tau)| \leq \eta$ for $\tau \in J_n$.

(a) $\Rightarrow$ (e): Assume that $F(\sigma_n + i\tau_n) = 0$ with $\sigma_n \to \infty$ for some $\epsilon > 0$. Pick $\sigma > 0$. Rewriting $F(s)$ as

$$(3.2) \qquad F(s) = 1 + T(s) + \frac{2e^{-\epsilon s}}{1 - e^{-\epsilon s}},$$

we see that $|1 + T(\sigma_n + i\tau_n)| \leq 4\exp(-\epsilon\sigma_n)$ for all sufficiently large n. By Lemma 3.4(b) we infer that

$$|1 + T(\sigma + i\tau_n)| \leq (\sigma_n/\sigma)^3(1 - |T(\sigma_n + i\tau_n)|^2) + |1 + T(\sigma_n + i\tau_n)| \to 0$$

as $n \to \infty$, and (e) holds.

(f) $\Rightarrow$ (b): Fix $\epsilon > 0$ and choose $\sigma_0 > 0$ arbitrarily large so that $2\exp(-\epsilon\sigma_0) < 1$ and $\epsilon\sigma_0 > 24\pi$. By assumption (f) we have $\tau_n \to \infty$ so that $T(\sigma_0 + i\tau_n) \to -1$. Using Lemma 3.4(a), we infer that $T(\sigma_0 + i\tau) \to -1$ as $\tau \to \infty$ in the union of intervals $\bigcup[\tau_n - 2\pi\epsilon^{-1}, \tau_n + 2\pi\epsilon^{-1}]$. Take n so large that

$$|T(\sigma_0 + i\tau) + 1| < \frac{1}{2}\exp(-\epsilon\sigma_0)$$

for $\tau \in [\tau_n - 2\pi\epsilon^{-1}, \tau_n + 2\pi\epsilon^{-1}]$. This interval contains an interval $J = [2\pi k\epsilon^{-1}, 2\pi(k+1)\epsilon^{-1}]$ for some positive integer k.

Next we show that there exists some $\sigma_1 > \sigma_0$ such that for all $\tau \in J$,

$$(3.3) \qquad |T(\sigma_1 + i\tau)| < 1 - 4\exp(-\epsilon\sigma_1)[1 - \exp(-\epsilon\sigma_1)]^{-1}.$$

Assume the contrary, i.e., for all $\sigma > \sigma_0$, there exists $\tau \in J$ with

$$|T(\sigma + i\tau)| \geq 1 - 4\exp(-\epsilon\sigma)[1 - \exp(-\epsilon\sigma)]^{-1}.$$

By Lemma 3.4(b),

$$1 - |T(\sigma_0 + i\tau)|^2 \leq (1 - |T(\sigma + i\tau)|)(1 + |T(\sigma + i\tau)|)(\sigma/\sigma_0)^3$$
$$\leq 8(\sigma/\sigma_0)^3 \exp(-\epsilon\sigma)[1 - \exp(-\epsilon\sigma)]^{-1} \to 0$$

as $\sigma \to \infty$. But this is impossible since $|T(\sigma_0 + i\tau)|$ is bounded away from 1 for $\tau \in J$.

We will now use the argument principle to show that $F(s)$ admits a unique zero in the rectangle $Q = \{\sigma_0 \leq \sigma \leq \sigma_1, \tau \in J\}$. For $s = \sigma_0 + i\tau$, $\tau \in J$, express $F(s)$ as in (3.2). Since $|1 + T(s)| < \frac{1}{2}\exp(-\epsilon\sigma_0)$ and $|2\exp(-\epsilon s)[1 - \exp(-\epsilon s)]^{-1}| \geq \exp(-\epsilon\sigma_0)$ for such s,

$$\left| \arg F(s) - \arg \frac{2\exp(-\epsilon s)}{1 - \exp(-\epsilon s)} \right| < \frac{\pi}{6} \qquad (\text{modulo } 2\pi)$$

and $F(s) \neq 0$ for $s = \sigma_0 + i\tau$, $\tau \in J$. Since $2\exp(-\epsilon s)[1 - \exp(-\epsilon s)]^{-1}$ winds around the origin exactly once in the counterclockwise direction as τ decreases from $2(k+1)\pi\epsilon^{-1}$ to $2k\pi\epsilon^{-1}$, $s = \sigma_0 + i\tau$, the index of 0 with respect to the image of ∂Q under F is 1 provided that we can show that $\Re F(s) > 0$ when s lies on the other three edges of ∂Q. To see that this is the case, when $s = \sigma_1 + i\tau$ ($\tau \in J$), use (3.3) to obtain

$$|T(s)| + |2\exp(-\epsilon s)[1 - \exp(-\epsilon s)]^{-1}| \leq |T(s)| + 2\exp(-\epsilon\sigma_1)[1 - \exp(-\epsilon\sigma_1)]^{-1} < 1$$

for $s = \sigma_1 + i\tau, \tau \in J$. When $s = \sigma + i\tau$ with $\tau = 2k\pi\epsilon^{-1}$ or $2(k+1)\pi\epsilon^{-1}$, then $|T(s)| < 1$ and

$$(1 + e^{-\epsilon s})(1 - e^{-\epsilon s})^{-1} = (1 + e^{-\epsilon\sigma})(1 - e^{-\epsilon\sigma})^{-1} > 1,$$

so $\Re F(s) > 0$ for such s. Thus $F(s) = 0$ has exactly one solution in Q by the argument principle. ∎

Proof of Corollary 3.2: By Lemma 2.7, $\lambda = i\omega$ is an eigenvalue of $\mathbf{A}$ if and only if $T(i\omega) = -1$. In the proof of Theorem 3.1 part (e) $\Rightarrow$ (d), we see that if (e) holds, then for any given $\eta \in (0, 1/2)$, there exist arbitrarily long subintervals J_n of $[0, \infty)$ such that $|T(i\tau) + 1| < \eta$ for τ in these intervals. Since $T(i\tau)$ rotates around the unit circle in the clockwise direction as τ increases, condition (e) cannot hold if $\limsup_{n\to\infty}(\omega_n - \omega_{n-1}) < \infty$. ∎

4. Relation to the unstabilized problem

We are going to relate the loss of wellposedness phenomenon to properties of the transfer function of the unstabilized open loop problem. We will need the spectral resolution of $\mathbf{A}$, thus let $\Phi_n = (\phi_n, \tilde{\phi}_n)$ ($n \in \{1, 2, \cdots\}$) be an orthonormal basis of eigenvectors of $\mathbf{A}$ to the eigenvalues λ_n (satisfying $T(\lambda_n) = -1$). Moreover we fix some $\mu > 0$. To avoid special cases we assume throughout this section that $\lambda = 0$ is not an eigenvalue of $\mathbf{A}$. This nonessential assumption merely means that (2.1) with (NS) admits no nontrivial steady state solutions.

Lemma 4.1. *There exists some* $\mathbf{x} = (x, \tilde{x}) \in \mathbf{X}$ *with* $\mu x - \tilde{D}\tilde{x} = 0$, $\mu \tilde{x} - Dx = 0$, *and* $Px = 1$.

Proof: Take arbitrary $z \in \operatorname{dom} D$, $\tilde{z} \in \operatorname{dom} \tilde{D}$ with $Pz = 1$ and put $\mathbf{x} = (z, \tilde{z}) - (\mu - \mathbf{A})^{-1}(\mu z - \tilde{D}\tilde{z}, \mu \tilde{z} - Dz)$. ∎

Lemma 4.2.

(a) *The operator mapping* $(z, \tilde{z})$ *into* $\tilde{P}\tilde{z}$ *is relatively bounded with respect to* $\mathbf{A}$.

(b) $\sum_n |\tilde{P}\tilde{\phi}_n / \lambda_n|^2$ *is finite.*

(c) *For all* n, $\tilde{P}\tilde{\phi}_n = -(\mu + \lambda_n)\langle \Phi_n, \mathbf{x} \rangle$.

Proof: With $\mathbf{x}$ as in the previous Lemma $\tilde{P}\tilde{z} = -\langle \tilde{z}, Dx \rangle - \langle \tilde{D}\tilde{z}, x \rangle$, which is evidently relatively bounded.

(b) is just (a) expressed by spectral resolution.

To prove (c) consider

$$(\mu + \lambda_n)\langle \Phi_n, \mathbf{x} \rangle = \langle \lambda_n \Phi_n, \mathbf{x} \rangle + \langle \Phi_n, \mu \mathbf{x} \rangle =$$
$$\langle \tilde{D}\tilde{\phi}_n, x \rangle + \langle D\phi_n, \tilde{x} \rangle + \langle \phi_n, \tilde{D}\tilde{x} \rangle + \langle \tilde{\phi}_n, Dx \rangle =$$
$$-\tilde{P}\tilde{\phi}_n \overline{Px} - P\phi_n \overline{\tilde{P}\tilde{x}} = -\tilde{P}\tilde{\phi}_n. \quad ∎$$

In general, for (UO) there occurs a loss of smoothness from control to observation:

Proposition 4.3. *Given* $f \in \mathbf{W}^{2,1}$ *with* $f(0) = 0$, *there exist unique* $u \in \mathbf{W}^{1,1}_{loc}([0, \infty), X)$ *and* $\tilde{u} \in \mathbf{W}^{1,1}_{loc}([0, \infty), \tilde{X})$ *satisfying* (2.1) *with* $Pu(t) = f(t)$, $u(0) = 0$, *and* $\tilde{u}(0) = 0$. *Moreover,* $y(t) = \tilde{P}\tilde{u}(t)$ *is defined everywhere and continuous in* t.

The functions u, $\tilde{u}$, *and* y *have exponential growth, and their Laplace transforms satisfy* $\hat{y}(s) = V(s)\hat{f}(s)$ *with*

$$V(s) = \frac{T(s) - 1}{T(s) + 1} \; .$$

Proof: In X consider

$$(u, \tilde{u})(t) = f(t)(x, \tilde{x}) + \int_0^t (\mu f(s) - f'(s))\mathbf{S}(t - s)\mathbf{x}\,ds \; .$$

By smoothness of f we have that

$$(u - fx, \tilde{u} - f\tilde{x}) \in \mathbf{W}^{1,1}_{loc}([0, \infty), \mathbf{X}) \cap \mathbf{C}([0, \infty), \operatorname{dom} \mathbf{A})$$

and satisfies the equation

$$(u(t) - f(t)x, \tilde{u}(t) - f(t)\tilde{x})' = \mathbf{A}(u(t) - f(t)x, \tilde{u}(t) - f(t)\tilde{x}) + (\mu f - f')(t)\mathbf{x} ,$$

from which we deduce (2.1). As Pz vanishes for $(z, \tilde{z}) \in \text{dom } \mathbf{A}$, we have

$$Pu(t) = f(t)Px = f(t)$$

. As the operator $(z, \tilde{z}) \to \tilde{P}\tilde{z}$ is relatively bounded with respect to $\mathbf{A}$, $y(t) = \tilde{P}\tilde{u}(t)$ exists and is continuous. Standard semigroup arguments imply that all functions have exponential growth, so that their Laplace transforms exist. (2.1) implies

$$s\hat{u}(s) = \tilde{D}\hat{\tilde{u}}(s),$$
$$s\hat{\tilde{u}}(s) = D\hat{u}(s),$$

thus

$$\hat{f}(s) + \hat{y}(s) = P\hat{u}(s) + \tilde{P}\hat{\tilde{u}}(s)$$
$$= T(s)(P\hat{u}(s) - \tilde{P}\hat{\tilde{u}}(s)) = T(s)(\hat{f}(s) - \hat{y}(s)),$$

hence $\hat{y}(s) = V(s)\hat{f}(s)$. ∎

Let $\mathbf{L}_\sigma^2 = \{f \in \mathbf{L}_{loc}^2 : e^{-\sigma t}f(t) \in \mathbf{L}^2\}$ with a suitable norm. We say that (2.1) with (UO) is wellposed for $\mathbf{L}_\sigma^2$ controls and $\mathbf{L}_\sigma^2$ observations, if the operator mapping $f \in \mathbf{W}^{2,1}$ with $f(0) = 0$ into y admits a continuous extension from $\mathbf{L}_\sigma^2$ into $\mathbf{L}_\sigma^2$. We remark that one can show that problem (2.1) with (UO) is wellposed for $\mathbf{L}_\sigma^2$ controls and observations iff it is wellposed for local $\mathbf{L}^2$ controls and observations. Here local $\mathbf{L}^2$ control and observation wellposedness means that for all $f \in \mathbf{W}^{2,1}$ with $f(0) = 0$, the input-output map satisfies $\|y\|_{\mathbf{L}^2(0,T)} \leq M(T)\|f\|_{\mathbf{L}^2(0,T)}$. The proof that wellposedness for local $\mathbf{L}^2$ controls and observations implies $\mathbf{L}_\sigma^2$ wellposedness uses techniques similar to those used by Curtain in [5]. (See especially Lemma 2.3.) Moreover, (UO) is $\mathbf{L}_\sigma^2$ wellposed for one $\sigma > 0$ iff it is so for all $\sigma > 0$.

Theorem 4.4. (2.1) with (UO) is wellposed for $\mathbf{L}_\sigma^2$ controls and $\mathbf{L}_\sigma^2$ observations iff none of the assertions (a) to (f) of Theorem 3.1 holds.

Proof: Since V is the transfer function for (UO), Plancherel's Theorem implies that (UO) is $\mathbf{L}_\sigma^2$ wellposed iff $V(\sigma + i\tau)$ is bounded for $\tau \in (-\infty, \infty)$. This is equivalent to the condition that T is bounded away from -1 on the same set. In this case, Theorem 3.1 (f) cannot hold. Conversely, if Theorem 3.1 (e) is false, then T is bounded away from -1 on the vertical line $\sigma + i\tau$, $\tau \in (-\infty, \infty)$. ∎

We will now compute the transfer function V in terms of the spectral resolution of $\mathbf{A}$.

Lemma 4.5.
$$V(s) = \tilde{P}\tilde{x} + \sum_n \frac{s - \mu}{(\mu - \lambda_n)(s - \lambda_n)} |\tilde{P}\tilde{\phi}_n|^2.$$

Proof: Consider the solution $\mathbf{u} = (u, \tilde{u})$ according to Proposition 4.3 with $f(t) = te^{\mu t}$, i.e.

$$\mathbf{u}(t) = te^{\mu t}\mathbf{x} - \int_0^t e^{\mu s}\mathbf{S}(t - s)\mathbf{x}ds \ .$$

Using the spectral resolution of $\mathbf{S}$ and Lemma 4.2(c), we obtain

$$\langle \mathbf{u}(t) - te^{\mu t}\mathbf{x}, \Phi_n \rangle = \frac{e^{\mu t} - e^{\lambda_n t}}{(\mu - \lambda_n)^2} \overline{\tilde{P}\tilde{\phi}_n} \ ;$$

consequently

$$\tilde{P}\tilde{u}(t) = te^{\mu t}\tilde{P}\tilde{x} + \sum_n \frac{e^{\mu t} - e^{\lambda_n t}}{(\mu - \lambda_n)^2} |\tilde{P}\tilde{\phi}_n|^2 \ .$$

(Lemma 4.2(b) guarantees that the sum converges.) Taking Laplace transforms and dividing by the Laplace transform of $te^{\mu t}$ we obtain the desired formula. ∎

Lemma 4.6.
$$T'(\lambda_n) = \frac{2}{|\tilde{P}\tilde{\phi}_n|^2}$$

Proof: As $T = (1 + V)/(1 - V)$, we have $T'(s) = 2V'(s)/(1 - V(s))^2$ for $\Re s > 0$. Taking the limit for $s \to \lambda_n$ and using the previous lemma we obtain

$$T'(\lambda_n) = \lim_{s \to \lambda_n} \frac{2 \sum_m \frac{1}{(s - \lambda_m)^2} |\tilde{P}\tilde{\phi}_m|^2}{(1 - \tilde{P}\tilde{x} - \sum_m \frac{s - \mu}{(\mu - \lambda_m)(s - \lambda_m)} |\tilde{P}\tilde{\phi}_m|^2)^2} = \frac{2}{|\tilde{P}\tilde{\phi}_n|^2} \ . \quad ∎$$

As an immediate consequence we have the following characterization of the loss of well-posedness case:

Theorem 4.7. *The assertions (a) to (f) of Theorem 3.1 hold if and only if the set $\{\tilde{P}\tilde{\phi}_n : n = 1, 2 \cdots\}$ is unbounded.*

Physically, $|\tilde{P}\tilde{\phi}_n|^2$ measures how quickly the system picks up energy when it is forced at resonance frequency $\frac{\lambda_n}{2\pi i}$:

Proposition 4.8. *The unstabilized open loop system (2.1) with (UO) and forcing function $f(t) = e^{\lambda_n t}$ admits a solution $\mathbf{u} = (u, \tilde{u})$ such that $\tilde{P}\tilde{u}(t) = \alpha e^{\lambda_n t} - |\tilde{P}\tilde{\phi}_n|^2 te^{\lambda_n t}$ with suitable α.*

Proof: Put

$$\mathbf{u}(t) = e^{\lambda_n t}\mathbf{x} - \sum_{k \neq n} \frac{\lambda_n - \mu}{\lambda_n - \lambda_k} e^{\lambda_n t}\langle \mathbf{x}, \Phi_k \rangle \Phi_k + [(\mu - \lambda_n)te^{\lambda_n t} + e^{\lambda_n t}]\langle \mathbf{x}, \Phi_n \rangle \Phi_n \ .$$

Since $\sum_{k \neq n} |\lambda_k \frac{\lambda_n - \mu}{\lambda_n - \lambda_k} e^{\lambda_n t} \langle \mathbf{x}, \Phi_k \rangle|^2$ converges, we infer that $\mathbf{u}(t) - e^{\lambda_n t} \mathbf{x} \in \text{dom } \mathbf{A}$, thus $P\mathbf{u}(t) = e^{\lambda_n t} P\mathbf{x} = e^{\lambda_n t}$. Using that $(\tilde{D}\tilde{\phi}_k, D\phi_k) = \lambda_k(\phi_k, \tilde{\phi}_k)$ and $(\tilde{D}\tilde{x}, Dx) = \mu(x, \tilde{x})$, we can check (2.1) by straightforward computation. Finally

$$\tilde{P}\tilde{u}(t) = e^{\lambda_n t}[\tilde{P}\tilde{x} - \sum_{k \neq n} \frac{\lambda_n - \mu}{\lambda_n - \lambda_k} \langle \mathbf{x}, \Phi_k \rangle \tilde{P}\tilde{\phi}_k + \tilde{P}\tilde{\phi}_n \langle \mathbf{x}, \Phi_n \rangle]$$

$$+ (\mu - \lambda_n) t e^{\lambda_n t} \langle \mathbf{x}, \Phi_n \rangle \tilde{P}\tilde{\phi}_n = \alpha e^{\lambda_n t} - t e^{\lambda_n t} \tilde{P}\tilde{\phi}_n \overline{\tilde{P}\tilde{\phi}_n}$$

by Lemma 4.2 (c). ∎

5. Examples

In this section we apply the criteria developed in Sections 3 and 4 to determine whether feedback delay causes a loss of wellposedness in some model mechanical stabilization problems.

EXAMPLE 5.1 (EULER-BERNOULLI BEAM). The boundary control problem

$$(5.1) \qquad w_{tt}(z,t) + w_{zzzz}(z,t) = 0, \qquad 0 < z < 1, \quad t > 0,$$

$$(5.2) \qquad w(0,t) = w_z(0,t) = 0, \qquad t > 0,$$

$$(5.3) \qquad w_{zzz}(1,t) = 0, \qquad t > 0,$$

$$(5.4) \qquad w_{zz}(1,t) = -k w_{tz}(1,t), \qquad t > 0, \quad k > 0,$$

models an Euler-Bernoulli beam, clamped at the end $z = 0$, and stabilized by an applied bending moment that is negatively proportional to angular velocity at the end $z = 1$. (All the physical constants have been normalized to 1.) It has recently been shown [4] (see also [3]) that this problem is uniformly exponentially stable whenever k in (5.4) is positive.

Problem (5.1) - (5.4) may be recast in the abstract form (2.1) with (SF) by taking $X = \tilde{X} = \mathbf{L}^2([0,1]; \mathbf{R})$,

$$(5.5) \qquad u(t) = w_{zz}(\cdot, t), \quad \tilde{u}(t) = w_t(\cdot, t),$$

$$Dx = -x''$$

defined on

$$\text{dom } D = \{ x \in \mathbf{W}^{2,2} : x'(1) = 0 \},$$
$$\tilde{D}x = \tilde{x}''$$

defined on

$$\text{dom } \tilde{D} = \{\, \tilde{x} \in \mathbf{W}^{2,2} : \tilde{x}(0) = \tilde{x}'(0) = 0 \,\},$$

and

$$Px = \frac{1}{\sqrt{k}}x(1), \quad \tilde{P}\tilde{x} = -\sqrt{k}\tilde{x}'(1).$$

It is easy to check that the Green's identity (2.2) holds.

Tedious but routine calculations yield that the eigenvalues λ_n of the operator $\mathbf{A}$ defined in Proposition 2.1 are given by $\lambda_{\pm n} = \pm i\omega_n$, where ω_n are the positive roots of the characteristic equation

$$(5.6) \qquad\qquad\qquad 1 + \cosh\sqrt{\omega}\cos\sqrt{\omega} = 0.$$

Observe that $\omega_n \sim (n - \frac{1}{2})^2\pi^2$ as $n \to \infty$, so that Corollary 3.2 does not apply. The normalized eigenfunction corresponding to $\lambda_n = i\omega_n$ is $\Phi_n = (\phi_n, \tilde{\phi}_n)$ where

$$\phi_n = c_n i\{(\cosh\sqrt{\omega_n}z + \cos\sqrt{\omega_n}z) - \beta_n(\sinh\sqrt{\omega_n}z + \sin\sqrt{\omega_n}z)\},$$
$$\tilde{\phi}_n = c_n\{(\cos\sqrt{\omega_n}z - \cosh\sqrt{\omega_n}z) + \beta_n(\sinh\sqrt{\omega_n}z - \sin\sqrt{\omega_n}z)\},$$

with

$$\beta_n = (\sinh\sqrt{\omega_n} - \sin\sqrt{\omega_n})(\cosh\sqrt{\omega_n} + \cos\sqrt{\omega_n})^{-1},$$

and where c_n is chosen to make $\|\Phi_n\| = 1$. Suppressing the subscripts n, we see that $\beta = \beta_n$ may be estimated as

$$\beta = (1 - 2e^{-\sqrt{\omega}}\sin\sqrt{\omega} - e^{-2\sqrt{\omega}})(1 + 2e^{-\sqrt{\omega}}\cos\sqrt{\omega} + e^{-2\sqrt{\omega}})^{-1}$$
$$= 1 - 2e^{-\sqrt{\omega}}(\sin\sqrt{\omega} + \cos\sqrt{\omega}) + O(e^{-2\sqrt{\omega}}) \quad (\omega = \omega_n \to \infty).$$

Then evaluating the integrals in the definition of $\|\Phi_n\|$ and using the above estimate for $\beta = \beta_n$, we find, after some calculation, that $c_n \to 1/\sqrt{2}$ $(n \to \infty)$. It follows that

$$|\tilde{P}\tilde{\phi}| = \left|c\sqrt{k\omega}\left\{\frac{2(\cosh\sqrt{\omega}\sin\sqrt{\omega} + \sinh\sqrt{\omega}\cos\sqrt{\omega})}{\cosh\sqrt{\omega} + \cos\sqrt{\omega}}\right\}\right|$$
$$\sim \sqrt{2k\omega} \quad \text{as} \quad \omega = \omega_n \to \infty,$$

where we have again suppressed the subscripts n. By Lemma 4.6

$$T'(i\omega_n) = 2/|\tilde{P}\tilde{\phi}_n|^2 \sim 1/k\omega_n \to 0$$

as $n \to \infty$. Hence, by Theorem 3.1, any positive time delay in (5.4) causes a loss of wellposedness in problem (5.1) - (5.4).

We remark that Salamon [**13**, pp 429-430] has observed that the unstabilized open loop problem corresponding to (5.1)-(5.4), that is, (5.1) and (5.2) with control $w_{zz}(1,t) = f(t)$ and observation $y(t) = w_{tz}(1,t)$, is ill-posed using local $\mathbf{L}^2$ controls and observations in a sense that is similar to that discussed in Section 4. Also, loss of wellposedness due to delays occurs in the analogue of problem (5.1)-(5.4) for a viscoelastic Euler-Bernoulli beam even when the viscoelastic damping at high frequencies is very significant [9].

Next, in place of problem(5.1)-(5.4), we consider (5.1), (5.2) together with the stabilizing scheme acting only through shearing force

$$(5.7) \qquad\qquad w_{zz}(1,t) = 0, \qquad t > 0,$$

$$(5.8) \qquad\qquad w_{zzz}(1,t) = kw_t(1,t), \qquad t > 0, \quad k > 0,$$

It has also been shown in [4] that (5.1), (5.2), (5.7), (5.8) is uniformly exponentially stable. The abstract problem (2.1), (SF) corresponding to (5.1), (5.2), (5.7), (5.8) is given by taking $X, \tilde{X}, u, \tilde{u}$ and $\tilde{D}$ as in the first part of Example 5.1, and setting

$$Dx = -x''$$

on

$$\operatorname{dom} D = \{\, x \in \mathbf{W}^{2,2} : x(1) = 0\},$$
$$Px = \frac{1}{\sqrt{k}} x'(1),$$

and

$$\tilde{P}\tilde{x} = \sqrt{k}\tilde{x}(1)$$

. It is easy to see that the eigenvalues of the corresponding operator $\mathbf{A}$ are still given by $\lambda_\pm = \pm i\omega_n$ where ω_n are the positive roots of (5.6). Calculations similar to those sketched above show that for this problem $|\tilde{P}\tilde{\phi}_n| \sim \sqrt{2k}$ as $n \to \infty$. (The essential difference from the previous calculation is that now there is no spatial derivative in the definition of $\tilde{P}$.) Hence,

$$T'(i\omega_n) = 2/|\tilde{P}\tilde{\phi}_n|^2 \to 1/k$$

as $n \to \infty$, so, by Theorem 3.1, time delays in (5.8) do not cause a loss of wellposedness in problem (5.1), (5.2), (5.7), (5.8).

EXAMPLE 5.2 (TORSIONAL VIBRATIONS). The boundary control problem for the wave equation

$$(5.9) \qquad \begin{aligned} w_{tt}(z,t) &= w_{zz}(z,t), \qquad 0 < z < 1, \quad t > 0, \\ w(0,t) &= 0, \quad w_z(1,t) = -kw_t(1,t) \quad (k > 0) \quad t > 0, \end{aligned}$$

models the stabilization of torsional vibrations in an elastic rod (all physical constants have been set equal to 1) that is clamped at $z = 0$, with a feedback torque negatively proportional to angular velocity at $z = 1$. It is well known (see, e.g., [2]) that problem

(5.9) is uniformly exponentially stable. The abstract form (2.1), (SF) of problem (5.9) is gotten by taking

$$X = \tilde{X} = \mathbf{L}^2([0,1]; \mathbf{R}),$$
$$u(t) = w_z(\cdot, t),$$
$$\tilde{u}(t) = w_t(\cdot, t),$$
$$Dx = x'$$

defined on

$$\operatorname{dom} D = \mathbf{W}^{1,2},$$
$$\tilde{D}\tilde{x} = \tilde{x}'$$

defined on

$$\operatorname{dom} \tilde{D} = \{ \tilde{x} \in \mathbf{W}^{1,2} : \tilde{x}(0) = 0 \},$$
$$Px = \frac{1}{\sqrt{k}} x(1),$$

and

$$\tilde{P}\tilde{x} = -\sqrt{k}\tilde{x}(1).$$

It is easy to check that Hypothesis D holds, and to show that the eigenvalues of the operator $\mathbf{A}$ defined in Proposition 2.1 are given by $\lambda_{\pm n} = \pm(n - \frac{1}{2})\pi i$, $n = 1, 2, \dots$. Thus, by Corollary 3.2, time delays in the feedback in problem (5.9) do not cause a loss of wellposedness.

We remark that Grimmer, Lenczewski and Schappacher [10] have developed a semigroup setting in which it is shown that a certain class of one-dimensional hyperbolic initial-boundary value problems which have boundary conditions involving delays are wellposed. The framework developed in [10] includes the wave equation with delayed boundary data as well as the analogous problem for the Timoshenko beam.

REFERENCES

[1] L. Brillouin, "Wave Propagation and Group Velocity," Academic Press, New York, 1960.

[2] G. Chen, *A note on the boundary stabilization of the wave equation*, SIAM J. Control and Optimization **19** (1981), 106–113.

[3] G. Chen, M. C. Delfour, A. M. Krall and G. Payre, *Modeling, stabilization and control of serially connected beams*, SIAM J. Control and Optimization **25** (1987), 526–546.

[4] G. Chen, S.G. Krantz, D.W. Ma, C.E. Wayne, and H.H. West, *The Euler-Bernoulli beam equation with boundary energy dissipation*, in "Operator Methods for Optimal Control Problems," S.J. Lee, ed., Marcel Dekker, New York, 1988, pp. 67–96.

[5] R.F. Curtain, *Well-posedness of infinite-dimensional linear systems in time and frequency domain*. TW-287 University of Groningen, preprint, to appear.

[6] R. Datko, *Not all feedback stabilized hyperbolic systems are robust with respect to small time delays in their feedbacks*, SIAM J. Control and Optimization **26** (1988), 697–713.

[7] R. Datko, *A rank-one perturbation result on the spectra of certain operators*, (to appear).

[8] R. Datko, J. Lagnese and M.P. Polis, *An example on the effect of time delays in boundary feedback of wave equations*, SIAM J. Control and Optimization **24** (1986), 152–156.

[9] W. Desch, K.B. Hannsgen, Y. Renardy and R.L. Wheeler, *Boundary stabilization of an Euler-Bernoulli beam with viscoelastic damping*, in "Proc. 26th IEEE Conference on Decision and Control," Los Angeles, CA, 1987, pp. 1792–1795.

[10] R.C. Grimmer, R. Lenczewski and W. Schappacher, *Wellposedness of hyperbolic equations with delay in the boundary conditions*, in "Proc. Trends in Semigroup Theory and Applications," Trieste, Italy, 1987 (to appear).

[11] K.B. Hannsgen, Y. Renardy and R.L. Wheeler, *Effectiveness and robustness with respect to time delays of boundary feedback stabilization in one-dimensional viscoelasticity*, SIAM J. Control and Optimization **26** (1988), 1200–1234.

[12] K.B. Hannsgen and R.L. Wheeler, *Time delays and boundary feedback stabilization in one-dimensional viscoelasticity*, in "Proc. Third International Conference on Distributed Parameter Systems," F. Kappel, K. Kunisch and W. Schappacher eds., Springer Lecture Notes in Control and Information Sciences, vol. 102, 1987, pp. 136–152.

[13] D. Salamon, *Infinite dimensional linear systems with unbounded control and observation: a functional analytic approach*, Trans. Amer. Math. Soc. **300** (1987), 383–431.

Wolfgang Desch
Institut für Mathematik
Karl-Franzens-Universität Graz
Brandhofgasse 18
A-8010 Graz
Austria

Robert L. Wheeler
Department of Mathematics
Virginia Polytechnic Institute and State University
Blacksburg, Virginia 24061-0123
USA

International Series of
Numerical Mathematics, Vol. 91
© 1989 Birkhäuser Verlag Basel

Parameter identification problems
in single-phase and two-phase flow

Richard E. Ewing and Tao Lin

Departments of Mathematics
University of Wyoming

Abstract. We discuss some difficulties in the solution of parameter identification problems in single-phase and two-phase flow. On the basis of the physics, some of the parameter identification problems of two-phase flow may be reduced to problems of single-phase flow – for example, if the parameters depend only on rock properties. Instead of the commonly used least squares method, a direct numerical method that we have developed is used for a one-dimensional single-phase model problem. Using this method, not only can we solve the model problem in a much shorter time than the least squares method requires, but we can also obtain an error estimate about the computation. Several numerical examples show the properties of this method.

Keywords. Parameter identification, inverse problem, PDE, numerical method.

1980 *Mathematics subject classifications*: Primary 35R30, 93C20; Secondary 93B30.

1. Introduction

Applications of single-phase and multi-phase flow in porous media arise in both hydrology and petroleum recovery areas. Hydrology deals with problems governing our drinking water supply, water necessary for agriculture and food supplies, water sources for industrial use and energy generation, and water for natural resource maintenance. Water management efforts, including the control of aquifer quality and quantity, are essential to our continuing society in light of contamination problems and the need to maintain minimal water table levels in our major aquifers. Large-scale mathematical models have been developed to monitor drainage areas and surface water supply in order to control the depletion and regeneration of aquifers and to minimize their pollution from industrial wastes, accidents, and saltwater intrusion.

The purpose of mathematical reservoir simulation models, in general, and petroleum applications, in particular, is to enable us to optimize the recovery of valuable minerals and hydrocarbon from permeable underground reservoirs. To do this, one must be able to predict the performance of the reservoir under various exploitation or production schemes. A series of models is constructed to yield information about, and understanding of, the complex chemical, physical, and fluid flow processes which accompany different recovery methods. Once an accurate model is obtained to describe the various production strategies, optimal control techniques can be utilized to maximize the profit functional which includes minimizing expenses of the recovery techniques, subject to oil price constraints and other important current economic considerations.

In all of these application areas, the mathematical models utilize various soil, rock, and fluid properties to describe the physics of fluid flow of one or more phases in porous media in various regimes. These models must describe and predict the results of the flow processes sufficiently well to aid in decision processes. The accuracy of the prediction capabilities of the models depends upon the accuracy of the mathematical model, the physical parameters put into the model, the numerical discretization, and the numerical solution process. In this paper, we emphasize the parameter identification problem.

The process of determining unknown physical parameter values, such as porosity and permeability, to be used in a mathematical reservoir model to give the best fit to measured production or observation well flow history, is commonly called "history matching" in petroleum applications. Optimally, one would like to have an automatic routine for history matching which is applicable to simulators of varying complexity and which can determine a set of parameters giving a good match to production data within a reasonable amount of time and with reasonable computational effort. One would also like to have some indication, through error estimates or statistical confidence intervals, of how accurately the parameters have been estimated, to indicate the reliability of predictions obtained from a simulator using these parameters.

The mathematical problem in an inhomogeneous reservoir requires an infinite number of parameters to obtain a complete solution, while computation with a reservoir simulator allows only a finite parameter specification. The most straightforward approach for specifying general properties in a finite difference or finite element simulator which utilizes a spatial grid network is to allow porosity and permeability to vary independently within each grid block or element. In field scale simulations using up to 25,000 grid blocks, this could require an algorithm for determining 50,000 or more unknowns simultaneously. While potentially minimizing the modeling error, this technique generates an extremely difficult and ill-conditioned optimization problem. Therefore, the feature of the history-matching problem in reservoir simulation that distinguishes it from parameter estimation problems in other fields of science and engineering is the large dimensionality of both the system state and the unknown parameters. To reduce both the statistical uncertainty and the computational complexity of the problem, one must decrease the number of unknowns and, where possible, utilize any additional available information.

In order to address the ill-posedness of the parameter estimation problem, we must obtain as accurate an initial guess as possible for the least squares history-matching process. Prior geological information from the reservoir under study and also from reservoirs of the same type should be utilized where possible. Kriging techniques from geostatistics [1–4] have been developed to statistically interpolate data, measured at the wells, throughout the reservoir. In this paper, we discuss in detail another method for taking measured well data and marching out into the reservoir to generate more accurate initial guesses for the full least squares minimization problem.

For multiphase flow problems, there are many more unknown parameters, and some of these parameters have a nonlinear dependence upon the unknown variables, such as phase saturations. This greatly complicates the history-matching problem. Although, in practice, an output least squares formulation is usually used to estimate all the variables simultaneously, this yields an extremely ill-posed problem.

In fact, an experiment was run at a short course at a major oil company concerning history matching with their multiphase simulator. A set of output data was given to a class of thirty who were asked to find parameters to fit the production data via the simulator.

Many "matches" were obtained yielding quite small least squares functionals but which involved widely different choices of the input parameters. This is a strong indication that better methods for history matching are needed.

In this paper, we emphasize the need to split the problem up, isolating the effects of the different parameters where possible, and to identify the parameters separately. This can frequently be done in two-phase oil/gas problems. Often, the reservoir pressure is sufficiently high that any gas present is dissolved in the oil phase. As oil is pumped from production wells, the pressure declines in certain parts of the reservoir and, after the bubble-point pressure is reached, gas begins to form a separate phase. In the initial time frame of the process, we have a single-phase (oil) flow regime and the complications of multi-phase flow are not present. During single-phase flow, the methods described in this paper can be used to identify the porosity and permeability; these are rock properties that do not depend upon multi-phase flow. The highly nonlinear capillary pressure curves and relative permeability curves can be approximated from cores and reservoir fluids in laboratory core floods once the total permeabilities and porosities are known. Descriptions of parameter estimation techniques for these nonlinear functions have appeared in the literature [5,6]. Once all of these initial guesses have been obtained, we can return to the full nonlinear least squares history matching to fine tune the parameters in a computationally feasible way.

In Section 2, we present the equations for two-phase flow problems and then the simplified single-phase flow case. Also, the general least squares history-matching method is specified. In Section 3, the direct perturbation method for obtaining estimates for the single-phase case is presented. The stability analysis and error estimates for our approach are also presented in Section 3. Section 4 contains numerical examples and conclusions.

2. Flow equations and discussion of the least squares method

The equations describing two phase, immiscible, incompressible displacement in a horizontal porous medium are given by

$$(1) \qquad \phi \frac{\partial S_w}{\partial t} - \nabla \cdot \left(k \frac{k_{rw}}{\mu_w} \nabla p_w \right) = q_w, \quad \mathbf{x} \in \Omega, \quad t \in J,$$

$$(2) \qquad \phi \frac{\partial S_o}{\partial t} - \nabla \cdot \left(k \frac{k_{ro}}{\mu_o} \nabla p_o \right) = q_o, \quad \mathbf{x} \in \Omega, \quad t \in J,$$

where the subscripts w and o refer to water and oil, respectively, and S_i is the saturation, p_i is the pressure, k_{ri} is the relative permeability, μ_i is the viscosity, and q_i is the external flow rate, each with respect to the ith phase.

The saturation constraint is given by

$$(3) \qquad S_w + S_o = 1.$$

From Equation (3) we see that one of the saturations can be eliminated. Let $S = S_w = 1 - S_o$. The pressure between the two phases is described by the capillary pressure

$$p_c(S) = p_o - p_w.$$

Note that $\dfrac{dp_c}{dS} \leq 0$.

The unknown distributed parameters in (1) and (2), which must be determined via field measurements and history matching, are the spatially varying porosity $\phi(\mathbf{x})$ and permeability $k(\mathbf{x})$. The viscosities can be determined from reservoir fluids. Finally, the nonlinear functions $k_{rw}(S_w)$, $k_{ro}(S_w)$, and $p_c(S_w)$ can be estimated from core floods in the laboratory.

The parameter estimation problem is to determine distributed parameters $\phi^*(\mathbf{x})$ and $k^*(\mathbf{x})$ such that, if these parameters are used in a numerical simulator for approximating the pressures p from (1) and (2), the predicted pressures at the well bore, obtained from the simulator, are "as close as possible" to the measured pressure values, in some sense. Part of the formulation of the problem involves how the "closeness" of the measured and calculated data should be defined. If $p^*(\mathbf{x}_i, t_n; \phi, k)$, $i = 1, \cdots, N_w$, $n = 1, \cdots, N_o$, denote the measured pressures at the well bores $\mathbf{x}_i$, $i = 1, \cdots, N_w$, at the N_o time observations t_n, $n = 1, \cdots, N_o$, and $P^*(\mathbf{x}_i, t_n; \hat{\phi}, \hat{k})$ denote the results of the simulation at $\mathbf{x}_i$ and time t_n using parameter estimates $\hat{\phi}$ and $\hat{k}$, then a typical objective functional used to optimize closeness is given by

$$(4) \qquad J = \sum_{i=1}^{N_w} \sum_{n=1}^{N_o} \left| p^*(\mathbf{x}_i, t_n; \phi, k) - P^*(\mathbf{x}_i, t_n; \hat{\phi}, \hat{k}) \right|^2.$$

The parameter identification problem can be solved theoretically by minimizing the above functional or a regularized version of it.

There are several difficulties in implementing the least squares approach. The first is that if many temporal observations are used, J may not have a unique, easily obtainable minimum to define the "optimal" estimates for the distributed parameters ϕ^* and k^*. For example, if the initial condition for p is constant and $q = 0$ in (1) and (2), then p^* and P^* should be equal, J would be totally insensitive to ϕ and k, and no unique solution to the problem exists. Similarly, if q is very small, the minimization problem governed by (4) would be highly ill-conditioned, since J would still be highly insensitive to variations in ϕ and k. One way to overcome the insensitivity of J is to design flow regimes, often termed pressure draw-down tests, which give sufficient change in the observables so that the minimization of the functional J is a better-conditioned problem.

The second difficulty in implementing the least squares method is that it is usually very time consuming. This is due to the nonlinearity of the least squares approach, even though the PDE involved may be linear. The available algorithms for least squares problems are usually gradient-like iterative procedures. In every iteration, one has to solve a PDE in order to update the functional or its gradient, and invert a linear system with a typically full matrix and many unknowns. Moreover, we have no idea about how many iterations we have to perform in order to reduce the functional value to a required tolerance. This means that we may have to solve certain PDE's and a very large linear system with a full matrix many times; hence, a least squares method is usually very time consuming, especially for large-scale identification problems.

Third, the general local convergence orders of the available algorithms for least squares methods are usually less than two, and the global convergence orders are quite difficult to estimate. Hence, the initial guess is crucial: a bad guess could cause divergence of the computations.

Due to the extreme difficulty in estimating all of the parameters simultaneously in the two-phase model (1) and (2), we want to split the identification problem into easier pieces. According to the physics explanation in the last section, the single-phase regime is often present in the early stages of multiphase flow and the data collected in this stage can be used to identify some rock properties, like the porosity and permeability, which do not depend on multi-phase flow. Also, single-phase flow is important in its own right in hydrology applications. For these reasons, we only discuss a simple, single-phase flow model in detail in the present paper, and hope to discuss how to apply these results to the complete parameter identification problems of two-phase flow in a later paper.

The simplest three-dimensional model for a single-phase flow regime incorporates three-dimensional variation of the permeabilities or transmissivities as a fucntion $a(\mathbf{x}) = k(\mathbf{x})/\mu$, for $\mathbf{x} = (x, y, z)$, in the differential equation

$$(5) \qquad \phi c \frac{\partial p}{\partial t} = \nabla \cdot a(\mathbf{x}) \nabla p + q(\mathbf{x}, t),$$

with associated initial and boundary conditions on p based on noisy measurements of $p(\mathbf{x}, t; a(\mathbf{x}))$, denoted by $p^*(\mathbf{x}_i, t_n; a(\mathbf{x}))$, at the wells $(\mathbf{x}_i, y_i)$, $i = 1, \cdots, N_w$, and for times t_n, $n = 1, \cdots, N_0$. In (5), $q(\mathbf{x}, t)$ accounts for withdrawal or injection of fluid within the reservoir and can model a regional source or sink, as in a spatially leaky acquifer, or point sources and sinks to treat injection or production wells. Thus, $q(\mathbf{x}, t)$ in (1) can be rewritten as

$$(6) \qquad q(\mathbf{x}, t) = \hat{q}(\mathbf{x}, t) + \sum_{i=1}^{N_w} q_i \delta(\mathbf{x} - \mathbf{x}_i),$$

where q_i, $i = 1, \cdots, N_w$, denote the volumetric flow rates ast the wells, positive for injection wells and negative for production wells.

If we use (5) to model a radial flow toward a well in a radially homogeneous medium, we end up, via a change of variables, with the following one-dimensional parabolic equation:

$$(7) \qquad u_t = (a(x)u_x)_x + f(x, t), \quad 0 < x < 1, \ t > 0,$$

$$(8) \qquad u(x, 0) = g(x), \quad 0 < x < 1,$$

$$(9) \qquad u(0, t) = \varphi_0(t), \ \text{and/or} \ u(1, t) = \varphi_1(t), \quad t > 0.$$

In this application, the internal measurements at the wells are transformed into boundary measurements at $x = 0$ and $x = 1$. When the communication between wells is appropriate, we then consider the specification of data at both boundaries. For this simple single phase flow model, we consider the parameter identification problem: Identify $a(x)$ in (7) from the following overspecified data at boundaries:

$$(10) \qquad u_x(0, t) = \psi_0(t), \ \text{and/or} \ u_x(1, t) = \psi_1(t), \quad t > 0.$$

The model problem specified in (7)–(10) is formulated as a Dirichlet problem with over-specified Neumann data. This corresponds to flowing wells with a constant bottom hole pressure constraint and then measuring the flow rates. A similar problem could be stated as a Neumann problem with overspecified Dirichlet data. This corresponds to flowing the well at a constant rate constraint and then measuring the bottom hole pressure. Although these are different physical and mathematical problems, the direct numerical procedure discussed in this paper treats them in exactly the same manner, since we use Cauchy initial data at the wells to march away from the data-bearing position of the line $x = 0$ and/or $x = 1$.

This model problem can surely be solved by the least squares approach; however, based on the observations of the least squares method, we want to develop a direct, noniterative numerical method to solve it. Some direct numerical methods have been developed for the steady-state problem [7]; for the time dependent problem above, we use a space-marching idea to derive the numerical scheme. The space-marching idea has been applied to the ill-posed heat conduction problem [8] and the parameter identification problem for hyperbolic equations [9]. The space-marching concept is to develop a numerical scheme to compute the parabolic equations in the direction of the x space variable. Directly applying this idea leads to an unstable scheme due to the parabolic property of the PDE [8]. In order to obtain a stable scheme, we introduce some regularization via perturbation, i.e., by adding a small term $\epsilon^2 u_{tt}$ in (7), and then develop a direct numerical scheme for the following modified problem:

(11) $$\epsilon^2 u_{tt} + u_t = (a(x)u_x)_x + f(x,t), \quad 0 < x < 1, \quad t > 0,$$

(12) $$u(x,0) = g(x), \quad 0 < x < 1,$$

(13) $$u(0,t) = \varphi_0(t), \quad u(1,t) = \varphi_1(t), \quad t > 0,$$

(14) $$u_x(0,t) = \psi_0(t), \quad u_x(1,t) = \psi_1(t), \quad t > 0,$$

(15) $$u_t(x,0) = (a(x)u_x(x,0))_x + f(x,0), \quad 0 < x < 1.$$

This stabilization, termed hyperbolic regularization, has been used for solving non-characteristic Cauchy problems [10,11] and some parameter identification problems [12–14] of parabolic equations. Since equation (11) is hyperbolic, the x and t variables play symmetric roles; hence, if we consider the x variable as a temporal variable, then we can easily find a numerical scheme for (11) by marching in the direction of the x-variable without violating stability. Moreover, the hyperbolic property of (11) enables the scheme to march not only forward but also backward in the x-variable. Thus, the scheme can march from both boundaries. The boundary conditions (13) and the extra data (14) at the boundaries then naturally become initial conditions for (11) with respect to the new time variable x. Since the computation will be carried out on a strip, $0 \le x \le 1, t > 0$, the original initial condition (8) becomes the boundary condition for (11) with respect to the new spacial variable t. Then (15) is an extra condition for (11) at the new spatial boundary, which enables us to form an extra equation for computing $a(x)$. The technical details for developing the scheme are presented in the next section.

3. A marching scheme and its analysis

Let h, k be the discretization steps for the x and t variables, respectively, and let u_j^n, f_j^n, g_j, and $a_{j+\frac{1}{2}}$ be the approximatons of $u(jh, nk)$, $f(jh, nk)$, $g(jh)$, and $a\left((j+\frac{1}{2})h\right)$, respectively. Then (11) can be approximated by the following well-known difference equation:

(16)
$$\epsilon^2 \frac{u_j^{n+1} - 2u_j^n + u_j^{n-1}}{k^2} + \frac{u_j^{n+1} - u_j^{n-1}}{2k}$$
$$= \frac{1}{h^2} \left[a_{j+\frac{1}{2}} \left(u_{j+1}^n - u_j^n \right) - a_{j-\frac{1}{2}} \left(u_j^n - u_{j-1}^n \right) \right] + f_j^n.$$

(15) can be approximated by

(17)
$$\frac{u_j^1 - g_j}{k} = \frac{1}{h^2} \left[a_{j+\frac{1}{2}} \left(g_{j+1} - g_j \right) - a_{j-\frac{1}{2}} (g_j - g_{j-1}) \right] + f_j^0.$$

In order to march in a positive direction along the x-axis, we solve (16) for u_{j+1}^n:

(18)
$$u_{j+1}^n = u_j^n + \frac{1}{a_{j+\frac{1}{2}}} \left\{ a_{j-\frac{1}{2}} \left(u_j^n - u_{j-1}^n \right) \right.$$
$$\left. + h^2 \left[\epsilon^2 \frac{u_j^{n+1} - 2u_j^n + u_j^{n-1}}{k^2} + \frac{u_j^{n+1} - u_j^{n-1}}{2k} - f_j^n \right] \right\}.$$

Next, we solve (17) for $a_{j+1+\frac{1}{2}}$:

(19)
$$a_{j+1+\frac{1}{2}} = \frac{1}{g_{j+2} - g_{j+1}} \left\{ a_{j+\frac{1}{2}} (g_{j+1} - g_j) \right.$$
$$\left. + h^2 \left[\frac{u_{j+1}^1 - g_{j+1}}{k} - f_{j+1}^0 \right] \right\}.$$

(18) and (19) define a complete difference scheme except for $j = 1$, since (18) and (19) are all two level equations in j. To compute u_1^n, we can use the following simple approximation:

(20)
$$u_1^n = \varphi_0(nk) + h \cdot \psi_0(nk).$$

To implement (18)–(20) on a computer, we make the following assumptions:

Assumption 1. $a_{\frac{1}{2}} = a\left(\frac{1}{2}h\right)$ is known.

Assumption 2. $g(x) \in C^2[0, 1]$ and $\exists \, d_0 > 0$ and $d_1 > 0$, satisfying

(21)
$$d_0 \leq |g'(x)| \leq d_1, \quad \forall x \in [0, 1].$$

Assumption 1 is used to start the computation from the left boundary and Assumption 2 is used to avoid the possible singularity induced by the term $1/(g_{j+2} - g_{j+1})$ in (19). Assumption 2 is related to the parameter identifiability. In the steady state case, Kunisch

[15] used a weighted seminorm approach to get rid of this strong "uniform flow" assumption. We will discuss how to eliminate or weaken this "strong" assumption for parabolic equations in a forthcoming paper.

The scheme for marching in the positive x-axis direction can be described as follows. Suppose $1 = Mh$:

    1. Compute u_1^n by (20) for $n = 1, \cdots, M - 1$.
    2. Compute $a_{1+\frac{1}{2}}$ by (19).

Then for $j \geq 2$,

    3. Compute u_j^n by (18) for $M - j$.
    4. Compute $a_{j+\frac{1}{2}}$ by (19).
    5. If $j < M - 1$, go to 3 for the next j; otherwise, stop.

A scheme for marching in the direction of the negative x-axis from the right boundary can be derived similarly and uses the other data, ϕ_1 and ψ_1. In computational practice, we may be able to march in only one direction; for example, we may have extra data only at one boundary. The computational results indicate that marching from two directions is usually better than from only one direction (see Example 4).

In order to investigate the error and stability, we introduce

$$(22) \qquad\qquad V(x,t) = u_x(x,t), \quad W = \epsilon u_t(x,t).$$

It is easy to see that (11) and (15) are equivalent to

$$(23) \qquad\qquad \epsilon V_t = W_x,$$

$$(24) \qquad\qquad \epsilon W_t = (aV)_x - \frac{1}{\epsilon} W + f,$$

$$(25) \qquad\qquad \frac{1}{\epsilon} W(x,0) = (a(x)V(x,0))_x + f(x,0),$$

and that (16), (17) are equivalent to

$$(26) \qquad\qquad \frac{W_{j+1}^n - W_j^n}{h} = \epsilon \frac{V_{j+\frac{1}{2}}^{n+1} - V_{j+\frac{1}{2}}^n}{k},$$

$$(27) \qquad \frac{a_{j+1+\frac{1}{2}} V_{j+1+\frac{1}{2}}^{n+1} - a_{j+\frac{1}{2}} V_{j+\frac{1}{2}}^{n+1}}{h} = \frac{\epsilon}{k}\left(W_{j+1}^{n+1} - W_{j+1}^n\right) + \frac{1}{\epsilon} W_{j+1}^{n+1} - f_{j+1}^{n+1},$$

$$(28) \qquad \frac{a_{j+1+\frac{1}{2}}(g_{j+2} - g_{j+1}) - a_{j+\frac{1}{2}}(g_{j+1} - g_j)}{h^2} = \frac{1}{\epsilon} W_{j+1}^0 - f_{j+1}^0.$$

We let

$$(29) \qquad\qquad \tilde{V}_{j+\frac{1}{2}}^n = V\left(\left(j + \frac{1}{2}\right)h, nk\right),$$

$$(30) \qquad\qquad \tilde{W}_j^n = W(jh, nk),$$

$$(31) \qquad\qquad \tilde{a}_{j+\frac{1}{2}} = a\left(\left(j + \frac{1}{2}\right)h\right),$$

and define the error functions as follows:

$$(32) \qquad \Delta V^n_{j+\frac{1}{2}} = \tilde{V}^n_{j+\frac{1}{2}} - V^n_{j+\frac{1}{2}},$$

$$(33) \qquad \Delta W^n_j = \tilde{W}^n_j - W^n_j,$$

$$(34) \qquad \Delta a_{j+\frac{1}{2}} = \tilde{a}_{j+\frac{1}{2}} - a_{j+\frac{1}{2}}.$$

In this section, we analyze only the right marching scheme; a similar analysis can be applied to the left marching scheme. Since the scheme is carried out in a triangular region, we can assume that

$$(35) \qquad \begin{aligned} \Delta V^n_{j+\frac{1}{2}} &= 0, \quad n \ge M - j + 1, \\ \Delta W^n_j &= 0, \quad n \ge M - j + 1. \end{aligned}$$

where

$$(36) \qquad M = \frac{1}{h}.$$

We now introduce some norms:

$$(37) \qquad \left\| \Delta V_{j+\frac{1}{2}} \right\|^2 = \sum_{n=1}^{M-j} \left| \Delta V^n_{j+\frac{1}{2}} \right|^2,$$

$$(38) \qquad \| \Delta W_j \|^2 = \sum_{n=0}^{M-j} \left| \Delta W^n_j \right|^2,$$

and

$$(39) \qquad \| \Delta_j \|_k^2 \equiv \left\| \Delta V_{j+\frac{1}{2}} \right\|^2 k + \| \Delta W_j \|^2 k + \left| \Delta a_{j+\frac{1}{2}} \right|^2.$$

By expanding V, W, and a at the grid points we have

$$(40) \qquad \frac{\tilde{W}^n_{j+1} - \tilde{W}^n_j}{h} = \epsilon \frac{\tilde{V}^{n+1}_{j+\frac{1}{2}} - \tilde{V}^n_{j+\frac{1}{2}}}{k} + O(h) + O(k),$$

$$(41) \qquad \frac{\tilde{a}_{j+1+\frac{1}{2}} \tilde{V}^{n+1}_{j+1+\frac{1}{2}} - \tilde{a}_{j+\frac{1}{2}} \tilde{V}^{n+1}_{j+\frac{1}{2}}}{h} = \frac{1}{\epsilon} \tilde{W}^{n+1}_{j+1} - f^{n+1}_{j+1} + O(h^2),$$

$$(42) \qquad \frac{a_{j+1+\frac{1}{2}}(g_{j+2} - g_{j+1}) - a_{j+\frac{1}{2}}(g_{j+1} - g_j)}{h^2} = \frac{1}{\epsilon} \tilde{W}^0_{j+1} - f^0_{j+1} + O(h^2).$$

Subtracting (26) from (40), (27) from (41), and (28) from (42), we then obtain the equations satisfied by the error functions:

(43)
$$\frac{\Delta W_{j+1}^n - \Delta W_j^n}{h} = \frac{\epsilon}{k}\left(\Delta V_{j+\frac{1}{2}}^{n+1} - \Delta V_{j+\frac{1}{2}}^n\right) + O(h) + O(k),$$

(44)
$$\frac{1}{h}\left[\left(\Delta a_{j+1+\frac{1}{2}}\tilde{V}_{j+1+\frac{1}{2}}^{n+1} - \Delta a_{j+\frac{1}{2}}\tilde{V}_{j+\frac{1}{2}}^{n+1}\right) + \left(a_{j+1+\frac{1}{2}}\Delta V_{j+1+\frac{1}{2}}^{n+1} - a_{j+\frac{1}{2}}\Delta V_{j+\frac{1}{2}}^{n+1}\right)\right]$$
$$= \frac{\epsilon}{k}\left(\Delta W_{j+1}^{n+1} - \Delta W_{j+1}^n\right) + \frac{1}{\epsilon}\Delta W_{j+1}^{n+1} - \frac{\epsilon}{k}\left(\tilde{W}_{j+1}^{n+1} - \tilde{W}_{j+1}^n\right) + O(h^2),$$

(45)
$$\frac{\Delta a_{j+1+\frac{1}{2}}(g_{j+2} - g_{j+1}) - \Delta a_{j+\frac{1}{2}}(g_{j+1} - g_j)}{h^2} = \frac{1}{\epsilon}\Delta W_{j+1}^0 + O(h^2).$$

The identification problem is in fact nonlinear even though the involved partial differential equation is linear. Like most analysis for the numerical scheme of nonlinear problems, our stability and error analysis will be based on some assumptions on the numerical solution.

Assumption 3. There exist positive constants M_1 and M_2 such that

(46)
$$M_1 \le \min_j a_{j+\frac{1}{2}} \le \max_j a_{j+\frac{1}{2}} \le M_2,$$
$$\max_j |W_j^0| \le M_2.$$

For the solution of the PDE (1) we make the following assumption:

Assumption 4:
$$u(x,t) \in C^2\left([0,1] \times [0,T]\right).$$

Our analysis is based on an energy inequality in which the energy is defined as

(47)
$$T_j = \left|a_{j+\frac{1}{2}}\right|^2 \left\|\Delta V_{j+\frac{1}{2}}\right\|^2 k + a_{j+\frac{1}{2}}\|\Delta W_j\|^2 k + k\left\|\tilde{V}_{j+\frac{1}{2}}\right\|^2 \left|\Delta a_{j+\frac{1}{2}}\right|^2$$
$$- \epsilon h \sum_{n=0} a_{j+\frac{1}{2}}\Delta V_{j+\frac{1}{2}}^{n+1}\left(\Delta W_j^{n+1} - \Delta W_j^n\right)$$
$$+ 2k \sum_{n=0} a_{j+\frac{1}{2}}\Delta a_{j+\frac{1}{2}}\Delta V_{j+\frac{1}{2}}^{n+1}\tilde{V}_{j+\frac{1}{2}}^{n+1} + \frac{g_{j+1} - g_j}{h}\epsilon^2 \tilde{V}_{j+\frac{1}{2}}^0 \left|\Delta a_{j+\frac{1}{2}}\right|^2.$$

To determine stability and error estimates, we need the following lemmas.

Lemma 1. There exists a constant M_3 such that $\forall k, h > 0$

(48)
$$\max_j \left\|\tilde{V}_{j+\frac{1}{2}}\right\|^2 k \le M_3.$$

Proof: The lemma follows from (29) and (37) since

$$V(x,t) \in C\left([0,1] \times [0,T]\right).$$

Lemma 2. *There exist positive constants M_4, M_5, and M_6, such that when h is small enough and*

$$(49) \qquad \frac{\epsilon h}{k} < \min(M_4, M_1, 1),$$

we have

$$(50) \qquad M_5\|\Delta_j\|_k^2 \leq T_j \leq M_6\|\Delta_j\|_k^2, \quad j \geq 0,$$

where M_4, M_5 and M_6 depend on M_1, M_2, M_3, d_0, and d_1, and will be identified in the proof.

Proof:

$$
\begin{aligned}
T_j &\leq |a_{j+1}|^2 \left\|\Delta V_{j+\frac{1}{2}}\right\|^2 k + a_{j+\frac{1}{2}}\|\Delta W_j\|^2 k + \left\|\tilde{V}_{j+\frac{1}{2}}\right\| k \left|\Delta a_{j+\frac{1}{2}}\right|^2 \\
&\quad + \frac{\epsilon h}{k}\left|a_{j+\frac{1}{2}}\right|\left(\|\Delta W_j\|^2 k + \left\|\Delta V_{j+\frac{1}{2}}\right\|^2 k\right) \\
&\quad + \left\|\tilde{V}_{j+\frac{1}{2}}\right\|^2 k \left|\Delta a_{j+\frac{1}{2}}\right|^2 + \left|a_{j+\frac{1}{2}}\right|^2 \left\|\Delta V_{j+\frac{1}{2}}\right\|^2 k \\
&\quad + \frac{g_{j+1} - g_j}{h}\epsilon^2 \left|\tilde{V}_{j+\frac{1}{2}}^0\right| \left|\Delta a_{j+\frac{1}{2}}\right|^2 \\
&\leq M_6\|\Delta_j\|_k^2.
\end{aligned}
$$

On the other hand,

$$
\begin{aligned}
T_j &\geq \left|a_{j+\frac{1}{2}}\right|^2 \left\|\Delta V_{j+\frac{1}{2}}\right\|^2 k + a_{j+\frac{1}{2}}\|\Delta W_j\|^2 k + \left\|\tilde{V}_{j+\frac{1}{2}}\right\|^2 k \left|\Delta a_{j+\frac{1}{2}}\right|^2 \\
&\quad - \frac{\epsilon h}{k}\left|a_{j+\frac{1}{2}}\right|\left(\alpha_1 \left\|\Delta V_{j+\frac{1}{2}}\right\|^2 k + \beta_1 \|\Delta W_j\|^2 k\right) \\
&\quad - \left(\alpha_2 \left|a_{j+\frac{1}{2}}\right|^2 \left\|\Delta V_{j+\frac{1}{2}}\right\|^2 k + \beta_2 \left\|\tilde{V}_{j+\frac{1}{2}}\right\|^2 k \left|\Delta a_{j+\frac{1}{2}}\right|^2\right) \\
&\quad + \frac{g_{j+1} - g_j}{h}\epsilon^2 \tilde{V}_{j+\frac{1}{2}}^0 \left|\Delta a_{j+\frac{1}{2}}\right|^2 \\
&= \left(\left|a_{j+\frac{1}{2}}\right|^2 - \frac{\epsilon h}{k}\left|a_{j+\frac{1}{2}}\right|\alpha_1 - \alpha_2 \left|a_{j+\frac{1}{2}}\right|^2\right) \left\|\Delta V_{j+\frac{1}{2}}\right\|^2 k \\
&\quad + \left(a_{j+\frac{1}{2}} - \frac{\epsilon h}{k}\left|a_{j+\frac{1}{2}}\right|\beta_1\right) \|\Delta W_j\|^2 k \\
&\quad + \left((1 - \beta_2)\left\|\tilde{V}_{j+\frac{1}{2}}\right\|^2 k + \frac{g_{j+1} - g_j}{h}\epsilon^2 \tilde{V}_{j+\frac{1}{2}}^0\right) \left|\Delta a_{j+\frac{1}{2}}\right|^2,
\end{aligned}
$$

where α_i and β_i are positive and $\alpha_i \cdot \beta_i = 1$, $i = 1,2$. Using Assumption 2,

$$\frac{g_{j+1} - g_j}{h} \epsilon^2 \tilde{V}^0_{j+\frac{1}{2}} > \frac{\epsilon^2 d_0}{2}$$

for h small enough, and we can find a $\beta_2 > 1$, such that

(51) $\qquad (1 - \beta_2) \left\| \tilde{V}_{j+\frac{1}{2}} \right\|^2 k + \frac{g_{j+1} - g_j}{h} \epsilon^2 \tilde{V}^0_{j+\frac{1}{2}} \geq \frac{\epsilon^2 d_0}{2} - (1 - \beta_2) M_3 > 0.$

Since the corresponding α_2 satisfies

$$\alpha_2 < 1,$$

we can easily find α_1, β_1, and M_4 such that

(52) $\qquad \left| a_{j+\frac{1}{2}} \right|^2 - \frac{\epsilon h}{k} \left| a_{j+\frac{1}{2}} \right| \alpha_1 - \alpha_2 \left| a_{j+\frac{1}{2}} \right|^2 > 0,$

(53) $\qquad a_{j+\frac{1}{2}} - \frac{\epsilon h}{k} \left| a_{j+\frac{1}{2}} \right| \beta_1 > 0,$

for

$$\frac{\epsilon h}{k} < \min(M_4, M_1, 1).$$

Then

$$T_j \geq M_5 \|\Delta_j\|^2_k,$$

and the lemma is proved.

Lemma 3.

$$\sum_{n=0} \left(\Delta W^{n+1}_{j+1} - \Delta W^n_j \right) \left(\Delta a_{j+1+\frac{1}{2}} \tilde{V}^{n+1}_{j+1+\frac{1}{2}} + \Delta a_{j+\frac{1}{2}} \tilde{V}^{n+1}_{j+\frac{1}{2}} \right)$$

(54) $\qquad \leq -\Delta a_{j+1+\frac{1}{2}} \sum_{n=0} \Delta W^n_{j+1} \left(\tilde{V}^{n+1}_{j+1+\frac{1}{2}} - \tilde{V}^n_{j+\frac{1}{2}} \right) - \Delta a_{j+1+\frac{1}{2}} \Delta W^0_{j+1} \tilde{V}^0_{j+1+\frac{1}{2}}$

$$-\Delta a_{j+\frac{1}{2}} \sum_{n=0} \Delta W^n_{j+1} \left(\tilde{V}^{n+1}_{j+\frac{1}{2}} - \tilde{V}^n_{j+\frac{1}{2}} \right) - \Delta a_{j+\frac{1}{2}} \Delta W^0_{j+1} \tilde{V}^0_{j+\frac{1}{2}}.$$

Proof: The lemma follows from (35) and discretized integration by parts.

Lemma 4. *If (49) is satisfied, then there exists a constant M_8 which depends on $u(x,t)$ and $g(x)$ as follows:*

(55) $\qquad M_8 = M_8 \left(\max_{x,t} |u_{xt}(x,t)|, \max_{x,t} |u_{tt}(x,t)|, d_1, \max_x |g''(x)| \right),$

such that, for t and h small enough,

$$(56) \quad \left| \frac{\epsilon h}{k} \Delta a_{j+1+\frac{1}{2}} \sum_{n=0} \Delta W_{j+1}^n \left(\tilde{V}_{j+1+\frac{1}{2}}^{n+1} - \tilde{V}_{j+1+\frac{1}{2}}^n \right) \right|$$
$$\leq \epsilon h M_8 \left(\|\Delta W_{j+1}\|^2 + M \left| \Delta a_{j+1+\frac{1}{2}} \right|^2 \right),$$

$$(57) \quad \left| \frac{\epsilon h}{k} \Delta a_{j+\frac{1}{2}} \sum_{n=0} \Delta W_{j+1}^n \left(\tilde{V}_{j+\frac{1}{2}}^{n+1} - \tilde{V}_{j+\frac{1}{2}}^n \right) \right| \leq \epsilon h M_8 \left(\|\Delta W_{j+1}\|^2 + M \left| \Delta a_{j+\frac{1}{2}} \right|^2 \right),$$

$$(58) \quad \left| \epsilon h \Delta a_{j+1+\frac{1}{2}} \Delta W_{j+1}^0 \left(\tilde{V}_{j+\frac{1}{2}}^0 - \tilde{V}_{j+1+\frac{1}{2}}^0 \right) \right| \leq h M_8 \left(k \|\Delta W_{j+1}\|^2 + k \left| \Delta a_{j+1+\frac{1}{2}} \right|^2 \right),$$

$$\left(a_{j+1+\frac{1}{2}} - a_{j+\frac{1}{2}} \right) \|\Delta W_{j+1}\|^2 \leq M_8 h \|\Delta W_{j+1}\|^2,$$

$$(59) \quad \left| \frac{g_{j+2} - 2g_{j+1} + g_j}{h} \epsilon^2 \tilde{V}_{j+\frac{1}{2}}^0 \Delta a_{j+\frac{1}{2}} \left(\Delta a_{j+1+\frac{1}{2}} + \Delta a_{j+\frac{1}{2}} \right) \right|$$
$$\leq M_8 \epsilon^2 h \left(\left| \Delta a_{j+1+\frac{1}{2}} \right|^2 + \left| \Delta a_{j+\frac{1}{2}} \right|^2 \right),$$

$$(60) \quad \left| \frac{g_{j+2} - g_{j+1}}{h} \epsilon^2 \left(\tilde{V}_{j+1+\frac{1}{2}}^0 - \tilde{V}_{j+\frac{1}{2}}^0 \right) \left| \Delta a_{j+1+\frac{1}{2}} \right|^2 \right| \leq M_8 \epsilon^2 h \left| \Delta a_{j+1+\frac{1}{2}} \right|^2,$$

$$(61) \quad \left| \frac{g_{j+2} - 2g_{j+1} + g_j}{h} \epsilon^2 \tilde{V}_{j+\frac{1}{2}}^0 \left| \Delta a_{j+\frac{1}{2}} \right|^2 \right| \leq M_8 \epsilon^2 h \left| \Delta a_{j+\frac{1}{2}} \right|^2,$$

$$\left| \frac{\epsilon h}{k} \sum_{n=0} \left(\tilde{W}_{j+1}^{n+1} - \tilde{W}_{j+1}^n \right) A_j^n \right|$$

(62)
$$\leq \epsilon h M_8 \left(\left\| \Delta V_{j+1+\frac{1}{2}} \right\|^2 + M \left| a_{j+1+\frac{1}{2}} \right|^2 + \left\| \Delta V_{j+\frac{1}{2}} \right\|^2 + M \left| a_{j+\frac{1}{2}} \right|^2 \right.$$
$$\left. + \left\| \tilde{V}_{j+1+\frac{1}{2}} \right\|^2 + M \left| \Delta a_{j+1+\frac{1}{2}} \right|^2 + \left\| \tilde{V}_{j+\frac{1}{2}} \right\|^2 + M \left| \Delta a_{j+\frac{1}{2}} \right|^2 \right),$$

where M is defined by (36), and

(63) $A_j^n = a_{j+1+\frac{1}{2}} \Delta V_{j+1+\frac{1}{2}}^{n+1} + a_{j+\frac{1}{2}} \Delta V_{j+\frac{1}{2}}^{n+1} + \Delta a_{j+1+\frac{1}{2}} \tilde{V}_{j+1+\frac{1}{2}}^{n+1} + \Delta a_{j+\frac{1}{2}} \tilde{V}_{j+\frac{1}{2}}^{n+1}.$

Proof: The proof follows simply by applying Assumptions 2 and 4, Taylor's expansion, and Cauchy's inequality.

Lemma 5. $\Delta a_{j+\frac{1}{2}}$'s satisfy

(64)
$$\frac{g_{j+2} - g_{j+1}}{h} \frac{\epsilon^2}{k} \tilde{V}_{j+1+\frac{1}{2}}^0 \left| \Delta a_{j+1+\frac{1}{2}} \right|^2 - \frac{g_{j+1} - g_j}{h} \frac{\epsilon^2}{k} \tilde{V}_{j+\frac{1}{2}}^0 \left| \Delta a_{j+\frac{1}{2}} \right|^2$$
$$= \frac{\epsilon h}{k} \tilde{V}_{j+\frac{1}{2}} \Delta W_{j+1}^0 \left(\Delta a_{j+1+\frac{1}{2}} + \Delta a_{j+\frac{1}{2}} \right)$$
$$+ \frac{g_{j+2} - 2g_{j+1} + g_j}{h} \frac{\epsilon^2}{k} \tilde{V}_{j+\frac{1}{2}}^0 \Delta a_{j+\frac{1}{2}} \left(\Delta a_{j+1+\frac{1}{2}} + \Delta a_{j+\frac{1}{2}} \right)$$
$$+ \frac{g_{j+2} - g_{j+1}}{h} \frac{\epsilon^2}{k} \left(\tilde{V}_{j+1+\frac{1}{2}}^0 - \tilde{V}_{j+\frac{1}{2}}^0 \right) \left| \Delta a_{j+1+\frac{1}{2}} \right|^2$$
$$+ \frac{g_{j+2} - 2g_{j+1} + g_j}{h} \frac{\epsilon^2}{k} \tilde{V}_{j+\frac{1}{2}} \left| \Delta a_{j+\frac{1}{2}} \right|^2 + O(h^2) \frac{h\epsilon^2}{k} \tilde{V}_{j+\frac{1}{2}}^0 \left(\Delta a_{j+1+\frac{1}{2}} + \Delta a_{j+\frac{1}{2}} \right).$$

Proof: From (45), we have

$$\Delta a_{j+1+\frac{1}{2}} - \Delta a_{j+\frac{1}{2}} = \frac{h^2 \Delta W_{j+1}^0}{\epsilon(g_{j+2} - g_{j+1})} - \frac{g_{j+2} - 2g_{j+1} + g_j}{g_{j+2} - g_{j+1}} \Delta a_{j+\frac{1}{2}} + \frac{O(h^2)h^2}{g_{j+2} - g_{j+1}},$$

and, by simple algebra, we obtain (64).

On the basis of these lemmas, we can state the theorem about the stability and error estimates as follows.

Theorem 6. *If Assumptions 1–4 are true, h and k satisfy (49), and*

$$(65) \qquad \frac{k}{h} \leq M_{13}$$

for some nonzero M_{13}, then there exist nonzero constants M_9 and M_{10} such that:

$$(66) \qquad \begin{aligned} \|\Delta_{j+1}\|_k^2 &\leq \left(M_9 + M_{10} \left(\frac{1}{\epsilon} + 1 + O(k) + O(h) \right) h \right) e^{M_{10}\left(\frac{1}{\epsilon} + 1 + O(k) + O(h)\right)} \\ &\quad \times \left(\|\Delta_0\|_k^2 + O(k) + O(h) + \epsilon^2 \right). \end{aligned}$$

Proof: Multiplying (43) by $ha_{j+\frac{1}{2}} \left(\Delta W_{j+1}^n + \Delta W_j^n \right)$ and adding hA_j^n times (44), we have

$$
\begin{aligned}
(67) \quad & \left(a_{j+\frac{1}{2}} \left| \Delta W_{j+1}^n \right|^2 + \left| a_{j+1+\frac{1}{2}} \right|^2 \left| \Delta V_{j+1+\frac{1}{2}}^{n+1} \right|^2 + \left| \tilde{V}_{j+1+\frac{1}{2}} \right|^2 \left| \Delta a_{j+1+\frac{1}{2}} \right|^2 \right) \\
& - \left(a_{j+1} \left| \Delta W_j^n \right| + \left| a_{j+\frac{1}{2}} \right|^2 \left| \Delta V_{j+\frac{1}{2}}^{n+1} \right|^2 + \left| \tilde{V}_{j+\frac{1}{2}} \right|^2 \left| \Delta a_{j+\frac{1}{2}} \right|^2 \right) \\
& = \frac{\epsilon h}{k} \Big[a_{j+\frac{1}{2}} \Delta V_{j+\frac{1}{2}}^{n+1} \Delta W_j^n - a_{j+\frac{1}{2}} \Delta V_{j+\frac{1}{2}}^n \Delta W_j^n + a_{j+1+\frac{1}{2}} \Delta V_{j+1+\frac{1}{2}}^{n+1} \Delta W_{j+1}^{n+1} \\
& \quad - a_{j+1+\frac{1}{2}} \Delta V_{j+1+\frac{1}{2}}^{n+1} \Delta W_{j+1}^n + a_{j+\frac{1}{2}} \left(\Delta V_{j+\frac{1}{2}}^{n+1} \Delta W_{j+\frac{1}{2}}^{n+1} - \Delta V_{j+\frac{1}{2}}^n \Delta W_{j+1}^n \right) \\
& \quad + \left(\Delta W_{j+1}^{n+1} - \Delta W_{j+1}^n \right) \left(\Delta a_{j+1+\frac{1}{2}} \tilde{V}_{j+1+\frac{1}{2}}^{n+1} + \Delta a_{j+\frac{1}{2}} \tilde{V}_{j+1+\frac{1}{2}}^{n+1} \right) \Big] \\
& \quad - \frac{\epsilon h}{k} \left(\tilde{W}_{j+1}^{n+1} - \tilde{W}_{j+1}^n \right) A_j + \frac{h}{\epsilon} \Delta W_{j+1}^{n+1} A_j^n - 2 a_{j+1+\frac{1}{2}} \Delta a_{j+1+\frac{1}{2}} \Delta V_{j+1+\frac{1}{2}}^{n+1} \tilde{V}_{j+1+\frac{1}{2}}^{n+1} \\
& \quad + 2 a_{j+\frac{1}{2}} \Delta a_{j+\frac{1}{2}} \Delta V_{j+\frac{1}{2}}^{n+1} \tilde{V}_{j+\frac{1}{2}}^{n+1} + h O(h^2) A_j^n \\
& \quad + h (O(k) + O(h)) \left(\Delta W_{j+1}^{n+1} + \Delta W_j^n \right) a_{j+\frac{1}{2}}.
\end{aligned}
$$

We add (64) to the result of summing (67) over n. Then multiplying this result by k and using Lemma 3 and the fact that

$$\sum_{n=0} \left(\Delta V_{j+\frac{1}{2}}^{n+1} \Delta W_{j+1}^{n+1} - \Delta V_{j+\frac{1}{2}}^n \Delta W_{j+1}^n \right) = 0,$$

we obtain:

$$T_{j+1} = T_j - \frac{\epsilon h}{k}\left[\Delta a_{j+1+\frac{1}{2}}k\sum_{n=0}\Delta W_{j+1}^n\left(\tilde{V}_{j+1+\frac{1}{2}}^{n+1}-\tilde{V}_{j+1+\frac{1}{2}}^n\right)\right.$$

$$\left. +\Delta a_{j+\frac{1}{2}}k\sum_{n=0}\Delta W_{j+1}^n\left(\tilde{V}_{j+\frac{1}{2}}^{n+1}-\tilde{V}_{j+\frac{1}{2}}^n\right)\right]$$

$$+\epsilon h\Delta a_{j+1+\frac{1}{2}}\Delta W_{j+1}^0\left(\tilde{V}_{j+\frac{1}{2}}^0-\tilde{V}_{j+1+\frac{1}{2}}^0\right) - \frac{\epsilon h}{k}k\sum_{n=0}\left(\tilde{W}_{j+1}^{n+1}-\tilde{W}_j^n\right)A_j^n$$

$$+\frac{hk}{\epsilon}\sum_{n=0}\Delta W_{j+1}^{n+1}A_j^n + k\left(a_{j+1+\frac{1}{2}}-a_{j+\frac{1}{2}}\right)||\Delta W_{j+1}||^2$$

(68)

$$+\frac{g_{j+2}-2g_{j+1}+g_j}{h}\epsilon^2\tilde{V}_{j+\frac{1}{2}}\Delta a_{j+\frac{1}{2}}\left(\Delta a_{j+1+\frac{1}{2}}+\Delta a_{j+\frac{1}{2}}\right)$$

$$+\frac{g_{j+2}-g_{j+1}}{h}\epsilon^2\left(\tilde{V}_{j+1+\frac{1}{2}}^0-\tilde{V}_{j+\frac{1}{2}}^0\right)\left|\Delta a_{j+1+\frac{1}{2}}\right|^2$$

$$+\frac{g_{j+2}-2g_{j+1}+g_j}{h}\epsilon^2\tilde{V}_{j+\frac{1}{2}}^0\left|\Delta a_{j+\frac{1}{2}}\right|^2 + O(h^2)hk\sum_{n=0}A_j^n$$

$$+(O(k)+O(h))hk\sum_{n=0}\left(\Delta W_{j+1}^{n+1}+\Delta W_j^n\right)a_{j+\frac{1}{2}}$$

$$+O(h^2)\epsilon^2 h\tilde{V}_{j+\frac{1}{2}}^0\left(\Delta a_{j+1+\frac{1}{2}}+\Delta a_{j+\frac{1}{2}}\right).$$

Then, by applying the inequalities in Lemma 4, we have

(69)
$$T_{j+1}\leq T_j + M_{11}\left(\frac{1}{\epsilon}+1+O(k)+O(h)\right)h\left(||\Delta_{j+1}||_k^2+||\Delta_j||_k^2\right)$$
$$+(O(k)+O(h)+\epsilon^2)hM_{12},$$

where

(70) $$M_{11}=M_{11}(M_2,M_3,M_8,d_1),$$

(71) $$M_{12}=M_{12}(M_2,M_3,M_8,d_1).$$

Using Lemma 2 in (69),

$$M_5||\Delta_{j+1}||_k^2\leq T_{j+1}$$

$$\leq T_0 + M_{11}\left(\frac{1}{\epsilon}+1+\Delta O(k)+O(h)\right)\left(2\sum_{m=0}^{j+1}||\Delta_m||_k^2\right)h$$
$$+(O(k)+O(h)+\epsilon^2)M_{12}$$

$$\leq M_6||\Delta_0||_k^2 + M_{11}\left(\frac{1}{\epsilon}+1+O(k)+O(h)\right)\left(2\sum_{m=0}^{j+1}||\Delta_m||_k^2\right)h$$
$$+(O(k)+O(h)+\epsilon^2)M_{12},$$

so that

$$\|\Delta_{j+1}\|_k^2 \le M_9 \left(\|\Delta_0\|_k + O(k) + O(h) + \epsilon^2 \right)$$
$$+ M_{10} \left(\frac{1}{\epsilon} + 1 + O(k) + O(h) \right) h \sum_{m=0}^{j+1} \|\Delta_m\|_k^2.$$

By applying Gronwall's inequality to (72), we obtain (66).

Remark 1. The requirement that $d_0 > 0$ is satisfied in (21) is very crucial. If $g'(x_0) = 0$ for some $x_0 \in [0, 1]$, then, for some j, the term $1/(g_{j+2} - g_{j+1})$ in (19) will result in singularity so that the computation blows up. Second, we may not be able to find M_4 such that (50) in Lemma 2 holds for some nonzero M_5 and M_6, and then we cannot obtain the results in Theorem 6.

Remark 2. (66) tells us not only the stability but also the error bounds of the scheme. Moreover, it tells us that we may get very bad results if we let $\epsilon \to 0$, since term $e^{\frac{M_{10}}{\epsilon}}$ will blow up. On the other hand, we definitely cannot use a very large ϵ, since ϵ^2 will then be large, and equation (11) will be completely different from (7). We can observe these situations in the numerical experiments in the next section. Hence, how to choose an optimal ϵ according to the data is a very attractive question, which we are still working on.

4. Numerical examples

In this section we use the scheme developed in the last section to solve the identification problem for parabolic equations for a variety of unknown functions $a(x)$. Even though the analysis of Section 3 requires some smoothness assumptions for $a(x)$, we also demonstrate, through Example 4, that the method works well for a discontinuous $a(x)$. The scheme is not only successful, but also very fast. Since there is no iteration procedure involved, it is, in fact, as fast as computing the parabolic equations once in the same region.

Example 1. Solve for $a(x)$ in

$$u_t = (a(x)u_x)_x - (x^2 + 2x)e^{x+t}, \quad 0 < x < 1, \quad t > 0,$$
$$u(0, t) = e^t, \quad u(1, t) = e^{1+t}, \quad t > 0,$$
$$u(x, 0) = e^x, \quad 0 < x < 1,$$

with the following extra data only at the left boundary:

$$u_x(0, t) = e^t.$$

In this case we can only use the right marching scheme, since no extra data is available at the right boundary. The true solution is $a(x) = 1 + x^2$; it is plotted with the numerical solution in Fig. 1. The numerical result is very satisfactory.

Example 2. Needless to say, we can develop the marching scheme without using perturbation; for example, just by letting $\epsilon = 0$ in the scheme we will get a scheme without perturbation. But this will usually lead to an unstable scheme since we have to compute the heat equation in the x direction. To show this, we use the scheme and its modified version with $\epsilon = 0$ to compute $a(x)$ in

$$u_t = (a(x)u_x)_x - e^{x-t}(20\cos(20x) + \sin(20x) + 2.5),$$
$$u(0,t) = e^{-t}, \quad u(1,t) = e^{1-t},$$
$$u(x,0) = e^x,$$

with the extra data

$$u_x(0,t) = e^{-t}.$$

The true solution is $a(x) = 1.5 + \sin(20x)$. The numerical results from both schemes are plotted in Fig. 2 together with the true solution. Note that the result from the scheme without perturbation blows up in the right half of interval $[0,1]$ while the result from the scheme with perturbation is very good in the whole interval $[0,1]$.

Example 3. As we mentioned in the last section, choosing a suitable ϵ is very important in practice. To show this, we use the scheme with a different ϵ to compute $a(x)$ in

$$u_t = (a(x)u_x) - \frac{x+201}{100}e^{x-t},$$
$$u(0,t) = e^{-t}, \quad u(1,t) = e^{1-t},$$
$$u(x,0) = e^x,$$

with extra data

$$u_x(0,t) = e^{-t}.$$

The true solution is $a(x) = 1 + \frac{x}{100}$. The results are plotted in Fig. 3, where we can easily see that the numerical result represented by Curve B is much better than the others, even though it is by no means optimal.

Example 4. If we have extra data at both boundaries, we will certainly get better computational results by using schemes marching in both directions. To show this, we use the right-marching scheme and the two-direction-marching scheme to solve for a non-smooth $a(x)$ in

$$u_t = (a(x)u_x) + f,$$
$$u(0,t) = e^t, \quad u(1,t) = e^{1+t},$$
$$u(x,0) = e^x,$$

with extra data

$$u_x(0,t) = e^t, \quad u_x(1,t) = e^t,$$

where

$$f(x,t) = \begin{cases} \dfrac{1}{2}e^{x+t}, & 0 \le x < \dfrac{1}{3}, \\[2mm] -\left(2 + \left(3x - \dfrac{1}{2}\right)\right)e^{x+t}, & \dfrac{1}{3} \le x < \dfrac{1}{2}, \\[2mm] \left(\dfrac{3}{2}x + \dfrac{3}{4}\right)e^{x+t}, & \dfrac{1}{2} \le x < \dfrac{2}{3}, \\[2mm] \dfrac{1}{4}e^{x+t}, & \dfrac{2}{3} \le x \le 1. \end{cases}$$

The true solution is piecewise linear:

$$a(x) = \begin{cases} \dfrac{1}{2}, & 0 \le x < \dfrac{1}{3}, \\[2mm] 3x - \dfrac{1}{2}, & \dfrac{1}{3} \le x < \dfrac{1}{2}, \\[2mm] -\dfrac{3}{2}x + \dfrac{7}{4}, & \dfrac{1}{2} \le x < \dfrac{2}{3}, \\[2mm] \dfrac{3}{4}, & \dfrac{2}{3} \le x \le 1. \end{cases}$$

The results are plotted in Fig. 4, in which Curve C, computed by using only the data at the left boundary, has larger error at points close to the right end of $[0,1]$.

Example 5. In applications, the available data are usually contaminated by random noise. How the random error in the data affects the computational results of the scheme in the last section is a very interesting question. Since the stability of the scheme proved in the last section is with respect to the data on the boundary, the random error coming from the boundary data should not have very much influence on the computational results. On the other hand, the scheme is more sensitive to the random error from the initial data, since, loosely speaking, the difference of the initial data is used in the denominator of (19). To see these, we apply the scheme to the following example which has data with random noise:

$$u_t = (a(x)u_x)_x - e^{x-t}(20\cos(20x) + \sin(20x) + 2.5) + magn_3 ran_3,$$

$$u(0,t) = e^{-t} + magn_1 ran_1, \quad u(1,t) = e^{1-t},$$

$$u(x,0) = e^x + magn_4 ran_4,$$

$$u_x(0,t) = e^{-t} + magn_2 ran_2,$$

where ran_i for $i = 1,2,3,4$ denote the random errors which are generated by the Vax Fortran random function; $magn_i$ for $i = 1,2,3,4$ are the magnitudes of the random error. Since marching from the left boundary to the right does need the data on the right boundary, we do not add any random noise to the data on the right boundary. In the first computation, we let $magn_{1-3} = 0.0001$, $magn_4 = 0.0$; this means that there is no random error in the initial data. The result is plotted in Fig. 5. Then, we let $magn_{1-3} = 0.0$, $magn_4 = 0.0001$; this means that the random error comes only from the initial data, and we plot the result in Fig. 6. According to these plots, the result from the first computation is better than that from the second computation. In other words, the random error in the initial data has more influence on the scheme.

Based on Section 3 and the numerical experiments in this section, we make the following conclusions:

1. Direct numerical methods are usually faster than iterative methods. They may be as fast as computing the PDE once in the region.

2. Via the direct computation methods, one can obtain estimates for the difference between the true and computed parameters. Similar convergence rates have been obtained in stationary cases for output least squares methods under special assumptions about availability of data [15,16], but are difficult to obtain in more general situations.

3. The least squares method is very general; it can be applied to almost all parameter identification problems. On the other hand, in order to find a suitable direct method, one has to understand the physical process, its mathematical model, and the advantage of the special properties of its mathematical model very well.

4. Direct numerical methods and least squares methods can be applied at the same time to a parameter identification problem to obtain better results. In this case, the results from the direct method may be a very good initial guess for the least squares method, and the least squares method can fine-tune the results from the direct method in a computationally feasible way.

Acknowledgments. This research was supported in part by U.S. Army Research Office Contract No. DAAG29–84–K–0002, by U.S. Air Force Office of Scientific Research Contract No. AFOSR–85–0117, by National Science Foundation Grant No. DMS–8504360, by Office of Naval Research Contract No. N00014–88–K–0370, and by the Institute for Scientific Computation through NSF Grant No. RII–8610680.

Figure 1.

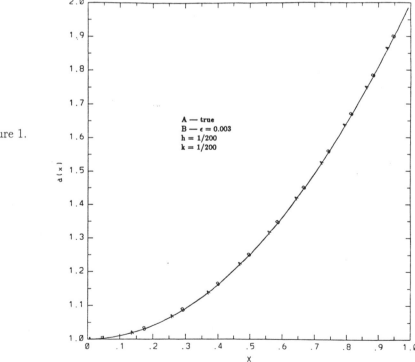

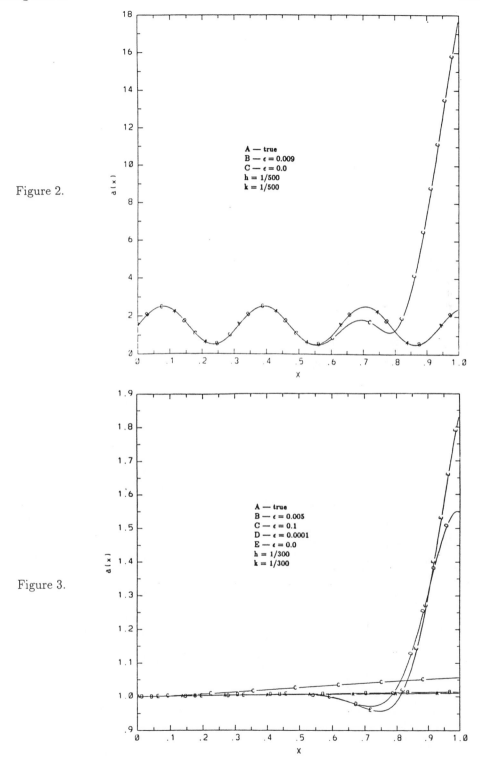

Figure 2.

Figure 3.

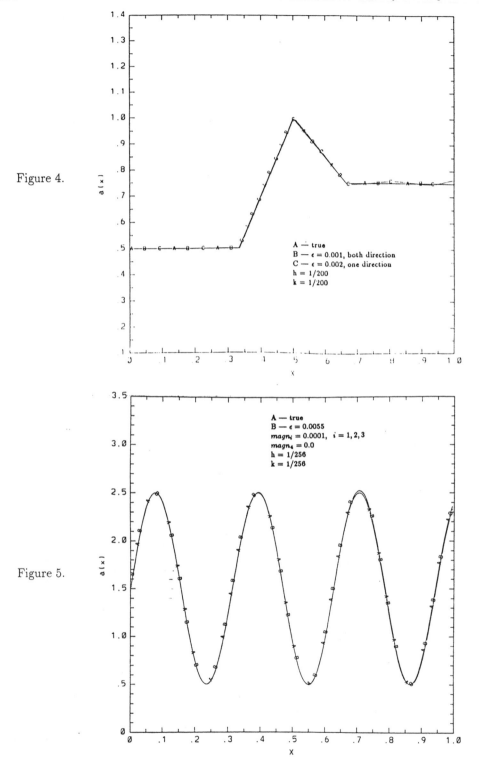

Figure 4.

Figure 5.

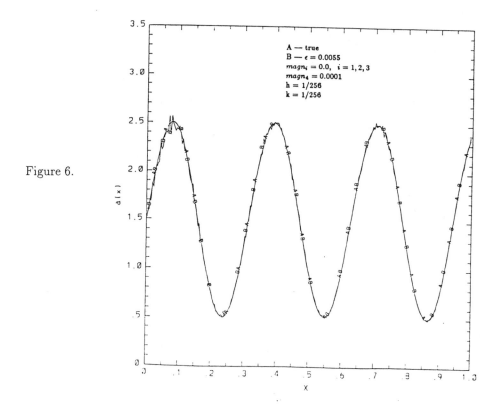

Figure 6.

REFERENCES

[1] P.K. Kitanidis and E.G. Vomvoris, *A geostatistical approach to the inverse problem in groundwater modeling (steady state) and one-dimensional simulations,* Water Resour. Res. **19(3)** (1983), 677–690.

[2] W.W-G. Yeh, Y.S. Yoon, and K.S. Lee, *Aquifer parameter identification with kriging and optimum parameterization,* Water Resour. Res. **19(1)** (1983), 225–233.

[3] G. Dagan, *Stochastic modeling of groundwater flow by unconditional and conditional probabilities, 1. Conditional simulation and the direct problem,* Water Resour. Res. **18(4)** (1982), 813–833.

[4] J.P. Delhomme, *Spatial variability and uncertainty in groundwater flow parameters: A geostatistical approach,,* Water Resour. Res. **15(2)** (1979), 269–280.

[5] G. Chavent, M. Dupuy, and P. Lemonnier, *History matching by use of optimal control theory,,* Trans., AIME, **259**, 74–86.

[6] G. Chavent, G. Cohen, and M. Espy, *Determination of relative permeabilities and capillary pressures by an automatic adjustment method,* SPE **9237**. presented at the 55th Annual Fall Tech. Conf. and Exhib. of SPE of AIME, Dallas, September 21–24, 1980.

[7] G.R. Richter, Numerical identification of a spatially varying diffusion coefficient, Math. Comp. **36** (1981), 375–386.

[8] J.V. Beck, B. Blackwell, and C.R. St. Clair, Jr.,, "Inverse Heat Conduction," Wiley, 1985.

[9] R.E. Ewing and T. Lin, *A direct method for parameter estimation in a hyperbolic partial differential equation,* Proc. of the 27th IEEE Conference on Decision and Control.

[10] C.F. Weber, Int. J. Heat Mass Transfer **24** (1981), 1783–1792.

[11] L. Eldén, *Modified equation for approximating the solution of Cauchy problem for the heat equation*, in "Inverse and Ill-Posed Problems," Academic Press, 1987, pp. 345–350.

[12] R.E. Ewing, T. Lin, and R. Falk, Inverse and ill-posed problems in reservoir simulation, in "Inverse and Ill-Posed Problems," Academic Press, 1987, pp. 483–497.

[13] T. Lin and R.E. Ewing, *A direct numerical method for an inverse problem for the heat equation via hyperbolic perturbation*, Proc. of the 27th IEEE Conference on Decision and Control.

[14] T. Lin and R.E. Ewing, *A note on source term identification for parabolic equations*, in this proceedings.

[15] K. Kunisch, *Rate of convergence for the estimation of a coefficient in a two point boundary value problem*, Inverse and Ill-Posed Problems (1987), 499–511, Academic Press.

[16] R.S. Falk, *Error estimates for the numerical identification of a variable coefficient*, Math. Comp. **40** (1983), 537–546.

Richard E. Ewing and Tao Lin
Departments of Mathematics
Petroleum Engineering, and Chemical Engineering
University of Wyoming
Laramie, Wyoming 82071
USA

International Series of
Numerical Mathematics, Vol. 91
© 1989 Birkhäuser Verlag Basel

Semigroup theory in linear viscoelasticity: weakly and strongly singular kernels ·

R.H. Fabiano and K. Ito

Center for Control Sciences
Brown University

Abstract. We discuss an integro-partial differential equation which arises in the theory of linear viscoelasticity. Previous well-posedness results are extended to include the case of strongly singular kernels. In addition, for the case of a weakly singular kernel and a finite delay, we show that the associated solution semigroup is differentiable.

Keywords. Linear viscoelasticity, strong singular kernels, differentiable semigroup.

1980 *Mathematics subject classifications*: 34K30, 35B65, 45K05

1. Introduction

We consider the following integro-differential equation in the Hilbert space H:

$$(1.1) \qquad \ddot{u}(t) + A\left[Eu(t) + \int_{-r}^{0} a(\theta)\dot{u}(t+\theta)\,d\theta\right] = f(t).$$

Here A is a positive definite, self-adjoint unbounded operator on H, and $f(t)$ is a locally integrable H-valued function. In addition, E is a positive constant (related to the stiffness in applications to linear viscoelasticity), and r satisfies $0 \le r \le \infty$. We assume that the function $a(\theta)$ satisfies $a(\theta) = \int_{-r}^{\theta} g(\xi)\,d\xi$, and we refer to the function $g(\theta)$ as the "history" kernel for (1.1). The kernel g will be further discussed later. Equations of this type arise is the modeling of viscoelastic beams and plates. Due to the great potential for applications to problems in modeling, control, and stabilization of large flexible space structures, there has been much interest recently in such equations (see [BF],[BFW],[Da1],[Da2],[Da3],[DM],[HRW], [HW],[MW] and the references therein for a sample of work in this area). We have been interested in the problem of developing a state space formulation for (1.1) which is suitable for applications to parameter estimation and optimal control. Another problem of interest is to understand the dissipative mechanism (internal damping in applications to structures) in (1.1), and it has been observed that this is related to the behavior of the kernel (in particular, the nature of the singularity of $g(\theta)$ at $\theta = 0$). In this paper, we will consider the following cases:

CASE 1   These are kernels which satisfy
  (WS.a) $g(\theta) > 0$ for $\theta \in (-r, 0)$,
  (WS.b) $g(\theta) \in L^1(-r, 0)$,
  (WS.c) $g(\theta) \in H^1(-r, -\alpha)$ for all $\alpha > 0$, and $g'(\theta) \ge 0$ for $\theta \in (-r, 0)$.

·This research was supported in part by the National Science Foundation under NSF Grant MCS-8504316 and by the Air Force Office of Scientific Research under Contract F49620-86-C-0111.

This class includes the weakly singular kernel $g(\theta)$ of the form

$$g(\theta) = c \frac{e^{\gamma\theta}}{|\theta|^p},$$

for $0 \leq p < 1$, $c > 0$, and $\gamma > 0$.

CASE 2   These are strongly singular kernels which satisfy
    (SS.a) $a(-r) = 0$, and $a(\theta) \in L^1(-r,0) \cap H^2(-r,\alpha)$ for all $\alpha > 0$,
    (SS.b) $g(\theta) = a'(\theta) > 0$ for $\theta \in (-r,0)$,
    (SS.c) $g'(\theta) \geq 0$ on $(-r,0)$.

In [FI] we discussed well-posedness results and an approximation framework for a state space formulation for (1.1) for the weakly singular kernel of Case 1. Actually, the following equivalent formulation of (1.1) is considered in [FI] (the equations (1.1) and (1.2) are equivalent for weakly singular kernels):

$$(1.2) \qquad \ddot{u}(t) + A\left[\hat{E}u(t) - \int_{-r}^{0} g(\theta)u(t+\theta)\,d\theta\right] = f(t),$$

where $\hat{E} = E + a(0)$. In Section 2 we extend the well-posedness results of [FI] to the case of a strongly singular kernel. The development of an approximation framework for this case will be discussed in a forthcoming paper. Finally, in Section 3 we show that, for a finite delay in the weakly singular case, the solution semigroup is differentiable. Because this regularity result implies that the spectrum is contained in a certain logarithmic sector (see [Pa], for example), it is a consequence that this model predicts damping rates which increase with mode frequency. This appears to be consistent with experimental observations of viscoelastic structures.

Now we will discuss notation and the state space formulation which will be used in this paper. Assume that V and H are Hilbert spaces and $V \subset H$ with continuous dense injection. Let V^* denote the strong dual space of V. We identify H with its dual, so that $V \subset H = H^* \subset V^*$. The dual product $\langle \phi, \psi \rangle_{V,V^*}$ on $V \times V^*$ is the unique extension by continuity of the scalar product $\langle \phi, \psi \rangle_H$ of H restricted to $V \times H$. Consider a symmetric sesquilinear form σ on V such that

$$(1.3) \qquad \begin{aligned} |\sigma(u,v)| &\leq C\,|u|_V\,|v|_V &&\text{for}\quad u,v \in V, \\ \operatorname{Re}\sigma(u,u) &\geq \omega\,|u|_V^2 &&\text{for}\quad u \in V, \end{aligned}$$

where $C, \omega > 0$. Let $A \in L(V, V^*)$ be defined by

$$(1.4) \qquad \sigma(u,v) = \langle Au, v \rangle_{V^*,V} \qquad \text{for all } u,v \in V.$$

It then follows from [T, Theorem 2.2.3] that the restriction of A on H with

$$\operatorname{dom} A = \{u \in V : \quad Au \in H\},$$

where we will use the same symbol A for such a restriction, defines a positive definite and self-adjoint operator on H, dom $A^{\frac{1}{2}} = V$, and

$$\sigma(u,v) = \left\langle A^{\frac{1}{2}}u, A^{\frac{1}{2}}v \right\rangle_H \qquad \text{for all } u, v \in V.$$

Thus, V can be equipped with the scalar product $\langle u, v \rangle_V = \sigma(u,v)$. Next, following the ideas found in [**Wa**] and [**Da3**], let $W = L^2_g(-r, 0; V)$ be the Hilbert space of all V-valued square integrable functions on the measure space $([-r, 0]$, Borel sets, $g\, d\theta)$, equipped with norm

$$|w|^2_W = \int_{-r}^0 g(\theta)|w(\theta)|^2_V \, d\theta.$$

Let Z denote the Hilbert space $V \times H \times W$ equipped with norm

$$|(u,v,w)|^2_Z = E|u|^2_V + |v|^2_H + \int_{-r}^0 g(\theta)|w(\theta)|^2_V \, d\theta.$$

Next, we define the operator $D : \text{dom}\, D \subset W \longrightarrow W$ on the domain

$$\text{dom}\, D = \left\{ w \in W : \begin{array}{l} w \text{ is locally absolutely continuous,} \\ w(0) = 0, \text{ and } \dfrac{dw}{d\theta} \in W \end{array} \right\},$$

by

$$Dw = \frac{d}{d\theta} w.$$

Now, again following [**Da3**] and [**Wa**], we introduce the state functions

$$v(t) = \dot{u}(t),$$
$$w(t, \theta) = u(t) - u(t + \theta).$$

Observe that (1.1) can now be written as

$$\ddot{u}(t) + A\left[Eu(t) - \int_{-r}^0 a(\theta)Dw(t)\, d\theta\right] = f(t),$$

where we used that $\dot{u}(t + \theta) = \frac{d}{d\theta} u(t + \theta) = -D\, w(t)$. Following an integration by parts, this may be rewritten as

$$\ddot{u}(t) + A\left[Eu(t) + \int_{-r}^0 g(\theta)w(t)\, d\theta\right] = f(t).$$

Observing this, and setting $z(t) = (u(t), v(t), w(t))$, one can formulate (1.1) as the following abstract Cauchy problem on Z:

(1.5) $\dot{z}(t) = \mathcal{A}z(t) + (0, f(t), 0).$

Here $\mathcal{A}$ is defined on the domain

$$\operatorname{dom} \mathcal{A} = \left\{ \begin{pmatrix} u \\ v \\ w \end{pmatrix} \in Z : \begin{array}{l} w \text{ is locally absolutely continuous,} \\ D\,w + v \in W, \quad v \in V, \quad w(0) = 0 \\ Eu + \displaystyle\int_{-r}^{0} g(\theta) w(\theta)\, d\theta \in \operatorname{dom} A \end{array} \right\},$$

by

$$\mathcal{A}(u, v, w) = \left(v, -A\left(Eu + \int_{-r}^{0} g(\theta) w(\theta)\, d\theta \right), D\,w + v \right).$$

This state space formulation for (1.1) arises naturally from energy considerations (see e.g., [**Wa**]) and has proven to be useful for applications to parameter estimation and optimal control problems for viscoelastic models. We will discuss these applications in future reports (see also [**BF**] and [**BFW**]).

2. Well-posedness for strongly singular kernels

Our purpose in this section is to show the well-posedness of the Cauchy problem (1.5). To do this we argue that the operator $\mathcal{A}$ satisfies the hypotheses of the Lumer-Phillips Theorem (see [**Pa**]) and hence generates a contraction semigroup.

Theorem 2.1. *Suppose the kernel g satisfies SS.a - SS.c. Then the operator $\mathcal{A}$ is the infinitesimal generator of a contraction semigroup $T(t)$ on Z.*

PROOF: First, we will show that $(u, v, w) \in \operatorname{dom} \mathcal{A}$ implies that $\int_{-r}^{0} g(\theta) |w(\theta)|_V\, d\theta$ exists. To see this, note that $w(\theta) = -\int_{\theta}^{0} Dw(\xi)\, d\xi$, so that for $R < r$ we have

$$\int_{-R}^{0} g(\theta) |w(\theta)|_V\, d\theta = \int_{-R}^{0} g(\theta) \left| \int_{\theta}^{0} (Dw(\xi) + v)\, d\xi + \theta v \right|_V d\theta$$

$$\leq \int_{-R}^{0} g(\theta) \left(\int_{\theta}^{0} \frac{d\xi}{g(\xi)} \right)^{\frac{1}{2}} \left(\int_{-R}^{0} g(\xi) |Dw(\xi) + v|_V^2\, d\xi \right)^{\frac{1}{2}} d\theta$$

$$+ \int_{-R}^{0} |\theta| g(\theta)\, d\theta\, |v|_V$$

$$\leq \left(R \int_{-r}^{0} |\theta| g(\theta)\, d\theta \right)^{\frac{1}{2}} |Dw + v|_W + \int_{-R}^{0} |\theta| g(\theta)\, d\theta\, |v|_V.$$

Now, $\int_{-r}^{0} -\theta g(\theta)\, d\theta = \int_{-r}^{0} \int_{-r}^{\theta} g(\xi)\, d\xi\, d\theta = \int_{-r}^{0} a(\theta)\, d\theta < \infty$, so that $|\theta|\, g \in L_1(-r, 0)$. Since $\left(\int_{-r}^{-R} g(\theta) |w(\theta)|_V\, d\theta \right)^2 \leq \int_{-r}^{-R} g(\theta)\, d\theta\, |w|_W^2$, it follows that $\int_{-r}^{0} g(\theta) |w(\theta)|_V\, d\theta$ exists. The rest of the proof proceeds as follows. We will show that $\mathcal{A}$ is dissipative, and that $(\lambda I - \mathcal{A}) \operatorname{dom} \mathcal{A} = Z$ for $\operatorname{Re} \lambda > 0$. Then denseness of $\operatorname{dom} \mathcal{A}$ follows from ([**Pa**, Theorem 1.4.6]), and we may conclude from the Lumer-Phillips Theorem that $\mathcal{A}$ is the infinitesimal generator of a contraction semigroup. To continue, we now show the dissipativeness of $\mathcal{A}$. For $z = (u, v, w) \in \text{dom}\,\mathcal{A}$,

$$(2.1) \quad \langle \mathcal{A}z, z \rangle_Z = E\langle u, v \rangle_V - \left\langle v, Eu + \int_{-r}^0 g(\theta)w(\theta)\,d\theta \right\rangle_V$$

$$+ \int_{-r}^0 g(\theta)\langle Dw + v, w \rangle_V\,d\theta$$

$$= -\left\langle v, \int_{-r}^0 g(\theta)w(\theta)\,d\theta \right\rangle_V + \int_{-r}^0 g(\theta)\langle Dw + v, w \rangle_V\,d\theta.$$

Let us consider for $0 < \epsilon < R < r$,

$$(2.2) \quad \int_{-R}^{-\epsilon} g(\theta)\langle Dw + v, w \rangle_V\,d\theta = \frac{1}{2}g(-\epsilon)|w(\epsilon)|_V^2 - \frac{1}{2}g(-R)|w(-R)|_V^2$$

$$- \frac{1}{2}\int_{-R}^{-\epsilon} g'(\theta)|w(\theta)|_V^2\,d\theta$$

$$+ \left\langle \int_{-R}^{-\epsilon} g(\theta)w(\theta)\,d\theta, v \right\rangle_V$$

$$\leq \frac{1}{2}g(-\epsilon)|w(\epsilon)|_V^2 + \left\langle \int_{-R}^{-\epsilon} g(\theta)w(\theta)\,d\theta, v \right\rangle_V.$$

Since $w(-\epsilon) = -\int_{-\epsilon}^0 (Dw + v)\,d\xi + \epsilon v$, it follows that

$$(2.3) \quad g(-\epsilon)|w(-\epsilon)|_V^2 \leq g(-\epsilon)\int_{-\epsilon}^0 \frac{d\theta}{g(\theta)}\int_{-\epsilon}^0 g(\theta)|Dw + v|_V^2\,d\theta + \epsilon^2 g(-\epsilon)|v|_V^2$$

$$\leq \int_{-\epsilon}^0 g(\theta)|Dw + v|_V^2\,d\theta + \epsilon^2 g(-\epsilon)|v|_V^2$$

$$\longrightarrow 0, \qquad \text{as } \epsilon \longrightarrow 0^+.$$

Thus, as $\epsilon \to 0^+$ and $R \to r^-$, (2.1)- (2.3) imply that $\langle \mathcal{A}z, z \rangle_Z \leq 0$ for all $z \in \text{dom}\,\mathcal{A}$, and hence $\mathcal{A}$ is dissipative. Finally, we will show that for $\text{Re}\,\lambda > 0$, $(\lambda I - \mathcal{A})\text{dom}\,\mathcal{A} = Z$. That is, given $(\phi, \psi, h) \in Z$, we must determine $(u, v, w) \in \text{dom}\,\mathcal{A}$ so that $(\lambda I - \mathcal{A})(u, v, w) = (\phi, \psi, h)$. This last equation is written as

$$(2.4) \qquad\qquad \lambda u - v = \phi,$$

$$(2.5) \qquad\qquad \lambda v + A\left(Eu + \int_{-r}^0 g(\theta)w(\theta)\,d\theta\right) = \psi,$$

$$(2.6) \qquad\qquad \lambda w - (Dw + v) = h.$$

Computing formally for now, from (2.6) we obtain

$$(2.7) \qquad\qquad w(\theta) = \int_\theta^0 e^{\lambda(\theta-\xi)}(v + h(\xi))\,d\xi$$

$$= \frac{1}{\lambda}(1 - e^{\lambda\theta})v + \int_\theta^0 e^{\lambda(\theta-\xi)}h(\xi)\,d\xi.$$

From (2.4) we have

(2.8) $$v = \lambda u - \phi.$$

Substituting (2.7) and (2.8) into (2.5), we obtain the following equation for u:

$$\lambda^2 u + A\left[Eu + \int_{-r}^0 g(\theta)\left(\int_\theta^0 e^{\lambda(\theta-\xi)}(\lambda u - \phi + h(\xi))\,d\xi\right)d\theta\right] = \psi + \lambda\phi,$$

or

(2.9) $$\left[\lambda^2 + A\left(E + \int_{-r}^0 g(\theta)(1 - e^{\lambda\theta})\,d\theta\right)\right]u$$

$$= \psi + \lambda\phi + \int_{-r}^0 g(\theta)\int_\theta^0 e^{\lambda(\theta-\xi)}A(\phi - h(\xi))\,d\xi\,d\theta,$$

where we used $\int_\theta^0 \lambda e^{\lambda(\theta-\xi)}\,d\xi = 1 - e^{\lambda\theta}$. We would like to show that (u, v, w) defined by (2.7), (2.8) and (2.9) satisfies $(u, v, w) \in \text{dom}\,\mathcal{A}$ and $(\lambda I - \mathcal{A})(u, v, w) = (\phi, \psi, h)$. Note that the right hand side of (2.9) belongs to V^* and, if we define

$$\Delta(\lambda) = \lambda^2 I + A\left(E + \int_{-r}^0 g(\theta)(1 - e^{\lambda\theta})\,d\theta\right),$$

then $\Delta(\lambda) \in L(V, V^*)$. Consider, for fixed $\lambda \in \mathbf{C}$, the following sesquilinear form μ on V:

$$\mu(u, v) = \langle\Delta(\lambda)u, v\rangle_{V^*,V}$$

$$= \lambda^2\langle u, v\rangle_H + \left(E + \int_{-r}^0 g(\theta)(1 - e^{\lambda\theta})\,d\theta\right)\sigma(u, v) \qquad \text{for u,v } \in V.$$

Then, there is a constant c so that

$$|\mu(u, v)| \le c|u|_V\,|v|_V \qquad \text{for u,v } \in V,$$

and, for $\lambda = a + b\mathrm{i}$ with $a > 0$,

(2.10) $$\text{Re}\,\mu(u, u) \ge (a^2 - b^2)|u|_H^2 + E|u|_V^2 \qquad \text{for all } u \in V.$$

Hence, it follows from [**Ta**] that if $\text{Re}\,\lambda \ge 0$, then $\Delta^{-1}(\lambda) \in L(V^*, V)$. Thus,

$$u = \Delta^{-1}(\lambda)\left(\psi + \lambda\phi + \int_{-r}^0 g(\theta)\int_\theta^0 e^{\lambda(\theta-\xi)}A(\phi - h(\xi))\,d\xi\,d\theta\right),$$

satisfies (2.9) and $u \in V$. Also, v defined by (2.8) satisfies $v \in V$. To see that $w(\theta)$ defined by (2.7) is in W, note that $\int_{-r}^0 \frac{1}{|\lambda|^2}|1 - e^{\lambda\theta}|^2 g(\theta)\,d\theta < \infty$ since $|\theta|g(\theta) \in L^1(-r, 0)$. Further, for finite r,

$$\int_{-r}^0 g(\theta)\left|\int_\theta^0 e^{\lambda(\theta-\xi)}h(\xi)\,d\xi\right|_V^2 d\theta \le \int_{-r}^0\left(\int_\theta^0 \frac{g(\theta)}{g(\xi)}\,d\xi\right)\left(\int_{-r}^0 g(\xi)|h(\xi)|_V^2\,d\xi\right)d\theta$$

$$\le \frac{1}{2}r^2|h|_W.$$

For $r = \infty$, we need to consider $\int_{-\infty}^0 g(\theta) |\sigma(\theta)|_V^2 \, d\theta$, where $\sigma(\theta) = \int_\theta^0 e^{\lambda(\theta-\xi)} h(\xi) \, d\xi$. We note that one can show that the operator D is dissipative on W using exactly the same arguments as above (i.e. (2.1)-(2.3)). Next, let $\sigma^R(\theta) = \int_\theta^0 e^{\lambda(\theta-\xi)} h^R(\xi) \, d\xi$, where $h^R(\theta) = \chi_{[-R,0)} \cdot h(\theta)$ for $R > 0$ (Here $\chi_{[-R,0)}$ is the usual characteristic function). Then, $(\lambda I - D)\sigma^R = h^R$ and $\sigma^R \in W$. In fact,

$$\int_{-R}^0 g(\theta) |\sigma^R(\theta)|_V^2 \, d\theta \leq \frac{1}{2} R^2 |h|_W^2$$

and

$$\int_{-\infty}^{-R} g(\theta) |\sigma^R(\theta)|_V^2 \leq \int_{-\infty}^{-R} g(\theta) \left| e^{\lambda(R+\theta)} \int_{-R}^0 e^{-\lambda(R+\xi)} h^R(\xi) \, d\xi \right|_V^2 d\theta$$

$$\leq \int_{-\infty}^{-R} e^{2(\operatorname{Re}\lambda)(R+\theta)} \left(\int_{-R}^0 \frac{g(\theta)}{g(\xi)} d\xi \right) \left(\int_{-R}^0 g(\xi) |h(\xi)|_V^2 \, d\xi \right) d\theta$$

$$\leq \frac{R}{2\operatorname{Re}\lambda} |h|_W^2 \, .$$

Thus, $\sigma^R \in \operatorname{dom} D$, and since D is dissipative, $\operatorname{Re}\lambda |\sigma^R|_W^2 \leq |h|_W |\sigma^R|_W$, so that $|\sigma^R|_W \leq \frac{1}{\operatorname{Re}\lambda} |h|_W$. Hence $\{\sigma^R\}_{R=-k}$, $k = 1, 2, 3 \ldots$, is a Cauchy sequence in W and its unique limit $\sigma \in W$ is given by $\sigma(\theta) = \int_\theta^0 e^{\lambda(\theta-\xi)} h(\xi) \, d\xi$. From these bounds we see that $w(\theta) \in W$.

It is now straightforward to see that $(u, v, w) \in \operatorname{dom}\mathcal{A}$ and $(\lambda I - \mathcal{A})(u, v, w) = (\phi, \psi, h)$, and the proof is complete. ∎

Corollary 2.2. *If r is finite, then $\lambda \in \rho(\mathcal{A})$ if and only if $\Delta^{-1}(\lambda) \in L(V^*, V)$. In general, the resolvent $(\lambda I - \mathcal{A})^{-1}$ is given by $(\lambda I - \mathcal{A})^{-1}(\phi, \psi, h) = (u, v, w) \in Z$ with*

$$u = \Delta^{-1}(\lambda) \left(\psi + \lambda\phi + \int_{-r}^0 g(\theta) \int_\theta^0 e^{\lambda(\theta-\xi)} A(\phi - h(\xi)) \, d\xi \, d\theta \right),$$

$$v = -E\Delta^{-1}(\lambda) A\phi + \lambda\Delta^{-1}(\lambda) \left(\psi - \int_{-r}^0 g(\theta) \int_\theta^0 e^{\lambda(\theta-\xi)} Ah(\xi) \, d\xi \, d\theta \right),$$

$$w = \int_\theta^0 e^{\lambda(\theta-\xi)} (v + h(\xi)) \, d\xi.$$

3. Differentiability of the solution semigroup

In this section we consider the Cauchy problem (1.5) for the case in which r is finite and the kernel $g(\theta)$ is weakly singular— that is, $g(\theta)$ satisfies (WS.a)-(WS.c). We show that under certain additional assumptions (namely, A1 and A2 below) the solution semigroup for this problem is differentiable. We first require the estimates given in the following two lemmas.

Lemma 3.1. *Assume $g(\theta)$ satisfies*

(A1)
$$\int_{-r}^0 g'(\theta) \left(\int_\theta^0 \frac{d\xi}{g(\xi)} \right)^{\frac{1}{2}} d\theta < \infty.$$

Then for each $\omega \in \mathbf{R}$ and $h \in W$, $\kappa(\theta) = (i\omega I - D)^{-1}h$ exists and there exists a positive constant M_1 such that

$$(3.1) \qquad \left| i\omega \int_{-r}^{0} g(\theta)\kappa(\theta)\,d\theta \right|_{V} \le M_1 |h|_{V}.$$

PROOF: Let $\kappa(\theta) = \int_{\theta}^{0} e^{i\omega(\theta - \xi)} h(\xi)\,d\xi$. Then from (2.7) and the proof of Theorem 2.1, it follows that $\kappa \in W$ and satisfies $i\omega\kappa - D\kappa = h$ with $\kappa(0) = 0$. Thus,

$$i\omega \int_{-r}^{0} g(\theta)\kappa(\theta)\,d\theta = \int_{-r}^{0} g(\theta)D\kappa\,d\theta - \int_{-r}^{0} g(\theta)h(\theta)\,d\theta.$$

In order to get a bound for this expression, let us consider for $0 < \epsilon < r$,

$$(3.2) \qquad \int_{-r}^{-\epsilon} g(\theta)D\kappa\,d\theta = -g(-r)\kappa(-r) + g(-\epsilon)\kappa(-\epsilon) - \int_{-r}^{-\epsilon} g'(\theta)\kappa(\theta)\,d\theta.$$

We will estimate the three terms on the right hand side separately. First,

$$\begin{aligned}
|g(-r)\kappa(-r)|_{V} &= \left| g(-r) \int_{-r}^{0} e^{i\omega(-r-\xi)} h(\xi)\,d\xi \right|_{V} \\
&\le (g(-r))^{\frac{1}{2}} \left(\int_{-r}^{0} \frac{g(-r)}{g(\xi)}\,d\xi \right)^{\frac{1}{2}} \left(\int_{-r}^{0} g(\theta)|h|_{V}^{2}\,d\theta \right)^{\frac{1}{2}} \\
&\le (rg(-r))^{\frac{1}{2}} |h|_{W}.
\end{aligned}$$

Second,

$$\begin{aligned}
|g(-\epsilon)\kappa(-\epsilon)|_{V} &\le (\epsilon\, g(-\epsilon))^{\frac{1}{2}} \left(\int_{-\epsilon}^{0} g(\theta)|h|_{V}^{2}\,d\theta \right)^{\frac{1}{2}} \\
&\longrightarrow 0 \qquad \text{as } \epsilon \longrightarrow 0^{+},
\end{aligned}$$

where we use $0 \le \epsilon\, g(-\epsilon) \le \int_{-\epsilon}^{0} g(\theta)\,d\theta$. Finally, the third term is bounded by

$$\left| \int_{-r}^{-\epsilon} g'(\theta)\kappa(\theta)\,d\theta \right|_{V} \le \int_{-r}^{-\epsilon} g'(\theta) \left(\int_{\theta}^{-\epsilon} \frac{d\xi}{g(\xi)} \right)^{\frac{1}{2}} |h|_{W}\,d\theta.$$

Since $\epsilon > 0$ is arbitrary, the estimate (3.1) follows from (A1). ∎

REMARK 3.2 Suppose that $g(\theta) = g_0(\theta)/|\theta|^p$, $0 \le p < 1$, where $g_0 \in C^1[-r, 0]$, and $g_0 \ge c > 0$ on $[-r, 0]$. Then it follows that g satisfies (A1). In fact, since $g'(\theta) = g_0'(\theta)/|\theta|^p - pg_0(\theta)/|\theta|^{p+1}$, and $\int_{\theta}^{0} \frac{d\xi}{g(\xi)} \le \frac{|\theta|^{p+1}}{c(1+p)}$, we have

$$\int_{-r}^{0} g'(\theta) \left(\int_{\theta}^{0} \frac{d\xi}{g(\xi)} \right)^{\frac{1}{2}} d\theta \le \int_{-r}^{0} \left(\frac{1}{c(1+p)} \right)^{\frac{1}{2}} |\theta|^{\frac{1}{2}(p+1)} \left(\frac{g_0'}{|\theta|^p} + \frac{g_0}{|\theta|^p} \right) d\theta,$$

where $|\theta|^{\frac{1}{2}(p+1)}|\theta|^{-p-1} = |\theta|^{-\frac{1}{2}(p+1)} \in L^1(-r, 0)$.

Lemma 3.3. *If $\phi \in V$ and $w \in \mathbf{R}$, then*

$$\left| iw \int_{-r}^{0} g(\theta) \int_{\theta}^{0} e^{iw(\theta-\xi)} \phi \, d\xi \, d\theta \right|_{V} \leq 2|\phi|_{V} \int_{-r}^{0} g(\theta) \, d\theta.$$

PROOF: Since $iw \int_{\theta}^{0} e^{iw(\theta-\xi)} \, d\xi = 1 - e^{iw\theta}$, we have

$$\left| iw \int_{-r}^{0} g(\theta) \int_{\theta}^{0} e^{iw(\theta-\xi)} \phi \, d\xi \, d\theta \right|_{V} \leq \left| \int_{-r}^{0} (1 - e^{iw\theta}) g(\theta) \, d\theta \right| |\phi|_{V}$$

$$\leq 2|\phi|_{V} \int_{-r}^{0} g(\theta) \, d\theta.$$

We now prove the main result of this section.

Theorem 3.4. *Assume that the self-adjoint operator A has the spectral resolution $A = \int_{\mu_0}^{\infty} \mu^2 dE(\mu)$, $\mu_0 > 0$. Let $\Delta(\omega, \mu) = -\omega^2 + \mu^2 \int_{-r}^{0} g(\theta) e^{i\omega\theta} d\theta$ for $(\omega, \mu) \in \mathbf{R}^2$ and suppose that there exists a positive constant M such that*

$$(\text{A2}) \qquad |\Delta^{-1}(\omega, \mu)\omega| \leq \frac{M}{|\omega|^p} \qquad and \qquad |\Delta^{-1}(\omega, \mu)\frac{\mu^2}{\omega}| \leq \frac{M}{|\omega|^p},$$

for all $|\omega|, \mu \geq \mu_0$ and some p satisfying $0 < p < 1$. Then the solution semigroup $T(t)$ generated by $\mathcal{A}$ is differentiable for $t \geq t_0$ for some $t_0 > 0$. One can take $t_0 = 3(\frac{r}{p})$.

PROOF: The idea of the proof is to obtain appropriate bounds on the operators $(\lambda I - \mathcal{A})^{-1}$ and $\lambda e^{\lambda t}(\lambda I - \mathcal{A})^{-1}$ in a certain logarithmic sector of the left half-plane. One may then use arguments similar to those of ([**Pa**, p. 54–59]) to conclude that $T(t)$ is differentiable for $t > t_0$. To proceed, we first consider the resolvent on the imaginary axis. Let $z = (\phi, \psi, h) \in Z$ and set $(u, v, w) = (i\omega I - \mathcal{A})^{-1}(\phi, \psi, h) \in Z$ for $\omega \in \mathbf{R}$. Then we will show that there exists $M_2 > 0$ such that

$$|u|_{V} \leq \frac{M_2}{|\omega|^p} |z|_{Z} \qquad and \qquad |v|_{H} \leq \frac{M_2}{|\omega|^p} |z|_{Z}.$$

From Corollary 2.2 with $\lambda = i\omega$ we have

$$|u|_{V} \leq \left(\int_{\mu_0}^{\infty} |\Delta^{-1}(\omega, \mu)\mu|^2 \langle dE(\mu)\psi, \overline{\psi} \rangle_{H} \right)^{\frac{1}{2}} + \left(\int_{\mu_0}^{\infty} |\Delta^{-1}(\omega, \mu)\omega|^2 \mu^2 \langle dE(\mu)\phi, \overline{\phi} \rangle_{H} \right)^{\frac{1}{2}}$$

$$+ \left(\int_{\mu_0}^{\infty} |\Delta^{-1}(\omega, \mu)\mu^2|^2 \mu^2 \langle dE(\mu)\zeta, \overline{\zeta} \rangle_{H} \right)^{\frac{1}{2}},$$

where $\zeta = \int_{-r}^{0} g(\theta) \int_{\theta}^{0} e^{i\omega(\theta-\xi)}(h(\xi) - \phi) \, d\xi \, d\theta$. Thus, from Lemmas 3.1 and 3.2 and the fact that $|\phi|_{V}^2 = \int_{0}^{\infty} \mu^2 \langle dE(\mu)\phi, \overline{\phi} \rangle_{H}$ for $\phi \in V$, we have

$$|u|_{V} \leq \left[\sup|\Delta^{-1}(\omega, \mu)\mu| \right] |\psi|_{H} + \left[\sup|\Delta^{-1}(\omega, \mu)\omega| \right] |\phi|_{V}$$

$$+ \left[\sup|\Delta^{-1}(\omega, \mu)\frac{\mu^2}{\omega}| \right] (2|\phi|_{V} + M_1|h|_{W}),$$

where the sup is over $\mu \geq \mu_0$. Note that $|\Delta^{-1}(\omega, \mu)\mu|^2 = |\Delta^{-1}(\omega, \mu)\omega| \, |\Delta^{-1}(\omega, \mu)\frac{\mu^2}{\omega}| \leq (M/|\omega|^p)^2$. Hence there exists $M_2 > 0$ such that $|u|_V \leq \frac{M_2}{|\omega|^p} |z|_Z$. Similarly,

$$
\begin{aligned}
|v|_H &\leq E\left[\sup|\Delta^{-1}(\omega, \mu)|\right] |\phi|_V + \left[\sup|\Delta^{-1}(\omega, \mu)\omega|\right] |\psi|_H \\
&\quad + M_1\left[\sup|\Delta^{-1}(\omega, \mu)\mu|\right] |h|_W \\
&\leq \frac{M_2}{|\omega|^p} |z|_Z.
\end{aligned}
$$

We cannot obtain a similar estimate for the third component w of the resolvent $(i\omega I - \mathcal{A})^{-1}$. Instead, for $\lambda = \alpha + i\omega$, $\alpha \leq 0$, consider the equation $(\lambda I - \mathcal{A})(u, v, w) = z = (\phi, \psi, h)$. This equation is equivalently written as

$$
(3.3) \qquad \left(\begin{bmatrix} I & 0 \\ 0 & I \end{bmatrix} + \alpha \begin{bmatrix} \Delta_{11} & \Delta_{12} \\ \Delta_{21} & \Delta_{22} \end{bmatrix}\right)\begin{pmatrix} \zeta \\ w \end{pmatrix} = (i\omega I - \mathcal{A})^{-1} z,
$$

where $\zeta = (u, v) \in V \times H$ and Δ_{ij} denotes the i, j component of the operator $(i\omega I - \mathcal{A})^{-1}$ corresponding to the partition of the state space Z by $Z = (V \times H) \times W$. We have just shown that

$$
(3.4) \qquad |\Delta_{11}|_{L(V \times H)} \leq \frac{M_2}{|\omega|^p} \qquad \text{and} \qquad |\Delta_{12}|_{L(W, V \times H)} \leq \frac{M_2}{|\omega|^p}.
$$

Now note that w satisfies $\lambda w - Dw = h + v$ with $w(0) = 0$. Thus,

$$
\begin{aligned}
(3.5) \qquad w(\theta) &= \int_\theta^0 e^{\lambda(\theta - \xi)}(h(\xi) - v)\, d\xi \\
&= \int_\theta^0 e^{\lambda(\theta - \xi)} h(\xi)\, d\xi + (1 - e^{\lambda\theta})(u - \frac{\phi}{\lambda}) \\
&= \bar{\Delta}_{22} h + \bar{\Delta}_{21}(u - \frac{\phi}{\lambda}),
\end{aligned}
$$

where the operators $\bar{\Delta}_{22} : W \longrightarrow W$ and $\bar{\Delta}_{21} : V \longrightarrow W$ are defined as indicated. We have

$$
\begin{aligned}
(3.6) \qquad |\bar{\Delta}_{22} h|_W^2 &= \left|\int_\theta^0 e^{\lambda(\theta - \xi)} h(\xi)\, d\xi\right|_W^2 \\
&= \int_{-r}^0 g(\theta) \left|\int_\theta^0 e^{\lambda(\theta - \xi)} h(\xi)\, d\xi\right|_V^2 d\theta \\
&\leq \int_{-r}^0 g(\theta) \int_\theta^0 e^{2\alpha(\theta - \xi)} \frac{d\xi}{g(\xi)} \left(\int_{-r}^0 g(\xi)|h(\xi)|_V^2\, d\xi\right) d\theta \\
&\leq \int_{-r}^0 \int_{-r}^\xi \frac{g(\theta)}{g(\xi)} e^{2\alpha(\theta - \xi)}\, d\theta\, d\xi \, |h|_W^2 \\
&\leq \int_{-r}^0 \frac{1}{2\alpha}(e^{2\alpha(-r - \xi)} - 1)\, d\xi\, |h|_W^2 \\
&\leq \frac{1}{4\alpha^2} e^{-2\alpha r} |h|_W^2
\end{aligned}
$$

and

$$(3.7) \qquad |\bar{\Delta}_{21}\phi|^2_W = \int_{-r}^0 g(\theta)|1 - e^{\lambda\theta}|^2 \, |\phi|^2_V \, d\theta$$

$$\leq \left(\int_{-r}^0 g(\theta) \, d\theta \right) e^{-2\alpha r} |\phi|^2_V.$$

From (3.3) and (3.5) we have

$$(3.8) \quad (I + \alpha\Delta_{11} + \alpha\Delta_{12}\bar{\Delta}_{21}P_V)\zeta = P_{V\times H}(i\omega I - A)^{-1} + \alpha\Delta_{12}\bar{\Delta}_{21}\left(\frac{\phi}{\lambda}\right) - \alpha\Delta_{12}\bar{\Delta}_{22}h,$$

where $P_V : V \times H \longrightarrow V$ is the canonical orthogonal projection defined by $P(u,v) = u$, and $P_{V\times H} : Z \longrightarrow V \times H$ is similarly the canonical projection defined by $P_{V\times H}(u,v,w) = (u,v)$. From (3.4) and (3.7), we have

$$|\alpha\Delta_{11} + \alpha\Delta_{12}\bar{\Delta}_{21}P_V|_{L(V\times H)} \leq |\alpha| \left(\frac{M_2}{|\omega|^p} + \frac{M_2}{|\omega|^p} e^{|\alpha|r} \right) \leq \frac{1}{2},$$

if $(\operatorname{Re}\lambda)r \geq a - p\log|\omega|$ for appropriately chosen $a > 0$. Thus, for such a λ, (3.8) posesses a unique solution ζ which satisfies

$$|\zeta|_{V\times H} \leq 2 \left(\frac{M_2}{|\omega|^p} |z|_Z + \frac{M_2}{|\omega|^p} e^{|\alpha|r} |\phi|_V + \frac{M_2}{2|\omega|^p} e^{|\alpha|r} |h|_W \right) \leq M_3 |z|_Z,$$

for $|\omega| \geq 1$ and some positive constant M_3. Similarly, from (3.5), (3.6), and (3.7) we have

$$|w|_W \leq e^{|\alpha|r} \left(|u|_V + \frac{1}{|\lambda|} |\phi|_V \right) + \frac{1}{2|\alpha|} e^{|\alpha|r} |h|_W \leq M_4 |\omega|^p |z|_Z.$$

Since (ζ, w) solves (3.3), we obtain

$$|(\lambda I - A)^{-1}|_{L(Z)} \leq M_4 |\omega|^p,$$

for all $a - p\log|\omega| \leq (\operatorname{Re}\lambda)r \leq 0$ and $|\omega| \geq 1$. Observe that for $\lambda = \sigma + i\tau$ with $\sigma = a - p\log|\tau|$, we have

$$|\lambda e^{\lambda t}(\lambda I - A)^{-1}| \leq C|\tau|e^{\operatorname{Re}\lambda\, t + p\log|\tau|} \leq Ce^{\frac{a}{r}t}|\tau|^{1+p-p\frac{t}{r}},$$

for some positive constant C.

With these estimates we may apply arguments similar to those of [Pa, p. 54–59], and the result follows. ∎

REMARK 3.5 Suppose $g(\theta) = ce^{\gamma\theta}/|\theta|^p$, $0 < p < 1$, $\gamma > 0$. In this case, one can show that conditions (A1) and (A2) are satisfied for the same p, so that Theorem 3.4 applies.

Corollary 3.6. *Using the same notation as in Theorem 3.4, suppose that there exists a positive constant M such that*

$$(\text{A3}) \qquad |\Delta^{-1}(\omega,\mu)\omega| \leq M \qquad \text{and} \qquad |\Delta^{-1}(\omega,\mu)\frac{\mu^2}{\omega}| \leq M,$$

for all $|\omega| \geq \frac{1}{2}\sqrt{E}\mu_0$ and $\mu \geq \mu_0$. Then the solution semigroup $T(t)$ generated by $\mathcal{A}$ is uniformly exponentially stable.

PROOF: First note that the third component w of the resolvent $(i\omega I - \mathcal{A})^{-1}$ satisfies

$$|w|_W \leq \frac{1}{2}r^2|h|_W + 2\int_{-r}^0 g(\theta)\,d\theta(|u|_V + r|\phi|_V),$$

for all w. From (2.10), for $\lambda = i\omega$ we have Re $\mu(u,u) \geq -\omega^2|u|_H + E|u|_V^2$ for all $u \in V$. Thus, by Corollary 2.2, $\lambda = i\omega \in \rho(\mathcal{A})$ for $|\omega| \leq \frac{1}{2}\sqrt{E}\mu_0$. Hence, using exactly the same arguments as in the proof of Theorem 3.4, one can prove that there exists $M_5 > 0$ such that $|(i\omega I - \mathcal{A})^{-1}| \leq M_5$ for all ω. Since $T(t)$ is a contraction semigroup, the corollary follows from ([**Pr**, Cor. 4]). ∎

REMARK 3.7   Suppose $g(\theta) = ce^{\gamma\theta}$, $\gamma, c > 0$. Then the condition (A3) is satisfied. Moreover, in this case Corollary 3.6 holds for the case $r = \infty$ since $\int_{-\infty}^0 \langle Dw, w\rangle_V\, g(\theta)\, d\theta \leq -\gamma|w|_W^2$ for all $w \in$ dom D.

REFERENCES

[**BF**] Burns, J. A., and Fabiano, R. F., *Modeling and approximation for a viscoelastic control problem*, in "Distributed Parameter Systems," Lecture Notes in Control and Information Sciences, Springer, 1987, pp. 23–39.

[**BFW**] Banks, H. T., Fabiano, R. H., and Wang, Y., *Estimation of Boltzmann damping coefficients in beam models*, LCDS/CCS Report No. 88-13. (July 1988), Brown University, Providence.

[**Da1**] Dafermos, C. M., *Asymptotic stability in viscoelasticity*, Archive for Rational Mechanics and Analysis **37** (1970), 297–308.

[**Da2**] Dafermos, C. M., *An abstract Volterra equation with applications to linear viscoelasticity*, J. Differential Equations, **7** (1970), 554–569.

[**Da3**] Dafermos, C. M., *Contraction semigroups and trend to equilibrium in continuum mechanics*, in "Applications of Methods of Functional Analysis to Problems in Mechanics," Lecture Notes in Mathematics, Springer, 1976, pp. 295–306.

[**DM**] Desch, W. and Miller, R. K., *Exponential stabilization of Volterra integrodifferential equations in Hilbert space*, J. Differential Equations **70** (1987), 366–389.

[**FI**] Fabiano, R. H. and Ito, K., *Semigroup theory and numerical approximation for equations in linear viscoelasticity*, LCDS/CCS Report No. 88-12. (July 1988), Brown University, Providence.

[**HRW**] Hannsgen, K. B., Renardy, Y., and Wheeler, R. L., *Effectiveness and robustness with respect to time delays of boundary feedback stabilization in one-dimensional viscoelasticity*, SIAM J. Control Optim. **26** (1988), 1200–1234.

[**HW**] Hannsgen, K. B. and Wheeler, R. L., *Time delays and boundary feedback stabilization in one-dimensional viscoelasticity*, in "Distributed Parameter Systems," Lecture Notes in Control and Information Sciences, Springer, 1987, pp. 136–152.

[**MW**] MacCamy, R. C. and Wong, J. S. W., *Stability theorems for some functional differential equations*, Trans. A.M.S. **164** (1972), 1–37.

[Pa] Pazy, A., "Semigroups of Linear Operators and Applications to Partial Differential Equations," Springer, New York, 1983.

[Pr] Prüss, J., *On the spectrum of C_o- semigroups*, Trans. Am. Math. Soc. **284** (1984), 847–857.

[R] Russell, D. L., *On mathematical models for the elastic beam with frequency proportional damping*, in "Control and Estimation in Distributed Parameter Systems," SIAM Frontiers in Applied Math (to appear).

[T] Tanabe, H., "Equations of Evolution," Pitman, London, 1979.

[Wa] Walker, J. A., "Dynamical Systems and Evolution Equations," Plenum Press, New York, 1980.

R.H. Fabiano and K. Ito
Center for Control Sciences
Division of Applied Mathematics
Brown University
Providence, RI 02912
USA

International Series of
Numerical Mathematics, Vol. 91
© 1989 Birkhäuser Verlag Basel

Constancy of the Hamiltonian
in infinite dimensional control problems

H.O. Fattorini*

Department of Mathematics
University of California

Abstract. We consider optimal control problems for systems described by the abstract autonomous quasilinear abstract parabolic differential equation

$$y'(t) = Ay(t) + f(y(t), u(t))$$

with an integral cost functional. We show that the maximum principle always includes constancy of the Hamiltonian. For "large" target sets, the Hamiltonian is identically zero for free time problems.

Keywords. Pontryagin's maximum principle, Hamiltonian, Hamiltonian equations, optimal controls.

1980 *Mathematics subject classifications*: 93E20, 93E25

1. Introduction

Consider an autonomous ordinary differential control system

$$(1.1) \qquad y'(t) = f(y(t), u(t)) \qquad (t \geq 0)$$

with control constraint

$$(1.2) \qquad u(t) \in U,$$

cost functional

$$(1.3) \qquad y_0(t, u) = \int_0^t f_0(y(s, u), u(s)) ds,$$

and initial and target condition

$$(1.4) \qquad y(0, u) = y^0, \quad y(\bar{t}, u) = \bar{y}.$$

respectively. We denote by $y(t, u)$ the trajectory of (1.1) corresponding to the control $u(\cdot)$; the initial condition y^0 is fixed). The time $\bar{t}$ may or may not be fixed in advance. Pontryagin's maximum principle for this control problem (see [10]) may be stated as follows. Define $\tilde{y} = (y_0, y)$, $\tilde{f}(y, u) = (f_0(y, u), f(y, u))$, $\tilde{z} = (z_0, u)$ (all vectors in $\mathbf{R} \times \mathbf{R}^m$) and let

$$(1.5) \qquad H(\tilde{y}, \tilde{z}, u) = \langle \tilde{z}, \tilde{f}(y, u) \rangle = z_0 f_0(y, u) + \langle z, f(y, u) \rangle$$

*This work was supported in part by the NSF under grant DMS 87-01877

be the Hamiltonian of the system. The Hamiltonian equations are

(1.6) $\tilde{y}'(t, u) = \text{grad}_z H(\tilde{y}(t, u), \tilde{z}(t, u), u(t)),$

(1.7) $\tilde{z}'(t, u) = -\text{grad}_z H(\tilde{y}(t, u), \tilde{z}(t, u), u(t)).$

Equation (1.6) is just a restatement of (1.1) plus the cost functional equation (1.3). Under suitable smoothness hypotheses on all functions involved, and assuming that an optimal control $\bar{u}(t)$ exists, Pontryagin's maximum principle holds: if $y(t, \bar{u})$ is the optimal trajectory then there exists a vector $\tilde{z} = (z_0, z) \in \mathbf{R}^{m+1}$, $(z_0, u) \neq 0$, $z_0 \leq 0$ such that the function $\tilde{y}(t, \bar{u}) = (y_0(t, \bar{u}), y(t, \bar{u}))$ and the solution $\tilde{z}(t, \bar{u})$ of (1.7) corresponding to the optimal control $\bar{u}(t)$ satisfying the final condition

(1.8) $\tilde{z}(\bar{t}) = \tilde{z} = (z_0, z)$

satisfy

(1.9) $H(\tilde{y}(t, \bar{u}), \tilde{z}(t, \bar{u}), \bar{u}(t)) = \max_{v \in U} H(\tilde{y}(t, \bar{u}), \tilde{z}(t, \bar{u}), v)$

almost everywhere in $0 \leq t \leq \bar{t}$. Moreover, we have

(1.10) $H(\tilde{y}(t, \bar{u}), \tilde{z}(t, \bar{u}), \bar{u}(t)) = \text{constant a.e. in} \quad 0 \leq t \leq \bar{t}.$

The stronger result

(1.11) $H(\tilde{y}(t, \bar{u}), \tilde{z}(t, \bar{u}), \bar{u}(t)) = 0 \text{ a.e. in} \quad 0 \leq t \leq \bar{t}.$

holds when the terminal time $\bar{t}$ is not fixed.

  We consider in this paper a generalization of (1.10) and (1.11) to systems described by certain quasilinear equations in Hilbert spaces

(1.12) $y'(t) = Ay(t) + f(y(t), u(t)),$

where A is the infinitesimal generator of a strongly continuos semigroup $S(t)$ and $f(t, u)$ is a suitable nonlinear term (hypotheses are precised in §2). The set-up for the maximum principle is that of [3], [4], [5], [8], [9]. Certain technical difficulties appear in the infinite dimensional case, beginning with the definition of the Hamiltonian,

(1.13) $H(\tilde{y}, \tilde{z}, u) = \langle \tilde{z}, \tilde{A}, \tilde{y} + \tilde{f}(y, u) \rangle = z_0 f_0(y, u) + \langle z, Ay + f(y, u) \rangle$

where $\tilde{A}$ is the operator in $\mathbf{R} \times E$ defined by $\tilde{A}(y_0, y) = (0, Ay)$, with domain $D(\tilde{A}) = \mathbf{R} \times D(A)$. In (1.13), y should be restricted to $D(A)$,)or z should be restricted to $D(A^*)$). However, there exist systems (notably those described by hyperbolic equations, see [7]) where it is not necessarily true that $y(t, u) \in D(A)$ or that $z(t, u) \in D(A^*)$ anywhere, so that an expression like (1.13) may not make sense. Thus our extension of (1.10) and (1.11) will only be valid for certain systems. Note that these technical difficulties do not appear in the generalization of the "dependence–on–v" part (1.9) of the maximum principle, since expressions such as $\langle z(t, \bar{u}), Ay(t, \bar{u}) \rangle$ are independent of v and can thus be eliminated from both sides. See [5] for details.

2. Existence and constancy of the Hamiltonian

By definition, a *solution* of the initial problem

$$(2.1) \qquad y'(t) = Ay(t) + f(y(t), u(t)) \qquad (0 \le t \le T),$$
$$(2.2) \qquad y(0) = y^0$$

is a continuous solution of the integrated version

$$(2.3) \qquad y(t) = S(t)y^0 + \int_0^t S(t - \sigma)f(y(\sigma), u(\sigma))d\sigma \qquad (0 \le t \le T),$$

where $S(\cdot)$ is the semigroup generated by A. We assume that the control set U is bounded and that

(F) $f(y, u)$ has a Fréchet derivative $\partial_y f(y, u)$ with respect to y and f (resp. $\partial_y f$) is continuous (resp. strongly continuous) and bounded on bounded subsets of $E \times U$.

Under these hypotheses, (2.3) can be solved by successive approximations in some interval $0 \le t \le T_1 < T$. If we can prove an *a priori* bound

$$(2.4) \qquad \|y(t)\| \le C \qquad (0 \le t \le T_0)$$

for any solution of (2.3) in an arbitrary interval $[0, T_0]$, where C does not depend on T_0 or on $u \in W(0, T; U)$, then (2.1) - (2.2) (or rather (2.3)) has a unique solution in $0 \le t \le T$. How to prove (2.4) depends on the particular equation under study: see [3] for quasilinear parabolic equations and [4] for the quasilinear hyperbolic case.

We note that the assumptions on f are also satisfied by an arbitrary translation, thus we may assume that $0 \in \rho(A)$, so that A^{-1} exists and is bounded.

We shall consider a cost functional of the form (1.3). The assumptions on f_0 are similar to those for f:

(F_0) $f_0(y, u)$ has a Fréchet derivative $\partial_y f_0(y, u)$ with respect to y and f_0 (resp. $\partial_y f_0$) is continuous (resp. strongly continuous) and bounded on bounded subsets of $E \times U$.

We begin by showing that, for certain systems defined by (2.1) the Hamiltonian (1.13) can be defined.

Let $g(t)$ be a continuous E-valued function. We say that $g(t)$ is *almost everywhere differentiable* in $0 \le t \le T$ with derivative $g'(\cdot) = h(\cdot)$ if there exists $h(\cdot) \in L^1(0, T; E)$ such that

$$g(t) = \int_0^t h(s)ds + g(0)$$

a.e. in $0 \le t \le T$. Of course, the definition implies that the derivative (defined in the usual way) exists almost everywhere and equals $h(\cdot)$.

In finite dimensional spaces, $g(\cdot)$ is just an absolutely continuous vector function.

Lemma 2.1. *Assume that, either (a) $y^0 \in D(A)$, $f(y, u) \in D(A)$ and $Af(y, u)$ is continuous in y and u, or (b) $y^0 \in D(A)$ and A generates an analytic semigroup $S(t)$. Then* *every solution of (2.3) is a.e. differentiable, belongs to $D(A)$ and satisfies (2.1) almost everywhere. Furthermore, we have*

$$(2.5) \qquad\qquad y'(\cdot), Ay(\cdot) \in L^p(0, T; E)$$

for $1 \le p \le \infty$ in case (a), for $1 \le p < \infty$ in case (b).

Under assumptions (a), $Af(y(\cdot), u(\cdot))$ is strongly measurable for $u(\cdot)$ strongly measurable and $y(\cdot)$ continuous. The same is true of $f(y, u)$ because of (F) thus the result is classical. Under assumptions (b), $f(y(\cdot), u(\cdot))$ is strongly measurable and bounded and the result follows from [1] (see also [3] for comments and other uses of the main result in [1]).

The Hamiltonian equations are again (1.6) and (1.7). The "second coordinate" of (1.6) is (2.1). The first coordinate is

$$(2.6) \qquad\qquad y_0'(t, u) = f_0(y(t, u), u(t))$$

which restates (1.3). Since the Hamiltonian does not depend of y_0, the first coordinate of equation (1.7) is $z_0'(t, u) = 0$, which simply means that z_0 is constant. The second coordinate is

$$(2.7) \qquad z'(t) = -(A^* + \partial_y f(y(t, u), u(t))^*)z(t) - z_0 \partial_y f_0(y(t, u), u(t)).$$

We note that the Cauchy problem for this equation is well posed backwards rather thatn forward (due to the negative sign on A^*). Solutions of (2.7) in $0 \le t \le \bar{t}$ corresponding to an arbitrary *final* condition

$$(2.8) \qquad\qquad z(\bar{t}) = z$$

are defined in the same way as solutions of (2.1) - (2.2). It is important to known when $z(t, u)$ is a genuine (rather than a weak) solution of (2.7). This is the case under the assumptions in the following result, which is an analogue of Lemma 2.1.

Lemma 2.2. *Assume that, either (a) $z \in D(A^*)$, $\partial_y f(y, u)^* E \in D(A^*)$ and $A^* \partial_y f(y, u)^*$ is strongly continuous in y and u, and that $\partial_y f_0(y, u) \in D(A^*)$ and $A^* \partial_y f_0(y, u)$ is continuous, or that (b) $z \in D(A^*)$ and A^* (or, equivalently, A) generates an analytic semigroup. Then every solution of (2.7) is a.e. differentiable, belongs to $D(A)$ and satisfies (2.7) almost everywhere. Furthermore, we have*

$$(2.9) \qquad\qquad z'(\cdot), A^* z(\cdot) \in L^p(0, T)$$

for $1 \le p \le \infty$ in case (a), for $1 \le p < \infty$ in case (b).

The proof is essentially the same as that of Lemma 2.1. We shall assume from now on without explicit mention that, besides (F) and (F_0), the assumptions of both Lemma 2.1 and Lemma 2.2 are satisfied: in particular, this will be the case if $y^0 \in D(A)$, $z \in D(A^*)$ and A generates an analytic semigroup.

We study below the differentiability of the Hamiltonian.

Lemma 2.3. Let $y(\cdot)$ (resp. $z(\cdot)$) be an arbitrary solution of (2.1) (resp. (2.7)). Then the function $\langle z(\cdot), Ay(\cdot)\rangle = \langle A^*z(\cdot), y(\cdot)\rangle$ is absolutely continuous in $0 \le t \le T$ with derivative

$$(2.10) \qquad \langle z(t), Ay(t)\rangle' = \langle z'(t), Ay(t)\rangle + \langle A^*z(t), y'(t)\rangle$$

Proof: Let $R_\mu = \mu R(\mu; A)$. Under the hypotheses in force the function $\langle z(t), R_\mu Ay(t)\rangle$ is absolutely continuous in $0 \le t \le T$ with derivative $\langle z(t), R_\mu Ay(t)\rangle' = \langle z'(t), AR_\mu y(t)\rangle + \langle A^*(R_\mu)^*z(t), y'(t)\langle$ in the set (of full measure) where $y(t)$ and $z(t)$ are both differentiable. In other words,

$$(2.11) \qquad \begin{aligned} \langle z(t), R_\mu Ay(t)\rangle = &\int_0^t \{\langle z'(s), R_\mu Ay(s)\rangle + \langle (R_\mu)^*A^*z(s), y'(s)\rangle\}ds \\ &+ \langle z(0), R_\mu Ay_0\rangle \qquad (0 \le t \le T). \end{aligned}$$

We use now that $\|R_\mu\|$ is bounded and that $R_\mu u \to u$, $(R_\mu)^*u \to u$ $(u \in E)$ as $\mu \to +\infty$ and that $y'(\cdot)$, $Ay(\cdot)$, $z'(\cdot)$ and $A^*z(\cdot)$ belong to $L^2(0,T;E)$ ((2.5) and (2.9)). This, combined with the dominated convergence theorem yields

$$(2.12) \qquad \begin{aligned} \langle z(t), Ay(t)\rangle = &\int_0^t \{\langle z'(s), Ay(s)\rangle + \langle A^*z(s), y'(s)\rangle\}ds \\ &+ \langle z(0), Ay_0\rangle \qquad (0 \le t \le T). \end{aligned}$$

This ends the proof of Lemma 2.3.

When A generates an analytic semigroup, the conclusions of Lemma 2.1, Lemma 2.2 and Lemma 2.3 can be suitably extended to the case where y^0 and z are arbitrary elements of E; it suffices to apply the results in intervals $\epsilon < t < T$ where $\epsilon > 0$.

We say that τ is a *regular point* of a solution $y(\cdot)$ of (2.1) if $y'(t)$ exists at $t = \tau$, $y(\tau) \in D(A)$ and (2.1) is satisfied at τ. Same definition for (2.7).

Lemma 2.4. Let $y(t,u)$ (resp. $z(t,u)$) be a solution of (2.1) (resp. 2.7)) corresponding to the same control $u(t)$. Let τ be a regular point of both $y(t,u)$ and $z(t,u)$ and such that $y_0'(\tau, u)$ exists and satisfies (2.6) and (2.10) holds. Then, if $D_t = d/dt$, we have

$$(2.13) \qquad D_t\big|_{t=\tau} H(z(t,u), y(t,u), u(t)) = 0$$

Proof: Under the assumptions on τ, $\langle z(t,u), Ay(t,u)\rangle$ is differentiable at $t = \tau$ with derivative $D_t\big|_{t=\tau}\langle z(t,u), Ay(t,u)\rangle = \langle z'(\tau, u), Ay(\tau, u)\rangle + \langle A^*z(\tau, z), y'(\tau, u)\rangle$. Accordingly,

$$(2.14) \qquad \begin{aligned} D_t\big|_{t=\tau} H&(y(t,u), z(t,u), u_0) = z_0\langle \partial_y f_0(y(\tau, u), u(\tau)), y'(\tau, u)\rangle \\ &+ \langle z'(\tau, u), Ay(\tau, u) + f(y(\tau, u), u(\tau))\rangle \\ &+ \langle A^*z(\tau, u), y'(\tau, u)\rangle + \langle z(\tau, u), \partial_y f(y(\tau, u), u(\tau))y'(\tau, u)\rangle \\ &= z_0\langle \partial_y f_0(y(\tau, u), u(\tau)), Ay(\tau, u) + f(y(\tau, u), u(\tau))\rangle \\ &+ \langle -(A^* + \partial_y f(y(\tau, u), u(\tau))^*)z(\tau, u), Ay(\tau, u) + f(y(\tau, u), u(\tau))\rangle \\ &+ \langle -z_0\partial_y f_0(y(\tau, u), u(\tau)), Ay(\tau, u) + f(y(\tau, u), u(\tau))\rangle \\ &+ \langle A^*z(\tau, u), Ay(\tau, u) + f(y(\tau, u), u(\tau))\rangle \\ &+ \langle z(\tau, u), \partial_y f(y(\tau, u), u(\tau))(Ay(\tau, u) + f(y(\tau, u), u(\tau)))\rangle = 0. \end{aligned}$$

This ends the proof of Lemma 2.4.

Lemma 2.5. *Let X be a metric space, U an arbitrary set,*

(2.15) $g(\xi, u) : X \times U \to \mathbf{R}$

a function such that

(2.16) $g(\xi) = \sup_{u \in U} g(\xi, u) < \infty$

for all $\xi \in X$, and let $g(\xi, u)$ be Lipschitz continuous in ξ uniformly with respect to u,

(2.17) $|g(\xi', u) - g(\xi, u)| \le K d(\xi', \xi)$ $(\xi, \xi' \in X, u \in U)$.

Then $g(\xi)$ is Lipschitz continuous in ξ (with the same constant K).

Proof: Let ξ, ξ' be elements of X. Let u, u' be elements of U such that $g(\xi, u) \ge g(\xi) - \epsilon d(\xi, \xi')$, $g(\xi', u') \ge g(\xi') - \epsilon d(\xi, \xi')$. We have $g(\xi, u') \le g(\xi)$, $g(\xi', u) \le g(\xi')$, so that

$$-K d(\xi, \xi') \le g(\xi', u') \le g(\xi) - g(\xi') + \epsilon d(\xi, \xi')$$

and

$$g(\xi) - g(\xi') - \epsilon d(\xi, \xi') \le g(\xi, u) - g(\xi', u) \le K d(\xi, \xi').$$

Since $\epsilon > 0$ is arbitrary, the result follows.

We prove (1.10) below. Constancy of the Hamiltonian does not depend on the control $\bar{u}(t)$ being an actual optimal control, but only on the fact that the maximum principle (1.9) is satisfied.

Theorem 2.6 (Constancy of the Hamiltonian). *Assume that the conditions on Lemma 2.1 hold, and let $\bar{u}(t)$ be an optimal control, or more generally a control satisfying the maximum principle (1.9). Then*

(2.18) $H(\tilde{y}(t, \bar{u}), \tilde{z}(t, \bar{u}), \bar{u}(t)) = constant$ $(0 \le t \le T)$.

Proof: Excising the unbounded term, we may write (1.9) in the following way:

(2.19) $\langle \tilde{z}(t, u), f(\tilde{y}(t, u), \bar{u}(t)) \rangle = \max_{v \in U} \langle \tilde{z}(t, \bar{u}), f(\tilde{y}(t, \bar{u}), v) \rangle$

almost everywhere in $0 \le t \le \bar{t}$. We use Lemma 2.5 with $\xi = (z, y) \in X$, where X is a bounded subset of $E \times E$ and $g(\xi, u) = g(z, y, u) = \langle z, f(y, u) \rangle$; obviously, $g(\xi, u)$ is Lipschitz continuous in ξ uniformly with respect to $u \in E$. It follows then that

$$g(z, y) = \max_{u \in U} \langle z, f(y, u) \rangle$$

is Lipschitz continuous in (z, y), that is, $|g(z', y') - g(z, y)| \le K(\|z' - z\| + \|y' - y\|)$. Accordingly,

$$|\langle z(t', \bar{u}), f(y(t', \bar{u}), \bar{u}(t')) \rangle - \langle z(t, \bar{u}), f(y(t, \bar{u}), \bar{u}(t)) \rangle|$$
$$\le K(\|z(t', \bar{u}) - z(t, \bar{u})\| + \|y(t', \bar{u}) - y(t, \bar{u})\|)$$

which shows that $\langle z(\cdot, \bar{u}), f(y(\cdot, \bar{u}), \bar{u}(\cdot)) \rangle$ (thus, in view of Lemma 2.4, the Hamiltonian $H(\tilde{y}(\cdot, \bar{u}), \tilde{z}(\cdot, \bar{u}), \bar{u}(\cdot))$ is absolutely continuous in $0 \le t \le \bar{t}$. Let τ be a point where both terms of $H(\tilde{y}(\cdot; \bar{u}), \tilde{z}(\cdot, \bar{u}), \bar{u}(\cdot))$ are differentiable. For $t > \tau$ we have

$$(t - \tau)^{-1} \{ H(\tilde{y}(t, \bar{u}), \tilde{z}(t, \bar{u}), \bar{u}(t)) - H(\tilde{y}(\tau, \bar{u}), \tilde{z}(\tau, \bar{u}, \bar{u}(\tau)) \}$$
$$(t - \tau)^{-1} \{ H(\tilde{y}(t, \bar{u}), \tilde{z}(t, \bar{u}), \bar{u}(t)) - H(\tilde{y}(\tau, \bar{u}), \tilde{z}(\tau, \bar{u}, \bar{u}(\tau)) \}$$

so that, by Lemma 2.3, $D_t|_{t=\tau} H(\tilde{y}(t, \bar{u}), \tilde{z}(t, \bar{u}), \bar{u}(t)) \ge 0$. Arguing in the same way with $t < \tau$ we deduce $D_t|_{t=\tau} H(\tilde{y}(t, \bar{u}), \tilde{z}(t, \bar{u}), \bar{u}(t)) \ge 0$. This ends the proof of Theorem 2.5.

3. The maximum principle in infinite dimensional spaces.

The results in §2 on constancy of the Hamiltonian depend only on the maximum principle (1.9) and not on the way used to show it. In contrast, to show (1.11), some details of the underlying theory are necessary.

A version of the maximum principle for control problems for input–output systems $u(\cdot) \to y(\cdot, u)$ has been established in [3], [4], [5]; the target condition may be of the form (1.4) or, more generally

$$(3.1) \qquad\qquad\qquad\qquad y(\bar{t}, u) \in Y$$

where Y is a subset of E. This approach to the maximum principle has been decisively generalized in [8], [9], where various necessary assumptions in [5] are removed, and, more significantly, the problem under study is generalized (and simplified) to an *infinite dimensional nonlinear programming problem*. We reproduce below (Theorem 3.1) the (particular case of) the main result [9] as it applies to the present situation. To do this, a few definitions are necessary.

Let $\{\Delta_n, n = 1, 2, \ldots\}$ be a sequence of sets in the Hilbert space E. Following [5] we say that $\{\Delta_n\}$ has *finite codimension* in E if and only if there exists a subspace H with $\dim H^{bot} < \infty$ such that

$$\bigcap_{n \geq 1} \Pi_H(\overline{\text{conv}}(\Delta_n))$$

has nonempty interior in H, where Π_H is the projection of E into H.

We recall the notion of *contingent cone* $K_Y(\bar{y})$ to a set Y at $\bar{y} \in Y$; $w \in E$ belong to $K_Y(\bar{y})$ if and only if there exists a sequence $\{h_k\}$ of positive numbers with $h_k \to 0$ and a sequence $\{y_k\} \subset Y$ with

$$(y_k - \bar{y})/h_k \to w \text{ as } k \to \infty$$

(see [2] for additional details).

Let $\bar{u}(\cdot)$ be an optimal control, $\bar{t}$ the time (or one of the times) at which the optimal trajectory $y(t; \bar{u})$ hits Y optimally. The hitting point $y(\bar{t}, \bar{u} =\in Y$ will be called $\bar{y}$. We denote by $W(0, \bar{t}; U)$ the *control space* of all admissible controls (that is, strongly measurable functions in $0 \leq t \leq \bar{t}$ satisfying the control constraint (1.2)). Let $\tilde{u}$ be an arbitrary control in $W(0, \bar{t}; U)$. Consider the control system

$$(3.2) \qquad\qquad z'(s) = (A + \partial_y f(s, y(s, \tilde{u}), \tilde{u}(s)))z(s) + v(s) \qquad (0 \leq s \leq \bar{t})$$
$$(3.3) \qquad\qquad z(0) = 0$$

(which we may call the *linearized system* associated with (2.1) - (2.2)). The control space for (3.2) - (3.3) consists of all functions of the form

$$(3.4) \qquad\qquad v(s) = f(s, y(s, \tilde{u}), u(s)) - f(s, y(s, \tilde{u}), \tilde{u}(s))$$

withe $u(\cdot) \in W(0, \bar{t}; U)$. The *reachable set* of all elements of the form $z(\bar{t})$, $z(\cdot)$ a solution of (3.2) - (3.3) for some v of the form (3.4) will be called $R(\bar{t}; \tilde{u})$.

In the result below, we assume that the control space $W(0, \bar{t}; U)$ is endowed with the distance

$$(3.5) \qquad\qquad d(u(\cdot), v(\cdot)) = \text{meas}\,\{t; 0 \leq t \leq \bar{t}; u(t) \neq v(t)\}.$$

The target set Y is assumed to be closed.

Theorem 3.1. *Assume that for every sequence $\{u^n\} \subset W(0, \bar{t}; U)$ with $d(u^n, \bar{u}) \to 0$ fast enough and for every sequence $\{y^n\} \subset Y$ with $y^n \to \bar{y}$ fast enough there exists $\rho > 0$ such that the sequence*

$$(3.6) \qquad \{\Delta_n\} = \{\overline{\mathrm{conv}}\,(K_Y(y^n) \cap B(0, \rho)) - \overline{\mathrm{conv}}\,R(\bar{t}; u^n)\}$$

has finite codimension in E, where $B(0, \rho)$ is the ball of center 0 and radius ρ. Then there exists a vector $(z_0, z) \in \mathbf{R} \times E$ such that

$$(3.7) \qquad\qquad\qquad (z_0, z) \neq 0, \quad z_0 \leq 0$$

such that the maximum principle (1.9) is satified, where H is defined by (1.13), $\tilde{y}(t, u) = (y_0(t, \bar{u}), y(t, \bar{u}))$ and $\tilde{z}(t, \bar{u}) = (z_0, z(t))$, with $z(\cdot)$ the solution of (2.7) with final condition (2.8).

Remark 3.2: In all known cases where condition (3.6) is satisfied (see [9]), one of the sequences

$$(3.8) \qquad\qquad \{\Delta_n^t\} = \{\overline{\mathrm{conv}}\,(N_Y(y^n) \cap B(0, \rho))\}$$
$$(3.9) \qquad\qquad \{\Delta_n^c\} = \{\overline{\mathrm{conv}}\,R(\bar{t}; u^n)\}$$

has finite codimension (which clearly implies that (3.6) has finite codimension).

Finite codimension of (3.8) is clearly a *controllability* condition for the linearized system (3.2) - (3.3). This kind of property is hard to come by and will *never* be satisfied under the conditions of Lemma 2.2; in fact, if (a) holds then $R(\bar{t}, \tilde{u}) \in D(A)$, while if (b) is satisfied then $R(\bar{t}, \tilde{u} \in D((-A)^\alpha)$ for all $\alpha < 1$. Thus, we shall directly assume that (3.8) is of finite codimension. This limits us to "large" target sets; for instance, (3.8) will be satisfied for a ball in a subspace of finite codimension but not when Y reduces to a point.

4. Vanishing of the Hamiltonian

The maximum principle, as proved in [3], [4], [9] does not take advantage of the fact that the terminal time $\bar{t}$ is not fixed. To prove (1.11) we shall use a "distortion–of–time" trick due to Clarke [2, p. 153]. Here (in contrast with §2) we have to assume that $\bar{u}(t)$ is an actual optimal control in the interval $0 \leq t \leq \bar{t}$ rather than just an extremal control (that is, a control satisfying the maximum principle (1.9)).

Associated with (2.1) – (2.2) we consider the auxiliary control system

$$(4.1) \qquad\qquad y'(\tau) = (1 + u_0(\tau))Ay(\tau) + (1 + u_0(\tau))f(y(\tau), u(\tau))$$
$$(4.2) \qquad\qquad y(0) = y^0$$

in the optimal interval $0 \leq \tau \leq \bar{t}$. The controls here are of the form $\tilde{u}(\tau) = (u_0(\tau), u(\tau))$, where $u_0(\cdot)$ is a measurable scalar function with $|u_0(\tau)| \leq \alpha$ a.e. in $0 \leq \tau \bar{t}$ ($\alpha < 1$ fixed) and $u(\cdot) \in W(0, \bar{t}; U)$; in other words, the control set for this system is

$$(4.3) \qquad\qquad \tilde{u} = [-\alpha, \alpha] \times U \subseteq \mathbf{R} \times F.$$

Treatment of the initial value problem (4.1) - (4.2) is essentially the same as that of (2.1) - (2.2), with $S(t - \sigma)$ in the integral equation (2.3) replaced by the solution operator $S(\tau, \sigma) = S(t(\tau) - t(\sigma))$ of the homogeneous equation $u'(\tau) = (1 + u_0(\tau))Au(\tau)$, where

$$(4.4) \qquad t(\tau) = \int_0^\tau (1 + u_0(\sigma))d\sigma \qquad (0 \leq \tau \leq \bar{t}).$$

Accordingly, the basic integral equation is

$$(4.5) \quad y(\tau) = S(t(\tau))y^0 + \int_0^\tau S(t(\tau) - t(\sigma))(1 + u_0(\sigma))f(y(\sigma), u(\sigma))d\sigma \qquad (0 \leq \tau \leq \bar{t})$$

or, in terms of the inverse function $\tau(t)$ which is Lipschitz continuous with derivative $\tau'(t) = (1 + u_0(t))^{-1}$ a.e. in $0 \leq t \leq t(\bar{t})$,

$$(4.6) \qquad y(\tau(t)) = S(t)y^0 + \int_0^t S(t - \eta)f(y(\tau(\eta)), u(\tau(\eta)))d\eta \qquad (0 \leq \tau \leq t(\bar{t})).$$

The treatment of (4.5) or of its equivalent version (4.6) is the same as that of (2.1) thus we omit the details. We use the notation $y(t, \tilde{u}) = y(t, u_0, u)$ to indicate the trajectory corresponding to the control $\tilde{u} = (u_0, u)$. We consider the optimal control problem for (4.1) - (4.2) with cost functional

$$(4.7) \qquad \begin{aligned} y_0(\tau, u_0, u) &= \int_0^\tau (1 + u_0(\sigma))f_0(y(\sigma), u(\sigma))d\sigma \\ &= \int_0^{t(\tau)} f_0(y(\tau(\eta)), u(\tau(\eta)))d\eta, \end{aligned}$$

target condition (3.1) and *fixed time* $\tau = \bar{t}$. We shall identify the new optimal control problems as (P^*); the original problem will be called (P).

Lemma 4.1. Let $\bar{u}(\cdot)$ be an optimal control for (P) in $0 \leq t \leq \bar{t}$. Then $(0, \bar{u}(t))$ is an optimal control for (P^*).

Proof: Assume this is not the case. Let $(\hat{u}_0(\cdot), \hat{u}(\cdot))$ be an admissible control for (P^*) such that the target condition (3.1) is satisfied and such that $y_0(\bar{t}, \hat{u}_0(\cdot), \hat{u}(\cdot)) < y_0(\bar{t}, 0, \bar{u}(\cdot)) = m$. Define $t(\tau)$ from $\hat{u}_0(\cdot)$ by (4.4) and let $u(t) = \hat{u}(\tau(t))$ in $0 \leq t \leq t(\bar{t})$. Obviously, $u(\cdot)$ is an admissible control for (P) whose corresponding trajectory satisfies the target condition (3.1); also, $y_0(t\bar{t}, u(\cdot)) = y_0(\bar{t}, \hat{u}_0(\cdot), \hat{u}(\cdot)) < y_0(\bar{t}, 0, \bar{u}(\cdot))$, which contradicts the free-time optimality of the control $\bar{u}(\cdot)$ for problem (P). This ends the proof.

In what follows, we shall use the maximum principle for the control system (4.1) – (4.2). Although systems of this type are not explicitly treated in [5] or [9], the theory can be extended easily. The linearized system corresponding to $\tilde{u}(\cdot) = (\hat{u}_0(\cdot), \hat{u}(\cdot)) \in W(0, \bar{t}; \tilde{U})$ is

$$(4.8) \qquad \tilde{z}' = \{(1 + \hat{u}_0(s))(A + \partial_y f(s, y(s, \hat{u}_0, \hat{u}), \hat{u}(s)))\}z(s) + \tilde{v}(s),$$
$$(4.9) \qquad \tilde{z}(0) = 0$$

where now the control space consists of all functions of the form

$$\tilde{v}(s) = (u_0(s) - \hat{u}_0(s))Ay(s, \hat{u}_0, \hat{u})$$
$$+(1 + u_0(s))f(s, y(s, \hat{u}_0, \hat{u}), u(s)) - (1 + \hat{u}_0(s))f(s, y(s, \hat{u}_0, \hat{u}), \hat{u}(s)).$$

(4.10)

Again, we denote by $R(\bar{t}, \hat{u}_0, \hat{u})$ the reachable set of (4.8) – (4.9). It is possible to show that the assumptions of Theorem 3.1 will hold in this case if they hold for (2.1) – (2.2). However, this is not necessary since (due to the reasons explained in Remark 3.2) we shall use the sequence (3.8) instead of (3.6).

Theorem 4.2. *Assume that the sequence (3.8) is of finite codimension in E for some $\rho > 0$. Let $(\bar{u}_0(\cdot), \bar{u}(\cdot)) \in W(0, \bar{t}; \tilde{U})$ be an optimal control for (4.1) – (4.2). Then the maximum principle (1.9) holds.*

In the case of interest to us, the optimal control will be of the form $(0, \bar{u}(\cdot))$. Thus, the function $z(t, \bar{u})$ in the maximum principle obeys the same equation (2.7) pertaining to the system (2.1) – (2.2).

We write the maximum principle in Hamiltonian form. We have

$$\tilde{H}(\tilde{y}, \tilde{z}, \tilde{u}) = \langle \tilde{z}, (1 + u_0)\tilde{A}\tilde{y} + (1 + u_0)\tilde{f}(y, \tilde{u}) \rangle$$
$$= z_0(1 + u_0)f_0(y, u) + \langle z, (1 + u_0)(Ay + f(y, u)) \rangle$$
$$= (1 + u_0)\{z_0 f_0(y, u) + \langle z, Ay + f(y, u) \rangle\} = (1 + u_0)H(\tilde{y}, \tilde{z}, u).$$

We know that $(0, \bar{u}(\cdot))$ is an optimal control for problem (P^*) in the interval $0 \le \tau \le \bar{t}$. For this optimal control we have $u_0 = 0$, so that $\tilde{H}(\tilde{y}, \tilde{z}, \tilde{u}) = H(\tilde{y}, \tilde{z}, \tilde{u})$ and the Hamiltonian equations are exactly the same examined at the beginning of §2. Accordingly, we have

$$H(\tilde{y}(t, \bar{u}), \tilde{z}(t, \bar{u}), \bar{u}(t))$$
$$= \max_{v \in U, |u_0| \le \alpha} (1 + u_0)H(y(t, \bar{u}), \tilde{z}(t, \bar{u}), v).$$

Setting $v = \bar{u}(t)$ on the right hand side we obtain the desired result:

Theorem 4.3. *Let $\bar{u}(t)$ be an optimal control for problem (P), and let the assumptions of Theorem 4.2 be satisfied. Then there exists a vector $(z_0, z) \in \mathbf{R} \times E$ such that*

$$(z_0, z) \neq 0, \quad z_0 \le 0$$

such that the maximum principle (1.9) is satisfied, where H is defined by (1.13), $\tilde{y}(t, u) = (y_0(t, \bar{u}), y(t, \bar{u}))$ and $\tilde{z}(t, \bar{u}) = (z_0, z(t, \bar{u}))$, with $z(\cdot, \bar{u})$ the solution of (2.7) with final condition (2.8). Moreover, (1.11) holds.

REFERENCES

[1] De Simon, L., *Un'applicazione della teoria degli integrali singolari allo studio delle equazioni differenziali astratte del primo ordine*, Rend. Sem. Mat. Univ. Padova **34** (1964), 205–223.
[2] Clarke, F., "Optimization and nonsmooth analysis," Wiley-Interscience, New York, 1983.

[3] Fattorini, H.O., *The maximum principle for nonlinear nonconvex systems in infinite dimensional spaces*, in "Lecture Notes in Control and Information Sciences, vol. 75, Proceedings of 2nd Vorau Conference on Distributed Parameter Systems (July 1964)," Springer, 1986, pp. 162–178.

[4] Fattorini, H.O., *The maximum principle for nonlinear nonconvex systems with set targets*, in "Proceedings of the 24th IEEE Conference on Decision and Control," Fort Lauderdale, December 1985.

[5] Fattorini, H.O., *A unified theory of necessary conditions for nonlinear nonconvex systems*, Appl. Math. Optim. **15** (1987), 141–185.

[6] Fattorini, H.O., *Optimal control of nonlinear systems: convergence of suboptimal controls I*, Proceedings of Special Session on Operator Methods for Optimal Control Problems, Annual AMS Meeting, New Orleans, January 1986, in "Lecture Notes in Pure and Applied Mathematics, Vol. 108," Marcel Dekker, New York, 1987, pp. 159–199.

[7] Fattorini, H.O., *Optimal control of nonlinear systems: convergence of suboptimal controls II*, Proceedings of the IFIP Working Conference on Control Problems for Systems Describing by Partial Differential Equations and Applications, Gainesville, February 1986, in "Lecture Notes in Control and Information Sciences, Vol. 97," Springer, 1987, pp. 230–246.

[8] Fattorini, H.O. and Frankowska, H., *Necessary conditions for infinite dimensional control problems*, in "Proceedings of 8th International Conference on Analysis and Optimization of Systems." Antibes-Juan Les Pins, June 1988

[9] Fattorini, H.O. and Frankowska, H., *Necessary conditions for infinite dimensional control problems*.

[10] Pontryagin, L.S., Boltyanskii, V.G., Gamkrelidze, R.V. and Mischenko, E.F., "The Mathematical Theory of Optimal Processes," (Russian), Goztekhizdat, Moscow, 1961. English translation: Interscience, New York, 1962

H.O. Fattorini
Department of Mathematics
University of California
Los Angeles, CA 90024
USA

International Series of
Numerical Mathematics, Vol. 91
© 1989 Birkhäuser Verlag Basel

Identification of objects of extreme conductivity
by boundary measurements

Avner Friedman and Michael Vogelius

Institute for Mathematics and its Applications
University of Minnesota

Department of Mathematics
University of Maryland

Keywords. Inverse problems, conductivity imaging, cracks

1980 *Mathematics subject classifications*: 35R30, 35J25

1. Introduction

The problem of how to determine internal conductivity profiles from boundary measurements has recently received considerable attention in the literature. The practical applications range from non-destructive testing, [7], to groundwater seepage monitoring, [18], and medical tomography, [2,8]. It is usually assumed that the (static, direct current) voltage potentials satisfy the differential equation

$$(1) \qquad \nabla \cdot (\gamma(x)\nabla u) = 0 \quad \text{in} \quad \Omega \subset \mathbf{R}^n, \quad n \geq 2,$$

where $\gamma(x)$ is the positive, real-valued conductivity, to be determined. The information, based upon which it is sought to determine $\gamma(x)$ consists of knowledge of currents $\gamma \frac{\partial u}{\partial \nu}$ and corresponding voltage potentials u at the boundary, $\partial\Omega$.

Let Λ_γ denote the linear operator from $H^{1/2}(\partial\Omega)$ to $H^{-1/2}(\partial\Omega)$, which takes Dirichlet-to Neumann-data:

$$\Lambda_\gamma(\phi) = \gamma \frac{\partial u}{\partial \nu}|_{\partial\Omega} \quad \text{with} \quad \nabla \cdot (\gamma \nabla u) = 0 \quad \text{in} \quad \Omega, \quad u = \phi \quad \text{on} \quad \partial\Omega.$$

Complete knowledge of Λ_γ is known to determine the function γ uniquely under quite general assumptions (cf. [9,10,11,13,16,17]). It is also known that the mapping $\Lambda_\gamma \to \gamma$ has a modulus of continuity of logarithmic type, provided γ is a priori known to belong to a bounded set in some Sobolev space, [1]. From a practical point of view results like these have the disadvantage that they require complete knowledge of the linear operator Λ_γ; furthermore an estimate of logarithmic type is somewhat disappointing as far as continuous dependence is concerned. Results like these do not explain the apparent practical success of various numerical algorithms to recover γ from only partial knowledge of Λ_γ (cf. [2,12,14,19,20]). To help obtain a better understanding of the basis for such algorithms it becomes relevant to analyze interesting classes of conductivities, where the spatial variation is further restricted, [3,4]. In particular this applies to classes that are not of a simple discrete parameter type, but which nonetheless permit identification through finitely many measurements and which have associated continuous dependence estimates of a Lipschitz- or a Hölder-type. Two challenging theoretical questions then become:

(i) exactly how many measurements are needed to identify conductivities within a certain class, and

(ii) for what characteristics associated with the particular type of conductivity does one obtain the best form of continuous dependence.

We have recently studied these questions in connection with inhomogeneities of extreme conductivity imbedded in a known reference conductor. We have studied two different kinds of geometric situations corresponding to

(i) finitely many small, well separated n-dimensional inhomogeneities in an n-dimensional reference medium, cf. [5] (mines).

(ii) an inhomogeneity in the form of a curve in a two dimensional reference medium, cf. [6] (a crack).

It turns out that there is a significant qualitative difference between the results we obtain for these two cases. The first case requires only measurements of one current and its corresponding voltage potential on the boundary, and it is possible to prove asymptotic continuous dependence estimates of Lipschitz-type for various characteristics of the inhomogeneities, such as location, relative size or orientation. In the second case two boundary measurements are necessary and sufficient for identification, and it is much more difficult to obtain continuous dependence estimates for the location or shape of the crack. We can prove a partial Lipschitz estimate when the crack is a priori known to be a line segment. In the following two sections we give a brief review of our results; for the case of a crack we furthermore provide an outline of the proof of the identifiability theorem.

2. Small inhomogeneities

We assume that each inhomogeneity has the form $z_k + \epsilon \rho_k Q_k B$, where ϵ and ρ_k are positive numbers, Q_k is a rotation around the origin and B is a bounded domain in $\mathbf{R}^n$ with

$$0 \in B \quad \text{and} \quad \partial B \text{ of type } C^{2+\beta}, \text{ for some } 0 < \beta < 1.$$

The points $\{z_k\}_{k=1}^K$ belong to Ω and satisfy:

$$|z_k - z_j| \geq d > 0 \quad, \quad \forall j \neq k, \text{ and}$$
$$\text{dist}(z_k, \partial\Omega) \geq d > 0 \quad, \quad \forall k.$$

The parameter ϵ determines the common, small length scale of the inhomogeneities and the ρ_k,

$$d \leq \rho_k \leq D$$

determine their relative size. Let $\omega_\epsilon = \cup_{k=1}^K (z_k + \epsilon \rho_k Q_k B)$ denote the entire collection of inhomogeneities. The voltage potential u_ϵ satisfies

(2) $$\nabla \cdot (\gamma \nabla u_\epsilon) = 0 \text{ in } \Omega \setminus \omega_\epsilon, \quad \gamma \frac{\partial u_\epsilon}{\partial \nu} = \psi \text{ on } \partial\Omega,$$

with the constraints

(3.a) $u_\epsilon = constant$ in $z_k + \epsilon\rho_k Q_k B$, for inhomogeneities that have infinite conductivity.

(3.b) $\gamma\frac{\partial u_\epsilon}{\partial \nu} = 0$ on $\partial(z_k + \epsilon\rho_k Q_k B)$, for inhomogeneities that have conductivity zero.

The constraints (3.a) and (3.b) represent the limiting cases of the equation (1) when $\gamma = +\infty$ or $\gamma = 0$ in $z_k + \epsilon\rho_k Q_k B$ respectively. The function $\gamma \in C^{2+\beta}$ here (and in the following) denotes the known reference conductivity. The inverse problem is to infer information about the the collection of inhomogeneities $\omega_\epsilon = \cup_{k=1}^K (z_k + \epsilon\rho_k Q_k B)$ from knowledge of $u_\epsilon|_{\partial\Omega}$ corresponding to one fixed boundary current ψ.

A main component of our analysis is the derivation of an asymptotic formula for the voltage potential u_ϵ. We shall not here give any details of this derivation, but only state that for the case where all inhomogeneities have infinite conductivity

(4)
$$u_\epsilon(y) = -\epsilon^n \sum_{k=1}^K \rho_k^n \gamma(z_k)\nabla_z N(z_k, y) \cdot Q_k A Q_k^T \nabla U(z_k)$$
$$+ U(y) + \epsilon^n \eta(\epsilon, y, \{Q_k\}, \{\rho_k\}, \{z_k\}),$$

with

$$\eta(\epsilon, y, \{Q_k\}, \{\rho_k\}, \{z_k\}) \to 0 \text{ as } \epsilon \to 0, \text{ uniformly in } y, \{Q_k\}, \{\rho_k\}, \text{ and } \{z_k\},$$

provided $\text{dist}(y, z_k) \geq c > 0$, $1 \leq k \leq K$. Here N is the so-called Neumann function, i.e., $N(\cdot, y)$ solves

$$-\nabla \cdot (\gamma\nabla N) = \delta_y \quad \text{in } \Omega$$
$$\gamma\frac{\partial N}{\partial \nu} = -\frac{1}{|\partial\Omega|} \quad \text{on } \partial\Omega$$

with

$$\int_{\partial\Omega} N ds = 0.$$

The positive definite symmetric matrix A, which occurs in (4) is the so-called "polarization tensor"; it depends on B only, and it models the effect of a single inhomogeneity (in the shape of B) imbedded in a uniform electric field, cf. [15]. U is the reference voltage potential, it depends on the known reference conductivity γ and the prescribed boundary current ψ through the solution of

$$\nabla \cdot (\gamma\nabla U) = 0 \quad \text{in } \Omega,$$
$$\gamma\frac{\partial U}{\partial \nu} = \psi \quad \text{on } \partial\Omega.$$

The functions u_ϵ and U are normalized by $\int_{\partial\Omega} u_\epsilon ds = \int_{\partial\Omega} U ds = 0$. A similar formula holds for inhomogeneities with conductivity zero. For a more detailed discussion of these formulae and the tensor A we refer the reader to [5]. We shall assume that the reference voltage potential U has no stationary points, i.e.,

$$\nabla U(x) \neq 0 \text{ for all } x \in \Omega.$$

In the remainder of this paper Γ always denotes a non-empty open subset of $\partial\Omega$. The formula (4) immediately leads to the following asymptotic "identifiability result" in the case when all the rotations, Q_k, are known apriori (for simplicity assume $Q_k = id$, $1 \leq k \leq K$).

Corollary 1. *Consider the voltage potentials u_ϵ and u'_ϵ corresponding to inhomogeneities $\omega_\epsilon = \cup_{k=1}^{K}(z_k + \epsilon\rho_k B)$ and $\omega'_\epsilon = \cup_{k=1}^{K'}(z'_k + \epsilon\rho'_k B)$ for fixed z_k, ρ_k, z'_k, ρ'_k and variable ϵ. If $\gamma\frac{\partial u_\epsilon}{\partial\nu} = \gamma\frac{\partial u'_\epsilon}{\partial\nu} = \psi$ on $\partial\Omega$ and $\lim_{\epsilon\to 0}\epsilon^{-n}\|u_\epsilon - u'_\epsilon\|_{L^\infty(\Gamma)} = 0$, then $K = K'$ and, after appropriate reordering $z_k = z'_k$ and $\rho_k = \rho'_k$, $1 \leq k \leq K$*

The corollary, as formulated above depends on knowledge of the limiting behaviour of u_ϵ and u'_ϵ as ϵ approaches zero; it may be alternately formulated in terms of knowledge of only two potentials, but then the exact locations and relative sizes are of course only determined asymptotically in ϵ.

Corollary 1'. *Given any function $\delta(\epsilon)$, $\lim_{\epsilon\to 0}\delta(\epsilon) = 0$, there exist a constant ϵ_0 and a function $\mu(\epsilon)$, $\lim_{\epsilon\to 0}\mu(\epsilon) = 0$, such that if u_ϵ and u'_ϵ, $0 < \epsilon < \epsilon_0$ are any two voltage potentials corresponding to inhomogeneities $\omega_\epsilon = \cup_{k=1}^{K}(z_k + \epsilon\rho_k B)$ and $\omega'_\epsilon = \cup_{k=1}^{K'}(z'_k + \epsilon\rho'_k B)$ respectively, with*

(5.a) $\gamma\frac{\partial u_\epsilon}{\partial\nu} = \gamma\frac{\partial u'_\epsilon}{\partial\nu} = \psi$ on $\partial\Omega$, and
(5.b) $\epsilon^{-n}\|u_\epsilon - u'_\epsilon\|_{L^\infty(\Gamma)} < \delta(\epsilon)$,

then

(5.c) $K = K'$ and, after appropriate reordering,
(5.d) $|z_k - z'_k| + |\rho_k - \rho'_k| \leq \mu(\epsilon)$, $1 \leq k \leq K$.

The constant ϵ_0 and the function μ depend on δ, d, D, Ω, Γ, B, γ and ψ, but are otherwise independent of the two sets of inhomogeneities.

Outline of proof of Corollary 1': Suppose the conclusions (5.c) and (5.d) are not true. Then there must exist a function $\delta_0(\epsilon)$, $\lim_{\epsilon\to 0}\delta_0(\epsilon) = 0$, and sequences K_ϵ, K'_ϵ, $z_{k,\epsilon}$, $z'_{k,\epsilon}$, $\rho_{k,\epsilon}$ and $\rho'_{k,\epsilon}$ indexed by $\epsilon \to 0$ such that either $|K_\epsilon - K'_\epsilon| \geq 1$ or $\sum |z_{k,\epsilon} - z'_{k,\epsilon}| + |\rho_{k,\epsilon} - \rho'_{k,\epsilon}| \geq c_0 > 0$ independent of ordering (and ϵ). By extraction of a subsequence we obtain that $K_\epsilon = K$, $K'_\epsilon = K'$, $z_{k,\epsilon} \to z_k$, $z'_{k,\epsilon} \to z'_k$, $\rho_{k,\epsilon} \to \rho_k$, $\rho'_{k,\epsilon} \to \rho'_k$ and

(6)   either $K \neq K'$ or $\sum |z_k - z'_k| + |\rho_k - \rho'_k| \geq c_0 > 0$ independent of ordering.

Because of the assumptions (5.a), (5.b) and the fact that $\lim_{\epsilon\to 0}\delta_0(\epsilon) = 0$ it now follows immediately from the expansion (4) (and unique continuation) that

$$\sum_{k=1}^{K}\rho_k^n\gamma(z_k)\nabla_z N(z_k,y) \cdot A\nabla U(z_k)$$

$$=\sum_{k=1}^{K'}(\rho'_k)^n\gamma(z'_k)\nabla_z N(z'_k,y) \cdot A\nabla U(z'_k),$$

and this contradicts the statement (6). ∎

It is slightly more difficult to determine the rotations Q_k, in the case when the relative sizes, ρ_k, are a priori known; if A is completely isotropic or Q_k is a rotation in a subspace on which A is isotropic, then

$$Q_k A Q_k^T \nabla U(z_k)$$

is independent of Q_k, and we cannot expect any result about identifiability of Q_k based on boundary measurements of

(7)
$$\sum_{k=1}^{K} \rho_k^n \gamma(z_k) \nabla_z N(z_k, y) \cdot Q_k A Q_k^T \nabla U(z_k).$$

We need an extra non-degeneracy condition. Such a condition may be formulated in full generality, but for simplicity of presentation we consider here only $n = 2$, in which case the relevant condition is that A not be isotropic.

Corollary 2. *Assume $n = 2$ and that A is not isotropic. Consider the voltage potentials u_ϵ and u'_ϵ corresponding to inhomogeneities $\omega_\epsilon = \cup_{k=1}^{K}(z_k + \epsilon Q_k B)$ and $\omega'_\epsilon = \cup_{k=1}^{K'}(z'_k + \epsilon Q'_k B)$ for fixed z_k, z'_k, Q_k, Q'_k and variable ϵ. If $\gamma \frac{\partial u_\epsilon}{\partial \nu} = \gamma \frac{\partial u'_\epsilon}{\partial \nu} = \psi$ on $\partial\Omega$ and $\lim_{\epsilon \to 0} \epsilon^{-2}\|u_\epsilon - u'_\epsilon\|_{L^\infty(\Gamma)} = 0$, then $K = K'$ and, after appropriate reordering, $z_k = z'_k$ and $Q_k = \pm Q'_k$, $1 \leq k \leq K$*

Corollary 2 may be formulated in terms of measurements of only two voltage potentials in a way analogous to Corollary 1'.

It is easy to see that we cannot asymptotically identify $\{\rho_k\}$ and $\{Q_k\}$ at the same time: indeed a simple calculation with positive definite symmetric 2×2 matrices shows that

> if A is not isotropic, and Q is any rotation $\neq \pm id$ then there
>
> exist $x \neq 0, \rho > 1$ such that $QAQ^T x = \rho A x$.

Given any $y \neq 0$ we may find a rotation Q' and a positive number λ such that $x = \lambda (Q')^T y$, and therefore $\lambda Q' Q A Q^T (Q')^T y = \lambda \rho Q' A (Q')^T y$. We consequently conclude that

> if A is not isotropic, Q is any rotation $\neq \pm id$ and y is any vector, then there
>
> exist a rotation Q' and a scalar $\rho > 1$ such that $Q' Q A (Q'Q)^T y = \rho Q' A (Q')^T y$.

This makes it obvious that $\{\rho_k\}$ and $\{Q_k\}$ may not be determined simultaneously from knowledge of boundary values of the expression (7) for one fixed boundary current ψ.

Since we consider here only the case $n = 2$, any rotation Q_k is uniquely characterized by a single angle, θ_k. The real strength of the formula (4) is not brought out in the asymptotic results of the previous corollaries. By considering the Fréchet derivative of (7) with respect to the parameters $\{z_k\}$, $\{\rho_k\}$ and $\{\theta_k\}$ it is possible to obtain two much stronger continuous dependence results. One of these results provide a Lipschitz estimate for the deviation in $\{z_k\}$ and $\{\rho_k\}$, with fixed $\{\theta_k\}$, in terms of the deviation in a single boundary voltage measurement. The other result provides a Lipschitz estimate for the deviation in $\{z_k\}$ and $\{\theta_k\}$, with fixed $\{\rho_k\}$, in terms of the deviation in a single boundary voltage measurement. From the previous discussion it is clear that it is not possible to obtain a Lipschitz estimate for all three quantities simultaneously, in terms of a single boundary voltage measurement. The result with fixed $\{\rho_k\} = \{1\}$ states

Theorem 1. *Assume $n = 2$ and that A is not isotropic. There exists constants δ_0, ϵ_0 and a function $\eta(\epsilon)$, $\lim_{\epsilon \to 0} \eta(\epsilon) = 0$, such that if u_ϵ and u'_ϵ, $0 < \epsilon < \epsilon_0$ are any two voltage potentials corresponding to inhomogeneities $\omega_\epsilon = \cup_{k=1}^{K}(z_k + \epsilon Q_k B)$ and $\omega'_\epsilon = \cup_{k=1}^{K'}(z'_k + \epsilon Q'_k B)$ respectively, with*

(8.a) $\gamma \frac{\partial u_\epsilon}{\partial \nu} = \gamma \frac{\partial u'_\epsilon}{\partial \nu} = \psi$ on $\partial\Omega$, and
(8.b) $\epsilon^{-2} \| u_\epsilon - u'_\epsilon \|_{L^\infty(\Gamma)} < \delta_0$,

then

(8.c) $K = K'$ and, after approppriate reordering,
(8.d) $|z_k - z'_k| + \min_{j \in \mathbf{Z}} |\theta_k - \theta'_k + j\pi| \leq C\epsilon^{-2} \| u_\epsilon - u'_\epsilon \|_{L^\infty(\Gamma)} + \eta(\epsilon), \; 1 \leq k \leq K$.

The constants ϵ_0, δ_0 and C and the function η depend on d, D, Ω, Γ, B, γ and ψ, but are otherwise independent of the two sets of inhomogeneities.

We note that A is isotropic for other than those domains which are invariant under all rotations (balls): if B, with appropriate choice of origin, is invariant under $Q(\theta_0)$ for just a single angle $0 < \theta_0 < 2\pi$, $\theta_0 \neq \pi$, then it follows from a simple computation with 2×2 matrices that the matrix A is isotropic. As a consequence Corollary 2 and Theorem 1, for example, do not apply if the inhomogeneities are in the shape of squares.

2. Cracks

Consider the case of a non self-intersecting C^2 curve, σ, of infinite or zero conductivity imbedded in a known two dimensional reference conductor (σ is the image of a twice continuously differentiable map $\sigma(t)$ from an interval $[a, b]$, $a \neq b$, into Ω and $\sigma(s) \neq \sigma(t) \; \forall \; s \neq t$). The voltage potential satisfies

$$(9) \qquad\qquad \nabla \cdot (\gamma \nabla u_\sigma) = 0 \text{ in } \Omega \setminus \sigma, \quad \gamma \frac{\partial u_\sigma}{\partial \nu} = \psi \text{ on } \partial\Omega$$

with the constraint

(10.a) $u_\sigma = constant$ on σ, if σ has infinite conductivity.
(10.b) $\gamma \frac{\partial u_\sigma}{\partial \nu} = 0$ on σ, if σ has conductivity zero (this is usually referred to as a crack).

The inverse problem is to determine the shape and location of σ from measurements of $u_\sigma|_\Gamma$. The constraints (10.a) and (10.b) are equivalent by duality; the only difference is that a fixed Neumann- condition on $\partial\Omega$ translates into a fixed Dirichlet-condition on $\partial\Omega$ and vice versa (and the reference conductivity becomes γ^{-1}). We shall state and give an outline of the proof of an identifiability result in the case of a curve of infinite conductivity and with fixed Neumann condition on $\partial\Omega$. This result is qualitatively different from the results in the previous section since it requires two sets of measurements.

To see that one measurement is not sufficient to determine σ, let U denote the solution to

$$\nabla \cdot (\gamma \nabla U) = 0 \text{ in } \Omega, \quad \gamma \frac{\partial U}{\partial \nu} = \psi \text{ on } \partial\Omega.$$

Let σ be any curve entirely lying on a level curve for U, then it is clear that $U = u_\sigma$ (since σ is perfectly conducting). Thus, for any fixed boundary current ψ, there is a continuum of curves which produces the same measurement for $u_\sigma|_{\partial\Omega}$.

We now proceed to construct two currents $\psi^{(1)}$ and $\psi^{(2)}$ which together with the corresponding voltage potentials $u_\sigma^{(1)}|_\Gamma$ and $u_\sigma^{(2)}|_\Gamma$ are sufficient to uniquely determine any perfectly conducting C^2 curve σ.

Let P and Q be two points on $\partial\Omega$. As the first boundary current we take

$$\psi^{(1)} = \begin{cases} 1, & \text{on } \widehat{PQ}; \\[2mm] -\dfrac{|\widehat{PQ}|}{|\widehat{QP}|}, & \text{on } \widehat{QP}. \end{cases}$$

As the second boundary current we take

$$\psi^{(2)} = \delta_P - \delta_Q,$$

where δ denotes a Dirac-delta function. We let $u_\sigma^{(i)}$, $i = 1, 2$ denote the solution to (9) with the constraint (10.a) on σ. We note that the solution corresponding to $\psi^{(2)}$ is not of finite energy; it is possible to construct L^∞ counterparts to $\psi^{(2)}$ such that the identifiability theorem remains true (for curves inside compact subsets of Ω, cf. [6]).

Theorem 2. *Assume that γ and $\partial\Omega$ are real analytic. Let σ and σ' be any two C^2 curves in Ω which do not self-intersect. If $u_\sigma^{(i)} = u_{\sigma'}^{(i)}$, $i = 1, 2$ on $\Gamma \subset \partial\Omega$ then $\sigma = \sigma'$.*

We do allow the curves considered here to be empty (formally if $a > b$), therefore Theorem 1.1 also asserts that the presence of a curve can be detected by two boundary measurements.

By duality this theorem also shows that it is possible to identify a curve of conductivity zero using two particular fixed boundary voltage potentials (the indefinite integrals of $\psi^{(1)}$ and $\psi^{(2)}$). It quite easy to use ideas similar to those presented here to construct two boundary potentials $\phi^{(1)}$ and $\phi^{(2)}$ so that the equivalent of Theorem 2 holds with these boundary potentials fixed and measurements consisting of $\gamma\frac{\partial u_\sigma}{\partial\nu}|_\Gamma$ (for a σ with infinite conductivity), cf. [6]. For the proof of the latter theorem it is sufficient to assume that $\partial\Omega$ is C^2. By duality this theorem also takes care of the case when σ has conductivity zero and fixed boundary currents are prescribed on $\partial\Omega$.

Outline of proof of Theorem 2, cf. [6]: Let O denote the open set separated from $\partial\Omega$ by the union of the two curves σ and σ'. A simple argument of unique continuation gives that $u_\sigma^{(i)} = u_{\sigma'}^{(i)}$ in $\Omega \setminus (O \cup \sigma \cup \sigma')$, $i = 1, 2$, and that the $u_\sigma^{(i)}$ and $u_{\sigma'}^{(i)}$, $i = 1, 2$ are all piecewise constant on ∂O (which is made up of parts of σ and σ'). Since each of these functions is continuous it follows that $u_\sigma^{(1)}$, say, is constant on ∂O. The constant value is the same as taken on by $u_\sigma^{(1)}$ on σ. It follows now immediately from the maximum principle that $u_\sigma^{(1)}$ is constant on O, and consequently that O is empty. We therefore get that $u_\sigma^{(i)} = u_{\sigma'}^{(i)}$ in $\Omega \setminus (\sigma \cup \sigma')$ and since each is a continuous function it follows that $u_\sigma^{(i)} = u_{\sigma'}^{(i)}$, $i = 1, 2$ in all of Ω.

The proof proceeds by contradiction. Assume that $\sigma \neq \sigma'$. Without loss of generality we may assume that there is a curve $\rho \subset \sigma'$, which does not intersect σ. The functions $u_\sigma^{(i)}$, $i = 1, 2$ are constant on ρ and analytic in a neighborhood of ρ. By an analytic continuation argument we may extend the curve ρ at both ends so that $u_\sigma^{(i)}$, $i = 1, 2$, are simultaneously constant on the extended curve ρ' (special consideration has to be given to stationary points of $u_\sigma^{(i)}$, cf. [6]). The continuation process can be repeated until the extended curve reaches $\partial\Omega$ or σ (it cannot self-intersect). The maximally extended curve cannot have both endpoints on σ, since this would contradict the maximum principle; it must therefore have at least have one endpoint, x_0, on $\partial\Omega$, the other endpoint may be on $\partial\Omega$ or σ.

We now examine the point x_0 more closely, using information about the boundary data $\psi^{(i)}$, $i = 1, 2$. It is not difficult to see that $|u_\sigma^{(2)}(x)| \to +\infty$ as $x \to P($ or $Q)$; thus the curve ρ' (an equipotential curve for $u_\sigma^{(2)}$) cannot meet $\partial\Omega$ at P or Q, i.e., $x_0 \neq P$ and Q.

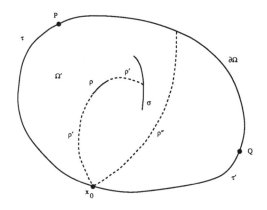

Figure 1.

Since $\gamma \frac{\partial u_\sigma^{(1)}}{\partial \nu} = \psi^{(1)}$ has value $+1$ or $-|\widehat{PQ}|/|\widehat{QP}|$ at x_0, and since ρ' is also an equipotential curve for $u_\sigma^{(1)}$ it follows that ρ' does not meet $\partial\Omega$ at an angle of $\pi/2$ at x_0. Since $\gamma \frac{\partial u_\sigma^{(2)}}{\partial \nu}(x_0) = \psi^{(2)}(x_0) = 0$ it now follows that the entire gradient, $\nabla u_\sigma^{(2)}(x_0)$, vanishes, i.e. x_0 is a stationary point for $u_\sigma^{(2)}$. Due to the stationarity of the point x_0, the fact that ρ' is not normal to $\partial\Omega$ and well known facts about the nodal structure of analytic, γ-harmonic functions it now follows that there is another curve $\rho'' \subset \Omega$ which meets $\partial\Omega$ at x_0, along which $u_\sigma^{(2)}$ is constant and which is not tangential to ρ' or $\partial\Omega$ at x_0 (this is the only place where we need that the boundary $\partial\Omega$ is analytic). The curve ρ'' may be extended until it hits σ or $\partial\Omega$ at the other end (it cannot self-intersect). The curves ρ' and ρ'' cannot intersect in Ω or both have their second endpoint on σ, since this would violate the maximum principle. The situation is therefore schematically as shown in figure 1 (the other endpoint of ρ' could also lie on $\partial\Omega$, but this is a somewhat simpler situation, which we leave to the reader). As shown in figure 1 the points P and Q must lie on the two different curves τ and

τ' that ρ'' separates $\partial\Omega$ into, otherwise we would get that $u_\sigma^{(2)}$ is constant in an open subset of Ω, an obvious contradiction. Consider now the domain Ω' with boundary $\tau \cup \rho' \cup \sigma \cup \rho''$, as shown in figure 1. Let c_0 denote the common constant value $u_\sigma^{(2)}$ attains on ρ'', ρ' and σ. Based on knowledge of the nodal structure of $u_\sigma^{(2)}$ near x_0, it is not difficult to see that $u_\sigma^{(2)}$ takes values below as well as above c_0 in a Ω'-neighborhood of x_0. The minimum of $u_\sigma^{(2)}$ in Ω' is thus strictly less than c_0. From the strong maximum principle we get that the minimum of $u_\sigma^{(2)}$ cannot be attained on τ (where $\frac{\partial u_\sigma^{(2)}}{\partial\nu}$ is non-negative); the minimum must therefore be attained on ρ'', ρ' or σ, in which case the value is c_0 – a contradiction. If the point Q had been on τ we would have considered the maximum of $u_\sigma^{(2)}$ instead. We conclude that our initial assumption $\sigma \neq \sigma'$ is not valid. $\blacksquare$

The question of continuous dependence of the location and shape of σ on the measurements $u_\sigma^{(i)}|_\Gamma$ is quite difficult, and at this point we only have partial results. Based on the uniqueness result and a compactness argument it is easy to see that if $\sigma_k \subset K \subset\subset \Omega$ are bounded in $C^{2+\alpha}[0,1]$, have lengths that are bounded away from zero and are bounded away from self-intersection then

$$\sum_{i=1}^{2} \|u_\sigma^{(i)} - u_{\sigma_k}^{(i)}\|_{L^\infty(\Gamma)} \to 0 \text{ implies that } \sigma_k \to \sigma(\text{in } C^2).$$

The crucial and difficult problem is to estimate the distance between σ and σ' in terms of the difference in the corresponding boundary voltages (through either a Lipschitz- or a Hölder-estimate).

Consider the special case that the curves are known to be line segments $\subset K \subset\subset \Omega$ (with lengths bounded away from zero); in that case it is possible to show, [6], that

$$\text{dist}_\Omega(l_\sigma, l_{\sigma'}) \leq C \sum_{i=1}^{2} \|\frac{\partial}{\partial s}(u_\sigma^{(i)} - u_{\sigma'}^{(i)})\|_{L^\infty(\Gamma)},$$

where l_σ and $l_{\sigma'}$ denote the entire lines generated by the two line segments σ and σ' respectively, and

$$\text{dist}_\Omega(l_\sigma, l_{\sigma'}) = \max_{x \in l_\sigma \cap \bar{\Omega}} \min_{y \in l_{\sigma'} \cap \bar{\Omega}} |x - y|.$$

$\frac{\partial}{\partial s}$ denotes the tangential derivative along $\partial\Omega$. We do not obtain a Lipschitz estimate for the location of the crack σ along the line l_σ, i.e., we cannot very well estimate the length of the crack or the size of translations parallel to the crack itself.

Acknowledgements. The first author is partially supported by National Science Foundation Grant DMS–8722187; the second author is partially supported by National Science Foundation Grant DMS–8601490 and by ONR Contract N–00014–85–K0169.

REFERENCES

[1] G. Alessandrini, *Stable determination of conductivity by boundary measurements*, Applicable Analysis **27** (1988), 153–172.

[2] D.C. Barber and B.H. Brown, *Recent developments in applied potential tomography- APT*, in "Information Processing in Medical Imaging," S. Bacharach ed., pp. 106–121. Nijhoff, 1986.

[3] H. Bellout and A. Friedman, *Identification problem in potential theory*, Arch. Rat. Mech. Anal. **101** (1988), 143–160.

[4] A. Friedman, *Detection of mines by electric measurements*, SIAM J. Appl. Math. **47** (1987), 201–212.

[5] A. Friedman and M. Vogelius, *Identification of small inhomogeneities of extreme conductivity by boundary measurements: a continuous dependence result*, Arch. Rat. Mech. Anal. (to appear).

[6] A. Friedman and M. Vogelius, *Determining cracks by boundary measurements*, IMA preprint 476, 1989.

[7] H. Fukue and K. Wada, *Application of electric resistance probe method to non-destructive inspection*, Mitsubishi Heavy Industries Technical Review **19** (1982), 83–94.

[8] D. Gisser, D. Isaacson and J. Newell, *Electric current computed tomography and eigenvalues*, SIAM J. Appl. Math. (to appear).

[9] V. Isakov, *On uniqueness of recovery of a discontinuous conductivity coefficient*, Comm. Pure Appl. Math. **41** (1988), 865–877.

[10] R. Kohn and M. Vogelius, *Determining conductivity by boundary measurements*, Comm. Pure Appl. Math. **37** (1984), 289–298.

[11] R. Kohn and M. Vogelius, *Determining conductivity by boundary measurements II. Interior results*, Comm. Pure Appl. Math. **38** (1985), 643–667.

[12] I.-J. Lee, *Computational experience with a variational method for electrical impedance tomography*, Tech. Note BN-1076, Univ. of Maryland, 1988.

[13] A. Nachman, *Reconstructions from boundary measurements*, Annals of Math. **128** (1988), 531–576.

[14] F. Santosa and M. Vogelius, *A backprojection algorithm for electrical impedance imaging*, SIAM J. Appl. Math. (to appear).

[15] M. Schiffer and G. Szegö, *Virtual mass and polarization*, Trans. Amer. Math. Soc. **67** (1949), 130–205.

[16] J. Sylvester and G. Uhlmann, *A uniqueness theorem for an inverse boundary value problem in electrical prospection*, Comm. Pure Appl. Math. **39** (1986), 91–112.

[17] J. Sylvester and G. Uhlmann, *A global Uniqueness theorem for an inverse boundary value problem*, Annals of Math. **125** (1987), 153–169.

[18] A. Wexler and C.J. Mandel, *An impedance computed tomography algorithm and system for ground water and hazardous waste imaging*, Paper presented to 2nd Annual Canadian/American Conf. on Hydrology, Banff, June 1985.

[19] A. Wexler, B. Fry and M. Neumann, *Impedance-computed tomography algorithm and system*, Applied Optics **24** (1985), 3985–3992.

[20] T.J. Yorkey, J.G. Webster and W.J. Tompkins, *Comparing reconstruction algorithms for electrical impedance tomography*, IEEE Trans. Biomedical Eng. **BME-34** (1987), 843–852.

Avner Friedman
Institute for Mathematics and its Applications
University of Minnesota
Minneapolis, Minnesota 55455
USA

Michael Vogelius
Department of Mathematics
University of Maryland
College Park, Maryland 20742
USA

International Series of
Numerical Mathematics, Vol. 91
© 1989 Birkhäuser Verlag Basel

Some results for an infinite horizon control problem governed by a semilinear state equation

Fausto Gozzi

Scuola Normale Superiore Pisa

Abstract. We consider an infinite horizon control problem governed by a semilinear state equation depending on a small parameter ε. We prove that, for ε sufficiently small, the associated Hamilton-Jacobi equation has a unique strict solution. We give some stability property and, consequently, we can solve the control problem by the classical dynamic programming method.

Keywords. Optimal control, infinite horizon, Hamilton-Jacobi equation.

1980 *Mathematics subject classifications*: 93C20, 49C10, 49C20

1. Introduction

We are concerned with a dynamical system governed by the following semilinear state equation:

(SE)
$$\begin{cases} y' = Ay + f(\varepsilon, y) + u & \text{on } [0, +\infty); \\ y(0) = x & x \in H; \end{cases}$$

where $A : D(A) \subset H \to H$ is the infinitesimal generator of a strongly continuous semigroup in a Hilbert space H and ε is a real parameter. f is a smooth function from $\mathbf{R} \times H$ to H which goes uniformly to 0 (with its derivatives) when ε goes to 0.

We consider the following infinite horizon optimal control problem [P]:

(1.1)
$$\begin{cases} \text{Minimize the functional:} \\ J(x, u) = \int_0^{+\infty} \left[g(y(s)) + \frac{1}{2} |u(s)|^2 \right] ds \\ \text{over all controls } u \in L^2(0, +\infty; U), \\ \text{where } y \text{ is subject to the state equation } (SE). \end{cases}$$

Here g is a smooth convex functions from H to $\mathbf{R}$.

We treat this problem by the dynamic programming approach (see [7] for instance) that is by studying the Hamilton-Jacobi equation associated to the finite horizon problem:

(HJ)
$$\begin{cases} \phi_t(t, x) = -\frac{1}{2} |\phi_x(t, x)|^2 + \langle Ax + f(\varepsilon, x), \phi_x(t, x) \rangle + g(x) = 0 \\ \qquad\qquad \forall (t, x) \in [0, +\infty) \times D(A) \\ \phi(0, x) = 0 \qquad \forall x \in H. \end{cases}$$

Then we pass to the limit for $t \to +\infty$. This lead us to study also the stationary Hamilton-Jacobi equation

(HJS) $\qquad 0 = -\dfrac{1}{2} |\phi'(x)|^2 + \langle Ax + f(\varepsilon, x), \phi'(x) \rangle + g(x) \qquad \forall x \in D(A),$

If we set $\varepsilon = 0$, $g(t, x) = \frac{1}{2}\langle Mx, x\rangle$, then problem [P] reduces to the well known linear regulator problem, which has been extensively studied (see for instance [10]). However, for applications, the use of a linear state equation is very restrictive.

Moreover, if g and ϕ_0 are general convex functions and $\varepsilon = 0$, then the cost J is a convex function on $L^2(0, +\infty; H)$ and equation (HJ) admits a unique strict (see section 4) solution on $[0, +\infty)$ (see [3]). We remark that if $\varepsilon \neq 0$ is sufficiently small then J is still convex but the method used in [4] only gives the existence of a local strict solution of equation (HJ) in some interval $[0, t]$.

The Hamilton-Jacobi equations in infinite dimensions (HJ) and (HJS) have been extensively studied in many recent works (see [5] and [6] for instance) using the approach of viscosity solutions. The results of existence and uniqueness of viscosity solutions are obtained under very general assumptions; however the viscosity solutions are not sufficiently regular for our pourposes.

As a matter of fact, to apply the classical dynamic programming approach to infinite horizon control problem it is necessary to study the limit of the solution of (HJ), that is $\phi(t, x)$, when $t \to \infty$; we call this limit $\phi^\infty(x)$. We need a result of existence and regularity of this limit; so we can study the closed loop equation for the infinite horizon problem

(1.2) $\qquad \begin{cases} y'(s) = Ay(s) + f(\varepsilon, y(s)) - \phi_x^\infty(y(s)) & \text{on } [0, +\infty); \\ y(0) = x & x \in H. \end{cases}$

It is also possible to study this problem by variational methods, in particular when the semigroup $\{e^{tA}\}_{t \geq 0}$ is compact (see [1] for finite horizon, and [3]). We obtain the existence of a feedback optimal control, however the solution ϕ is still not regular and the closed loop equation cannot be directly solved. For other results in this direction see [12] and [13].

In a previous paper (see [8]) we analized the finite horizon problem and gave, under suitable hypotheses, an existence theorem for (HJ). More precisely we showed that, for every given $R > 0$ there exists $\bar{\varepsilon}(R) > 0$ such that for $|\varepsilon| \leq \bar{\varepsilon}(R)$ the equation (HJ) has a unique global regular solution (in a sense which we will precise in section 4) ϕ_ε in $[0, T] \times \Sigma_R$, where Σ_R is the closed ball in H with radius R.

Using this result it is possible to solve the finite horizon problem by the classical dynamic programming method.

Our goal is now to show that the infinite horizon problem can also be solved by this method.

First we show that the existence result we proved for (HJ) holds true in the infinite horizon case as well; then we study the limit of the solution as $t \to \infty$. Under some stability hypothesis we can show the existence and regularity of this limit. This enable us to solve the infinite horizon control problem.

More precisely we fix an upper bound r_0 for the norm of the initial state x, and show that the closed loop equation (1.2) has a unique mild solution y_ε with $\sup_{t \in [0, +\infty)} |y_\varepsilon(t)| \leq r_0$, and that there exists a unique optimal control u_ε such that $\sup_{t \in [0, +\infty)} |u_\varepsilon(t)| \leq M(r_0)$ where M is a given function $(0, +\infty) \to (0, +\infty)$ depending only on the data.

We assume that g is strictly convex and sufficiently regular (2 times differentiable) and A is strictly dissipative (a stability hypothesis). For explanations see section 3.

To prove the solvability of (HJ) with infinite horizon we consider the approximating equation:

$(HJ)_\alpha$
$$\begin{cases} \phi_t(t,x) = -\frac{1}{\alpha}(\phi(t,x) - \phi_\alpha(t,x)) + \langle Ax + f(\varepsilon,x), \phi_x(t,x) \rangle + g(x) \\ \qquad\qquad \forall (t,x) \in [0,+\infty) \times D(A) \\ \phi(0,x) = 0 \qquad \forall x \in H \end{cases}$$

where ϕ_α is the convex regularization of ϕ (see next section).

When $\varepsilon = 0$, the solution ϕ_ε^α of $(HJ)_\alpha$ is convex and it is possible to show that ϕ_ε^α converges to the solution of equation (HJ) (see [3]). We prove that if ε is sufficiently small then the solution ϕ_ε^α exists globally and is convex. Thus it is possible to show the existence of the limit as $\alpha \to 0$.

2. Notations and preliminary results

We begin by specifying the notations which we will use throughout this paper. Let H be a Hilbert space with norm $|\cdot|_H$ and scalar product $\langle \cdot, \cdot \rangle_H$. We denote by $\mathcal{L}(H)$ the Banach algebra of the linear bounded operators from H into H.

Now let X, Y be two Hilbert spaces. If $f : X \to Y$ is a k-times Fréchet differentiable function, we set, for every $r > 0$:

$$(2.1) \qquad |f|_{h,r} = \sup\{|f^{(h)}(x)| : |x| \le r\} \quad, \quad h = 0,..,k,$$

$$(2.2) \qquad \|f\|_{k,r} = \sup\left\{ \frac{|f^{(k)}(x) - f^{(k)}(y)|}{|x-y|} : |x|,|y| \le r, x \ne y \right\} \quad, \quad h = 0,..,k.$$

Note that, if f is a $n+1$-times Fréchet differentiable function, we have:

$$(2.3) \qquad \|f\|_{n,r} = |f|_{n+1,r}.$$

For $k \in \mathbf{N}$ we define:

$$(2.4) \qquad C^k(X,Y) = \Big\{ f : X \to Y, \; k-\text{times continuously Fréchet differentiable:}$$
$$|f|_{h,r} < +\infty \quad \forall h = 0,..,k, \, \forall r > 0 \Big\},$$

$$(2.5) \qquad C_{Lip}^k(X,Y) = \Big\{ f \in C^k(X,Y) : \|f\|_{k,r} < +\infty, \, \forall r > 0 \Big\}.$$

If $Y = \mathbf{R}$ we write $C^k(X)$ instead of $C^k(X,\mathbf{R})$.

We remark that $C^k(X,Y)$ and $C_{Lip}^k(X,Y)$ are Fréchet spaces with the seminorms $\sum_{h=0}^k |f|_{h,r}$ and $\|f\|_{k,r} + \sum_{h=0}^k |f|_{h,r}$ ($r \in \mathbf{N}$) respectively, and we say that $f_n \to f$ in $C^k(X,Y)$ if $|f_n - f|_{i,r} \to 0$ for $i = 0,..,k$ and for every $r > 0$.

Now let $\phi : [0,+\infty) \times H \to \mathbf{R}$ and $k \ge 1$. We say that $\phi \in B\big([0,+\infty); C^k(H)\big)$ if ϕ satisfies the following conditions:

i) $\displaystyle\sup_{t \in [0,+\infty)} |\phi(t,\cdot)|_{h,r} < +\infty$ \quad for $h = 0,..,k \; \forall r > 0,$

ii) $\phi : [0,+\infty) \times H \to \mathbf{R}$ is continuous,

iii) $\phi_x : [0,+\infty) \times H \to H$ is continuous;

and, if $k \geq 2$:

iv) $\dfrac{\partial^h \phi}{\partial x^h}$ is strongly continuous for $h = 2, .., k$, that is the map:

$$[0, +\infty) \times H \to H$$

$$(t, x) \to \frac{\partial^h \phi}{\partial x^h}(y_1, .., y_{h-1})$$

is continuous $\forall (y_1, .., y_{h-1}) \in H^{h-1}$.

Analogously, we say that $\phi \in B\left([0, +\infty); C^k_{Lip}(H)\right)$ if $\phi \in B\left([0, +\infty); C^k(H)\right)$ and:

$$\sup_{t \in [0,+\infty)} \|\phi(t, \cdot)\|_{k,r} < +\infty \quad \forall r > 0.$$

Moreover we say that $\phi_n \to \phi$ in $B([0, +\infty); C^k(H))$ if:

$$(2.6) \qquad \sup_{t \in [0,+\infty)} |\phi(t, \cdot) - \phi_n(t, \cdot)|_{i,r} \to 0 \quad \text{for } i = 0, .., k; \quad \forall r > 0.$$

Finally, setting $\Sigma_r = \{x \in H, |x| \leq r\}$, we similarly define the spaces $C^k(\Sigma_r)$, $C^k_{Lip}(\Sigma_r)$, $B([0, +\infty); C^k(\Sigma_r))$, $B([0, +\infty); C^k_{Lip}(\Sigma_r))$, and we set $|\phi|_{C^k(\Sigma_r)} = \sum_{i=0}^{k} |\phi|_{i,r}$, and $|\phi|_{C^k_{Lip}(\Sigma_r)} = |\phi|_{C^k(\Sigma_r)} + \|\phi\|_{k,r}$, for $\phi \in C^k(\Sigma_r)$ and $\phi \in C^k_{Lip}(\Sigma_r)$, respectively.

Now we consider the regularization of a convex function (see [3]) and recall its fundamental properties which we will use in the following.

We denote by K (resp. K_R) the set of all convex functions $\phi \in C^1(H)$ (resp. $C^1(\Sigma_R)$) such that $\phi'(0) = 0$.

For $\phi \in K$ we set:

$$(2.7) \qquad \phi_\alpha(x) = \min\left\{\phi(y) + \frac{1}{2\alpha}|x - y|^2; \ y \in H\right\} \quad \alpha \in \mathbf{R} - \{0\}, x \in H.$$

We remark that the minimum exists and is unique, due to the convexity of ϕ.

Moreover setting:

$$(2.8) \qquad x_\alpha = (1 + \alpha\phi')^{-1}(x) \qquad \alpha \in \mathbf{R}, x \in H$$

we have:

$$(2.9) \qquad \phi_\alpha(x) = \phi(x_\alpha) + \frac{\alpha}{2}|\phi'(x_\alpha)|^2.$$

We collect now the properties of ϕ_α we need, which are proved in [3].

Lemma 2.1. *Let* $\phi, \overline{\phi} \in K \cap C^2_{Lip}(H)$; *then* $\phi_\alpha, \overline{\phi}_\alpha \in K \cap C^2_{Lip}(H)$ *and the following estimates hold for every* $R > 0$:

a) $|\phi_\alpha|_{i,R} \leq |\phi|_{i,R} \qquad$ *for* $i = 0, 1, 2$;

b) $\|\phi_\alpha\|_{2,R} \leq \|\phi\|_{2,R}$;

c) $|\phi_\alpha - \overline{\phi}_\alpha|_{i,R} \leq |\phi - \overline{\phi}|_{i,R} \qquad$ *for* $i = 0, 1$;

d) $|\phi_\alpha - \overline{\phi}_\alpha|_{2,R} \leq |\phi - \overline{\phi}|_{2,R} + \alpha\|\overline{\phi}\|_{2,R}|\phi - \overline{\phi}|_{1,R}$.

Lemma 2.2. Let $\phi \in K \cap C^2_{Lip}(H)$; and set:

(2.10)
$$R_{\alpha,\phi}(x) = \frac{1}{\alpha}(\phi(x) - \phi_\alpha(x)) - \frac{1}{2}|\phi'(x)|^2$$

Then for $R > 0$ we have:

   i) $|R_{\alpha,\phi}|_{0,R} \le \alpha |\phi|^2_{1,R} |\phi|_{2,R}$;

   ii) $|R_{\alpha,\phi}|_{1,R} \le \alpha (|\phi|^2_{1,R} \|\phi\|_{2,R} + |\phi|_{2,R}|\phi|_{1,R})$.

We observe that the same statements hold if K is replaced by K_R.

Remark 2.3. It is not difficult to show that if $\phi, \overline{\phi} \in K \cap C^k_{Lip}(H)$ $(k \ge 3)$ then analogous estimates hold true for $|\phi_\alpha|_{i,R}$, $|\phi_\alpha - \overline{\phi}_\alpha|_{i,R}$ and $|R_{\alpha,\phi}|_{i-1,R}$, $i = 3, .., k$.

3. The control problem

We analize our control problem [P] in a more precise way.
Minimize

(1.1)
$$J(x,u) = \int_0^{+\infty} \left[g(y(s)) + \frac{1}{2}|u(s)|^2 \right] ds$$

over all $u \in L^2(0, +\infty; H)$, where $y(s, u, x) \in C([0, +\infty); H)$ is the unique "mild" solution of the state equation

(SE)
$$\begin{cases} y' = Ay + f(\varepsilon, x) + u & \text{on } [0, +\infty] \\ y(0) = x & x \in H. \end{cases}$$

Our hypotheses (as we said in the introduction) are the following:

   i) A: $D(A) \subset H \to H$ is the infinitesimal generator of a strongly continous semigroup in H and there exist $\omega > 0$ s.t.

$$\Re\langle Ax, x\rangle \le -\omega \, |x|^2 \quad \forall x \in D(A)$$

   that is A is a strictly dissipative operator on H.

   ii) $f(\varepsilon, \cdot) \in C^2_{Lip}(H, H)$, $\quad \forall \varepsilon \in \mathbf{R}$

   $f(\varepsilon, 0) = 0$, $\quad \dfrac{\partial f}{\partial x} f(\varepsilon, 0) = 0$, $\quad \forall \varepsilon \in \mathbf{R}$

   $f(\varepsilon, x) \xrightarrow[\varepsilon \to 0]{} 0$ uniformly on $C^2_{Lip}(H, H)$.

   iii) $g \in C^2_{Lip}(H)$

   $g(0) = 0 \quad g'(0) = 0$

   $\langle g''(x)z, z\rangle \ge \mu|z|^2 \forall x, z \in H$, $\quad \mu > 0$ fixed

   that is g is strictly convex on H.

Remark 3.1. We obtain the same results if we replace the assumption iii) by the following:

$iii)'$ $g \in B([0, +\infty); C^2_{Lip}(H))$

$$g(s,0) \equiv 0 \quad \frac{\partial g}{\partial x}(s,0) \equiv 0 \text{ on } [0, +\infty)$$

$$\left\langle \frac{\partial^2 g(s,x)}{\partial x^2} z, z \right\rangle \geq \mu |z|^2, \quad \forall x, z \in H, \quad \forall s \in \mathbf{R}^+, \quad \mu > 0, \text{ fixed}$$

that is g is strictly convex on H uniformly on $[0, +\infty)$.

By this hypothesis we can prove the following two statements about the state equation (SE) and the cost functional $J(x,u)$.

Proposition 3.2. Let $x \in H$, $u \in L^2(0, +\infty; H)$; Assume that i) — iii) hold true.
Let

(3.1)
$$r = 2 \left(|x| + \sup_{T \in (0, +\infty)} \int_0^T e^{-(T-s)\frac{w}{2}} |u(s)| \, ds \right) = r(|x|, |u|)$$

$$\leq 2 \left(|x|_H + (2/w)^{\frac{1}{2}} |u|_{L^2(0,+\infty;H)} \right)$$

then there exists $\varepsilon_0(r) > 0$ such that $\forall |\varepsilon| < \varepsilon_0$ (SE) has a unique "mild" solution $y(x,u)$ with the properties

a) $y \in C([0, +\infty), H) \cap L^2(0, +\infty; H)$

b) $|y(s,x,u)| \leq e^{-\frac{w}{2}s}|x| + \int_0^s e^{-\frac{w}{2}(s-\sigma)}|u(\sigma)| \, d\sigma \leq r$.

The proof is a standard application of the contraction mapping principle, Gronwall inequality and Fourier transform (see [9], [11] chapter 6 for similar results).

An immediate consequence is:

Proposition 3.3. For every $x \in H$ and $u \in L^2(0, +\infty; H)$ we have $J(x,u) < +\infty$.

Proof: It is enough to remember that (r taken as in Prop. 3.1)

(3.2)
$$g(y(s)) \leq |g|_{2,r} \frac{|y(s)|^2}{2}$$

(by the mean value theorem).
So we have

$$J(x,u) \leq \frac{1}{2} \left[|g|_{2,r} |y|^2_{L^2} + |u|^2_{L^2} \right] < +\infty$$

due to a).

Q.E.D.

Remark 3.4. The above proposition implies that under our hypotheses the set of admissible control is the whole space $L^2(0, +\infty; H)$.

4. The Hamilton Jacobi equation

As we said in the introduction we consider the Hamilton-Jacobi equation associated to the control problem [P] with finite horizon, that is:

(HJ)
$$\begin{cases} \phi_t(t,x) = -\frac{1}{2}|\phi_x(t,x)|^2 + \langle Ax + f(\varepsilon,x), \phi_x(t,x)\rangle + g(x) \\ \qquad \forall (t,x) \in [0,+\infty) \times D(A) \\ \phi(0,x) = 0. \end{cases}$$

We say that a function $\phi \in B\left([0,+\infty), C^1_{Lip}(\Sigma_R) \cap K_R\right)$ is a strict solution of (HJ) if $\phi(\cdot,x) \in C^1([0,+\infty]) \; \forall x \in D(A)$, and satisfy (HJ).

More generally, if ϕ is defined only on $[0,+\infty) \times \Sigma_R$ we say that ϕ is a strict solution on Σ_R if the conditions above hold true on this ball.

The following result holds:

Theorem 4.1. *For every $R > 0$ there exists $\bar{\varepsilon}(R) > 0$ such that $\forall |\varepsilon| < \bar{\varepsilon}(R)$, (HJ) has a unique strict solution $\phi_\varepsilon: [0,+\infty) \times \Sigma_R \to \mathbf{R}$. Moreover we have:*

(4.1)
$$\phi_\varepsilon(t,\cdot) \in C^2_{Lip}(\Sigma_R) \cap K_R \quad \forall t \geq 0,$$

(4.2)
$$\sup_{t \in [0,+\infty)} |\phi_\varepsilon(t,\cdot)|_{C^2_{Lip}(\Sigma_R)}.$$

Remark 4.2. The following proof also works when the initial data in (HJ) is a function $\phi_0 \in C^2_{Lip}(H) \cap K$. It is a simple generalisation. We study the case $\phi_0 = 0$ for simplicity, as this is what we need to solve the control problem [P].

Proof: The proof consists of several steps. It is essentially a refinement of the proof for the finite horizon problem.

I) Integral approximating problem.

We consider the approximating equation

$(HJ)_\alpha$
$$\begin{cases} \phi_t(t,x) = -\frac{1}{\alpha}\left(\phi(t,x) - \phi_\alpha(t,x)\right) + \langle Ax + f(\varepsilon,x), \phi_x(t,x)\rangle + g(x) \\ \varphi(0,x) = 0 \end{cases}$$

where the "bad" term $1/2\,|\phi_x(t,x)|^2$ is replaced by $\dfrac{\phi - \phi_\alpha}{\alpha}(t,x)$ (see section 1).

By the characteristic method we can write $(HJ)_\alpha$ in the following integral form

(IF)
$$\phi(t,x) = \int_0^t e^{-(t-s)/\alpha}\left[g(\zeta_\varepsilon(s,t,x)) + \frac{1}{\alpha}\phi_\alpha\left(s,\zeta_\varepsilon(s,t,x)\right)\right] ds \overset{def}{=} (\Gamma_\alpha^\varepsilon \phi)(t,x)$$

where $\zeta_\varepsilon(s,t,x)$ is the "mild" solution of the final value problem

(C)
$$\begin{cases} \zeta'_\varepsilon(s) = -A\zeta_\varepsilon(s) - f(\varepsilon,\zeta_\varepsilon(s)) \quad s \in [0,t] \\ \zeta_\varepsilon(t) = x. \end{cases}$$

First we want to show that there exists a solution for the approximating problem (IF). After we will study the convergence of this solution as $\alpha \to 0$.

We need some preliminary results

Lemma 4.3. *(Characteristics estimate)* We fix $R > 0$. Let $t \in [0, +\infty)$, $x \in \Sigma_R \subset H$, and $|\epsilon| \leq \varepsilon_0(R)$ as in Prop.3.1.

Then the Cauchy problem (C) has a unique "mild" solution $\zeta_\epsilon(\cdot, t, x)$ on $[0, t)$ and we have $\zeta_\epsilon(s, t, \cdot) \in C^2_{Lip}(H, H)$, $\forall s, t \in \mathbf{R}^+$, $s \leq t$.

Moreover the following estimates hold.

$$|\zeta_\epsilon(s, t, x)| \leq |x| e^{-\omega(t-s)/2}$$
$$|\zeta_\epsilon(s, t, \cdot)|_{1,R} \leq e^{-\omega(t-s)/2}$$
$$|\zeta_\epsilon(s, t, x)(z, z)|_H \leq |f(\epsilon, \cdot)|_{2,R} \int_s^t e^{-\omega(t-\sigma)/2} |\zeta_{\epsilon x}(\sigma, t, x)z|_H^2 \, d\sigma$$
$$\|\zeta_\epsilon(s, t, \cdot)\|_{2,R} \leq \int_s^t e^{-\omega(t-\sigma)/2} \left[3|f(\epsilon, \cdot)|_{2,R} + \|f(\epsilon, \cdot)\|_{2,R}\right] |\zeta_\epsilon(\sigma, t, \cdot)|_{1,R} \, d\sigma.$$

For the proof see [8].

II) Convexity estimates.

Lemma 4.4. Let $R > 0$, $C > 0$ be fixed. Let

$$(4.3) \qquad \mathcal{F}_C^R = \left\{ \phi \in B\left([0, +\infty), C^2_{Lip}(\Sigma_R) \cap K_R\right) \quad s.t. \quad \sup_{t \in [0, +\infty)} |\phi(t, \cdot)|_{1,R} \leq C \right\};$$

then there exists $\varepsilon(R, C) > 0$ s.t. $\forall |\varepsilon| \leq \varepsilon(R, C)$ the function $\Gamma_\alpha^\varepsilon$ is a convexity-preserving map, that is

$$\Gamma_\alpha^\varepsilon(\mathcal{F}_C^R) \subset B\left([0, +\infty), C^2_{Lip}(\Sigma_R) \cap K_R\right).$$

If C is sufficienty large, then we also have $\Gamma_\alpha^\varepsilon(\mathcal{F}_C^R) \subset \mathcal{F}_C^R$.

The proof is exactly the same of the one in [8], where $\varepsilon(R, C)$ is obtained independent from the horizon T in the case of strict dissipativity of A.

The estimate of the successive approximations for $\Gamma_\alpha^\varepsilon$ is more technical.

III) Successive approximations.

Lemma 4.5. Let $R > 0$ be fixed. We set $L = L(R) = \frac{2}{\omega} |g|_{C^2_{Lip}(\Sigma_R)}$. Consider the sequence $\{\phi^n\}_{n \in \mathbf{N}}$ in $B\left([0, +\infty), C^2_{Lip}(\Sigma_R) \cap K_R\right)$

$$(4.4) \qquad\qquad \phi^0 = \Gamma_\alpha^\varepsilon(0), \qquad \phi^{n+1} = \Gamma_\alpha^\varepsilon(\phi^n).$$

There exist $\varepsilon_1(R) > 0$ s.t. $\forall |\varepsilon| < \varepsilon_1(R)$, we have:

A) $\phi^n \in \mathcal{F}_L^R$ and also $\displaystyle\sup_{t \in [0, +\infty)} |\phi^n(t, \cdot)|_{C^2_{Lip}(\Sigma_R)} \leq L, \quad \forall n \in \mathbf{N}$;

B) $\displaystyle\sum_{n \geq 1} \sup_{t \in [0, +\infty)} |\phi^{n+1}(t, \cdot) - \phi^n(t, \cdot)|_{C^2_{Lip}(\Sigma_R)} < +\infty$;

C) $\Gamma_\alpha^\varepsilon$ is continous on $B\left([0, +\infty), C^2_{Lip}(\Sigma_R)\right)$.

Proof: The derivatives of $\Gamma_\alpha^\varepsilon(\phi)$ can be written as:

(4.5) $\quad (\Gamma_\alpha^\varepsilon \phi)_x(t,x) = \int_0^t e^{-(t-s)/\alpha} \zeta_{\varepsilon x}^*(s,t,x)[g'(\zeta_\varepsilon(s,t,x)) + \frac{1}{\alpha}\phi_{\alpha x}(s,\zeta_\varepsilon(s,t,x))]ds$

$(\Gamma_\alpha^\varepsilon \phi)_{xx}(t,x) =$

(4.6)
$= \int_0^t e^{-(t-s)/\alpha} \zeta_{\varepsilon x}^*(s,t,x)[g''(\zeta_\varepsilon(s,t,x)) + \frac{1}{\alpha}\phi_{\alpha xx}(s,\zeta_\varepsilon(s,t,x))]\zeta_{\varepsilon x}(s,t,x)ds$

$+ \int_0^t e^{-(t-s)/\alpha} \zeta_{\varepsilon xx}^*(s,t,x)[g'(\zeta_\varepsilon(s,t,x)) + \frac{1}{\alpha}\phi_{\alpha x}(s,\zeta_\varepsilon(s,t,x))]ds.$

Now we can begin the proof

A) We use a recurrence argument. First we have that A) holds true for the function $0 \in B([0,+\infty); C_{Lip}^2(\Sigma_R))$. After we suppose that A) holds for $\phi^0, \ldots, \phi^{n-1}$. Then we have

$$|\phi^n(t,\cdot)|_{0,R} \leq \int_0^t e^{-(t-s)/\alpha}\left[|g(\zeta_\varepsilon(s,t,x))|_{0,R} + \frac{1}{\alpha}|\phi_\alpha^{n-1}(s,\zeta_\varepsilon(s,t,\cdot))|_{0,R}\right] ds \quad \forall y \in \Sigma_R.$$

Now let $h \in K_R$ $h(0) = 0$; we have

(4.7) $$h(y) = h\left(\frac{Ry}{|y|}\frac{|y|}{R}\right) \leq \frac{|y|}{R}|h|_{0,R}.$$

It follows that (see lemma 4.3.)

(4.8) $$\left|h(\zeta_\varepsilon(s,t,\cdot))\right|_{0,R} \leq |h|_{0,R}e^{-\frac{\omega}{2}(t-s)}.$$

Now using the last inequality and also Lemma 2.1. we have:

(4.9) $$|\phi^n(t,\cdot)|_{0,R} \leq \int_0^t e^{-(t-s)/\alpha}e^{-\frac{\omega}{2}(t-s)}\left[|g|_{0,R} + \frac{1}{\alpha}|\phi^{n-1}(s,\cdot)|_{0,R}\right] ds$$

because $\phi_\alpha^{n-1}(s,\cdot)$ is convex. By iterating the above inequality we obtain:

(4.10)
$$|\phi^n(t,\cdot)|_{0,R} \leq \int_0^t e^{-(t-s)/\alpha}\left[|g|_{0,R}\sum_{k=0}^n \frac{(t-s)^k}{\alpha^k k!}\right]e^{-\frac{\omega}{2}(t-s)}ds$$

$$\leq \frac{2}{\omega}|g|_{0,R}.$$

In similar way we can write, using (4.5) and lemma 4.3

(4.11)
$$|\phi^n(t,\cdot)|_{1,R} \leq \int_0^t e^{-(t-s)/\alpha}e^{-\frac{\omega}{2}(t-s)}\left[|g|_{1,R} + \frac{1}{\alpha}|\phi^{n-1}(s,\cdot)|_{1,R}\right] ds$$

$$\leq \frac{2}{\omega}|g|_{1,R}.$$

Now we define

(4.12)
$$\left[\phi^n(t,\cdot)\right]_{2,\mathrm{R}} \overset{def}{=} |\phi^n(t,\cdot)|_{2,\mathrm{R}} + \|\phi^n(t,\cdot)\|_{2,\mathrm{R}}.$$

To evaluate this quantity we argue as follows: let $h \in C^2_{Lip}(\Sigma_{\mathrm{R}})$, $h(0) = 0$, $h'(0) = 0$, let $|z| \leq \mathrm{R}$. Then $|h(z)| \leq |h|_{1,\mathrm{R}}|z|$ and $|h'(z)| \leq |h|_{2,\mathrm{R}}|z|$, so

(4.13)
$$h'(\zeta_\epsilon(s,t,x)) \leq |h|_{2,\mathrm{R}}\mathrm{R}e^{-\frac{w}{2}(t-s)} \qquad \forall |x| \leq \mathrm{R}.$$

We apply this inequality to the functions g and $\phi_\alpha(s,\cdot)$ in (4.6) and we obtain that

(4.14)
$$\begin{aligned}
\left[\phi^n(t,\cdot)\right]_{2,R} &\leq \int_0^t e^{-(t-s)/\alpha}e^{-\omega(t-s)/2}\gamma(t,s)\left[|g|_{2,R} + \frac{1}{\alpha}|\phi^{n-1}(s,\cdot)|_{2,R}\right]ds \\
&+ \int_0^t e^{-(t-s)/\alpha}e^{-\omega(t-s)/2}\left[\|g\|_{2,R} + \frac{1}{\alpha}\|\phi^{n-1}(s,\cdot)\|_{2,R}\right]ds
\end{aligned}$$

where

$$\gamma(t,s) = c_1(\varepsilon,R,f)\frac{1}{\omega}\left(1 - e^{-\omega(t-s)/2}\right) + e^{-\omega(t-s)/2}$$

and

$$c_1(\varepsilon,R,f) = |f(\varepsilon,\cdot)|_{2,R}(3+4R) + R\|f(\varepsilon,\cdot)\|_{2,R}.$$

But, due to i), we can chose $\varepsilon_2(R) > 0$ such that

$$c_1(\varepsilon,R,f) \leq \omega \qquad \forall |\varepsilon| \leq \varepsilon_2(R).$$

So it follows that $\gamma(t,s) \leq 1$ and we can repeat the reasoning we used to prove (4.10) and (4.11) and obtain

(4.15)
$$\left[\phi^n(t,\cdot)\right]_{2,R} \leq \frac{2}{\omega}[g]_{2,R}.$$

This completes the proof of $A)$.

$B)$ We apply (4.13) to the function $(\phi^n_\alpha - \phi^{n-1}_\alpha)(s,\cdot)$ and we have

(4.16)
$$\begin{aligned}
|\phi^{n+1}(t,\cdot) &- \phi^n(t,\cdot)|_{C^1(\Sigma_R)} \leq \\
&\leq \int_0^t e^{-(t-s)/\alpha}\frac{1}{\alpha}|\phi^n(s,\cdot) - \phi^{n-1}(s,\cdot)|_{1,R}(R+1)e^{-\omega(t-s)/2}ds
\end{aligned}$$

by iterating it follows

(4.17)
$$\leq \frac{L(R+1)}{\alpha}\int_0^t e^{-\omega(t-s)/2}e^{-(t-s)/\alpha}\frac{(t-s)^n}{n!\alpha^n}ds \overset{def}{=} a_n(t)$$

and

$$(4.18) \qquad \sum_{n\geq 1} a_n(t) \leq \frac{2L(R+1)}{\alpha\omega}\left(1 - e^{-\omega t/2}\right) \leq \frac{2L(R+1)}{\alpha\omega}.$$

For the last estimate we have, using lemma 2.1. and the same method of A):

$$|\phi^{n+1}(t,\cdot) - \phi^n(t,\cdot)|_{2,R} \leq$$

$$(4.19) \qquad \leq \int_0^t e^{-(t-s)/\alpha} e^{-\omega(t-s)/2}\frac{1}{\alpha}\left[|\phi^n(s,\cdot) - \phi^{n-1}(s,\cdot)|_{2,R} + \right.$$

$$\left. +\alpha\|\phi^n(s,\cdot)\|_{2,R}|\phi^n(s,\cdot) - \phi^{n-1}(s,\cdot)|_{1,R}\right] ds$$

Now, using the previous estimates and iterating:

$$\leq \int_0^t e^{-(t-s)/\alpha} e^{-\omega(t-s)/2}\frac{(t-s)^n}{n!\alpha^n}\left[\frac{L}{\alpha} + L^2(n+1)\right] ds \stackrel{def}{=} b_n(t).$$

By adding up n the previous inequality we obtain:

$$\sum_{n\geq 1} b_n(t) \leq \int_0^t e^{-\omega(t-s)/2}\left[\frac{L}{\alpha} + L^2 + \frac{L^2}{\alpha}(t-s)\right] ds \leq \text{(by integrating)}$$

$$\frac{2L}{\alpha\omega} + \frac{2L^2}{\omega} + \frac{2L^2}{\alpha\omega} < +\infty$$

and B) is proved.

C) Let $\phi, \overline{\phi} \in B([0,+\infty); C^2_{Lip}(\Sigma_R) \cap K_R)$, and let:

$$D = \left\{\sup_{t\in[0,+\infty)} |\phi(t,\cdot)|_{C^2(\Sigma_R)}, |\overline{\phi}(t,\cdot)|_{C^2(\Sigma_R)}\right\}$$

Then we have by similar estimates:

$$\sup_{t\in[0,+\infty)} |\Gamma^\varepsilon_\alpha\phi(t,\cdot) - \Gamma^\varepsilon_\alpha\overline{\phi}(t,\cdot)|_{C^2(\Sigma_R)} \leq$$

$$\leq \frac{1}{\alpha}\int_0^t e^{-(t-s)/\alpha}(D+1)|\phi(s,\cdot) - \overline{\phi}(s,\cdot)|_{C^2(\Sigma_R)} ds$$

$$\leq (D+1)\sup_{t\in[0,+\infty)} |\phi(t,\cdot) - \overline{\phi}(t,\cdot)|_{C^2(\Sigma_R)}.$$

In fact: $\qquad \frac{1}{\alpha}\int_0^t e^{-(t-s)/\alpha} ds \leq 1 \quad \forall t \in [0,+\infty)$

and C) is proved.

Q.E.D.

Remark 4.6. Note that estimate A) does not depend on α. This fact will be used in the next section to prove the convergence of ϕ^α as $\alpha \to 0$.

An immediate consequence of lemma 4.5. is the following:

Corollary 4.7. *Under the Hypothesis i) — iii) we have that for every $R > 0$ there exists $\varepsilon_2(R) > 0$ such that $\forall |\varepsilon| \leq \varepsilon_2(R)$, $\exists \phi_\varepsilon^\alpha \in B([0,+\infty); C_{Lip}^2(\Sigma_R) \cap K_R)$ s.t.*

$$\phi^n \xrightarrow{n \to +\infty} \phi_\varepsilon^\alpha \quad in \ B\left([0,+\infty); C^2(\Sigma_R)\right)$$

and

$$\phi_\varepsilon^\alpha(t,x) = (\Gamma_\alpha^\varepsilon \phi_\varepsilon^\alpha)(t,x) \quad on \ [0,+\infty) \times \Sigma_R$$

so that ϕ_ε^α is also a strict solution of the equation $(HJ)_\alpha$.

IV) Convergence of ϕ_ε^α as $\alpha \to 0$.

Lemma 4.8. *We assume that i) — iii) hold. Let $R > 0$ be fixed and let $\varepsilon_2(R)$ be as in part III). Let $\phi_\varepsilon^\alpha \in B([0,+\infty); C^2(\Sigma_R) \cap K_R)$ be the strict solution of $(HJ)_\alpha$. Then there exists $\phi_\varepsilon \in B([0,+\infty); C_{Lip}^1(\Sigma_R) \cap K_R)$ such that:*

a) $\phi_\varepsilon^\alpha \xrightarrow{\alpha \to 0} \phi_\varepsilon$ *in* $B([0,+\infty); C^1(\Sigma_R))$,
b) ϕ_ε *is a strict solution of the Hamilton-Jacobi equation (HJ).*
c) $\forall t \in [0,+\infty)$, $\phi_\varepsilon(t,\cdot) \in C_{Lip}^2(\Sigma_R) \cap K_R$.

Proof: The proof is similar to the one of [3](p.38-41, 98-99) We use the estimate A) and Lemma 2.2.

Let $\alpha > 0$, $\beta > 0$. Then ϕ^β fulfils the equation:

$$(4.20) \quad \begin{cases} \phi_t^\beta + \frac{1}{\alpha}(\phi^\beta - \phi_\alpha^\beta) - \langle Ax + f(\varepsilon,x), \phi_x^\beta \rangle = g + R_{\alpha,\phi^\beta} + R_{\beta,\phi^\beta} \\ \qquad \forall (t,x) \in [0,+\infty) \times (D(A) \cap \Sigma_R) \\ \phi^\beta(0,x) = 0 \qquad \forall x \in H. \end{cases}$$

It follows, by using the integral form of (4.20) and the inequality (4.13):

$$|\phi^\beta(t,\cdot) - \phi^\alpha(t,\cdot)|_{C^1(\Sigma_R)} \leq \frac{1}{\alpha} \int_0^t e^{-(t-s)/\alpha} |\phi^\beta(s,\cdot) - \phi^\alpha(s,\cdot)|_{C^1(\Sigma_R)} ds$$

$$+ \int_0^t e^{-(t-s)/\alpha} e^{-\omega(t-s)/2}(R+1)(|R_{\alpha,\phi^\beta}(s,\cdot)|_{C^1(\Sigma_R} + |R_{\beta,\phi^\beta}(s,\cdot)|_{C^1(\Sigma_R)}) ds.$$

Now by lemma 2.2. and Gronwall's inequality we get:

$$|\phi^\beta(t,\cdot) - \phi^\alpha(t,\cdot)|_{C^1(\Sigma_R)} \leq (1+R) \int_0^t e^{-\omega(t-s)/2}(\alpha+\beta)\frac{3}{2}L^3 ds \leq \frac{3}{2\omega}(1+R)L^3(\alpha+\beta)$$

so a) is proved.

To show b) we remark that, if $x \in D(A)$, then $\phi^\alpha(\cdot,x) \in C^1([0,+\infty))$ (it follows from the definition of Γ), and:

$$\phi_t^\alpha = g + R_{\alpha,\phi^\alpha(t,\cdot)} - \frac{1}{2}|\phi_x^\alpha|^2 + \langle Ax + f(\varepsilon,x), \phi_x^\alpha \rangle$$

so if $x \in D(A)$, then $\phi_\varepsilon(\cdot,x)$ is differentiable and $\phi_t^\alpha(\cdot,x) \xrightarrow{\alpha \to 0} \phi_{\varepsilon,t}(\cdot,x)$ in $C([0,+\infty))$, and ϕ_ε is a strict solution.

To show c) we apply Ascoli's theorem exactly as in [3] (p.40-41).

Q.E.D.

V) Uniqueness.

This also is similar to the proof contained in [3](p.39, 99).

Let ϕ_1, ϕ_2 two strict solutions of (3.1). Then for $\alpha > 0$ we have:

(4.21)
$$\begin{cases} \phi_{i,t}(t,x) = -\frac{1}{\alpha}(\phi_i - \phi_{i,\alpha}) + \langle Ax + f(\varepsilon,x), \phi_{i,x}\rangle + g + R_{\alpha,\phi_i} \\ \qquad \forall(t,x)\in[0,+\infty)\times(D(A)\cap\Sigma_R) \qquad i=1,2 \\ \phi_i(0,x) = 0 \qquad \forall x \in H \end{cases}$$

By using the integral form (4.1) and applying Gronwall's lemma we obtain, using the same method seen in part IV):

$$|\phi_1(t,\cdot) - \phi_2(t,\cdot)|_{C^1(\Sigma_R)} \leq 3(1+R)\alpha L^3/\omega$$

and the uniqueness follows from the arbitrariness of α.

Now theorem 4.1 is proved setting $\bar{\varepsilon}(R) = \varepsilon_2(R)$ as in (4.19).

5. Stability

We discuss in this section the limit, when t goes to $+\infty$, of the solution $\phi_\varepsilon(t,x)$ of the Hamilton-Jacobi equation (HJ). The main result is theorem 5.1.

Theorem 5.1. *Let $r > 0$, let $R(r) = r + (2/\omega)^{\frac{1}{2}}L(r)$, ($L$ as in lemma 4.5), $|\varepsilon| \leq \bar{\varepsilon}(R)$. Then there exists the limit*

$$\lim_{t\to+\infty} \phi_\varepsilon(t,x) \overset{def}{=} \phi_\varepsilon^\infty(x) \qquad in\ C^1(\Sigma_r).$$

The function ϕ_ε^∞ is $C^1(\Sigma_r)$ and convex.

We begin by showing two useful lemmas.

Lemma 5.2. *(Monotonicity) Let $r > 0$ and let $R = R(r) > 0$ as above; for $|\varepsilon| < \bar{\varepsilon}(R)$ let $\phi_\varepsilon \in B([0,+\infty); C^2_{Lip}(\Sigma_R) \cap K_R)$ be the strict solution of (HJ).*

Then $\forall x \in \Sigma_r$, $\phi_\varepsilon(\cdot,x)$ is a non-decreasing function $\mathbf{R}^+ \to \mathbf{R}^+$.

Proof: Let $t > 0$. By the results contained in [8] we have that $\forall x \in \Sigma_r$ $\forall |\varepsilon| \leq \bar{\varepsilon}(R)$:

(5.1)
$$\phi_\varepsilon(t,x) = \inf_{u\in L^2(0,t;H)} \left\{ \int_0^t \left[g(y(s)) + \frac{1}{2}|u(s)|^2 \right] ds \right\}$$

where y is the mild solution of

$$\begin{cases} y' = Ay + f(\varepsilon,y) + u \\ y(0) = x. \end{cases}$$

Now let $t_1, t_2 \in \mathbf{R}^+$, $t_1 < t_2$. Let $x \in \Sigma_r$. Let (u^{t_2}, y^{t_2}) be the optimal pair (which exists and is unique due to results of [8]) for the problem associated to $\phi(t_2,x)$ as in (5.1). We have

(5.2)
$$\phi(t_2,x) - \phi(t_1,x) \geq \int_{t_1}^{t_2} \left[g(y^{t_2}(s)) + \frac{1}{2}|u^{t_2}(s)|^2 \right] \geq 0$$

because $g \geq 0$.

Q.E.D.

Lemma 5.3. *Under the hypotheses of lemma (5.2) we have*

(5.3)
$$\lim_{t_1 \to \infty, t_1 < t_2} |\phi(t_2, \cdot) - \phi(t_1, \cdot)|_{0,r} = 0.$$

Proof: Let (u^{t_1}, y^{t_1}) be an optimal pair for the problem associated to $\phi(t_1, x)$, $x \in \Sigma_r$. Let

$$\bar{u}^{t_2}(s) = \begin{cases} u^{t_1}(s) & 0 \le s < t_1 \\ 0 & t_1 \le s \le t_2 \end{cases}$$

and let $\bar{y}^{t_2}(s)$ be the corresponding solution of (SE). We have by monotonicity:

(5.4)
$$|\phi(t_2, x) - \phi(t_1, x)| \le \int_{t_1}^{t_2} |g(\bar{y}^{t_2}(s))| \, ds.$$

Now, using the definition of $\bar{y}^{t_2}(s)$ we have

(5.5)
$$|\bar{y}^{t_2}(s)| \le e^{-\frac{\psi}{2}(s - t_1)} |y^{t_1}(t_1)| \le |x| e^{-\frac{\psi}{2} s}$$

where the last inequality is due to the optimality of y^{t_1}, which implies, as seen in [8]

$$|y^{t_1}(\sigma)| \le |x| e^{-\frac{\psi}{2} \sigma}.$$

Then

$$|\phi(t_2, \cdot) - \phi(t_1, \cdot)|_{0,r} \le r |g|_{1,r} \int_{t_1}^{t_2} e^{-\frac{\psi}{2} s} \, ds$$

from which (5.3) follows.

Q.E.D.

Remark 5.4. From these two lemmas it follows that

$$\exists \lim_{t \to +\infty} \phi_\varepsilon(t, x) \overset{def}{=} \phi_\varepsilon^\infty(x)$$

uniformly and nondecreasing on Σ_r.

To complete the proof we have to show the convergence of the spatial derivative of $\phi_\varepsilon(t, x)$ when $t \to +\infty$. We argue as in [3](pages 40-41). Let $x, y \in \Sigma_{r/2}$, $|h| \le 1$, $t > 0$. Set

(5.6)
$$\psi^t(x, y, h) = \psi^t(h) = \phi_\varepsilon(t, x + hy)$$

by Lemma 4.5.-A) it follows

(5.7)
$$\psi^t(h) \le L(r)$$

and

$$\frac{\partial \psi^t}{\partial h}(h) = \langle y, \phi_{\varepsilon x}(t, x + hy) \rangle$$

from which

(5.8)
$$\left| \frac{\partial \psi^t}{\partial h}(h) \right| \le |y|L$$

$$\left| \frac{\partial \psi^t}{\partial h}(h) - \frac{\partial \psi^t}{\partial h}(k) \right| \le L|y|^2|h - k|.$$

So $\left\{ \frac{\partial \psi^t}{\partial h} \right\}_{t>0}$ is equicontinuous and by Ascoli-Arzelà theorem there exist a sequence $t_k \uparrow +\infty$ such that $\left\{ \frac{\partial \psi^{t_k}}{\partial h} \right\}_{k \in \mathbf{N}}$ is uniformly convergent in $[-1, 1]$. It follows

(5.9)
$$\begin{cases} \psi^{t_k}(h) \to \phi_\varepsilon^\infty(x + hy) \\ \frac{\partial \psi^{t_k}}{\partial h} \to \frac{d}{dh}\phi_\varepsilon^\infty(x + hy). \end{cases}$$

This clearly implies that ϕ_ε^∞ is Gateaux differentiable on $\Sigma_{r/2}$ and

(5.10)
$$\frac{\partial \psi^{t_k}}{\partial h}(0) \xrightarrow{k \to +\infty} \langle y, D_x \phi_\varepsilon^\infty(x) \rangle.$$

Now it is enough to prove the continuity of $D_x \phi_\varepsilon^\infty(x)$. Let $x, z \in \Sigma_{r/2}$.

$$\left| \frac{\partial \psi^{t_k}}{\partial h}(x, y, 0) - \frac{\partial \psi^{t_k}}{\partial h}(x, z, 0) \right| \le L|x - z||y|$$

due to the regularity of $\phi_\varepsilon(t_k, x)$. And from (5.10) we obtain for $k \uparrow +\infty$

(5.11)
$$| < y, D_x \phi_\varepsilon(x) - D_x \phi_\varepsilon(z) > | \le L|y||x - z|.$$

So $D_x \phi_\varepsilon^\infty$ is lipschitz continuous on $\Sigma_{r/2}$. It remains to show that

$$\lim_k \phi_\varepsilon(t_k, x) = \phi_\varepsilon^\infty(x) \qquad \text{on } C^1(\Sigma_r)$$

for every sequence $t_k \uparrow +\infty$. This is easy to prove by considering a subsequence and applying the Ascoli theorem again.

Finally it is not difficult to refine this proof and obtain the convergence on all the open ball $Int \, \Sigma_r$.

Q.E.D.

Now we recall that a function $\phi \in C^1(H)$ is a solution of the stationary Hamilton-Jacobi equation (HJS) if ϕ satisfy this equation for every $x \in D(A)$. We say that ϕ is a solution of (HJS) on a ball $\Sigma_r \subset H$ if the above condition are verified on this ball.

An important consequence of the theorem 5.4. is:

Corollary 5.5. *The function ϕ_ε is a solution of the stationary Hamilton-Jacobi equation* (HJS).

Proof: We have to prove that, for every $x \in D(A) \cap \Sigma_r$, (HJS) is verified. Consider the equation (HJ) and let $t \to +\infty$. The right hand side converges exactly to the right hand side of (HJS). This implies that:

$$\exists \lim_{t \to +\infty} \phi_{\varepsilon,t}(t,x) \stackrel{def}{=} a(x) \in \mathbf{R} \qquad \forall x \in D(A) \cap \Sigma_r.$$

But we have:

$$\phi_\varepsilon(t,x) \uparrow \phi_\varepsilon^\infty(x) < +\infty \qquad \forall x \in \Sigma_r$$

which yields $a(x) = 0 \quad \forall x \in D(A) \cap \Sigma_r$.

$$\text{Q.E.D.}$$

6. Solution of the control problem

Let $r > 0$ be fixed. Let $|x| \le r$, $|u|_{L^2} \le L(r)$.
Finally let $R = r + (2/\omega)^{\frac{1}{2}} L(r)$ and $|\varepsilon| \le \bar{\varepsilon}(R)$.
Then, as seen in section 3-4-5, the following statements hold true:

1) The state equation (SE) has a unique mild solution

$$y \in C([0,+\infty); H) \cap L^2(0,+\infty; H) \qquad \text{with } |y(t)| \le R.$$

2) The Hamilton-Jacobi equation has a unique strict solution

$$\phi_\varepsilon \in B([0,+\infty); C^1_{Lip}(\Sigma_R) \cap K_R).$$

3) There exist the limit on $C^1(\Sigma_r)$:

$$\lim_{t \to +\infty} \phi_\varepsilon(t,x) \stackrel{def}{=} \phi_\varepsilon^\infty.$$

Using these results we can solve the control problem [P] in a standard way.

Lemma 6.1. *Let $r, M > 0$ be fixed. Let $x \in \Sigma_r$, $u \in L^2(0,+\infty; H)$, with $|u|_{L^2} \le M$. Let $R = r + (2/\omega)^{\frac{1}{2}} M$ and $|\varepsilon| \le \bar{\varepsilon}(R)$. Let y be the mild solution of the Cauchy problem:*

$$(6.1) \qquad \begin{cases} y' = Ay + f(\varepsilon,y) + u & \text{on } [0,+\infty); \\ y(0) = x. \end{cases}$$

Then the following fundamental identity holds for every $(t,x) \in [0,+\infty) \times \Sigma_r$:

$$(6.2) \qquad \phi_\varepsilon(t,x) + \frac{1}{2}\int_0^t |u + \phi_{\varepsilon x}(t-s,y(s))|^2 ds = \int_0^t [g(s,y(s)) + \frac{1}{2}|u(s)|^2]ds.$$

The proof is standard (see for instance [3] p.51-52, [7], [10]).

Now we consider the closed loop equation:

$$(6.3) \quad \begin{cases} y'(s) = Ay(s) + f(\varepsilon, y(s)) - \phi_{\varepsilon x}^{\infty}(y(s)) & \text{on } [0, +\infty); \\ y(0) = x & x \in \Sigma_r. \end{cases}$$

We remark that, since $\phi_{\varepsilon,x}^{\infty}$ is a locally Lipschitz monotone operator on H, then equation (6.3) has a unique mild solution $y_\varepsilon \in C([0, +\infty); H) \cap L^2(0, +\infty; H)$. Furthermore, for the monotonicity of $\phi_{\varepsilon x}^{\infty}$ we have that:

$$(6.4) \quad |y_\varepsilon(s)| \leq |x| \leq r \quad \forall s \in [0, +\infty).$$

Hence we can state the following theorem:

Theorem 6.2. Let $r_0 > 0$ be fixed. There exists $\varepsilon_3(r_0) > 0$ such that for $|x| \leq r_0$, $|\varepsilon| \leq \varepsilon_3(r_0)$, problem [P] has a unique optimal control u_ε. Moreover u_ε is given by the feedback formula:

$$(6.5) \quad u_\varepsilon(s) = -\phi_{\varepsilon,x}^{\infty}(y_\varepsilon(s))$$

where $y_\varepsilon(s)$ is the solution of the closed loop equation (6.3).

Proof: Let:

$$M_0 = M_0(r_0) = \max \left\{ (2/\omega)^{\frac{1}{2}} r_0 L(r_0), L(r_0), 2L(r_0)^{\frac{1}{2}} \right\}.$$

First we observe that if $|u|_{L^2} > M_0$ then $J(u) > J(0)$, in fact:

$$J(x, 0) = \int_0^\infty g(y_1(s)) ds$$

where

$$(6.6) \quad y_1' = Ay_1 + f(\varepsilon, y_1), \qquad y_1(0) = x;$$

which implies

$$|y_1(s)| \leq |x| e^{-\omega s/2} \quad \forall \varepsilon \leq \bar{\varepsilon}(r_0)$$

and, by (4.7):

$$(6.7) \quad J(x,0) \leq |g|_{0,r_0} \int_0^\infty e^{-\omega s/2} ds \leq 2r_0 |g|_{0,r_0}/\omega \leq L(r_0) \leq$$
$$\leq J(x, u) \quad \text{as} \quad |u|_{L^2} > M_0.$$

So we only have to minimize J for $u \in L^2(0, +\infty; H)$, $|u|_{L^2} \leq M_0$.

Now taking (6.2) and letting $t \to \infty$, for ε sufficiently small, we have:

$$(6.8) \quad \phi_\varepsilon^{\infty}(x) + \frac{1}{2} \int_0^{+\infty} |u(s) + \phi_{\varepsilon x}^{\infty}(y(s))|^2 ds = J(x, u) \quad \forall x \in \Sigma_{r_0}, \forall |u| \leq M_0(r_0).$$

At this point we show that u_ε given by (6.5) is such that $|u_\varepsilon|_{L^2} \leq M_0(r_0)$. In fact in this case it easily follows by (6.8) that u_ε is the unique optimal control for [P].

We have:

$$|\phi_{\varepsilon x}^\infty(y_\varepsilon(s))| = \lim_{t \to +\infty} |\phi_{\varepsilon x}(t, y_\varepsilon(s))| \leq \qquad \text{by (4.13)}$$

$$\leq L(r_0)|y_\varepsilon(s)| \leq L(r_0)r_0 e^{-\omega s/2}$$

It follows:

$$|u_\varepsilon|_{L^2} \leq r_0 L(r_0)(2\omega)^{\frac{1}{2}} \leq M_0(r_0).$$

Q.E.D.

Remark 6.3. By using the same method of [8] it is now possible, to show the regularity of the optimal pair $(u_\varepsilon, y_\varepsilon)$ with respect to parameter ε.

More precisely, under some regularity assumption on the data, we can give a Taylor expansion for the value function and for the optimal pair about the point $\varepsilon = 0$ (see [8] for details).

7. An example

Parabolic systems. Let $H = H^1(0,1)$. We consider the parabolic state equation:

(7.1)
$$\begin{cases} y_t(t,x) = \Delta_x y(t,x) + \varepsilon f(y(t,x)) + u(t,x) & t \in [0,+\infty), \ x \in (0,1) \\ y(0,x) = y_0(x) \in H^1(0,1); \quad y(t,0) = 0 = y(t,1) & \forall t \in [0,+\infty). \end{cases}$$

We denote by A the operator on H defined by:

$$\begin{cases} D(A) = H^3(0,1) \cap H_0^1(0,1) \\ Ay = \Delta y. \end{cases}$$

The control u is a generic element of $L^2(0,+\infty; H)$, and f is a smooth function $\mathbf{R} \to \mathbf{R}$ such that, if we define:

$$F : H \to H; \quad F(y)(x) = f(y(x)) \ \forall x \in (0,1),$$

then we have $F \in C_{Lip}^2(H,H)$. For example we can take $f(z) = z^2$ and use the fact that $H^1(0,1)$ is an algebra.

We want minimize the cost:

(7.2)
$$J(y_0, u) = \int_0^{+\infty} [\frac{1}{2}|y(s)|_H^2 + \frac{1}{2}|u(s)|_H^2]ds$$

over all controls $u \in L^2(0,+\infty; H)$, where y is the mild solution of (7.1).

Now assumptions i) — iii) are verified, in particular:

$$\langle Ay, y \rangle_H \leq -C_0|y|_H^2 \qquad \forall y \in D(A),$$

where $C_0 > 0$.

If r_0 is the supremum of the norm of the initial state y_0, we take M_0 as in section 6, and we set $|\varepsilon| \leq \varepsilon_3(r_0)$.

Then by the theorem 3.1 and theorem 6.2 there exists a unique optimal pair $(u_\varepsilon, y_\varepsilon)$ for problem [P], and the following feedback formula holds:

$$u_\varepsilon(s) = -\phi_{\varepsilon x}^\infty(y_\varepsilon(s))$$

where ϕ_ε^∞ is the solution of the stationary Hamilton-Jacobi equation (HJS).

We can repeat the same example in higher dimensions setting $H = H^k(\Omega)$ where Ω is an open bounded subset of $\mathbf{R}^n$ with sufficiently smooth boundary $\partial\Omega$ and $k > n/2$ (so H is an algebra).

Acknowledgements. I thanks Prof. Fleming of Brown University for useful discussions.

REFERENCES

[1] V. Barbu, *Hamilton-Jacobi equations and non linear control problems*, to appear.

[2] V. Barbu and G. Da Prato, "Hamilton-Jacobi equations in Hilbert spaces," Pitman, London, 1983.

[3] V. Barbu and G. Da Prato, *Hamilton-Jacobi equations in Hilbert spaces; variational and semigroup approach*, Ann. Mat. Pura Appl., (IV) **CXLII** (1985), 303–349.

[4] V. Barbu, G. Da Prato and C. Popa, *Existence and uniqueness of the dynamic programming equation in Hilbert space*, Nonlinear Analysis, Theory, Methods and Appl. **7** (1983), 283–299.

[5] P. Cannarsa and G. Da Prato, *Nonlinear optimal control with infinite horizon for distributed parameter systems and stationary Hamilton-Jacobi equations*.

[6] M.G. Crandall and P.L. Lions, *Hamilton-Jacobi equations in infinite dimensions, Part I : Uniqueness of viscosity solutions*, J. Func. Anal. **62**, (1985), 379–396; *Part II : Existence of viscosity solutions*, to appear in J. Func. Anal.; *Part III*, to appear in J. Func. Anal.; *Part IV*, to appear..

[7] W.H. Fleming and R.W. Rishel, "Deterministic and stochastic optimal control," Springer, New York, 1975.

[8] F. Gozzi, *Some results for an optimal control problem with a semilinear state equation*, to appear.

[9] D. Henry, *Geometric theory of semilinear parabolic equations*, in "Lecture Notes in Math. 840," Springer-Verlag, 1981.

[10] J.L. Lions, "Optimal control of systems governed by partial differential equations," Springer, Wiesbaden, 1972.

[11] A. Pazy, "Semigroups of linear operators and applications to partial differential equations," Springer-Verlag, New York-Heidelberg- Berlin, 1983.

[12] Y.C. You, *A nonquadratic Bolza problem and a quasi-Riccati equation for distributed parameter systems*, SIAM J. on Control and Opt. **25** (1987), 905–920.

[13] G. Di Blasio, *Global Solutions for a class of Hamilton-Jacobi equations in Hilbert spaces*, Numer. Funct. Anal. and Optimiz. **8(3-4)** (1985-86), 261–300.

Fausto Gozzi
Scuola Normale Superiore
Piazza dei Cavalieri 7
I-56100 Pisa
Italy

International Series of
Numerical Mathematics, Vol. 91
© 1989 Birkhäuser Verlag Basel

Stabilization of the viscoelastic Timoshenko beam

Kenneth B. Hannsgen*

Department of Mathematics
Virginia Polytechnic Institute and State University

Abstract. The decay of lateral vibrations in a viscoelastic Timoshenko beam with boundary feedback is investigated by analysis of the spectrum.

Keywords. Boundary feedback, Timoshenko beam, viscoelasticity

1980 *Mathematics subject classifications*: 93D15, 45K05

1. Introduction

This report concerns the decay of lateral vibrations in a uniform Timoshenko beam that obeys viscoelastic stress-strain relations. We derive and analyze a characteristic equation that determines the decay rate when the beam is subjected to shear force and bending moment feedback at one end. In particular, we develop an asymptotic formula for the decay rates of high-frequency vibrations, in a case where shearing and bending vibrations are essentially uncoupled at high frequencies. These results extend joint work with W. Desch, Y. Renardy, and R.L. Wheeler (reported in [1, Section 3]), which gave a similar analysis in the elastic case.

2. Statement of the problem

We begin with the equations

$$
(1) \qquad
\begin{aligned}
\rho w_{tt} - G(w_x - \varphi)_x &= 0 \\
R\varphi_{tt} - E\varphi_{xx} - G(w_x - \varphi) &= 0
\end{aligned}
$$

$(0 < x < 1, 0 < t < \infty)$ [5,6] for the elastic case. Here ρ is the linear mass density and R the mass moment of inertia of the cross section. For a fixed geometry, E is (proportional to) Young's modulus and G is the shear modulus of the material. The unknown $w(x,t)$ denotes the lateral displacement of the beam; the slope w_x splits into the two terms φ, representing bending, and $w_x - \varphi$, the part due to shear.

For viscoelastic materials, we replace $E\varphi_{xx}$ by

$$
A_\infty \varphi_{xx}(x,t) + \frac{d}{dt} \int_0^\infty a(\tau)\varphi_{xx}(x,t-\tau)d\tau,
$$

where $A(t) = A_\infty + a(t)$ $(t > 0)$ is the relaxation modulus. Similarly, G is replaced by the shear modulus $B(t) = B_\infty + b(t)$. We assume that $A_\infty > 0$ and $B_\infty > 0$ (so the material is a solid) and that $a(t)$ and $b(t)$ are completely monotone on $(0,\infty)$ with

$$
\int_0^\infty [a(t) + b(t)]e^{\delta t} dt < \infty
$$

*Supported by the Air Force Office of Scientific Research under grant AFOSR-86-0085 for some $\delta > 0$. Then a and b have Laplace transforms

$$\hat{a}(\lambda) = \int_0^\infty e^{-\lambda t} a(t)dt, \quad \hat{b}(\lambda) = \int_0^\infty e^{-\lambda t} b(t)dt$$

$(Re\lambda \geq -\delta)$. Let $\hat{A}(\lambda) = A_\infty \lambda^{-1} + \hat{a}(\lambda), \hat{B}(\lambda) = B_\infty \lambda^{-1} + \hat{b}(\lambda)$. From the Stieltjes representations for $\hat{a}$ and $\hat{b}$ [7], one sees that each of $\hat{A}, \hat{B}$ has at most one zero (λ_A and λ_B), both of which must be real and negative. Let $\sigma_* = max\{-\delta, \lambda_A, \lambda_B\}$.

Now make the elastic-viscoelastic substitutions indicated above to obtain the viscoelastic version of (1):

$$\rho w_{tt}(x,t) - [B_\infty(w_x - \varphi)_x(x,t) + \frac{d}{dt}\int_0^\infty b(\tau)(w_x - \varphi)_x(x,t-\tau)d\tau] = 0$$

(2) $\quad R\varphi_{tt}(x,t) - [A_\infty \varphi_{xx}(x,t) + \frac{d}{dt}\int_0^\infty a(\tau)\varphi_{xx}(x,t-\tau)d\tau + B_\infty(w_x - \varphi)(x,t)$

$$+ \frac{d}{dt}\int_0^\infty b(\tau)(w_x - \varphi)(x,t-\tau)d\tau] = 0 \ (0 < x < 1, \quad 0 < t < \infty).$$

At the left boundary we prescribe

(3) $\qquad\qquad\qquad \varphi(0,t) = w(0,t) = 0 \qquad (0 < t < \infty).$

At $x = 1$, we recall that (analogously to the elastic case [1])

$$A_\infty \varphi_x(1,t) + \frac{d}{dt}\int_0^\infty a(\tau)\varphi_x(1,t-\tau)d\tau = M(t) = \text{applied bending moment},$$

$$B_\infty(w_x - \varphi)(1,t) + \frac{d}{dt}\int_0^\infty b(\tau)(w_x - \varphi)(1,t-\tau)d\tau = S(t) = \text{applied shearing force}.$$

We prescribe

(4)

$$S(t) = -\int_0^\infty w_t(1,t-\tau)dK_1(\tau)$$

$$M(t) = -\int_0^\infty \varphi_t(1,t-\tau)dK_2(\tau) \quad (0 < t < \infty),$$

where $K_j(0) = 0, K_j(t) = \kappa_j + \int_0^t k_j(\tau)d\tau (t > 0)$, with $\kappa_j \geq 0$ and $k_j \in L^1(0,\infty), j = 1,2$. These are the viscoelastic analogues of the stabilizing feedbacks studied, e.g., in [5].

3. The characteristic equation

First consider product solutions $w(x,t) = v(x)e^{\lambda t}, \varphi(x,t) = \psi(x)e^{\lambda t} \ (Re\lambda > -\delta)$ of (2), (3), (4), and assume that $\hat{k}_j(\lambda)$ exists, $j = 1, 2$. Substitution yields

(2′)

$$\rho\lambda^2 v - (B_\infty + \lambda\hat{b}(\lambda))(v' - \psi)' = 0$$

$$R\lambda^2\psi - (A_\infty + \lambda\hat{a}(\lambda))\psi'' - (B_\infty + \lambda\hat{b}(\lambda))(v' - \psi) = 0 \ ('= d/dx, 0 < x < 1)$$

(3')
$$v(0) = \psi(0) = 0,$$

(4')
$$(A_\infty + \lambda\hat{a}(\lambda))\psi'(1) = -\lambda(\kappa_2 + \hat{k}_2(\lambda))\psi(1),$$
$$(B_\infty + \lambda\hat{b}(\lambda))(v' - \psi)(1) = -\lambda(\kappa_1 + \hat{k}_1(\lambda))v(1).$$

Since (2') has constant coefficients, its solutions are smooth, so we can differentiate and eliminate v algebraically to get

(2'')
$$\psi'''' - \lambda\left(\frac{\rho}{\hat{B}(\lambda)} + \frac{R}{\hat{A}(\lambda)}\right)\psi'' + \frac{\rho\lambda}{\hat{A}(\lambda)}(1 + \frac{R\lambda}{\hat{B}(\lambda)})\psi = 0 \ (0 < x < 1)$$

(3'')
$$[A_\infty + \lambda\hat{a}(\lambda)]\psi'''(0) - R\lambda^2\psi'(0) = \psi(0) = 0,$$

(4'')
$$\psi'(1) + \frac{\kappa_2 + \hat{k}_2(\lambda)}{\hat{A}(\lambda)}\psi(1) = 0$$
$$\rho\lambda\psi''(1) - \rho R\frac{\lambda^2}{\hat{A}(\lambda)}\psi(1) + (\kappa_1 + \hat{k}_1(\lambda))(\psi'''(1) - \frac{R\lambda}{\hat{A}(\lambda)}\psi'(1)) = 0,$$

with the obvious interpretations at $\lambda = 0$. Equation (2'') has a fundamental set of solutions, analytic in $\lambda, \lambda \in \Lambda \equiv \{Re\lambda > \sigma_*\}$. Imposing the boundary conditions (3'') and (4'') on the general solution of (2'') yields a linear equation

$$M(\lambda)c = 0$$

in $\mathbf{C}^4$, with coefficient matrix $M(\lambda)$ analytic in Λ. Nontrivial solutions exist if and only if

(5)
$$\Delta(\lambda) \equiv det M(\lambda) = 0.$$

As in [3,4], if $\Delta(\lambda) \neq 0$ in a half-plane $\{Re\lambda \geq \sigma_1 > \sigma_*\}$, then for suitable initial and history data, the system (2), (3), (4) has a solution that decays to zero like $e^{\sigma_1 t}$ as $t \to \infty$. Briefly, one derives a formal representation for the solution by treating the tails $\int_t^\infty$ in (2) and (4) as forcing terms and taking Laplace transforms. The resulting boundary value problem (with complex parameter λ, the transform variable) can be solved by an integral formula involving a Green's function, which is analytic for λ in Λ except at zeros of Δ. Estimates on the Green's function now permit one to justify the existence and decay of the solution in a suitable sense, whose precise nature depends on the smoothness and compatibility properties of the data.

It is not clear what the best existence-uniqueness framework is for this initial data problem. Without an inertial mass at $x = 1$, we cannot fit this system into the recently developed well-posedness theories, such as [2]. The remainder of this report is devoted to an analysis of Equation (5).

4. Solutions of the characteristic equation for $Re\lambda \geq 0$

We note first that, under mild conditions on the functions K_j (permitting $K_j \equiv 0$, in particular), (5) has no solution with $Re\lambda \geq 0$. Suppose, in fact, that a nontrivial solution of $(2')$, $(3')$, $(4')$ exists. Multiply the first equation of $(2')$ by $\bar{v}$ and the second by $\bar{\psi}$, integrate from 0 to 1 in x, and add. After integrating by parts one gets

$$
(6) \quad \begin{aligned}
&\lambda[\rho\|v\|^2 + R\|\psi\|^2 + (\kappa_1 + \hat{k}_1(\lambda))|v(1)|^2 + (\kappa_2 + \hat{k}_2(\lambda))|\psi(1)|^2] \\
&= -\hat{B}(\lambda)\|v' - \psi\|^2 - \hat{A}(\lambda)\|\psi'\|^2
\end{aligned}
$$

($\|\cdot\| = L^2$ norm) or, when $\lambda = 0$, $B_\infty\|v' - \psi\|^2 + A_\infty\|\psi'\|^2 = 0$. If k_1 and k_2 are completely monotone (positive, decreasing, and convex would suffice), then $Re\lambda\hat{k}_j(\lambda) \geq 0$. Taking real parts in (6), we get $\|v' - \psi\| = \|\psi'\| = 0$, and $(3')$ then gives us $v \equiv \psi \equiv 0$. We have proved the following.

Proposition 1. *With $A(t)$ and $B(t)$ as above, if κ_1 and κ_2 are nonnegative and k_1 and k_2 are completely monotone, then (5) has no solution with $Re\lambda \geq 0$.*

5. Solutions of the characteristic equation for $Im\lambda \to \infty$

For $Re\lambda < 0$, numerical methods alone have been successful in locating solutions of (5). See [1] for the elastic case; [3] indicates that the viscoelastic case will be similar. As $Im\lambda \to \infty$ (or $-\infty$, since zeros occur in conjugate pairs) in $\Sigma \equiv \{-\delta < Re\lambda < 0\}$, on the other hand, asymptotic formulas for $\hat{A}$ and $\hat{B}$ can simplify (5) and permit one to locate its roots. In this section, we assume that

$$
(7) \qquad e^{\delta t}k_j(t) \in L^1(0,\infty), j = 1,2,
$$

so that $\hat{k}_j(\lambda) \to 0$ as $\lambda \to \infty$ in Σ; this can be weakened, of course, if one is interested in a smaller strip.

Since $|\hat{A}(\lambda)| + |\hat{B}(\lambda)| \to 0$ ($\lambda \to \infty$ in Σ), the auxiliary equation for $(2'')$ takes the form

$$
\{\eta^2 - \frac{\rho\lambda}{\hat{B}(\lambda)}[1 + o(1)]\}\{\eta^2 - \frac{R\lambda}{\hat{A}(\lambda)}[1 + o(1)]\} = 0,
$$

so there is a fundamental set $\{e^{\eta_j x}\}_{j=1}^4$ for $(2'')$ with

$$
(8) \qquad \begin{aligned}
\eta_1 &= -\eta_2 = (R\lambda/\hat{A}(\lambda))^{1/2} + o(1) \\
\eta_3 &= -\eta_4 = (\rho\lambda/\hat{B}(\lambda))^{1/2} + o(1) \quad (\lambda \to \infty \text{ in } \Sigma).
\end{aligned}
$$

(We take both square roots positive when $\lambda > 0$.) By [3, Lemmas 2.2 and 2.3],

$$
(9) \qquad \begin{aligned}
Re\, \eta_1 &\sim (R/A(0))^{1/2}(\sigma - RA'(0)/2A(0)) \\
Re\, \eta_3 &\sim (\rho/B(0))^{1/2}(\sigma - \rho B'(0)/2B(0))
\end{aligned}
$$

if, respectively, $A'(0+) > -\infty$ or $B'(0+) > -\infty$, while $Re\eta_1 \to \infty (Re\eta_3 \to \infty)$ if $A'(0+) = -\infty (B'(0+) = -\infty)$.

To approximate Δ, substitute the general solution of (2″) into (3″) and (4″), dropping the lower order terms $\hat{k}_j$. The result (cf. [5, (3.16-(3.17)]) is

$$-\frac{1}{\lambda}\Delta(\lambda) \sim det[m_{ij}(\lambda)] \quad (\lambda \to \infty \text{ in } \Sigma),$$

where

$$m_{1j} = 1, m_{2j} = \eta_j F_j, m_{3j} = (\hat{A}\eta_j + \kappa_2)e^{\eta_j}$$
$$m_{4j} = (\lambda + \kappa_1\rho^{-1}\eta_j)F_j e^{\eta_j} \quad (j = 1, 2, 3, 4),$$

with

$$F_1 = F_2 = R\lambda^2 - \lambda\hat{A}(\lambda)\eta_1^2 = O(\lambda\hat{A}(\lambda)) = o(\lambda)$$
$$F_3 = F_4 = R\lambda^2 - \lambda\hat{A}(\lambda)\eta_3^2 \sim \lambda^2[R - \rho\hat{A}(\lambda)/\hat{B}(\lambda)]$$

($\lambda \to \infty$ in Σ).

When F_3 dominates F_1, there is a simple leading order term for $\Delta(\lambda)$. Thus we assume throughout the following that

(10) $|R - \rho\hat{A}(\lambda)/\hat{B}(\lambda)| \geq q > 0$

when $\lambda \in \Sigma$ with $Im\lambda$ sufficiently large. (When $A(0+) + B(0+) < \infty$, this simply means that $\rho A(0+) \neq RB(0+)$.) The result is

$$\lambda^{-1}\Delta(\lambda) \sim C\eta_3 F_3^2[(\lambda\rho - \kappa_1\eta_3)e^{-\eta_3} + (\lambda\rho + \kappa_1\eta_3)e^{\eta_3}] \cdot [(\hat{A}\eta_1 - \kappa_2)e^{-\eta_1} + (\hat{A}\eta_1 + \kappa_2)e^{\eta_1}]$$
$$\sim C\eta_3 F_3^2(\hat{A}/R\lambda)^{1/2} \cdot [(\lambda\rho - \kappa_1\eta_3)e^{-\eta_3} + (\lambda\rho + \kappa_1\eta_3)e^{\eta_3}] \cdot [(\lambda R - \kappa_2\eta_1)e^{-\eta_1} + (\lambda R + \kappa_2\eta_1)e^{\eta_1}]$$

($\lambda \to \infty$ in Σ), where C is a nonzero constant. By Rouché's Theorem, the zeros of Δ are given asymptotically by those of the two bracketed factors in the last expression. Using (8) and (9), we see that these two families of zeros are determined to leading order, respectively, by $\hat{B}$ and κ_1 (the shearing parameters) and $\hat{A}$ and κ_2 (the bending parameters). Moreover, the bracketed factors take precisely the same form as the characteristic determinant for the viscoelastic wave equation [3, (2.10)] (with $\epsilon = I = 0, s \to \lambda, \beta \to \eta_j, k \to \kappa_i$). As in [3], we can then read off the asymptotic behavior of the zeros, using (8) and (9). The result is as follows.

Proposition 2. *Assume (7) and (10), in addition to the general conditions of Section 2. For $\lambda \in \Sigma$ and $Im\lambda$ sufficiently large, all solutions of (5) fall into two families, Z_1 and Z_2. Z_1 (resp. Z_2) is empty if $B'(0+)$ (resp. $A'(0+)$) is $-\infty$. If $B'(0+) > -\infty$, then Z_1 consists of a sequence $\{z_{1n}\}$ with $Imz_{1n} \to \infty$,*

$$2Rez_{1n} \to 2\sigma_1 \equiv \frac{\rho B'(0+)}{B(0+)} + \left(\frac{B(0+)}{\rho}\right)^{1/2} log \left|\frac{\kappa_1 - (\rho B(0+))^{1/2}}{\kappa_1 + (\rho B(0+))^{1/2}}\right|.$$

(Z_1 is empty if $\sigma_1 < -\delta$.)

When $A'(0+) > -\infty$, we get a similar sequence $\{z_{2n}\}$ for Z_2, with $\rho \to R, B \to A$, and $\kappa_1 \to \kappa_2$.

Thus the asymptotic decay rate for high frequency vibrations is the larger of $e^{\sigma_j t}(j = 1, 2)$ when $A'(0+) + B'(0+) > -\infty$; this generalizes the conclusion for the elastic case [1].

REFERENCES

[1] W. Desch, K.B. Hannsgen, Y. Renardy and R.L. Wheeler, *Boundary stabilization of an Euler-Bernoulli beam with viscoelastic damping*, Proc. 26th IEEE Conference on Decision and Control, Los Angeles, CA, 1987, 1792–1795.

[2] W. Desch and R.K. Miller, *Exponential stabilization of Volterra integral equations with singular kernel*, J. Integral Equations Appl. (to appear).

[3] K.B. Hannsgen, Y. Renardy and R.L. Wheeler, *Effectiveness and robustness with respect to time delays of boundary feedback stabilization in one-dimensional viscoelasticity*, SIAM J. on Control and Optimization **26** (1988) (to appear).

[4] K.B. Hannsgen and R.L. Wheeler, *Existence and decay estimates for boundary feedback stabilization of torsional vibrations in a viscoelastic rod*, Proc. C.I.R.M. Meeting on Volterra Integrodifferential Equations in Banach Spaces and Applications, Trento, Italy, 1987, Pitman. (to appear)

[5] J.V. Kim and Y. Renardy, *Boundary control of the Timoshenko beam*, SIAM J. on Control and Optimization **25** (1987), 1417–1429.

[6] S. Timoshenko, "Vibration Problems in Engineering," Van Nostrand, New York, 1955.

[7] D.V. Widder, "The Laplace Transform," Princeton Univ. Press, Princeton, N.J., 1946.

Kenneth B. Hannsgen
Department of Mathematics
Virginia Polytechnic Institute and State University
Blacksburg, VA 24061-0123
USA

International Series of
Numerical Mathematics, Vol. 91
© 1989 Birkhäuser Verlag Basel

Linear quadratic optimal control problem for linear systems with unbounded input and output operators: Numerical approximations [*]

K. Ito and H. T. Tran

Center for Control Sciences
Brown University

Abstract. We present an abstract approximation framework for the numerical treatment of Riccati operators for a class of linear infinite dimensional systems with unbounded input and output operators. As a simple application, we show how the results of the general convergence theory may be applied and hence solve the linear quadratic control problem for linear functional differential equations with point delays in the controls.

Keywords. Linear quadratic regulator problem in Hilbert space, feedback controls, hereditary control systems, averaging approximation.

1980 *Mathematics subject classifications*: 34K35, 49A22, 65N30

1. Introduction

For at least two decades now, there have been a number of interesting developments in the theory of linear quadratic regulator problems for infinite dimensional systems which involves unbounded input operator in the evolution equation and/or unbounded output operator in the quadratic performance index (e.g., see [Ic], [VK], [KL], [IP], [PS], [La], [FLT], [DLT] and the references cited there). As with the linear quadratic regulator for systems represented by ordinary differential equations, the solution to a Riccati equation defines the feedback structure of the optimal control law. Thus the main issue in this subject has been the study of existence and uniqueness of solutions of Riccati equations.

In this paper we present an approximation framework for solutions to Riccati equations for a class of infinite dimensional systems with unbounded input and output operators studied in Pritchard-Salamon [PS]. We will call it the Pritchard-Salamon class. Our study by no means is complete or comprehensive. It does not cover many of the important boundary control problems studied in e.g. [LT1], [LT2]. For such a problem we refer to [La1], [LT3] for approximation results. One of our motivations to use the Pritchard-Salamon framework is that it enables us to treat the control problem governed by delay differential equations with delays in control and observation (in a systematic manner). The Pritchard-Salamon class assumes the smoothness of the underlined semigroup and observation map (e.g. see the assumption (**H3**)). Because of the smoothness assumption, the Riccati solution has a smoothing property (i.e., $\Pi \in \mathcal{L}(X^*, X)$, see Theorems 2.2 and 2.3 for the notation). Thus an objective of our study is to develop an approximation theory

[*]This research was supported in part by the National Science Foundation under NSF Grant MCS-8504316 and by the Air Force Office of Scientific Research under Contract F49620-86-C-011.

for Riccati solutions that yields not only the strong convergence of approximating solutions in $\mathcal{L}(H)$-topology but also in the stronger topology $\mathcal{L}(X^*, X)$ for the Pritchard-Salamon class. In §2, a general convergence theory which can be used in computational techniques is developed in the context of the theory of infinite dimensional Riccati equations developed in [PS]. We then show in §3 that the abstract framework of §2 can be used for a numerical scheme based on the averaging approximation method to treat the infinite dimensional Riccati equations in connection with hereditary control problems involving point delays in the inputs.

2. A general approximation framework

In this section we present a general convergence framework that can be used to treat approximation techniques for the linear quadratic optimal control problem for the Pritchard-Salamon class [PS]. The results given here; in particular, for infinite time horizon problem, extend the approximation theory developed in [Gi], [BK], [I2] in which the input and output operators are assumed to be bounded to the unboundedness cases. Specifically, the general idea is to approximate the regulator problem by a sequence of finite dimensional state space problems, each embedded into the state spaces H and X of the original problem.

We assume throughout that H, U and Y are Hilbert spaces and we identify them with their duals, that X is a Hilbert space such that $X \subset H = H^* \subset X^*$ with continuous dense injections and where X^* denote the strong dual space of X. In a formal sense our basic model is a linear control system given by an abstract equation [PS], [IP]:

$$(2.1) \qquad \frac{d}{dt}x(t) = Ax(t) + Bu(t), \qquad x(0) = x_0,$$
$$y(t) = Cx(t), \qquad 0 \le t \le T,$$

in H where $u(\cdot) \in L^2(0,T;U)$, $y(\cdot) \in L^2(0,T;Y)$ and A is the infinitesimal generator of a strongly continuous semigroup $S(t)$ on H. We interpret equation (2.1) in the mild form, that is its solution $x(t)$ is given by the variation of constants formula

$$(2.2) \qquad x(t) = S(t)x_0 + \int_0^t S(t-s)Bu(s)\,ds, \qquad 0 \le t \le T.$$

Moreover, as in [PS], we assume the following to discuss the problem involving possible unboundedness of the operators B and C: namely, $B \in \mathcal{L}(U, X^*)$ and $C \in \mathcal{L}(W, Y)$ where W is a Hilbert space such that $W \subset H$ with continuous dense embedding. Therefore, in order to make the expression (2.2) precise and to allow for trajectories in all three spaces W, H and X^*, we assume the following hypothesis [PS]:

(H1) $S(t)$ is also a strongly continuous semigroup on W and X^*;

(H2) For any $u(\cdot) \in L^2(0,T;U)$, $\int_0^T S(T-s)Bu(s)\,ds \in W$ and there exists a constant $b > 0$ such that

$$\left\| \int_0^T S(T-s)Bu(s)\,ds \right\|_W \le b\|u\|_{L^2(0,T;U)};$$

(H3) There exists a constant $c > 0$ such that

$$\int_0^T \left\| CS(t)x_0 \right\|_Y^2 \, dt \leq c \|x_0\|_{X^*},$$

for every $x_0 \in W$;

(H4) $Z = \text{dom}_{X^*}(A) \subset W$ with a continuous dense embedding where the Hilbert space Z is endowed with the graph norm of A on X^*.

Associated with the control system (2.2) is the performance index

$$(2.3) \qquad J_T(u; x_0) = \langle x(T), Gx(T) \rangle_{X^*, X} + \int_0^T \left(\|Cx(t)\|_Y^2 + \|u(t)\|_U^2 \right) \, dt,$$

where $G \in \mathcal{L}(X^*, X)$ is a self-adjoint, non-negative operator. The following result gives a characterization of G which will be useful later.

Corollary 2.1. $G = (G^{1/2})^* G^{1/2}$ where $G^{1/2} = \mathcal{L}(X^*, H)$.

Proof: Let i be the canonical embedding from X into H, i.e., $\|ix\|_H = \|x\|_X$. Then $i^* : H \to X^*$ and $i^*i : X \to X^*$. Hence, from the assumption on the operator G, we have

$$\langle x, Gy \rangle_{X^*, X} = \langle Gx, y \rangle_{X, X^*}$$
$$= \langle i^* i Gx, y \rangle_{X^*}.$$

But, since $\langle x, Gy \rangle_{X^*, X} = \langle x, i^* i Gy \rangle_{X^*}$, $i^* i G \in \mathcal{L}(X^*)$ is a self-adjoint operator. From [**Ka**, p. 281], $i^* i G = \tilde{G} \tilde{G}$, where $\tilde{G} \in \mathcal{L}(X^*)$ is a self-adjoint, non-negative operator. Therefore,

$$\langle x, Gy \rangle_{X^*, X} = \langle x, i^* i Gy \rangle_{X^*}$$
$$= \langle \tilde{G}x, \tilde{G}y \rangle_{X^*}$$
$$= \langle (i^*)^{-1} \tilde{G}x, (i^*)^{-1} \tilde{G}y \rangle_H$$
$$= \langle x, \tilde{G}i^{-1}(i^*)^{-1} \tilde{G}y \rangle_{X^*}.$$

The result then follows with $G^{1/2} = (i^*)^{-1} \tilde{G}$. ∎

The following representation for the optimal control and Riccati operator is quite standard in the literature (e.g. see [**La1**], [**Gi**]). That is, from the variation of constants formula (2.2), the output function $y(t) = Cx(t)$ in $L^2(0, T; Y)$ can be written as

$$y(t) = Mx_0 + Lu \qquad \text{in } L^2(0, T; Y),$$

where the linear operators M and L are defined by

$$(2.4) \qquad (Mx_0)(t) = CS(t)x_0, \qquad \text{for } x_0 \in W,$$

$$(2.5) \qquad (Lu)(t) = C \int_0^t S(t - s) Bu(s) \, ds, \qquad \text{for } u \in L^2(0, T; U).$$

By **(H1)-(H4)**, M has a unique continuous extension to all of X^* which will be also denoted by $M \in \mathcal{L}(X^*, L^2(0,T;Y))$, and $L \in \mathcal{L}(L^2(0,T;U), L^2(0,T;Y))$. Therefore, the cost functional (2.3) can be written as

(2.6) $$J_T(u; x_0) = \|M_T x_0 + L_T u\|_H^2 + \|M x_0 + Lu\|_{L^2(0,T;Y)}^2 + \|u\|_{L^2(0,T;U)}^2,$$

where the bounded linear operators $L_T : L^2(0,T;U) \to H$ and $M_T : X^* \to H$ associated with the terminal constraint are defined by

(2.7) $$L_T u = G^{1/2} \int_0^T S(T-s)Bu(s)\, ds, \qquad \text{for } u \in L^2(0,T;U),$$

(2.8) $$M_T x_0 = G^{1/2} S(T) x_0, \qquad \text{for } x_0 \in X^*.$$

Theorem 2.2. *Assume that T is finite and $x_0 \in X^*$. The optimal control u_T^* that minimizes (2.6) is given by*

(2.9) $$u_T^* = -(I + L^*L + L_T^* L_T)^{-1}(L^*M + L_T^* M_T) x_0$$

and if we define the self-adjoint operator $\Pi_T \in \mathcal{L}(X^*, X)$ by

(2.10) $$\Pi_T = (M^*, M_T^*) \left(\begin{bmatrix} I & 0 \\ 0 & I \end{bmatrix} + \begin{bmatrix} L \\ L_T \end{bmatrix} [L^*, L_T^*] \right)^{-1} \begin{pmatrix} M \\ M_T \end{pmatrix},$$

then

$$\langle \Pi_T x_0, x_0 \rangle_{X,X^*} = J_T(u_T^*; x_0) = \min J_T(u; x_0).$$

Moreover, if we let $\Pi_T(t) = \Pi_{T-t}$, $t \le T$, then the optimal control is given by the feedback law

(2.11) $$u_T^*(t) = -B^* \Pi_T(t) x_T^*(t),$$

where $x_T^*(t)$ is the corresponding optimal trajectory.

Proof: To simplify our illustration we define the bounded linear operators $\tilde{M} = \begin{pmatrix} M \\ M_T \end{pmatrix}$:

$X^* \to L^2(0,T;Y) \times H$ and $\tilde{L} = \begin{pmatrix} L \\ L_T \end{pmatrix} : L^2(0,T;U) \to L^2(0,T;Y) \times H$. Then the performance index (2.6) can be equivalently written as

$$\begin{aligned}
J_T(u; x_0) &= \|\tilde{M} x_0 + \tilde{L} u\|_{L^2(0,T;Y) \times H}^2 + \|u\|_{L^2(0,T;U)}^2 \\
&= \langle \tilde{M} x_0, \tilde{M} x_0 \rangle + 2\langle \tilde{L}^* \tilde{M} x_0, u \rangle + \langle (I + \tilde{L}^* \tilde{L}) u, u \rangle \\
&= \langle \tilde{M} x_0, \tilde{M} x_0 \rangle - \langle (I + \tilde{L}^* \tilde{L})^{-1} \tilde{L}^* \tilde{M} x_0, \tilde{L}^* \tilde{M} x_0 \rangle \\
&\quad + \langle u + (I + \tilde{L}^* \tilde{L})^{-1} \tilde{L}^* \tilde{M} x_0, (I + \tilde{L}^* \tilde{L})(u + (I + \tilde{L}^* \tilde{L})^{-1} \tilde{L}^* \tilde{M} x_0) \rangle,
\end{aligned}$$

which is minimized at $u_T^* = -(I + \tilde{L}^*\tilde{L})^{-1}\tilde{L}^*\tilde{M}x_0$. Next, we evaluate the cost functional at the optimal control to obtain

$$
\begin{aligned}
J_T(u_T^*; x_0) &= \langle \tilde{M}x_0, \tilde{M}x_0 \rangle - \langle (I + \tilde{L}^*\tilde{L})^{-1}\tilde{L}^*\tilde{M}x_0, \tilde{L}^*\tilde{M}x_0 \rangle \\
&= \langle \tilde{M}^*\tilde{M}x_0, x_0 \rangle - \langle \tilde{M}^*\tilde{L}(I + \tilde{L}^*\tilde{L})^{-1}\tilde{L}^*\tilde{M}x_0, x_0 \rangle \\
&= \langle \tilde{M}^*\tilde{M}x_0, x_0 \rangle - \langle \tilde{M}^*(I + \tilde{L}\tilde{L}^*)^{-1}\tilde{L}\tilde{L}^*\tilde{M}x_0, x_0 \rangle,
\end{aligned}
$$

where we used the identity $\tilde{L}(I + \tilde{L}^*\tilde{L})^{-1} = (I + \tilde{L}\tilde{L}^*)^{-1}\tilde{L}$. Now, we have

$$
\begin{aligned}
J_T(u_T^*; x_0) &= \langle \tilde{M}^*(I - (I + \tilde{L}\tilde{L}^*)^{-1}\tilde{L}\tilde{L}^*)\tilde{M}x_0, x_0 \rangle \\
&= \langle \tilde{M}^*(I + \tilde{L}\tilde{L}^*)^{-1}\tilde{M}x_0, x_0 \rangle.
\end{aligned}
$$

The last assertion of the theorem is a consequence of Theorem 2.6 in [**PS**]. ∎

Next we consider the control problem of minimizing the performance index

$$
(2.12) \qquad J(u; x_0) = \int_0^\infty \left(\left\| Cx(t) \right\|_Y^2 + \left\| u(t) \right\|_U^2 \right) dt,
$$

subject to (2.1). The following theorem is a special case of the general results in [**PS**].

Theorem 2.3. *Assume that the pair (A, B) is stabilizable, meaning that there exists an operator $K \in \mathcal{L}(X^*, U)$ such that $A - BK$ generates an exponentially stable semigroup on X^* and that (A, C) is detectable in the sense that there exists an operator $F \in \mathcal{L}(Y, X^*)$ such that $A - FC$ generates an exponentially stable semigroup on X^*. Then there exists a unique non-negative, self-adjoint solution $\Pi \in \mathcal{L}(X^*, X)$ to the algebraic Riccati equation*

$$
(A^*\Pi + \Pi A - \Pi BB^*\Pi + C^*C)x = 0, \qquad \forall x \in Z.
$$

Moreover, the optimal control u^* that minimizes (2.12) has the following feedback form

$$
u^*(t) = -B^*\Pi T(t)x_0, \qquad x_0 \in X^*,
$$

where the strongly continuous closed loop semigroup generated by $A - BB^*\Pi$, $T(t)$ on X^*, is exponentially stable.

Next, we formulate a sequence of approximate regulator problems and present convergence results for the corresponding Riccati operators for the finite and infinite time horizon control problems. In our formulation, the Hilbert spaces X and H are fundamental. To this end, we consider a sequence of approximations $(Z^N, A^N, B^N, C^N; i^N, j^N)$ such that for $N = 1, 2, \ldots, Z^N$ is the finite dimensional Euclidean space, $A^N \in \mathcal{L}(Z^N)$, $B^N \in \mathcal{L}(U, Z^N)$, $C^N \in \mathcal{L}(Z^N, Y)$ and i^N, j^N are the injective linear maps

$$
i^N : Z^N \to H, \quad j^N : Z^N \to X.
$$

Let H^N be a sequence of finite dimensional linear subspaces of H which is isometrically isomorphic to Z^N by means of the embedding i^N, meaning that $H^N = \text{image}(i^N)$. On Z^N

we will always consider its norm to be induced from the H-norm, that is $\|z\|_N^2 = \|i^N z\|_H^2$ for $z \in Z^N$. Then it is obvious that the operators i^N and $(i^N)^*$ satisfy

(2.13) $(i^N)^* i^N = $ the identity on Z^N,

(2.14) $i^N (i^N)^* = P_H^N$, the orthogonal projection of H onto H^N.

Also, we denote by $X^N \subset X$ the image of the linear map j^N. In general, X^N is not contained in H^N. Note that if we identify Z^N with X^N in the sense that the norm $\|\cdot\|_{\tilde N}$ on Z^N is induced from one on X, meaning that $\|z\|_{\tilde N}^2 = \|j^N z\|_X^2$ for $z \in Z^N$, then we also have

(2.15) $k^N j^N = $ the identity on Z^N,

(2.16) $j^N k^N = P_X^N$, the orthogonal projection of X onto X^N,

where the linear map k^N is defined by $\langle k^N x, z \rangle_{\tilde N} = \langle x, j^N z \rangle_X$, $z \in Z^N$ and $x \in X$. For simplicity of our discussions, we assume that

(2.17)
$$C^N (i^N)^* = C P_H^N,$$
$$i^N B^N = P_H^N B.$$

Then the Nth approximate optimal control problem to (2.3) is as follows:

(2.18) Minimize $J_T^N(u; x_0) = \langle \xi^N(T), G^N \xi^N(T) \rangle_N + \int_0^T \left(\left\| C^N \xi^N(t) \right\|_Y^2 + \|u(t)\|_U^2 \right) dt,$

subject to

$$\frac{d}{dt} \xi^N(t) = A^N \xi^N(t) + B^N u(t), \qquad t > 0,$$
$$\xi^N(0) = (j^N)^* x_0 \in Z^N, \qquad x_0 \in X^*,$$

where the symmetric matrix G^N is defined by $j^N G^N (j^N)^* = P_X^N G (P_X^N)^*$. Then, from (2.13), (2.14) and (2.17),

$$y(t) = C^N \xi^N(t)$$
$$= C^N \left(e^{t A^N} (j^N)^* x_0 + \int_0^t e^{(t-s) A^N} B^N u(s) \, ds \right)$$
$$= C^N (i^N)^* i^N \left(e^{t A^N} (j^N)^* x_0 + \int_0^t e^{(t-s) A^N} (i^N)^* i^N B^N u(s) \, ds \right)$$
$$= C i^N e^{t A^N} (j^N)^* x_0 + C i^N \int_0^t e^{(t-s) A^N} (i^N)^* B u(s) \, ds$$
$$= (M^N x_0)(t) + (L^N u)(t) \in L^2(0, T; Y),$$

where the linear maps M^N and L^N, which approximate the corresponding operators defined by (2.4) and (2.5), are given by

(2.19) $(M^N x_0)(t) = C i^N e^{tA^N} (j^N)^* x_0, \quad x_0 \in X^*,$

(2.20) $(L^N u)(t) = C i^N \int_0^t e^{(t-s)A^N} (i^N)^* B u(s)\, ds, \quad u \in L^2(0, T; U).$

Also, by (2.15), (2.16) and the identity $k^N P_X^N = k^N$, we have

$$\langle \xi^N(T), G^N \xi^N(T) \rangle_N = \langle (j^N)^*(k^N)^* \xi^N(T), G^N (j^N)^*(k^N)^* \xi^N(T) \rangle_N$$
$$= \langle (k^N)^* \xi^N(T), j^N G^N (j^N)^*(k^N)^* \xi^N(T) \rangle_{X^*,X}$$
$$= \langle (k^N)^* \xi^N(T), G(k^N)^* \xi^N(T) \rangle_{X^*,X}.$$

Therefore, the performance index of the Nth approximate control problem (2.18) can be equivalently written as

$$J_T^N(u; x_0) = \| M_T^N x_0 + L_T^N u \|_H^2 + \| M^N x_0 + L^N u \|_{L^2(0,T;Y)}^2 + \| u \|_{L^2(0,T;U)}^2,$$

where the linear maps M_T^N and L_T^N approximating the corresponding operators (2.7) and (2.8) are given by

(2.21)
$$M_T^N x_0 = G^{1/2} (k^N)^* e^{TA^N} (j^N)^* x_0, \quad x_0 \in X^*$$

(2.22) $L_T^N u = G^{1/2} (k^N)^* \int_0^T e^{(T-s)A^N} (i^N)^* B u(s)\, ds, \quad u \in L^2(0, T; U).$

Consequently, by Theorem 2.2, the Nth optimal control to (2.18) is given by

(2.23) $u_T^N = -(I + (L^N)^* L^N + (L_T^N)^* L_T^N)^{-1} ((L^N)^* M^N + (L_T^N)^* M_T^N) x_0, \qquad x_0 \in X^*$

If we define the self-adjoint operator Π_T^N on Z^N by

(2.24) $j^N \Pi_T^N (j^N)^* x_0$

$$= ((M^N)^*, (M_T^N)^*) \left(\begin{bmatrix} I & 0 \\ 0 & I \end{bmatrix} + \begin{bmatrix} L^N \\ L_T^N \end{bmatrix} [(L^N)^*, (L_T^N)^*] \right)^{-1} \begin{pmatrix} M^N \\ M_T^N \end{pmatrix} x_0,$$

where $x_0 \in X^*$, then

$$\langle \Pi_T^N (j^N)^* x_0, (j^N)^* x_0 \rangle_N = J_T^N(u_T^N; x_0) = \min J_T^N(u; x_0).$$

Moreover, if we let $\Pi_T^N(t) = \Pi_{T-t}^N$, $t \leq T$, then $\Pi_T^N(t)$, $t \leq T$, satisfies the differential Riccati equation in Z^N

(2.25)
$$\frac{d}{dt} \Pi_T^N(t) + (A^N)^* \Pi_T^N(t) + \Pi_T^N(t) A^N$$
$$- \Pi_T^N(t) B^N (B^N)^* \Pi_T^N(t) + (C^N)^* C^N = 0,$$
$$\Pi_T^N(T) = G^N.$$

Hence, the following result is an immediate consequence of (2.23), (2.24) and Theorem 2.2 (e.g. see [**La**], [**IT**]).

Theorem 2.4. *Assume that T is finite and that the approximation is consistent, meaning that as $N \to \infty$,*

$$\|M^N x_0 - M x_0\|_{L^2(0,T;Y)} \to 0, \qquad \text{for all } x_0 \in X^*,$$

$$\|L^N u - L u\|_{L^2(0,T;Y)} \to 0, \qquad \text{for all } u \in L^2(0,T;U),$$

$$\|M_T^N x_0 - M_T x_0\|_H \to 0, \qquad \text{for all } x_0 \in X^*,$$

$$\|L_T^N u - L_T u\|_H \to 0, \qquad \text{for all } u \in L^2(0,T;U),$$

$$\|(M^N)^* y - M^* y\|_X \to 0, \qquad \text{for all } y \in L^2(0,T;Y),$$

$$\|(L^N)^* y - L^* y\|_{L^2(0,T;U)} \to 0, \qquad \text{for all } y \in L^2(0,T;Y),$$

$$\|(M_T^N)^* \phi - M_T^* \phi\|_X \to 0, \qquad \text{for all } \phi \in H,$$

$$\|(L_T^N)^* \phi - L_T^* \phi\|_{L^2(0,T;U)} \to 0, \qquad \text{for all } \phi \in H,$$

where the convergences are assumed to be uniform in T on bounded intervals. Then as $N \to \infty$ and for all $x_0 \in X^*$,

$$\|u_T^N - u_T^*\|_{L^2(0,T;U)} \to 0,$$

$$\|j^N \Pi_T^N (j^N)^* x_0 - \Pi_T x_0\|_X \to 0,$$

uniformly in T.

The above result concerns with the convergences of the Riccati operators and the optimal open loop controls. The following result shows the convergence of the feedback kernels, $j^N \Pi_T^N(t) B^N$, $t \leq T$.

Corollary 2.5. *Assume that for all $u \in U$, $\|(k^N)^*(i^N)^* B u - B u\|_{X^*} \to 0$ as $N \to \infty$, then*

$$\|j^N \Pi_T^N(t) B^N u - \Pi_T(t) B u\|_X \to 0,$$

for all $u \in U$, and the convergence is uniform in t for $t \in [0,T]$. Moreover, if B is of finite rank, i.e., $\dim(U) = $ finite, then the convergence is in the uniform operator topology.

Proof: By Theorem 2.4, we have

$$j^N \Pi_T^N(t)(j^N)^* B u \to \Pi_T(t) B u,$$

for all $u \in U$. But from (2.13)–(2.17),

$$j^N \Pi_T^N(t) B^N u = j^N \Pi_T^N(t)(i^N)^* B u$$
$$= j^N \Pi_T^N(t)(j^N)^*(k^N)^*(i^N)^* B u,$$

which shows the corollary. ∎

For the infinite time horizon control problem, the following fundamental convergence results extends the approximation framework pursued in [BK], [Gi], [I2] in which $B \in \mathcal{L}(U,H)$ and $C \in \mathcal{L}(H,Y)$ are assumed to the case where $B \in \mathcal{L}(U,X^*)$ and $C \in \mathcal{L}(W,Y)$.

Theorem 2.6. *Suppose that the assumptions in Theorems 2.3, 2.4 and Corollary 2.5 are satisfied and that for a positive integer N_0,*

(a) *(uniform stabilizability of (A^N, B^N)) $\exists$ constants $M_1 \geq 1$, $\omega_1 > 0$ independent of N and a sequence of operators $K^N \in \mathcal{L}(Z^N, U)$ such that $\sup \|j^N (K^N)^*\|_{\mathcal{L}(U,X)} < \infty$ and*

$$\|j^N e^{t(A^N - B^N K^N)^*} k^N \phi\|_X \leq M_1 e^{-\omega_1 t} \|\phi\|_X,$$

and

(b) *(uniform detectability of (A^N, C^N)) $\exists$ constants $M_2 \geq 1$, $\omega_2 > 0$ independent of N and a sequence of operators $F^N \in \mathcal{L}(Y, Z^N)$ such that $\sup \|(F^N)^* k^N\|_{\mathcal{L}(X,Y)} < \infty$ and*

$$\|j^N e^{t(A^N - F^N C^N)^*} k^N \phi\|_X \leq M_2 e^{-\omega_2 t} \|\phi\|_X,$$

for $N \geq N_0$ and $\phi \in X$. Then, as $N \to \infty$,

$$\|j^N \Pi^N (j^N)^* x_0 - \Pi x_0\|_X \to 0,$$

for all $x_0 \in X^*$, where Π is the unique, non-negative, self-adjoint solution of the algebraic Riccati equation, see Theorem 2.3, and $\Pi^N \in \mathcal{L}(Z^N)$ is a unique, non-negative, self-adjoint solution to the Nth approximate algebraic Riccati equation in Z^N:

$$(2.26) \qquad (A^N)^* \Pi^N + \Pi^N A^N - \Pi^N B^N (B^N)^* \Pi^N + (C^N)^* C^N = 0.$$

Moreover, for each $u \in U$,

$$\|j^N \Pi^N B^N u - \Pi B u\|_X \to 0,$$

as $N \to \infty$.

Proof: Appendix. ∎

We remark that if $\dim(U) = $ finite, then the sequence of feedback kernels $j^N \Pi^N B^N$ converges in the uniform operator topology to ΠB. Furthermore, the convergence results in Theorem 2.6 are stronger than those obtained in, e.g. [**BK**, Theorem 2.2], in the sense that $\Pi \in \mathcal{L}(X^*, X)$ where $X \subset H \subset X^*$.

In the next section we shall show how the results of our approximation framework may be applied and hence solve the linear quadratic control problem for linear functional differential equations with general delays in the state and inputs.

3. The linear regulator problem for retarded systems

We illustrate the application of our approximation results presented in the previous section with a linear quadratic optimal control problem involving linear retarded functional differential equations with delays in input. We consider the initial value problem

$$(3.1) \qquad \frac{d}{dt} x(t) = A_0 x(t) + A_1 x(t - r) + \int_{-r}^{0} A_{01}(s) x(t + s) \, ds$$
$$+ B_0 u(t) + B_1 u(t - r),$$

$$(3.2) \qquad y(t) = C_0 x(t),$$

where $x(t) \in R^n$, $u(t) \in R^m$, $y(t) \in R^p$ and A_0, A_1, B_0, B_1, C_0 are matrices of suitable dimensions. Furthermore, $A_{01}(\cdot)$ is an $n \times n$ matrix valued square integrable functions on $[-r, 0]$. It is well known that system (3.1) admits a unique solution $x(\cdot) \in L^2_{loc}(-r, \infty; R^n) \cap H^1_{loc}(0, \infty; R^n)$ for every input $u(\cdot) \in L^2_{loc}(0, \infty; R^m)$ and every initial condition of the form

$$(3.3) \qquad x(0) = \eta, \quad x(\theta) = \phi_1(\theta), \quad u(\theta) = \phi_2(\theta), \qquad \theta \in (-r, 0),$$

where $\phi = (\eta, \phi_1, \phi_2) \in R^n \times L^2(-r, 0; R^n) \times L^2(-r, 0; R^m)$. Moreover, this solution depends continuously on u and ϕ on compact intervals (e.g., see [PS], [BT], [DM]). The corresponding output $y(\cdot)$ is in $L^2_{loc}(0, \infty; R^p)$ and depends in this space continuously on u and ϕ.

Associated with the control problem (3.1)–(3.2) we consider the performance measure

$$(3.4) \qquad J(u; \phi) = \int_0^T \left(|C_0 x(t)|^2_{R^p} + |u(t)|^2_{R^m} \right) dt,$$

for given $\phi = (\eta, \phi_1, \phi_2) \in R^n \times L^2 \times L^2$, and $T \in (0, \infty]$.

3.1. A dual evolution formulation. The *classical* way of formulating (3.1) as a well-posed control problem defined on a Hilbert space is to introduce the state of system (3.1) at time $t \geq 0$ to be the triple $\tilde{z}(t) = \big(x(t), x(t + \theta), u(t + \theta)\big) \in R^n \times L^2(-r, 0; R^n) \times L^2(-r, 0; R^m)$ which completely describes the past history of the solution [Ic]. An alternative state concept [PS], [VK] is to introduce the state function to be

$$z(t) = \Big(x(t), A_1\chi_{[-r,0]}(\theta)x(t - r - \theta) + \int_{-r}^{\theta} A_{01}(\xi)x(t + \xi - \theta)\, d\xi$$
$$+ B_1\chi_{[-r,0]}(\theta)u(t - r - \theta)\Big) \in H,$$

where H denote the Hilbert space $R^n \times L^2(-r, 0; R^n)$ equipped with norm

$$(3.5) \qquad \|(\eta, \phi)\|^2_H = \|\eta\|^2_{R^n} + \int_{-r}^{0} \|\phi(\theta)\|^2_{R^n}\, d\theta.$$

We note that if $\mathcal{F} : H \times L^2(-r, 0; R^m) \to H$, denotes the so-called structual operator defined by (e.g., see [BM], [S1])

$$(3.6) \qquad \mathcal{F}(\eta, \phi_1, \phi_2) = \Big(\eta, A_1\chi_{[-r,0]}(\theta)\phi_1(-r - \theta) + \int_{-r}^{\theta} A_{01}(\xi)\phi_1(\xi - \theta)\, d\xi$$
$$+ B_1\chi_{[-r,0]}(\theta)\phi_2(-r - \theta)\Big),$$

then $z(t) = \mathcal{F}\tilde{z}(t)$ and it satisfies the following evolution equation in H

$$(3.7) \qquad \begin{aligned} \frac{d}{dt}z(t) &= A_T^* z(t) + B_T^* u(t), & t \geq 0, \\ z(0) &= \mathcal{F}(\eta, \phi_1, \phi_2) \in H, \\ y(t) &= C_T^* z(t), & t \geq 0, \end{aligned}$$

where the output operator $C_T^* \in \mathcal{L}(H, R^p)$ is defined by

$$(3.8) \qquad\qquad C_T^*(\eta, \phi) = C_0 \eta.$$

A_T^* and B_T^* are the dual operators of A_T and B_T where the linear operators A_T and B_T are associated with the transposed system

$$(3.9) \qquad \begin{aligned} \frac{d}{dt} w(t) &= A_0^T w(t) + A_1^T w(t-r) + \int_{-r}^{0} A_{01}^T(s) w(t+s)\, ds + C_0^T y(t) \\ u(t) &= B_0^T w(t) + B_1^T w(t-r) \end{aligned}$$

and are defined by [PS]

$$(3.10) \qquad \begin{aligned} \operatorname{dom}(A_T) &= \{(\eta, \phi) \in H \mid \phi \in H^1(-r, 0; R^n),\ \phi(0) = \eta\} \\ A_T(\eta, \phi) &= \left(A_0^T \eta + A_1^T \phi(-r) + \int_{-r}^{0} A_{01}^T(s) \phi(s)\, ds,\ \dot\phi \right) \end{aligned}$$

and

$$(3.11) \qquad\qquad B_T(\eta, \phi) = B_0^T \eta + B_1^T \phi(-r),$$

for $(\eta, \phi) \in X = \operatorname{dom}(A_T) \subset H$ where X is the Hilbert space equipped with norm

$$\|(\phi(0), \phi)\|_X^2 = \|\phi(0)\|_{R^n}^2 + \int_{-r}^{0} \|\dot\phi(s)\|_{R^n}^2\, ds.$$

Then, the cost functional (3.4) is equivalently written as

$$(3.12) \qquad\qquad J(u, z) = \int_0^T \left(|C_T^* z(t)|_{R^p}^2 + |u(t)|_{R^m}^2 \right) dt,$$

with $z = \mathcal{F}\phi$.

We note that $B_T^* \in \mathcal{L}(R^m, X^*)$ and it follows from [PS], [IP] that the hypotheses (H1)-(H4) in Section 2 are satisfied for the minimization problem (3.12) subject to (3.7). Here, $W = H$ and $Z = \operatorname{dom}_{X^*}(A_T^*) = H$.

Remarks. Above we formulate our problem (3.1) using the dual state space framework instead of the more natural *classical* state space concept. The advantage of doing so is that we employ a state space, $R^n \times L^2$, which is smaller than that the state space, $R^n \times L^2 \times L^2$, of the evolution equation for $\tilde{z}(t)$. Moreover, the dual formulation X is independent of problem data (comparing with the corresponding space V^* in [PS] for the natural state space formulation).

3.2. Averaging approximations. In this section we describe an approximation scheme for retarded systems which has been studied by [**BB**], [**Gi**], [**S2**], [**LM**] and many others. A basic idea of the following formulation of AVE scheme was first introduced in [**I1**]. To this end, for every positive integer N, let $t_j^N = -\frac{jr}{N}$, $j = 0, \ldots, N$, χ_j denote the characteristic function of $[t_j, t_{j-1})$ for $j = 1, 2, \ldots, N$ and B_j^N, $j = 0, \ldots, N$ denote the usual first order spline elements on the interval $[-r, 0]$ corresponding to the mesh $t_0^N, \ldots, t_N^N$,

$$
B_j^N(\theta) = \begin{cases} \frac{N}{r}(\theta - t_{j+1}), & \theta \in [t_{j+1}^N, t_j^N], \\ \frac{N}{r}(t_{j-1}^N - \theta), & \theta \in [t_j^N, t_{j-1}^N], \\ 0, & \text{elsewhere.} \end{cases}
$$

The injections $i^N : Z^N \to H$ are given by

(3.13)
$$
i^N z = \left(z_0, \sum_{i=1}^N z_i \chi_i \right),
$$

where Z^N is the Euclidean space $R^{n(N+1)}$ equipped with the induced norm described in Section 2 and $z = \text{col}(z_0, z_1, \ldots, z_N) \in Z^N$. The adjoint operator $(i^N)^* : H \to Z^N$, where $(i^N)^*(\eta, \phi) = z$, is then given by

(3.14)
$$
z_0 = \eta, \qquad z_j = \frac{N}{r} \int_{t_j^N}^{t_{j-1}^N} \phi(\theta) \, d\theta, \quad j = 1, \ldots, N.
$$

The injections $j^N : Z^N \to X$ are given by

(3.15)
$$
j^N z = \left(z_0, \sum_{i=0}^N z_i B_i^N(\theta) \right).
$$

The linear maps $k^N : X \to Z^N$ are given by

(3.16) $k^N(\phi(0), \phi) = \text{col}(\phi_0, \phi_1, \ldots, \phi_N)$, $\phi_j = \phi(t_j^N)$, $\quad j = 0, 1, \ldots, N$.

We next define the approximating operators $A_T^N : H \to H^N$ by

(3.17) $A_T^N(\eta, \phi) = \left(A_0^T z_0 + A_1^T z_N + \sum_{j=1}^N (A_{01}^T)_j z_j, \sum_{j=1}^N \frac{N}{r}(z_{j-1} - z_j)\chi_j \right)$,

where z_j, $j = 0, \ldots, N$, are defined by (3.14) and $(A_{01})_j = \int_{t_j^N}^{t_{j-1}^N} A_{01}(s) \, ds$. The following result establishes the stability of the averaging approximations with respect to the X−norm which is stronger than the H−norm obtained earlier in [**BB**], [**MT**].

Theorem 3.1. *There exists a constant ω independent of N such that*

$$\left\| j^N \left(e^{tA_T^N} z + \int_0^t e^{(t-s)A_T^N} (f(s), 0)\, ds \right) \right\|_X \leq e^{\omega t} \left(\| j^N z \|_X + \sqrt{3} \| f \|_{L^2(0,t;R^n)} \right).$$

Proof: For $z = \operatorname{col}(z_0, z_1, \ldots, z_N) \in Z^N$, let $e^{tA_T^N} z + \int_0^t e^{(t-s)A_T^N}(f(s), 0)\, ds = z(t) = \operatorname{col}(z_0(t), \ldots, z_N(t))$. Then $z(t)$ satisfies

(3.18)
$$\frac{d}{dt} z_0(t) = A_0^T z_0(t) + A_1^T z_N(t) + \sum_{j-1}^{N} (A_{01}^T)_j z_j(t) + f(t),$$

(3.19)
$$\frac{d}{dt} z_j(t) = \frac{N}{r} (z_{j-1}(t) - z_j(t)), \qquad j = 1, 2, \ldots, N.$$

Considering first equation (3.19), we obtain

$$\sum_{j=1}^{N} \langle \frac{d}{dt} z_j(t), \frac{d}{dt}(z_{j-1} - z_j(t)) \rangle = \sum_{j=1}^{N} \frac{N}{r} \langle z_{j-1}(t) - z_j(t), \frac{d}{dt}(z_{j-1}(t) - z_j(t)) \rangle,$$

where,

$$\text{RHS} = \frac{1}{2} \sum_{j=1}^{N} \frac{N}{r} \frac{d}{dt} |z_{j-1}(t) - z_j(t)|^2$$

$$= \frac{1}{2} \frac{d}{dt} \| j^N z(t) \|_X^2 - \frac{1}{2} \frac{d}{dt} |z_0(t)|^2,$$

where we used $\| j^N z(t) \|_X^2 = |z_0(t)|_{R^n}^2 + \sum_{j=1}^{N} \frac{N}{r} |z_{j-1}(t) - z_j(t)|_{R^n}^2$. And,

$$\text{LHS} \leq -\frac{1}{2} \sum_{j=1}^{N} |\frac{d}{dt} z_j(t)|^2 + \frac{1}{2} \sum_{j=1}^{N} |\frac{d}{dt} z_{j-1}(t)|^2$$

$$= \frac{1}{2} |\frac{d}{dt} z_0(t)|^2 - \frac{1}{2} |\frac{d}{dt} z_N(t)|^2.$$

Therefore, we obtain the following estimate

(3.20)
$$\frac{d}{dt} \| j^N z(t) \|_X^2 \leq \frac{d}{dt} |z_0(t)|^2 + |\frac{d}{dt} z_0(t)|^2.$$

Next, since $z_j(t) = z_0(t) - \int_{t_j^N}^{0} \frac{d}{d\theta}(j^N z(t))\, d\theta$, $j = 1, \ldots, N$, we have

$$|z_j(t)| \leq |z_0(t)| + \int_{t_j^N}^{0} |\frac{d}{d\theta}(j^N z(t))|\, d\theta$$

$$\leq |z_0(t)| + \sqrt{r} \left(\int_{-r}^{0} |\frac{d}{d\theta}(j^N z(t))|^2\, d\theta \right)^{1/2}.$$

Since both terms on the right-hand side are bounded above by $\|j^N z(t)\|_X$, we obtain

(3.21) $$|z_j(t)| \leq (1 + \sqrt{r})\|j^N z(t)\|_X, \qquad j = 1, \ldots, N.$$

Finally, considering the equation (3.18), we have

$$\langle \frac{d}{dt} z_0(t), z_0(t) \rangle = \langle A_0^T z_0(t), z_0(t) \rangle + \langle A_1^T z_N(t), z_0(t) \rangle$$
$$+ \sum_{j=1}^{N} \langle (A_{01}^T)_j z_j(t), z_0(t) \rangle + \langle f(t), z_0(t) \rangle.$$

But with use of Cauchy-Schwarz and estimate (3.21) we argue,

(3.22)
$$\frac{1}{2} \frac{d}{dt} |z_0(t)|^2 \leq |A_0^T| |z_0(t)|^2 + |A_1^T| |z_N(t)| |z_0(t)|$$
$$+ \sum_{j=1}^{N} \int_{t_j^N}^{t_{j-1}^N} |A_{01}(s)| ds |z_0(t)| |z_j(t)|$$
$$+ |z_0(t)| |f(t)|$$
$$\leq \omega_1 |z_0(t)| \|j^N z(t)\|_X + |z_0(t)| |f(t)|,$$

where $\omega_1 = \left(|A_0^T| + |A_1^T| + \|A_{01}\|_{L^1} \right)(1 + \sqrt{r})$. From the above estimate, we also obtain

$$\frac{1}{2} |\frac{d}{dt} z_0(t)|^2 \leq \omega_1^2 \|j^N z(t)\|_X^2 + |f(t)|^2.$$

Therefore, using (3.20), (3.22) and this inequality we have

(3.23) $$\frac{1}{2} \frac{d}{dt} \|j^N z(t)\|_X^2 \leq (\frac{1}{2} + \omega_1 + \omega_1^2) \|j^N z(t)\|_X^2 + \frac{3}{2} |f(t)|^2.$$

Hence, the assertion in the theorem is a consequence of Gronwall's inequality with $\omega = \frac{1}{2} + \omega_1 + \omega_1^2$. ∎

The following result is a direct consequence of Theorem 3.1 and (3.23).

Corollary 3.2. *The following statements hold.*

  (i) $j^N A_T^N k^N - \omega I$ *is dissipative on* X.
  (ii) *For every* $T > 0$, *there exists a positive constant* c, *dependent on* T *but independent of* N, *such that*

$$\left\| j^N \int_0^T e^{(T-s)A_T^N} (i^N)^* C_T y(s) \, ds \right\|_X \leq c \|y\|_{L^2(0,T;R^p)},$$

for all $y(\cdot) \in L^2(0, T; R^p)$ *and all* N.

Remarks. The dual statement of Corollary 3.2 is given by

   (i) For every $\phi \in H$,

(H3)^{N}
$$\|C_T^* i^N e^{t(A_T^N)^*} (j^N)^* \phi\|_{L^2(0,T;R^p)} \le c\|\phi\|_{X^*},$$

which is the corresponding hypothesis **(H3)** of Section 2 associated with the Nth approximate system to (3.7). Furthermore, from [**MT**, Theorem 4.5] we have

   (ii) For every $T > 0$, there exists a positive constant b, dependent on T but independent of N, such that for every $\phi \in X$,

(H2')^{N}
$$\|B_T i^N e^{t A_T^N} (i^N)^* \phi\|_{L^2(0,T;R^m)} \le b\|\phi\|_H.$$

The dual statement of which is

   (iii) For any $u \in L^2(0,T;R^m)$,

(H2)^{N}
$$\left\| i^N \int_0^T e^{(T-s)(A_T^N)^*} (i^N)^* B_T^* u(s)\, ds \right\|_H \le b\|u\|_{L^2(0,T;R^m)},$$

the corresponding hypothesis **(H2)** of Section 2 associated with the Nth approximate system to (3.7).

We now establish some elementary convergence results which will play an important role.

Lemma 3.3. For every $\phi \in H$ we have

$$\|(k^N)^*(i^N)^* \phi - \phi\|_{X^*} \le \frac{1}{\sqrt{2}} \left(\frac{r}{N}\right) \|\phi\|_H.$$

Proof: Let $\phi \in H$ and $\psi \in X$, then

$$|\langle (k^N)^*(i^N)^*\phi - \phi, \psi\rangle_{X^*,X}| = |\langle \phi, i^N k^N \psi - \psi\rangle_H|$$
$$\le \|\phi\|_H \|i^N k^N \psi - \psi\|_H.$$

Thus the lemma follows from the estimate $\|i^N k^N \psi - \psi\|_H \le \frac{1}{\sqrt{2}}\|\dot\psi\|_{L^2} \frac{r}{N}$, [**LM**, Proposition A2].

Corollary 3.4. Let $z = col(z_0, z_1, \ldots, z_N) \in Z^N$. There exists a constant $c_1 > 0$ such that

$$\|i^N A_T^N z - A_T j^N z\|_H \le c_1 \left(\frac{r}{N}\right) \|j^N z\|_X.$$

Proof: From (3.13), (3.14) and the characterizations of A_T, A_T^N, the L^2–component of $i^N A_T^N z - A_T j^N z$ is identically zero. Moreover, the R^n–component of $i^N A_T^N z$ and $A_T j^N z$ are given by

$$i^N A_T^N z\big|_{R^n} = \sum_{j=1}^{N} (A_{01}^T)_j z_j + A_0^T z_0 + A_1^T z_N$$

$$= \sum_{j=1}^{N} \int_{-r}^{0} A_{01}(s)\chi_j(s) z_j\, ds + A_0^T z_0 + A_1^T z_N$$

$$= \int_{-r}^{0} A_{01}(s)(i^N z)(s)\, ds + A_0^T z_0 + A_1^T z_N$$

and

$$A_T j^N z \big|_{R^n} = \int_{-r}^0 A_{01}(s)(j^N z)(s)\, ds + A_0^T z_0 + A_1^T z_N.$$

Since $i^N z = i^N k^N j^N z$, the result follows from Lemma 3.3. ∎

Remarks. By Corollary 3.4, without loss of generality, we will use the following relation

(3.24) $i^N A_T^N z = A_T j^N z,$

for the remainder of this section. Also, since the input function does not have distributed delay, it follows directly from (3.11), (3.13), (3.15) and (3.16) that

(3.25) $B_T j^N = B_T i^N$ and $B_T = B_T i^N k^N,$

which implies the condition in Corollary 2.5. Moreover, by Lemma 3.3 we have

(3.26) $\|(k^N)^* - i^N\|_{\mathcal{L}(Z^N, X^*)} \le \frac{1}{\sqrt{2}}\left(\frac{r}{N}\right).$

We turn to prove convergence results of the approximating semigroups.

Theorem 3.5. *The following statements hold.*

(i) *For all* $\phi = (\eta, \phi) \in H$

$$S_T(t)\phi = \lim_{N\to\infty} i^N e^{t A_T^N} (i^N)^* \phi,$$

$$S_T^*(t)\phi = \lim_{N\to\infty} i^N e^{t(A_T^N)^*} (i^N)^* \phi,$$

and the limits are uniform on every compact interval $[0, T]$.

(ii) *For all* $\phi = (\phi(0), \phi) \in X$

$$S_T(t)\phi = \lim_{N\to\infty} j^N e^{t A_T^N} k^N \phi,$$

and the limit is uniform on every compact interval $[0, T]$.

Proof: (i) These statements have been proved in [**BB**], [**Gi**].

(ii) For $\phi \in X = \mathrm{dom}\,(A_T)$, $\|\cdot\|_X$ is equivalent to the graph norm of A_T. From statement (i), without loss of generality, we will prove

$$\|A_T j^N e^{t A_T^N} k^N \phi - A_T S_T(t)\phi\|_H \to 0,$$

as $N \to \infty$ and for all $\phi \in X$. Now, by using the identities (3.24) and $(i^N)^* i^N = \mathrm{id}$, we obtain

$$A_T j^N e^{t A_T^N} k^N \phi = i^N A_T^N e^{t A_T^N} k^N \phi$$
$$= i^N e^{t A_T^N} (i^N)^* i^N A_T^N k^N \phi$$

and since $\phi \in \mathrm{dom}\,(A_T)$

$$A_T S_T(t)\phi = S_T(t)A_T\phi,$$

so that if we show $\|i^N A_T^N k^N \phi - A_T\phi\|_H \to 0$, as $N \to \infty$ and for all $\phi \in X$, then the result follows from statement (i). To this end we have

$$i^N A_T^N k^N \phi = \left(A_0^T \phi(0) + A_1^T \phi(-r) + \sum_{j=1}^{N}(A_{01}^T)_j \phi(t_j),\right.$$

$$\left.\sum_{j=1}^{N} \frac{N}{r}(\phi(t_{j-1}^N) - \phi(t_j^N))\chi_j\right)$$

and

$$A_T\phi = \left(A_0^T \phi(0) + A_1^T \phi(-r) + \int_{-r}^{0} A_{01}^T(s)\phi(s)\,ds,\ \dot\phi\right).$$

From [**BB**, Lemma 3.2], it was shown that

$$\sum_{j=1}^{N} \frac{N}{r}(\phi(t_{j-1}^N) - \phi(t_j^N))\chi_j \to \dot\phi \qquad \text{in } L^2(-r,0;R^n).$$

And since

$$\sum_{j=1}^{N}(A_{01}^T)_j\phi(t_j) = \sum_{j=1}^{N}\int_{t_j^N}^{t_{j-1}^N} A_{01}^T(s)\phi(t_j)\,ds = \int_{-r}^{0} A_{01}^T(s)\left[\sum_{j=1}^{N}\chi_j(s)\phi(t_j)\right]ds$$

$$= \int_{-r}^{0} A_{01}^T(s)(i^N k^N \phi)(s)\,ds,$$

we have from Lemma 3.3

$$\sum_{j=1}^{N}(A_{01}^T)_j\phi(t_j) \to \int_{-r}^{0} A_{01}^T(s)\phi(s)\,ds.$$

This concludes our proof. ∎

In all subsequent results, we will verify the convergence theory of Section 2 as it is applied to the specific retarded functional differential equations problem which is the focus of the present section. Turning first to Theorem 2.4, we will show that the statements on consistency of our approximation scheme hold. This guarantees the convergences of the Riccati operators and the optimal open loop controls for the finite time horizon problem. Moreover, by Lemma 3.3 and Corollary 2.5, the feedback kernels, $j^N \Pi_T^N(t)(B_T^N)^*$, $t \leq T$, converge to $\Pi_T(t)B_T^*$ in the uniform operator topology.

Lemma 3.6. *Let T be finite. Then as $N \to \infty$, the following statements hold.*

(i) $\|C_T^* i^N e^{t(A_T^N)^*}(j^N)^* \phi - C_T^* S_T^*(t)\phi\|_{L^2(0,T;R^p)} \to 0$, *for all $\phi \in X^*$.*

(ii) *For all $y \in L^2(0, T; R^p)$,*

$$\left\| j^N \int_0^T e^{(T-s)A_T^N}(i^N)^* C_T y(s)\, ds - \int_0^T S_T(T-s)C_T y(s)\, ds \right\|_X \to 0.$$

Proof: (i) By **(H3)^N**, i.e., $\|M^N\|_{\mathcal{L}(X^*, L^2(0,T;R^p))} \le c$, it suffices to prove this statement for all $\phi \in H$ which is dense in X^*. To this end, let $\phi \in H$ and by using the identity $(j^N)^*(k^N)^*(i^N)^*\phi = (i^N)^*\phi$, we obtain

$$M^N \phi - M\phi = M^N(\phi - (k^N)^*(i^N)^*\phi) + (C_T^* i^N e^{t(A_T^N)^*}(i^N)^*\phi - C_T^* S_T^*(t)\phi).$$

Since $C_T^* \in \mathcal{L}(H, R^p)$, the result follows from Theorem 3.5 and Lemma 3.3.

(ii) By Corollary 3.2, it suffices to prove the result for all $y(\cdot) \in C^1(0, T; R^p)$ with $y(0) = 0$. Without loss of generality [see remarks in the proof of Theorem 3.5 (ii)], we will prove

$$\left\| A_T j^N \int_0^T e^{(T-s)A_T^N}(i^N)^* C_T y(s)\, ds - A_T \int_0^T S_T(T-s)C_T y(s)\, ds \right\|_H \to 0.$$

From [Pa, pp. 4–5] and using the relation (3.24) we obtain

$$A_T j^N \int_0^T e^{(T-s)A_T^N}(i^N)^* C_T y(s)\, ds = i^N A_T^N \int_0^T e^{(T-s)A_T^N}(i^N)^* C_T y(s)\, ds$$

$$= -i^N (i^N)^* C_T y(T) + i^N \int_0^T e^{(T-s)A_T^N}(i^N)^* C_T \dot{y}(s)\, ds.$$

Similarly,

$$A_T \int_0^T S_T(T-s)C_T y(s)\, ds = -C_T y(T) + \int_0^T S_T(T-s)C_T \dot{y}(s)\, ds.$$

Since $C_T y = (C_0^T y, 0)$, we have $i^N (i^N)^* C_T = C_T$. Result then follows from Theorem 3.5 (i). ∎

Lemma 3.7. *Let T be finite. Then, as $N \to \infty$, the following statements hold.*

(i) $\|G^{1/2}(k^N)^* e^{T(A_T^N)^*}(j^N)^* \phi - G^{1/2} S_T^*(T)\phi\|_H \to 0$, *for all $\phi \in X^*$.*

(ii) $\|j^N e^{T A_T^N} k^N (G^{1/2})^* \phi - S_T(T)(G^{1/2})^* \phi\|_X \to 0$, *for all $\phi \in H$.*

Proof: Statement (i) follows by using similar arguments as in the proof of Lemma 3.6 (i), and statement (ii) is an immediate consequence of Theorem 3.5 (ii).

Lemma 3.8. *Let T be finite. Then, as $N \to \infty$, the following statements hold.*

(i) $\left\| C_T^* i^N \int_0^t e^{(t-s)(A_T^N)^*} (i^N)^* B_T^* u(s)\, ds - C_T^* \int_0^t S_T^*(t-s) B_T^* u(s)\, ds \right\|_{L^2(0,T;R^p)} \to 0$, *for all* $u \in L^2(0,T;R^m)$.

(ii) $\left\| B_T^N i^N \int_t^T e^{(s-t) A_T^N} (i^N)^* C_T y(s)\, ds - B_T \int_t^T S_T(s-t) C_T y(s)\, ds \right\|_{L^2(0,T;R^m)} \to 0$, *for all* $y \in L^2(0,T;R^p)$.

Proof: (ii) Follow from Theorem 3.5 (i).

(i) It suffices to show that for each $t > 0$,

$$(3.27) \qquad \left\| i^N \int_0^t e^{(t-s)(A_T^N)^*} (i^N)^* B_T^* u(s)\, ds - \int_0^t S_T^*(t-s) B_T^* u(s)\, ds \right\|_H \to 0.$$

By **(H2)**^{N}, it will be enough to show this statement for $u \in C^1(0,T;R^n)$ with $u(0) = 0$. Without loss of generality we also assume that $0 \in \rho(A_T^*)$. Then

$$A_T^* \int_0^t S_T^*(t-s) B_T^* u(s)\, ds = -B_T^* u(t) + \int_0^t S_T^*(t-s) B_T^* \dot{u}(s)\, ds,$$

which implies

$$\int_0^t S_T^*(t-s) B_T^* u(s)\, ds = -(A_T^*)^{-1} B_T^* u(t) + \int_0^t S_T^*(t-s)(A_T^*)^{-1} B_T^* \dot{u}(s)\, ds.$$

Similarly,

$$i^N \int_0^t e^{(t-s)(A_T^N)^*} (i^N)^* B_T^* u(s)\, ds = -i^N (A_T^{N*})^{-1} (i^N)^* B_T^* u(t)$$

$$+ i^N \int_0^t e^{(t-s)(A_T^N)^*} (A_T^{N*})^{-1} (i^N)^* B_T^* \dot{u}(s)\, ds.$$

From [**PS**, Proposition 5.13], we have $\text{dom}_{X^*}(A_T^*) = H$. Therefore $(A_T^*)^{-1} B_T^* \in H$. Thus, if we show

$$\left\| i^N (A_T^{N*})^{-1} (i^N)^* B_T^* - (A_T^*)^{-1} B_T^* \right\|_{\mathcal{L}(R^m, H)} \to 0,$$

as $N \to \infty$, then statement (i) follows from Theorem 3.5 (i). We have,

$$\left\| i^N (A_T^{N*})^{-1} (i^N)^* B_T^* - (A_T^*)^{-1} B_T^* \right\|_{\mathcal{L}(R^m, H)} = \left\| B_T i^N (A_T^N)^{-1} (i^N)^* - B_T (A_T)^{-1} \right\|_{\mathcal{L}(H, R^m)}.$$

Using (3.24) and (3.25), we obtain

$$B_T i^N (A_T^N)^{-1} (i^N)^* = B_T j^N (A_T^N)^{-1} (i^N)^*$$

$$= B_T (A_T)^{-1} A_T j^N (A_T^N)^{-1} (i^N)^*$$

$$= B_T (A_T)^{-1} i^N (i^N)^*$$

$$= B_T (A_T)^{-1} P_H^N.$$

Since, [**BB**], $P_H^N \phi \to \phi$, for $\phi \in H$, this completes the proof. ∎

Lemma 3.9. *Let T be finite. Then, as $N \to \infty$, the following statements hold.*

(i) $\|G^{1/2}(k^N)^* \int_0^T e^{(T-s)(A_T^N)^*}(i^N)^* B_T^* u(s)\, ds - G^{1/2} \int_0^T S_T^*(T-s) B_T^* u(s)\, ds\|_H \to 0$,
for all $u \in L^2(0, T; R^m)$.

(ii) $\|B_T i^N e^{tA_T^N} k^N (G^{1/2})^* \phi - B_T S_T(t)(G^{1/2})^* \phi\|_{L^2(0,T;R^m)} \to 0$, *for all $\phi \in H$.*

Proof: Statement (ii) follows from (3.25) and Theorem 3.5 (ii), and statement (i) follows from (3.26) and (3.27). ∎

Finally we discuss convergence results of Theorem 2.6. Since $C_T^* \in \mathcal{L}(H, R^p)$, the condition on uniform detectability of $((A_T^N)^*, (C_T^N)^*)$ has been proved in [**I1**]. To verify the statement of uniform stabilizability of $((A_T^N)^*, (B_T^N)^*)$ we show below that the approximating semigroups $j^N e^{t(A_T^N - K_T^N B_T^N)} k^N$ is uniformly differentiable. Then by using similar arguments as in [**LM**], it follows that if the semigroup generated by $A_T - K_T B_T$ is exponentially stable, then the approximating semigroups $j^N e^{t(A_T^N - K_T^N B_T^N)} k^N$ is also exponentially stable.

Theorem 3.10. *For all $N \in \mathcal{N}$ and some positive constant b, the approximating semigroups $j^N e^{t(A_T^N - K_T^N B_T^N)} k^N$ are differentiable for $t > b$ in the uniform operator topology, uniformly with respect to N.*

Proof: We will first observe two facts.

Fact 1. Let $\phi \in X$, then there exist positive constants b and M (independent of N) such that $\|j^N A_T^N e^{tA_T^N} k^N \phi\|_X \leq M\|\phi\|_X$, $t \geq b$ and for all N.

For it, note that by [**LM, Theorem 3.1**], $\|A_T^N e^{tA_T^N} z\|_H \leq M\|i^N z\|_H$ for all N and where t is sufficiently large. Since $\|\cdot\|_X$ is equivalent to the graph norm of A_T, without loss of generality we can consider

$$\|j^N A_T^N e^{tA_T^N} k^N \phi\|_X = \|A_T j^N e^{tA_T^N} A_T^N k^N \phi\|_H$$
$$= \|i^N A_T^N e^{tA_T^N} A_T^N k^N \phi\|_H \leq M\|i^N A_T^N k^N \phi\|_H,$$

where $z = i^N A_T^N k^N \phi = A_T j^N k^N \phi$ satisfies

$$\|z\|_H \leq \|A_T\|_{\mathcal{L}(X,H)} \|P_X^N \phi\|_N \leq c_1 \|\phi\|_X.$$

Fact 2. Let $\phi \in X$, then for every $T > 0$ there exists a positive constant c (dependent on T but independent of N) such that $\int_0^T |B_T i^N e^{sA_T^N} A_T^N k^N \phi|_{R^m}^2\, ds \leq c\|\phi\|_X^2$.

Fact 2 follows from $(\mathbf{H2'})^N$ and the above estimate.

Finally, we turn to prove the desired result in the theorem. That is we need to show that

$$\|j^N (A_T^N - K_T^N B_T^N) e^{t(A_T^N - K_T^N B_T^N)} k^N \phi\|_X \leq \hat{M}\|\phi\|_X,$$

for some sufficiently large t and where the positive constant $\hat{M}$ is independent of N. To this end, since $B_T^N = B_T i^N$ we have

$$j^N (A_T^N - K_T^N B_T^N) e^{t(A_T^N - K_T^N B_T^N)} k^N \phi = j^N (A_T^N - K_T^N B_T j^N) e^{tA_T^N} k^N \phi$$
$$- j^N (A_T^N - K_T^N B_T^N) \int_0^t e^{(t-s)(A_T^N - K_T^N B_T^N)} k^N j^N K_T^N B_T i^N e^{sA_T^N} k^N \phi\, ds.$$

Therefore, for sufficiently large t the first term $\leq \hat{M}_1 \|\phi\|_X$, by virtue of $j^N K_T^N \in \mathcal{L}(R^m, X)$, $B_T \in \mathcal{L}(X, R^m)$, Theorem 3.1 and Fact 1. The second term $\leq \hat{M}_2 \|\phi\|_X$ by Fact 2 and [**Ka**, pp. 499- 489]. This completes the proof. ∎

Appendix

We give here a proof of Theorem 2.6 in two steps.

Step 1. Suppose that $x_0 \in X^*$, then there exists a positive constant α (independent of N) such that $\min J^N(u; x_0) \leq \alpha \|x_0\|_{X^*}^2$.

Proof. Let $x_0 \in X^*$ and consider the minimization problem

$$\text{Minimize} \int_0^T (\|C^N \xi^N(t)\|_Y^2 + \|u(t)\|_U^2)\, dt,$$

subject to

$$\frac{d}{dt} \xi^N(t) = A^N \xi^N(t) + B^N u(t), \qquad t > 0,$$

$$\xi^N(0) = (j^N)^* x_0.$$

Suppose that $u(t) = -K^N \xi^N(t)$, then we have

$$\xi^N(t) = e^{t(A^N - B^N K^N)}(j^N)^* x_0$$

$$= e^{tA^N}(j^N)^* x_0 - \int_0^t e^{(t-s)A^N} B^N K^N e^{s(A^N - B^N K^N)}(j^N)^* x_0\, ds.$$

Hence, using (2.17) and (2.15) we obtain

$$\int_0^T \|C^N \xi^N(t)\|_Y^2\, dt \leq 2 \int_0^T \|C i^N e^{tA^N}(j^N)^* x_0\|_Y^2\, dt$$

$$+ 2 \int_0^T \|C i^N \int_0^t e^{(t-s)A^N}(i^N)^* B K^N (j^N)^*(k^N)^* e^{s(A^N - B^N K^N)}(j^N)^* x_0\, ds\|_Y^2\, dt,$$

where the first term on the right-hand side $\leq c_1 \|x_0\|_{X^*}^2$, by the uniform boundedness principle, (**H3**), (2.4), (2.19) and the first condition in Theorem 2.4. From (**H2**), (2.5), (2.20) and the second condition of Theorem 2.4, the second term can be estimated by

$$\leq 2c_2 \|j^N (K^N)^*\|_{\mathcal{L}(U,X)} \int_0^T \|(k^N)^* e^{s(A^N - B^N K^N)}(j^N)^* x_0\|_Y^2\, ds \leq c_3 \|x_0\|_X^2,$$

where we used the assumption (**a**) of the theorem. Thus for any $T > 0$, we have

(A1)
$$\int_0^T \|C^N \xi^N(t)\|_Y^2\, dt \leq c \|x_0\|_{X^*}^2,$$

where c is independent of N. Moreover, we have

$$\int_0^\infty \|C^N \xi^N(t)\|_Y^2 \, dt = \int_0^T \|C^N \xi^N(t)\|_Y^2 \, dt + \sum_{k=1}^\infty \int_0^T \|C^N \xi^N(t + kT)\|_Y^2 \, dt,$$

where, from (A1) the first term $\leq c\|x_0\|_{X^*}^2$, and from (A1) and assumption **(a)** the second term $\leq c\|x_0\|_{X^*}^2 \sum_{k=1}^\infty M_1^2 e^{-2\omega_1 kT}$. Let T be sufficiently large so that $M_1^2 e^{-2\omega_1 T} \leq 1/2$ then $\sum_{k=1}^\infty M_1^2 e^{-2\omega_1 kT} \leq 2$. Hence,

$$\text{(A2)} \qquad \int_0^\infty \|C^N \xi^N(t)\|_Y^2 \, dt \leq 3c\|x_0\|_{X^*}^2.$$

Also, since $u(t) = -K^N \xi^N(t) = -K^N (j^N)^* (k^N)^* e^{t(A^N - B^N K^N)} (j^N)^* x_0$, we obtain

$$\text{(A3)} \qquad \int_0^\infty \|u(t)\|_U^2 \, dt \leq \|j^N (K^N)^*\|_{\mathcal{L}(U,X)}^2 \frac{M_1^2}{2\omega_1} \|x_0\|_{X^*}^2.$$

Since $\min J^N(u; x_0) \leq \int_0^\infty (\|C^N \xi^N(t)\|_Y^2 + \|u(t)\|_U^2) \, dt$, the claim in Step 1 holds by estimates (A2) and (A3).

Now from [**PS**, Theorem 3.3], Step 1 implies that (2.26) has a non-negative, self-adjoint solution $\Pi^N \in \mathcal{L}(Z^N)$. Next, we will show uniqueness and convergence of Π^N.

Step 2. There exists positive constant β, independent of N, such that

$$\int_0^\infty \|(k^N)^* e^{t(A^N - B^N \hat{K}^N)} (j^N)^* x_0\|_{X^*}^2 \, dt \leq \beta \|x_0\|_{X^*}^2,$$

where $\hat{K}^N = (B^N)^* \Pi^N$.

Proof. Since *Step 1* implies $\|j^N \Pi^N (j^N)^*\|_{\mathcal{L}(X^*, X)} \leq \alpha$, the assumption and the proof of Corollary 2.5 show that $\|(k^N)^* B^N \hat{K}^N (j^N)^*\|_{\mathcal{L}(X^*)}$ is uniformly bounded in N. Thus, there exists positive constants M, ω (independent of N) such that

$$\text{(A4)} \qquad \|(k^N)^* e^{t(A^N - B^N \hat{K}^N)} (j^N)^*\|_{\mathcal{L}(X^*)} \leq M e^{\omega t}, \qquad t \geq 0.$$

Next, we observe that

$$A^N - B^N \hat{K}^N = A^N - F^N C^N + (F^N C^N - B^N \hat{K}^N).$$

By the variation of constants formula

$$e^{t(A^N - B^N (B^N)^* \Pi^N)} = e^{t(A^N - F^N C^N)} + \int_0^t e^{(t-s)(A^N - F^N C^N)} (F^N C^N - B^N (B^N)^* \Pi^N)$$
$$\times e^{s(A^N - B^N (B^N)^* \Pi^N)} \, ds, \qquad t \geq 0.$$

Let $\xi^N(t) = e^{t(A^N - B^N \hat{K}^N)}(j^N)^* x_0$. Then

$$\int_0^T \|(k^N)^* \xi^N(t)\|_{X^*}^2 \, dt \leq 2 \int_0^T \|(k^N)^* e^{t(A^N - F^N C^N)}(j^N)^* x_0\|_{X^*}^2 \, dt$$

$$+ 2 \int_0^T \left\| (k^N)^* \int_0^t e^{(t-s)(A^N - F^N C^N)}(j^N)^*(k^N)^*(F^N C^N - B^N \hat{K}^N) \right.$$

$$\left. \times \, e^{s(A^N - B^N \hat{K}^N)}(j^N)^* x_0 \, ds \right\|_{X^*}^2 \, dt.$$

By assumption **(b)**, the first term $\leq \frac{M_2^2}{\omega_2}\|x_0\|_{X^*}^2$. Now considering the second term in this sum, by Young's inequality we obtain

$$\leq 4\|(k^N)^* F^N\|_{\mathcal{L}(Y,X^*)}^2 \left(\int_0^T \|(k^N)^* e^{s(A^N - F^N C^N)}(j^N)^*\|_{X^*} \, ds\right)^2 \int_0^T \|C^N \xi^N(s)\|^2 \, ds$$

$$+ 4\|(k^N)^*(i^N)^* B\|_{\mathcal{L}(U,X^*)}^2 \left(\int_0^T \|(k^N)^* e^{s(A^N - F^N C^N)}(j^N)^*\|_{X^*} \, ds\right)^2 \int_0^T \|\hat{K}^N \xi^N(s)\|^2 \, ds.$$

From the assumption in Corollary 2.5 and the uniform boundedness principle, we have $\sup \|(k^N)^*(i^N)^* B\|_{\mathcal{L}(U,X^*)} < \infty$. Moreover, since

$$\int_0^\infty \left(\|C^N \xi^N(s)\|^2 + \|\hat{K}^N \xi^N(s)\|^2\right) ds = \langle \Pi^N (j^N)^* x_0, (j^N)^* x_0 \rangle \leq \alpha \|x_0\|_{X^*}^2.$$

Step 2 then follows from the assumption **(b)**. Hence by (A4), it follows from Datko's lemma **[Da]** that there exist constants $M_3 \geq 1$ and $\omega_3 > 0$ such that

$$\text{(A5)} \qquad \|(k^N)^* e^{t(A^N - B^N \hat{K}^N)}(j^N)^* x_0\|_X \leq M_3 e^{-\omega_3 t}\|x_0\|_{X^*}, \qquad t \geq 0.$$

Finally, to prove convergence of the Riccati operators we suppose that we choose $G = \gamma j_X$ where $\gamma = \max(\|\Pi\|_{\mathcal{L}(X^*,X)}, \alpha)$ and j_X is the canonical embedding from X^* into X. Then, for all $x \in X^*$,

$$\langle Gx, x \rangle \geq \langle \Pi x, x \rangle \qquad \text{and} \qquad \langle G(j^N)^* x, (j^N)^* x \rangle \geq \langle \Pi^N (j^N)^* x, (j^N)^* x \rangle.$$

Thus, we have

$$\langle \Pi x_0, x_0 \rangle \leq \langle \Pi_T(0) x_0, x_0 \rangle \leq \langle \Pi x_0, x_0 \rangle + \langle Gx(T), x(T) \rangle,$$

where $x(t) = T(t) x_0$, $x_0 \in X^*$. Thus, by Theorem 2.3,

$$\text{(A6)} \qquad 0 \leq \langle (\Pi_T(0) - \Pi) x_0, x_0 \rangle \leq \langle Gx(T), x(T) \rangle$$

$$\leq \gamma M_4^2 e^{-2\omega_4 T}\|x_0\|_{X^*}^2.$$

Similarly by (A5) we also have

$$\text{(A7)} \qquad 0 \leq \langle (j^N \Pi_T^N (j^N)^* - j^N \Pi^N (j^N)^*) x_0, x_0 \rangle \leq \gamma M_3^2 e^{-2\omega_3 T}\|x_0\|_{X^*}^2.$$

Therefore, we have

$$\|j^N \Pi^N (j^N)^* x_0 - \Pi x_0\|_X = \|j^N \Pi^N (j^N)^* x_0 - j^N \Pi_T^N (j^N)^* x_0\|_X$$

$$+ \|j^N \Pi_T^N (j^N)^* x_0 - \Pi_T x_0\|_X$$

$$+ \|\Pi_T x_0 - \Pi x_0\|_X.$$

Hence the strong convergence of $j^N \Pi^N (j^N)^*$ to Π in $\mathcal{L}(X^*, X)$ follows from (A6), (A7) and Theorem 2.4. ∎

References

[BB] H. T. Banks and J. Burns, *Hereditary control problems: Numerical methods based on averaging approximations*, SIAM J. Control and Opt. **16** (1978), 169–208.

[BK] H. T. Banks and K. Kunisch, *The linear regulator problem for parabolic systems*, SIAM J. Control and Opt. **22** (1984), 684–698.

[BM] C. Bernier and A. Manitius, *On semigroups in $R^n \times L^p$ corresponding to differential equations with delays*, Canada J. Math. **30** (1978), 897–914.

[BT] J. G. Borisovic and A. S. Turbabin, *On the Cauchy problem for linear non-homogeneous differential equations with retarded argument*, Soviet Math. Doklady **10** (1969), 401–405.

[Da] R. Datko, *Uniform asymptotic stability of evolutionary processes in a Banach space*, SIAM J. Num. Anal. **3** (1972), 428–445.

[DLT] G. Da Prato, I. Lasiecka and R. Triggiani, *A direct study of the Riccati equation arising in hyperbolic boundary control problems*, J. Diff. Eqs. **64** (1986), 26–47.

[DM] M. C. Delfour and A. Manitius, *The structural operator F and its role in the theory of retarded systems: Part I.*, J. Math. Anal. Appl. **73** (1980), 466–490.

[FLT] F. Flandoli, I. Lasiecka and R. Triggiani, *Algebraic Riccati equations with non smoothing observation arising in hyperbolic and Euler-Bernoulli boundary control problems*, Annali di Matematice Pure e Applicata, to appear.

[Gi] J. Gibson, *Linear-quadratic optimal control of hereditary differential systems: Infinite dimensional Riccati equations and numerical approximations*, SIAM J. Control and Opt. **21** (1983), 95–139.

[Ic] A. Ichikawa, *Quadratic control of evolution equations with delays in control*, SIAM J. Control and Opt. **20** (1982), 645–668.

[I1] K. Ito, *Finite dimensional compensators for infinite dimensional systems via Galerkin-type approximation*, LCDS/CCS Report (1986), Brown University, Providence, Rhode Island; submitted to SIAM J. Control and Opt..

[I2] K. Ito, *Strong convergence and convergence rates of approximating solutions for algebraic Riccati equations in Hilbert spaces*, Lecture Notes in Control and Information Sciences **102** (1987), Springer-Verlag, New York.

[IP] K. Ito and R. K. Powers, *Chandrasekhar equations for infinite dimensional systems: Part II. Unbounded input and output case*, ICASE Rept. No. 87–32 (1987), NASA Langley Research Center, Hampton, Virginia.

[IT] K. Ito and H. T. Tran, *Linear quadratic regulator problem for infinite dimensional linear systems with delays in control*, Proceedings of the 27th IEEE Conference on Decision and Control **3** (1988), 2012–2017.

[Ka] T. Kato, "Perturbation Theory for Linear Operators," Springer-Verlag, New York, 1966.

[KL] H. N. Koivo and E. B. Lee, *Controller synthesis for linear systems with retarded states and control variables and quadratic cost*, Automatica **8** (1972), 203–208.

[La] I. Lasiecka, *Unified theory for abstract parabolic boundary problems: A semigroup approach*, Appl. Math. Opt. **6** (1980), 287–333.

[La1] I. Lasiecka, *Approximations of Riccati equations for abstract boundary control systems - application to hyperbolic systems*, Numer. Funct. Anal. and Opt. 8, **3 and 4** (1985-1986), 207–243.

[LM] I. Lasiecka and A. Manitius, *Differentiability and convergence rates of approximating semigroups for retarded functional differential equations*, SIAM J. Num. Anal. **25** (1988).

[LT1] I. Lasiecka and R. Triggiani, *Dirichlet boundary control problem for parabolic equations with quadratic cost: analytically and Riccati's feedback synthesis*, SIAM J. Control and Opt. **21** (1983), 41–67.

[LT2] I. Lasiecka and R. Triggiani, *Riccati equations for hyperbolic partial differential equations with $L_2(0,T;L_2(\Gamma))$-Dirichlet boundary terms*, SIAM J. Control and Opt. **24** (1986), 884–925.

[LT3] I. Lasiecka and R. Triggiani, *The regulator problem for parabolic equations with Dirichlet boundary control part II: Galerkin approximation*, Appl. Math. Opt. **16** (1987), 187–216.

[MT] A. Manitius and H. T. Tran, *Numerical approximations for hereditary systems with input and output delays: Convergence results and convergence rates*, LCDS/CCS Report 88–14 (1988), Brown University, Providence, Rhode Island; submitted to SIAM J. Control and Opt..

[Pa] A. Pazy, "Semigroups of Linear Operators and Applications to Partial Differential Equations," Applied Mathematical Sciences 44, Springer-Verlag, New York, 1983.

[PS] A. J. Pritchard and D. Salamon, *The linear quadratic optimal control problem for infinite dimensional systems with unbounded input and output operators*, SIAM J. Control and Opt. **25** (1987), 121–144.

[S1] D. Salamon, "On Control and Observation of Neutral Systems," RNM 91, Pitman, London, 1984.

[S2] D. Salamon, *Structure and stability of finite dimensional approximations for functional differential equations*, SIAM J. Control and Opt. **23** (1985), 928–951.

[VK] R. B. Vinter and R. H. Kwong, *The finite time quadratic control problem for linear systems with state and control delays: An evolution equation approach*, SIAM J. Control and Opt. **19** (1981), 139–153.

K. Ito and H.T. Tran
Center for Control Sciences
Division of Applied Mathematics
Brown University
Providence, RI 02912
USA

International Series of
Numerical Mathematics, Vol. 91
© 1989 Birkhäuser Verlag Basel

On a class of moment problems for solving
minimum norm control problems

W. Krabs

Fachbereich Mathematik
Technische Hochschule Darmstadt

Abstract. We consider a class of infinite moment problems which consist of countably many pairs of equations (see (1.1)) for functions $f_j \in L^\infty[0, T]$, $j \in \mathbb{N}$. The aim consists of finding a sequence $f = (f_j)_{j \in \mathbb{N}}$ of such functions which satisfy these equations and the condition that

$$\delta_T(f) = \left(\sum_{j=1}^{\infty} (\text{ess sup}_{t \in [0,T]} |f_j(t)|)^2 \right)^{1/2}$$

is finite and as small as possible.

This problem can be split up into a sequence of moment problems for each f_j for which explicit solutions are derived.

It is also compared with the same problem where $\delta_T(f)$ is replaced by

$$\gamma_T(f) = \text{ess sup}_{t \in [0,T]} \left(\sum_{j=1}^{\infty} f_j(t)^2 \right)^{1/2}$$

which is less than or equal to $\delta_T(f)$ and necessary and sufficient conditions are given for a sequence f which minimizes $\delta_T(f)$ also to minimize $\gamma_T(f)$ and to satisfy $\gamma_T(f) = \delta_T(f)$. Finally, a generalization of the problem of minimizing $\gamma_T(f)$ is studied.

Keywords. Moment problems, vibrations, distributed control, boundary control, minimum norm problems.

1980 *Mathematics subject classifications*: 65K99, 93C20, 93B05

1. Introduction

Distributed control of vibrations, described by an abstract wave equation, from a given initial state to a position of rest within a given time interval $[0, T], T > 0$ leads to the problem of finding a sequence $(f_j)_{j \in \mathbb{N}}$ of functions $f_j \in L^\infty[0, T]$ which satisfy an infinite sequence of moment equations of the form

(1.1)
$$\int_0^T f_j(t) \cos \sqrt{\lambda_j} t \, dt = c_j^1,$$
$$\int_0^T f_j(t) \sin \sqrt{\lambda_j} t \, dt = c_j^2$$
$$\text{for all} \quad j \in \mathbb{N}$$

where the sequence $(\lambda_j)_{j \in \mathbb{N}}$ consists of positive reals, is increasing, and satisfies $\lim_{j \to \infty} \lambda_j = \infty$. The sequences $(c_j^1)_{j \in \mathbb{N}}$ and $(c_j^2)_{j \in \mathbb{N}}$ are in ℓ^2. In [5] it is shown how the problem of finding a sequence $f = (f_j)_{j \in \mathbb{N}}$ in $L^\infty[0, T]$ which satisfies (1.1) and which minimizes

$$(1.2) \qquad \gamma_T(f) = \operatorname{ess\,sup}_{t \in [0,T]} (\sum_{j=1}^{\infty} f_j(t)^2)^{1/2}$$

can be solved.

Here we consider the same problem with $\gamma_T(f)$ replaced by

$$(1.3) \qquad \delta_T(f) = (\sum_{j=1}^{\infty} (\operatorname{ess\,sup}_{t \in [0,T]} |f_j(t)|)^2)^{1/2}.$$

Let X be the space of all sequences $f = (f_j)_{j \in \mathbb{N}}$ with $f_j \in L^\infty[0, T]$ for all $j \in \mathbb{N}$ such that

$$\sum_{j=1}^{\infty} (\operatorname{ess\,sup}_{t \in [0,T]} |f_j(t)|)^2 < \infty.$$

It is easy to see that

$$(1.4) \qquad \gamma_T(f) \le \delta_T(f) \quad \text{for all} \quad f \in X$$

which implies $X \subseteq L^\infty([0, T], \ell^2)$ the latter being the space of all $f = (f_j)_{j \in \mathbb{N}}$ such that $\gamma_T(f)$ is finite. Further it can be shown that, for every $j \in \mathbb{N}$, there is exactly one solution of (1.1) of the form

$$f_j(t) = y_j^1 \cos \sqrt{\lambda_j} t + y_j^2 \sin \sqrt{\lambda_j} t$$

where

$$((y_j^1)^2 + (y_j^2)^2)^{1/2} \le \frac{2\sqrt{\lambda_j}}{\sqrt{\lambda_j}T - |\sin \sqrt{\lambda_j}T|} ((c_j^1)^2 + (c_j^2)^2)^{1/2}$$

$$\le \frac{2\sqrt{\lambda_1}}{\sqrt{\lambda_1}T - |\sin \sqrt{\lambda_1}T|} ((c_j^1)^2 + (c_j^2)^2)^{1/2}$$

(see [2]). This implies

$$|y_j^1| + |y_j^2| \le \sqrt{2}((y_j^1)^2 + (y_j^2)^2)^{1/2} \le \frac{2\sqrt{2\lambda_1}}{\sqrt{\lambda_1}T - |\sin \sqrt{\lambda_1}T|} ((c_j^1)^2 + (c_j^2)^2)^{1/2}$$

and hence

$$|f_j(t)| \le |y_j^1| + |y_j^2| \le \frac{2\sqrt{2\lambda_1}}{\sqrt{\lambda_1}T - |\sin \sqrt{\lambda_1}T|} ((c_j^1)^2 + (c_j^2)^2)^{1/2}$$

for all $j \in \mathbb{N}$. As a further consequence we have

$$\delta_T(f) \le \frac{2\sqrt{2\lambda_1}}{\sqrt{\lambda_1}T - |\sin \sqrt{\lambda_1}T|} (\sum_{j=1}^{\infty} (c_j^1)^2 + (c_j^2)^2)^{1/2}$$

which shows that the set of sequences $f = (f_j)_{j \in \mathbb{N}}$ in X which satisfy (1.1) and for which $\delta_T(f)$ is finite is non-empty.

In order to minimize $\delta_T(f)$ on the set of sequences $f = (f_j)_{j \in \mathbb{N}} \in X$ which satisfy (1.1) we have to determine, for

every $j \in \mathbb{N}$, a function $\hat{f}_j \in L^\infty[0,T]$ which satisfies

(1.5)
$$\int_0^T \hat{f}_j(t) \cos \sqrt{\lambda_j} t \, dt = c_j^1,$$

$$\int_0^T \hat{f}_j(t) \sin \sqrt{\lambda_j} t \, dt = c_j^2$$

and minimizes

(1.6)
$$\delta_T(\hat{f}_j) = \operatorname{ess\,sup}_{t \in [0,T]} |\hat{f}_j(t)|.$$

For the sequence $\hat{f} = (\hat{f}_j)_{j \in \mathbb{N}}$ it then follows that (1.1) is satisfied and that

(1.7)
$$\delta_T(\hat{f}) = \left(\sum_{j=1}^\infty \delta_T(\hat{f}_j)^2 \right)^{1/2}$$

is minimal.

Before we show how the sequence $\hat{f} = (\hat{f}_j)_{j \in \mathbb{N}}$ can be determined we discuss the question under which condition this sequence also solves the problem of minimizing (1.2) subject to (1.1) and satisfies

(1.8)
$$\left(\sum_{j=1}^\infty \delta_T(\hat{f}_j)^2 \right)^{1/2} = \operatorname{ess\,sup}_{t \in [0,T]} \left(\sum_{j=1}^\infty \hat{f}_j(t)^2 \right)^{1/2}.$$

From the results in [5] it follows that there are sequences $(y_j^1)_{j \in \mathbb{N}}$ and $(y_j^2)_{j \in \mathbb{N}}$ in ℓ^2 such that

(1.9)
$$\int_0^T \left(\sum_{j=1}^\infty (y_j^1 \cos \sqrt{\lambda_j} t + y_j^2 \sin \sqrt{\lambda_j} t)^2 \right)^{1/2} dt = 1$$

and

(1.10)
$$\sum_{j=1}^\infty c_j^1 y_j^1 + c_j^2 y_j^2 = \operatorname{ess\,sup}_{t \in [0,T]} \left(\sum_{j=1}^\infty \hat{f}_j(t)^2 \right)^{1/2}.$$

On using (1.1), (1.4), and (1.5) we therefore obtain from (1.8)

$$\sum_{j=1}^\infty c_j^1 y_j^1 + c_j^2 y_j^2 = \int_0^T \sum_{j=1}^\infty (y_j^1 \cos \sqrt{\lambda_j} t + y_j^2 \sin \sqrt{\lambda_j} t) \hat{f}_j(t) dt$$

$$\leq \int_0^T \sum_{j=1}^\infty |y_j^1 \cos \sqrt{\lambda_j} t + y_j^2 \sin \sqrt{\lambda_j} t| \delta_T(\hat{f}_j) dt$$

$$\leq \int_0^T \left(\sum_{j=1}^\infty (y_j^1 \cos \sqrt{\lambda_j} t + y_j^2 \sin \sqrt{\lambda_j} t)^2 \right)^{1/2} dt \left(\sum_{j=1}^\infty \delta_T(\hat{f}_j)^2 \right)^{1/2}$$

$$= \operatorname{ess\,sup}_{t \in [0,T]} \left(\sum_{j=1}^\infty \hat{f}_j(t)^2 \right)^{1/2}.$$

so that (1.10) implies the existence of some $\alpha = \alpha(t) \geq 0$, for every $t \in [0, T]$ with

$$\sum_{j=1}^{\infty}(y_j^1 \cos \sqrt{\lambda_j}t + y_j^2 \sin \sqrt{\lambda_j}t)^2 > 0,$$

such that

(1.11) $$|y_j^1 \cos \sqrt{\lambda_j}t + y_j^2 \sin \sqrt{\lambda_j}t| = \delta_T(\hat{f}_j)\alpha(t)$$

for all $j \in \mathbb{N}$ and all but finitely many $t \in [0, T]$. Further it follows, for every $j \in \mathbb{N}$, that

(1.12) $$\hat{f}_j(t) = \delta_T(\hat{f}_j)\text{sgn}(y_j^1 \cos \sqrt{\lambda_j}t + y_j^2 \sin \sqrt{\lambda_j}t)$$

for all but finitely many $t \in [0, T]$.
From (1.11) and (1.9) we deduce that

$$\int_0^T \alpha(t)dt = (\sum_{j=1}^{\infty} \delta_T(\hat{f}_j)^2)^{-1/2} = \delta_T(\hat{f})^{-1/2}$$

which implies

(1.13) $$\int_0^T |y_j^1 \cos \sqrt{\lambda_j}t + y_j^2 \sin \sqrt{\lambda_j}t|dt = \frac{\delta_T(\hat{f}_j)}{\delta_T(\hat{f})}$$

for all $j \in \mathbb{N}$.
Conversely, assume that $(\hat{f}_j)_{j \in \mathbb{N}}$ is a sequence in X which minimizes (1.3) subject to (1.1) and that there are two sequences $(y_j^1)_{j \in \mathbb{N}}$ and $(y_j^2)_{j \in \mathbb{N}}$ in ℓ^2 such that (1.9), and (1.12), (1.13) for all $j \in \mathbb{N}$ are satisfied. Then it is easy to see that

$$\sum_{j=1}^{\infty} c_j^1 y_j^1 + c_j^2 y_j^2 \leq \text{ess sup }_{t\in[0,T]}(\sum_{j=1}^{\infty} \hat{f}_j(t)^2)^{1/2}$$

and

$$\sum_{j=1}^{\infty} c_j^1 y_j^1 + c_j^2 y_j^2 = (\sum_{j=1}^{\infty} \delta_T(\hat{f}_j)^2)^{1/2}.$$

By virtue of (1.4) this implies (1.8) which is sufficient for $(\hat{f}_j)_{j \in \mathbb{N}}$ to minimize (1.2) subject to (1.1).

Result: A sequence $(\hat{f}_j)_{j \in \mathbb{N}}$ in X which minimizes (1.3) subject to (1.1) also minimizes (1.2) subject to (1.1) and satisfies (1.8), if and only if there are two sequences $(y_j^1)_{j \in \mathbb{N}}$ and $(y_j^2)_{j \in \mathbb{N}}$ in ℓ^2 such that (1.9), and (1.12), (1.13) for all $j \in \mathbb{N}$ are satisfied.

2. The minimization of (1.6) subject to (1.5)

To obtain a unified treatment for all $j \in \mathbb{N}$ we consider the problem of finding a function $\hat{f} \in L^{\infty}[0,T]$ which satisfies

$$(2.1) \qquad \int_0^T \hat{f}(t) \cos \sqrt{\lambda} t \, dt = c^1,$$

$$\int_0^T \hat{f}(t) \sin \sqrt{\lambda} t \, dt = c^2$$

and minimizes

$$(2.2) \qquad \delta_T(\hat{f}) = \operatorname{ess\,sup}_{t \in [0,T]} |\hat{f}(t)|.$$

In (2.1) λ is a positive real and $c^1, c^2 \in \mathbb{R}$ are such that $|c^1| + |c^2| > 0$.

From the results in [5] it follows that there is exactly one $\hat{f} \in L^{\infty}[0,T]$ which satisfies (2.1) and minimizes (2.2). It is of the form

$$(2.3) \qquad \hat{f}(t) = \delta_T(\hat{f}) \operatorname{sgn}(y^1 \cos \sqrt{\lambda} t + y^2 \sin \sqrt{\lambda} t)$$

for all but finitely many $t \in [0,T]$ where

$$(2.4) \qquad \delta_T(\hat{f}) = c^1 y^1 + c^2 y^2 > 0$$

and

$$(2.5) \qquad \int_0^T \operatorname{sgn}(y^1 \cos \sqrt{\lambda} t + y^2 \sin \sqrt{\lambda} t) \cos \sqrt{\lambda} t \, dt = \frac{c^1}{\delta_T(\hat{f})},$$

$$\int_0^T \operatorname{sgn}(y^1 \cos \sqrt{\lambda} t + y^2 \sin \sqrt{\lambda} t) \sin \sqrt{\lambda} t \, dt = \frac{c^2}{\delta_T(\hat{f})}.$$

Similar results can also be found in [1], Chap. 2 and references therein.

In order to determine y^1 and y^2 with (2.4), (2.5) and thus to obtain the function (2.3) which satisfies (2.1) and minimizes (2.2) we first assume that

$$(2.6) \qquad \sqrt{\lambda} T \in (0, \pi].$$

Then, for every pair $y^1, y^2 \in \mathbb{R}$ there is exactly one $t_0 \in [0, \frac{\pi}{\sqrt{\lambda}}]$ such that

$$\tan \sqrt{\lambda} t_0 = -\frac{y^1}{y^2}.$$

This shows that, under the assumption (2.6), the system (2.5) can be rewritten in the form

$$\hat{\mu}\sigma \int_0^{t_0} \cos \sqrt{\lambda} t \, dt - \hat{\mu}\sigma \int_{t_0}^T \cos \sqrt{\lambda} t \, dt = c^1,$$

$$\hat{\mu}\sigma \int_0^{t_0} \sin \sqrt{\lambda} t \, dt - \hat{\mu}\sigma \int_{t_0}^T \sin \sqrt{\lambda} t \, dt = c^2$$

where $\hat{\mu} = \delta_T(\hat{f})$ and $\sigma = +1$ or -1, or, equivalently,

$$\hat{\mu}\sigma(2\sin\sqrt{\lambda}t_0 - \sin\sqrt{\lambda}T) = c^1\sqrt{\lambda},$$
$$\hat{\mu}\sigma(\cos\sqrt{\lambda}T - 2\cos\sqrt{\lambda}t_0 + 1) = c^2\sqrt{\lambda}.$$

These two equations hold, if and only if

$$2c^2\sin\sqrt{\lambda}t_0 + 2c^1\cos\sqrt{\lambda}t_0 - c^2\sin\sqrt{\lambda}T - c^1\cos\sqrt{\lambda}T - c^1 = 0.$$

If we choose $\alpha \in (-\frac{\pi}{2}, \frac{\pi}{2})$ such that $\tan\alpha = \frac{c^1}{c^2}$, then the last equation becomes equivalent to

$$2\sin(\sqrt{\lambda}t_0 + \alpha) = \sin(\sqrt{\lambda}T + \alpha) + c^1$$

which yields

(2.7)
$$t_0 = \frac{1}{\sqrt{\lambda}}\arcsin(\frac{1}{2}\sin(\sqrt{\lambda}T + \alpha) + \frac{1}{2}c^1) - \frac{\alpha}{\sqrt{\lambda}}$$

with

(2.8)
$$\alpha = \arctan\frac{c^1}{c^2}.$$

Let $c^1 = 0$. Then $c^2 \neq 0$, $\alpha = 0$, and

(2.9)
$$t_0 = \frac{1}{\sqrt{\lambda}}\arcsin(\frac{1}{2}\sin\sqrt{\lambda}T).$$

Further it follows that $\cos\sqrt{\lambda}T - 2\cos\sqrt{\lambda}t_0 + 1 \neq 0$,

$$\delta_T(\hat{f}) = \hat{\mu} = \frac{|c^2|\sqrt{\lambda}}{|\cos\sqrt{\lambda}T - 2\cos\sqrt{\lambda}t_0 + 1|} = c^2 y^2,$$

and hence

(2.10)
$$y^2 = \frac{\sqrt{\lambda}\,\mathrm{sgn}c^2}{|\cos\sqrt{\lambda}T - 2\cos\sqrt{\lambda}t_0 + 1|}.$$

Finally,

(2.11)
$$y^1 = -y^2\tan\sqrt{\lambda}t_0.$$

Let $c^1 \neq 0$. Then t_0 is given by (2.7) with α by (2.8). Further $2\sin\sqrt{\lambda}t_0 - \sin\sqrt{\lambda}T \neq 0$ and

$$\delta_T(\hat{f}) = \hat{\mu} = \frac{|c_1|\sqrt{\lambda}}{|2\sin\sqrt{\lambda}t_0 - \sin\sqrt{\lambda}T|}$$
$$= (-c^1\tan\sqrt{\lambda}t_0 + c^2)y^2 > 0$$

which implies

$$(2.12) \qquad y^2 = \frac{1}{-c^1 \tan \sqrt{\lambda} t_0 + c^2} \times \frac{|c_1|\sqrt{\lambda}}{|2 \sin \sqrt{\lambda} t_0 - \sin \sqrt{\lambda} T|}.$$

Again y^1 is given by (2.11).
   Next we replace the assumption (2.6) by

$$(2.13) \qquad \sqrt{\lambda} T \in ((k-1)\pi, k\pi] \quad \text{for some} \quad k \in \mathbb{N}.$$

Then, for every pair $y^1, y^2 \in \mathbb{R}$ the zeros of the function $y^1 \cos \sqrt{\lambda} t + y^2 \sin \sqrt{\lambda} t$ in $[0, T]$ are given by

$$(2.14) \qquad t_0, t_1 = t_0 + \frac{\pi}{\sqrt{\lambda}}, \cdots, t_{k-1} = t_0 + (k-1)\frac{\pi}{\sqrt{\lambda}},$$

where t_0 is the unique solution of

$$\tan \sqrt{\lambda} t_0 = -\frac{y^1}{y^2} \quad \text{in} \quad [0, \frac{\pi}{\sqrt{\lambda}}].$$

This shows that, under the assumption (2.13), the system (2.5) can be rewritten in the form

$$\hat{\mu}\sigma \int_0^{t_0} \cos \sqrt{\lambda} t dt - \hat{\mu}\sigma \int_{t_0}^{t_1} \cos \sqrt{\lambda} t dt + -\cdots + (-1)^k \hat{\mu}\sigma \int_{t_{k-1}}^T \cos \sqrt{\lambda} t dt = c^1,$$

$$\hat{\mu}\sigma \int_0^{t_0} \sin \sqrt{\lambda} t dt - \hat{\mu}\sigma \int_{t_0}^{t_1} \sin \sqrt{\lambda} t dt + -\cdots + (-1)^k \hat{\mu}\sigma \int_{t_{k-1}}^T \sin \sqrt{\lambda} t dt = c^2,$$

where $\hat{\mu} = \delta_T(\hat{f})$ and $\sigma = +1$ or -1 or, equivalently,

$$\hat{\mu}\sigma(2k \sin \sqrt{\lambda} t_0 - (-1)^{k-1} \sin \sqrt{\lambda} T) = c^1 \sqrt{\lambda},$$
$$\hat{\mu}\sigma(1 - 2k \cos \sqrt{\lambda} t_0 + (-1)^{k-1} \cos \sqrt{\lambda} T) = c^2 \sqrt{\lambda}.$$

These two equations hold, if and only if

$$2kc^2 \sin \sqrt{\lambda} t_0 + 2kc^1 \cos \sqrt{\lambda} t_0 + (-1)^k c^2 \sin \sqrt{\lambda} T + (-1)^k c^1 \cos \sqrt{\lambda} T - c^1 = 0.$$

If we choose $\alpha \in (-\frac{\pi}{2}, \frac{\pi}{2})$ such that $\tan \alpha = \frac{c^1}{c^2}$, then the last equation becomes equivalent to

$$2k \sin(\sqrt{\lambda} t_0 + \alpha) = (-1)^{k-1} \sin(\sqrt{\lambda} T + \alpha) + c^1$$

which yields

$$(2.15) \qquad t_0 = \frac{1}{\sqrt{\lambda}} \arcsin(\frac{(-1)^{k-1}}{2k} \sin(\sqrt{\lambda} T + \alpha) + \frac{c^1}{2k}) - \frac{\alpha}{\sqrt{\lambda}}$$

with (2.8)

$$\alpha = \arctan \frac{c^1}{c^2}.$$

The computation of y^1 and y^2 is the same as under the assumption (2.6).

If $c^1 = 0$, then y^2 is given by

$$(2.16) \qquad y^2 = \frac{\sqrt{\lambda} \operatorname{sgn} c^2}{|(-1)^{k-1} \cos \sqrt{\lambda} T - 2k \cos \sqrt{\lambda} t_0 + 1|}$$

and

$$\delta_T(\hat{f}) = c^2 y^2.$$

If $c^1 \neq 0$, then y^2 is given by

$$(2.17) \qquad y^2 = \frac{1}{-c^1 \tan \sqrt{\lambda} t_0 + c^2} \times \frac{|c_1| \sqrt{\lambda}}{|2k \sin \sqrt{\lambda} t_0 - (-1)^{k-1} \sin \sqrt{\lambda} T|}$$

and

$$(2.18) \qquad \delta_T(\hat{f}) = \frac{|c_1| \sqrt{\lambda}}{|2k \sin \sqrt{\lambda} t_0 - (-1)^{k-1} \sin \sqrt{\lambda} T|}.$$

In both cases y^1 is given by

$$(2.19) \qquad y^1 = -y^2 \tan \sqrt{\lambda} t_0.$$

3. Application to a Vibrating String

We consider a vibrating string governed by the wave equation

$$(3.1) \qquad \begin{aligned} y_{tt}(t,x) - y_{xx}(t,x) &= f(t,x), \\ x \in (0,1), t &\in (0,T] \end{aligned}$$

for some $T > 0$ under the boundary conditions

$$(3.2) \qquad y(t,0) = y(t,1) = 0 \quad \text{for} \quad t \in [0,T].$$

The control function $f = f(t,x)$ can be chosen in $L^\infty([0,T], L^2(0,1))$.

We assume an initial state for $t = 0$ to be given by

$$(3.3) \qquad \begin{aligned} y(0,x) &= x(1-x), \quad y_t(0,x) = 0 \\ &\text{for all} \quad x \in [0,1]. \end{aligned}$$

The problem consists in the first place of finding $f \in L^\infty([0,T], L^2(0,1))$ such that the (generalized) solution $y = y(t,x)$ of (3.1), (3.2), (3.3) satisfies the end condition

$$(3.4) \qquad \begin{aligned} y(T,x) &= y_t(T,x) = 0 \\ &\text{for almost all} \quad x \in [0,1]. \end{aligned}$$

By virtue of the closed form (generalized) solution of (3.1), (3.2) , (3.3) this requirement can be shown to be equivalent to finding a sequence $(f_j)_{j \in \mathbb{N}}$ in $L^\infty([0, T], \ell^2)$ such that

(3.5) $$f_{2i} = 0 \quad \text{a.e. for all} \quad i \in \mathbb{N}$$

and

(3.6)
$$\int_0^T f_{2i-1}(t) \cos(2i - 1)\pi t \, dt = 0,$$
$$\int_0^T f_{2i-1}(t) \sin(2i - 1)\pi t \, dt = \frac{4\sqrt{2}}{[(2i - 1)\pi]^2}$$
$$\text{for all} \quad i \in \mathbb{N}.$$

If such a sequence can be found, then the (generalized) solution $y = y(t, x)$ of (3.1), (3.2), (3.3) which belongs to the control function

$$f(t, x) = \sqrt{2} \sum_{i=1}^{\infty} f_{2i-1}(t) \sin(2i - 1)\pi x$$

in (3.1) satisfies (3.4).

The infinite system, (3.6) of moment equations is of the form (1.1) where

$$
\begin{aligned}
c_j^1 &= 0 && \text{for all} \quad j \in \mathbb{N}, \\
c_{2i}^2 &= 0 && \text{for all} \quad i \in \mathbb{N}, \quad (3.5) \text{ is satisfied,} \\
c_{2i-1}^2 &= \frac{4\sqrt{2}}{[(2i - 1)\pi]^2} && \text{for all} \quad i \in \mathbb{N}, \quad \text{and} \\
\lambda_j &= (j\pi)^2 && \text{for all} \quad j \in \mathbb{N}.
\end{aligned}
$$

The problem (being introduced in Section 1) now consists of determining $\hat{f} = (\hat{f}_j)_{j \in \mathbb{N}} \in L^\infty([0, T], \ell^2)$ with (3.5), (3.6) such that

(3.7) $$\delta_T(\hat{f}) = \left(\sum_{i=1}^{\infty} \delta_T(\hat{f}_{2i-1})^2\right)^{1/2}$$

is minimized where, for every $i \in \mathbb{N}$,

(3.8) $$\delta_T(\hat{f}_{2i-1}) = \operatorname{ess\,sup}_{t \in [0,T]} |\hat{f}_{2i-1}(t)|.$$

By the results of Section 2 every $\hat{f}_{2i-1} \in L^\infty(0, T)$ is of the form

(3.9) $$\hat{f}_{2i-1}(t) = \delta_T(\hat{f}_{2i-1}) \times \begin{cases} -1 & \text{for} \quad t \in (0, t_0], \\ (-1)^\ell & \text{for} \quad t \in (t_0 + \ell, t_0 + \ell + 1] \\ & \text{and} \quad \ell = 0, \cdots, k - 2, \\ (-1)^{k-1} & \text{for} \quad t \in (t_0 + k - 1, T], \end{cases}$$

where $(2i-1)T \in (k-1, k]$ and

(3.10) $$\delta_T(\hat{f}_{2i-1}) = \frac{(2i-1)\pi |c_{2i-1}^2|}{|(-1)^{k-1} \cos(2i-1)\pi T - 2k \cos(2i-1)\pi t_0 + 1|},$$

(3.11) $$c_{2i-1}^2 = \frac{4\sqrt{2}}{[(2i-1)\pi]^2},$$

(3.12) $$t_0 = \frac{1}{(2i-1)\pi} \arcsin(\frac{(-1)^{k-1}}{2k} \sin(2i-1)\pi T).$$

For $T = 0.25$ we obtain the following numerical results:

$2i-1$	t_0	$\delta_T(\hat{f}_{2i-1})$	k
1	0.1150267	10.998155	1
3	0.03834224	0.380377327	1
5	0.01131341	0.161499164	2
7	0.00808101	0.070588754	2
9	0.00417784	0.047063394	3
11	0.00341823	0.028894152	3

As an approximation of $\delta_T(\hat{f})$ we consider, for every $N \in \mathbb{N}$, the quantity

$$\delta_T^{2N-1}(\hat{f}) = (\sum_{i=1}^{2N-1} \delta_T(\hat{f}_{2i-1})^2)^{1/2}.$$

For these quantities we obtain the following values:

$2N-1$	1	3	5
$\delta_{0.25}^{2N-1}(\hat{f})$	10.998155	11.004731	11.005916

$2N-1$	7	9	11
$\delta_{0.25}^{2N-1}(\hat{f})$	11.006142	11.006243	11.006281

For comparison we also give the corresponding approximations of the minimum of

$$\gamma_T(f) = \text{ess sup}_{t \in [0,t]}(\sum_{i=1}^{\infty} f_{2i-1}(t)^2)^{1/2}$$

subject to $(1.1)_{2i-1}$ for $j \in \mathbb{N}$ by the minimum value $\gamma_T^{2N-1}(\hat{f})$ of

$$\gamma_T^{2N-1}(f) = \operatorname{ess\,sup}_{t \in [0,T]} (\sum_{i=1}^{N} f_{2i-1}(t)^2)^{1/2}$$

subject to $(1.1)_{2i-1}$ for $i = 1, ..., N$ which have been calculated in [4]:

$2N-1$	1	3	5
$\gamma_{0.25}^{2N-1}(\hat{f})$	10.9982	11.0012	11.0014

4. A generalization

Boundary control of vibrations, described by a partial differential equation, from a given initial state to a position of rest within a given time interval $[0, T], T > 0$, leads to the problem of finding a vector function $w = (w_1, \cdots, w_m)^T \in L^{\infty}([0,T], \mathbb{R}^m)$ which satisfies an infinite sequence of moment equations of the form

(4.1)
$$\int_0^T \sum_{i=1}^m a_{ji} w_i(t) \cos \sqrt{\lambda_j} t \, dt = c_j^1,$$
$$\int_0^T \sum_{i=1}^m a_{ji} w_i(t) \sin \sqrt{\lambda_j} t \, dt = c_j^2$$

for all $j \in \mathbb{N}$

where the sequence $(\lambda_j)_{j \in \mathbb{N}}$ consists of positive reals, is increasing, and satisfies $\lim_{j \to \infty} \lambda_j = \infty$ and the sequences $(c_j^1)_{j \in \mathbb{N}}, (c_j^2)_{j \in \mathbb{N}}$ are in ℓ^2 (as in Section 1).

The real infinite matrix $(a_{ji})_{j \in \mathbb{N}}, i = 1, \cdots, m$ is such that for every $i = 1, \cdots, m$ the sequence $(a_{ji})_{j \in \mathbb{N}}$ is in ℓ^2.

In addition, for every $i = 1, \cdots, m$, the equations

(4.2)
$$\int_0^T w_i(t) dt = 0 \quad \text{and} \quad \int_0^T t w_i(t) dt = 0$$

have to be satisfied (see [3] and [4]).

As a joint generalization of the systems (1.1) and (4.1), (4.2) we consider the following infinite system of equations

(4.3)
$$\int_0^T (\sum_{i=1}^{\infty} b_{ji} w_i(t)) v_j(t) dt = c_j$$

for all $j \in \mathbb{N}$.

The sequence $(v_j)_{j \in \mathbb{N}}$ of functions $v_j \in L^1[0,T]$ for $j \in \mathbb{N}$ and the sequence $(c_j)_{j \in \mathbb{N}} \in \ell^2$ are given and the real matrix $(b_{ji})_{j, i \in \mathbb{N}}$ is such that, for every sequence $(w_i)_{i \in \mathbb{N}} \in$

$L^\infty([0,t], \ell^2)$, the corresponding sequence $(\sum_{i=1}^\infty b_{ji}w_i)_{j\in\mathbb{N}}$ has the property that $\sum_{i=1}^\infty b_{ij}w_i \in L^\infty[0,T]$ for every $j \in \mathbb{N}$.

The problem consists of finding a sequence $w = (w_i)_{i\in\mathbb{N}} \in L^\infty([0,T], \ell^2)$ which satisfies (4.3) and minimizes

$$(4.4) \qquad \beta_T(w) = \text{ess sup}_{t\in[0,T]}(\sum_{i=1}^\infty w_i(t)^2)^{1/2}.$$

In order to obtain approximate solutions this problem is replaced by the task of finding a vector function $w^M = (w_1, \cdots, w_M) \in L^\infty([0,T], \mathbb{R}^M)$ which satisfies the equations

$$(4.5) \qquad \int_0^T (\sum_{i=1}^M b_{ji}w_i(t))v_j(t)dt = c_j$$

for $j = 1, \cdots, N$ where $M, N \in \mathbb{N}$ are given integers and minimizes

$$(4.6) \qquad \beta_T^M(w^M) = \text{ess sup}_{t\in[0,T]}(\sum_{i=1}^M w_i(t)^2)^{1/2}.$$

As a dual problem we consider the problem of finding a vector $y = (y_1, \cdots, y_N)^T \in \mathbb{R}^N$ which satisfies

$$(4.7) \qquad \int_0^T (\sum_{i=1}^M (\sum_{j=1}^N y_j b_{ji}v_j(t))^2)^{1/2} dt \le 1$$

and maximizes

$$(4.8) \qquad \gamma_T^M(y) = \sum_{j=1}^N c_j y_j.$$

We assume that, for every $y \in \mathbb{R}^N$ and every $i \in \{1, \cdots, M\}$ the function $\sum_{j=1}^N y_j b_{ji}v_j$ has only finitely many zeros in $[0,T]$. Then it follows that both problems have solutions $\hat{w}^M \in L^\infty([0,T], \mathbb{R}^M)$ and $\hat{y} \in \mathbb{R}^N$, respectively, and that $\beta_T^M(\hat{w}^M) = \gamma_T^M(\hat{y})$. From this it follows that

$$\beta_T^M(\hat{w}^M) = \sum_{j=1}^N c_j\hat{y}_j = \int_0^T \sum_{i=1}^M (\sum_{j=1}^N \hat{y}_j b_{ji}v_j(t))\hat{w}_i(t)dt$$

$$\le \int_0^T (\sum_{i=1}^M (\sum_{j=1}^N \hat{y}_j b_{ji}v_j(t))^2)^{1/2}(\sum_{i=1}^M \hat{w}_i(t)^2)^{1/2} dt$$

$$\le \text{ess sup}_{t\in[0,T]}(\sum_{i=1}^M \hat{w}_i(t)^2)^{1/2} = \beta_T^M(\hat{w}^M)$$

which implies

$$(4.9) \qquad (\sum_{i=1}^{M} \hat{w}_i(t)^2)^{1/2} = \beta_T^M(\hat{w}^M)$$

for all but finitely many $t \in [0, T]$ and the existence of some $\alpha = \alpha(t)$, for every i and every $t \in [0, T]$ such that

$$(\sum_{j=1}^{N} \hat{y}_j b_{ji} v_j(t))^2 > 0,$$

for which we have

$$\hat{w}_i(t) = \alpha(t) \sum_{j=1}^{N} \hat{y}_j b_{ji} v_j(t).$$

This, in connection with (4.9), implies

$$(4.10) \qquad \hat{w}_i(t) = \gamma_T^M \sum_{j=1}^{N} \hat{y}_j b_{ji} v_j(t) / (\sum_{i=1}^{M} (\sum_{j=1}^{N} \hat{y}_j b_{ji} v_j(t))^2)^{1/2}$$

for every $i = 1, \cdots, M$ and all but finitely many $t \in [0, T]$.

Insertion into (4.5) gives

$$(4.11) \qquad \sum_{\ell=1}^{N} \int_0^T \frac{1}{\hat{\gamma}(t)} \sum_{i=1}^{M} b_{ji} b_{\ell i} v_j(t) v_\ell(t) dt \hat{y}_\ell = \frac{c_j}{\gamma_T^M(\hat{y})}$$
$$\text{for all} \quad j = 1, \cdots, N$$

where

$$(4.12) \qquad \hat{\gamma}(t) = (\sum_{i=1}^{M} (\sum_{j=1}^{N} \hat{y}_j b_{ji} v_j(t))^2)^{1/2}$$

(> 0 for all but finitely many $t \in [0, T]$).

Result: If $\hat{w}^M \in L^\infty([0, T], \mathbb{R}^M)$ minimizes (4.6) subject to (4.5), then every $\hat{w}_i \in L^\infty[0, T]$ is of the form (4.10) where $\gamma_T^M(\hat{y})$ is given by (4.8) and $\hat{y} \in \mathbb{R}^N$ solves (4.11), (4.12).

Conversely it can be shown that, if $\hat{y} \in \mathbb{R}^N$ is a solution of (4.11), (4.12) with $\gamma_T^N(\hat{y})$ being defined by (4.8) and if $\hat{w}^M \in L^\infty([0, T], \mathbb{R}^M)$ is defined by (4.10), then $\hat{w}^M$ minimizes (4.6) subject to (4.5).

In analogy to a procedure being described in [5] the system (4.11), (4.12) can be solved approximately by replacing the integrals on the left-hand sides of (4.11) by suitable quadrature formulae and by applying iteration.

REFERENCES

[1] Butkovskiy, A.G., and L.M. Pustyl'nikov, "Mobile Control of Distributed Parameter Systems," Ellis Horwood, 1987. (English Transl.)

[2] Krabs, W., *On Time-Minimal Distributed Control of Vibrating Systems Governed by an Abstract Wave Equation*, Appl. Math. Optim. **13** (1985), 137–149.

[3] Krabs, W., *On Boundary Controllability of One-Dimensional Vibrating Systems*, Math. Meth. in the Appl. Sc. **1** (1979), 322–345.

[4] Krabs, W., *Optimal Control of Processes Governed by Partial Differential Equations, Part II: Vibrations*, ZOR **26** (1982), 63–86.

[5] Krabs W. and Zheng, Yan, *Numerical Solution of Time-Minimal Problems of Distributed Control of Vibrations*, Appl. Math. Optim. (to appear).

W. Krabs
Fachbereich Mathematik
Technische Hochschule Darmstadt
Schlossgartenstrasse 7
D-6100 Darmstadt
West Germany

International Series of
Numerical Mathematics, Vol. 91
© 1989 Birkhäuser Verlag Basel

Asymptotic energy estimates for Kirchhoff plates subject to weak viscoelastic damping

John E. Lagnese

Department of Mathematics
Georgetown University

Abstract. Estimates on the rate of decay of the viscoelastic energy $E(t)$ are obtained for solutions of a dynamic Kirchhoff plate equation which is subject to both viscoelastic damping and to boundary dissipation induced through the action of bending and twisting moments and shear forces applied along a part of the edge of the plate, the remaining part being clamped. Assuming the relaxation kernal $D(s)$ exhibits an algebraic rate of decay of the forms $s^m[D(s) - D_\infty] \in L^1(0, \infty)$, and that the geometry of the plate is suitably restricted, it is shown that the energy satisfies $t^{m+1} E(t) \in L^1(0, \infty)$.

Keywords. Plate equations, viscoelastic damping, boundary controllability, energy estimates.

1. Formulation of the boundary value problem

Let Ω be a bounded, open connected set in $\mathbf{R}^2$ having a Lipschitz boundary Γ. We assume that $\Gamma = \bar{\Gamma}_0 \cup \bar{\Gamma}_1$ where Γ_0 and Γ_1 are relatively open, disjoint subsets of Γ with $\Gamma_1 \neq \emptyset$. We denote by $\nu = (\nu_1, \nu_2)$ the unit normal vector to Γ (when it exists) pointing out of Γ, and set $\tau = (-\nu_2, \nu_1)$, a unit positively oriented tangent vector to Γ. Let $t > 0$, and set

$$Q = \Omega \times (0, \infty), \ \ \Sigma = \Gamma \times (0, \infty), \ \ \Sigma_0 = \Gamma_0 \times (0, \infty); \ \ \Sigma_1 = \Gamma_1 \times (0, \infty).$$

We consider the following boundary value problem, which describes the small vibrations of a thin, homogeneous, isotropic, viscoelastic plate of uniform thickness h:

$$(1.1) \qquad \rho h w'' - \frac{\rho h^3}{12} \Delta w'' + D(0) \Delta^2 w + \int_{-\infty}^t D'(t - s) \Delta^2 w(s) ds = 0 \quad \text{in } Q,$$

$$(1.2) \qquad\qquad\qquad w = \frac{\partial w}{\partial \nu} = 0 \quad \text{on } \Sigma_0,$$

$$(1.3) \ \ D(0)[\Delta w + (1 - \mu) B_1 w] + \int_{-\infty}^t D'(t - s)[\Delta w(s) + (1 - \mu) B_1 w(s)] ds = -M_\tau \quad \text{on } \Sigma_1,$$

$$(1.4) \qquad \begin{aligned} &D(0)[\frac{\partial \Delta w}{\partial \nu} + (1 - \mu) \frac{\partial B_2 w}{\partial \tau}] + \int_{-\infty}^t D'(t - s)[\frac{\partial \Delta w(s)}{\partial \nu} + (1 - \mu) \frac{\partial B_2 w(s)}{\partial \tau}] ds \\ &\qquad - \frac{\rho h^2}{12} \frac{\partial w''}{\partial \nu} = -\frac{\partial}{\partial \tau} M_\nu - g_3 \quad \text{on } \Sigma_1, \end{aligned}$$

where $I = \partial/\partial t$ and where

$$B_1 w = 2\nu_1\nu_2 \frac{\partial^2 w}{\partial x \partial y} - \nu_1^2 \frac{\partial^2 w}{\partial y^2} - \nu_2^2 \frac{\partial^2 w}{\partial x^2},$$

$$B_2 w = (\nu_1^2 - \nu_2^2) \frac{\partial^2 w}{\partial x \partial y} + \nu_1\nu_2 \Big(\frac{\partial^2 w}{\partial y^2} - \frac{\partial^2 w}{\partial x^2} \Big).$$

In the above system, w denotes vertical deflection of the plate from its equilibrium position $w = 0$. The various parameters which appear have the following meanings:

  ρ : mass density;
  μ : viscoelastic Poisson's ratio (assumed constant);
  M_τ : bending moment about the tangent vector τ;
  M_ν : twisting moment about the normal vector ν;
  g_3 : shear force in the vertical direction;
  $D(t)$: viscoelastic flexural rigidity.

The reader is referred to [5, Chapter I] for a heuristic derivation of the model.

  To simplify the writing of the system (1.1) – (1.4), we make the change $t \to t\sqrt{D(0)/\rho h}$ in the time scale. Equations (1.1) – (1.4) are then brought to the forms

(1.5) $w''(t) - \gamma\Delta w''(t) + \Delta^2 w(t) + \Delta^2 \displaystyle\int_0^\infty D'(s)w(t-s)ds = 0 \quad \text{in } Q, \quad \gamma = \dfrac{h^2}{12},$

(1.6) $$w = \frac{\partial w}{\partial \nu} = 0 \quad \text{on } \Sigma_0,$$

(1.7)
$$\mathcal{B}_1 w(t) + B_1 \int_0^\infty D'(s)w(t-s)ds = v_1,$$

$$\mathcal{B}_2 w(t) - \gamma \frac{\partial w''(t)}{\partial \nu} + B_2 \int_0^\infty D'(s)w(t-s)ds = v_2 \quad \text{on } \Sigma_1,$$

where the kernal $D(t)$ satisfies $D(0) = 1$ and where we have introduced boundary operators $\mathcal{B}_1$, $\mathcal{B}_2$ defined by

$$\mathcal{B}_1 w = \Delta w + (1 - \mu)B_1 w,$$

$$\mathcal{B}_2 w = \frac{\partial \Delta w}{\partial \nu} + (1 - \mu)\frac{\partial B_2 w}{\partial \tau}.$$

"Initial" conditions for the system are

(1.8) $w(0+) = w^0, \quad w'(0+) = w^1, \quad w(-s) = \vartheta^0(s), \quad 0 < s < \infty.$

The functions v_1, v_2 are the *control variables* and are at our disposal. They will be chosen below as certain *feedbacks* in $\{w, w'\}$.

We shall assume that the function D satisfies

(1.9) $$D \in C^2([0,\infty)), \quad D(t) > 0, \quad D'(t) < 0, \quad D''(t) \geq 0.$$

Hypotheses (1.9) assure that the viscoelastic energy (defined below) is nonincreasing, that $D_\infty \doteq D(\infty)$ and $D'_\infty \doteq D'(\infty)$ both exist and that $D_\infty \geq 0$ and $D'_\infty = 0$. We shall suppose that

(1.10) $$D_\infty > 0.$$

Assumption (1.10) means that the material behaves likes an elastic solid at $t = +\infty$.

We now seek feedback laws for the controls v_1, v_2 which will induce further dissipation in the system. First, the "energy" of the system must be properly defined. The total energy can be expected to consist of two parts. One part involves the current kinetic and strain energies, and the other will involve the past history of strains. To obtain the appropriate expression, we first rewrite (1.5) and (1.7) in the forms

(1.11) $$w'' - \gamma \Delta w'' + D_\infty \Delta^2 w + \Delta^2 \int_0^\infty D'(s)[w_t(s) - w]ds = 0 \quad \text{in } Q,$$

(1.12)
$$D_\infty \mathcal{B}_1 w + \mathcal{B}_1 \int_0^\infty D'(s)[w_t(s) - w]ds = v_1,$$
$$D_\infty \mathcal{B}_2 w - \gamma \frac{\partial w''}{\partial \nu} + \mathcal{B}_2 \int_0^\infty D'(s)[w_t(s) - w]ds = v_2 \quad \text{on } \Sigma_1$$

where in (1.11) and (1.12) we have introduced the notation

$$w_t(s) = w(t - s).$$

We multiply (1.11) by w' and integrate the product over Ω to obtain

(1.13)
$$(w''(t), w'(t)) - \gamma(\Delta w''(t), w'(t)) + D_\infty(\Delta^2 w(t), w'(t))$$
$$+ (\Delta^2 \int_0^\infty D'(s)[w_t(s) - w(t)]ds, w'(t)) = 0,$$

where $(u, v) = \int_\Omega u(X)v(X)dX$, $X = \{x, y\}$.

We have

(1.14) $$(\Delta w'', w') = -(\nabla w'', \nabla w') + \int_{\Gamma_1} w' \frac{\partial w''}{\partial \nu} d\Gamma.$$

The third and fourth terms in (1.13) are transformed with the aid of the following *Green's formula* [4, p.206]:

(1.15) $$(\Delta^2 u, v) = a(u; v) + \int_\Gamma [(\mathcal{B}_2 u)v - (\mathcal{B}_1 u)\frac{\partial v}{\partial \nu}]d\Gamma,$$

where

$$
a(u; v) = \int_\Omega \left[\frac{\partial^2 u}{\partial x^2} \frac{\partial^2 v}{\partial x^2} + \frac{\partial^2 u}{\partial y^2} \frac{\partial^2 v}{\partial y^2} + \mu \left(\frac{\partial^2 u}{\partial x^2} \frac{\partial^2 v}{\partial y^2} + \frac{\partial^2 u}{\partial y^2} \frac{\partial^2 v}{\partial x^2} \right) \right.
$$

(1.16)

$$
\left. + 2(1 - \mu) \frac{\partial^2 u}{\partial x \partial y} \frac{\partial^2 v}{\partial x \partial y} \right] dX.
$$

Identity (1.15) is valid for all u, v in $H^4(\Omega)$ and for all $\mu \in \mathbf{R}$.

Use of (1.14) and (1.15) in (1.13) yields

$$
(w''(t), w'(t)) + \gamma(\nabla w''(t), \nabla w'(t)) + D_\infty a(w(t); w'(t))
$$

(1.17)

$$
+ \int_{\Gamma_1} [w'(t)(D_\infty \mathcal{B}_2 w(t) - \gamma \frac{\partial w''(t)}{\partial \nu}) - D_\infty \frac{\partial w'(t)}{\partial \nu} \mathcal{B}_1 w(t)] d\Gamma
$$

$$
+ \int_0^\infty D'(s)(\Delta^2 w_t(s) - \Delta^2 w(t), w'(t)) ds = 0.
$$

We have

$$
(1.18) \quad (\Delta^2 w_t(s) - \Delta^2 w, w') = (\Delta^2(w_t(s) - w), w' - w_t'(s)) + (\Delta^2(w_t(s) - w), w_t'(s)).
$$

The first term on the right side of (1.18) equals

$$
- a(w_t(s) - w; w_t'(s) - w')
$$

$$
- \int_{\Gamma_1} [(w_t'(s) - w')\mathcal{B}_2(w_t(s) - w) - \frac{\partial}{\partial \nu}(w_t'(s) - w')\mathcal{B}_1(w_t(s) - w)] d\Gamma
$$

$$
= -\frac{1}{2} \frac{\partial}{\partial t} a(w_t(s) - w)
$$

$$
- \int_{\Gamma_1} [(w_t'(s) - w')\mathcal{B}_2(w_t(s) - w) - \frac{\partial}{\partial \nu}(w_t'(s) - w')\mathcal{B}_1(w_t(s) - w)] d\Gamma.
$$

The second term on the right side of (1.18) equals

$$
a(w_t(s) - w; w_t'(s)) + \int_{\Gamma_1} [w_t'(s)\mathcal{B}_2(w_t(s) - w) - \frac{\partial}{\partial \nu}(w_t'(s))\mathcal{B}_1(w_t(s) - w)] d\Gamma
$$

$$
= -\frac{1}{2} \frac{\partial}{\partial s} a(w_t(s) - w)
$$

$$
+ \int_{\Gamma_1} [w_t'(s)\mathcal{B}_2(w_t(s) - w) - \frac{\partial}{\partial \nu}(w_t'(s))\mathcal{B}_1(w_t(s) - w)] d\Gamma.
$$

Therefore

$$
(\Delta^2 w_t(s) - \Delta^2 w, w') = -\frac{1}{2} \frac{\partial}{\partial t} a(w_t(s) - w) - \frac{1}{2} \frac{\partial}{\partial s} a(w_t(s) - w)
$$

$$
+ \int_{\Gamma_1} [w' \mathcal{B}_2(w_t(s) - w) - \frac{\partial w'}{\partial \nu} \mathcal{B}_1(w_t(s) - w)] d\Gamma.
$$

Consequently we have

$$\int_0^\infty D'(s)(\Delta^2 w_t(s) - \Delta^2 w, w')ds$$

(1.19)
$$= -\frac{1}{2}\frac{\partial}{\partial t}D'(s)a(w_t(s) - w)ds + \frac{1}{2}\int_0^\infty D''(s)a(w_t(s) - w)ds$$

$$+ \int_0^T \int_{\Gamma_1} [w'B_2(w_t(s) - w) - \frac{\partial w'}{\partial \nu}B_1(w_t(s) - w)]d\Gamma ds.$$

Substituting (1.19) into (1.17) and using the boundary conditions (1.7) yields

(1.20)
$$\frac{1}{2}\frac{\partial}{\partial t}[\|w'(t)\|^2 + \gamma\|\nabla w'(t)\|^2 + D_\infty a(w(t)) - \int_0^\infty D'(s)a(w_t(s) - w(t))ds]$$

$$= -\int_{\Gamma_1} [w'(t)v_2(s) - \frac{\partial w'(t)}{\partial \nu}v_1(t)]d\Gamma - \frac{1}{2}\int_0^\infty D''(s)a(w_t(s) - w(t))ds.$$

We define the *total energy* of the system (1.5) – (1.7) as

(1.21) $$E(t) = \frac{1}{2}[\|w'(t)\|^2 + \gamma\|\nabla w'(t)\|^2 + D_\infty a(w(t)) + \int_0^\infty |D'(s)|a(w_t(s) - w(t))ds].$$

After an integration in t, (1.20) may be written

(1.22)
$$E(t) = E(0) - \int_0^t \int_{\Gamma_1} [w'(t)v_2(t) - \frac{\partial w'(t)}{\partial \nu}v_1(t)]d\Gamma dt$$

$$- \frac{1}{2}\int_0^t \int_0^\infty D''(s)a(w_t(s) - w(t))dsdt.$$

The last term on the right side of (1.22) represents *viscoelastic damping*. We may introduce additional damping through an appropriate choice of v_1 and v_2 in the first integral on the right side of (1.22). To this end, we set

(1.23) $$v_1 = -\mathcal{F}_1(w'), \quad v_2 = -\mathcal{F}_2(w').$$

In (1.23), $\mathcal{F}_1, \mathcal{F}_2$ are differential operators chosen in such a way that

(1.24) $$\int_{\Gamma_1} [w'(t)v_2(t) - \frac{\partial w'(t)}{\partial \nu}v_1(t)]d\Gamma = \int_{\Gamma_1} F\{w'_x, w'_y, w'\} \cdot \{w'_x, w'_y, w'\}d\Gamma,$$

where $F = [f_{ij}]$ is a symmetric, *positive semi-definite* matrix on Γ_1. The first integral on the right side of (1.22) then represents *mechanical dissipation* in the system. It may be verified that (1.24) will hold if $\mathcal{F}_1$ and $\mathcal{F}_2$ are defined by

(1.25)
$$\mathcal{F}_1(\varphi) = (\nu_1^2 f_{11} + \nu_2^2 f_{22} + 2\nu_1\nu_2 f_{12})\frac{\partial \varphi}{\partial \nu}$$

$$+ [\nu_1\nu_2(f_{22} - f_{11}) + (\nu_1^2 - \nu_2^2)f_{12}]\frac{\partial \varphi}{\partial \tau} + (\nu_1 f_{13} + \nu_2 f_{23})\varphi,$$

$$\mathcal{F}_2(\varphi) = -f_{33}\varphi - (\nu_1 f_{13} + \nu_2 f_{23})\frac{\partial\varphi}{\partial\nu}$$

(1.26)
$$+ \frac{\partial}{\partial\tau}\{[\nu_1\nu_2(f_{22} - f_{11}) + (\nu_1^2 - \nu_2^2)f_{12}]\frac{\partial\varphi}{\partial\nu}$$

$$+ (\nu_1^2 f_{22} + \nu_2^2 f_{11} - 2\nu_1\nu_2 f_{12})\frac{\partial\varphi}{\partial\tau}\}.$$

In particular, if F is chosen to be diagonal with equal diagonal elements $f_{ii} = f_0$, then (1.25), (1.26) reduce to

$$\mathcal{F}_1(\varphi) = f_0\frac{\partial\varphi}{\partial\nu}, \quad \mathcal{F}_2(\varphi) = -f_0\frac{\partial}{\partial\tau}\left(f_0\frac{\partial\varphi}{\partial\tau}\right).$$

The purpose of this paper is to derive asymptotic estimates for the total energy (1.21) *of the system* (1.5) – (1.8), *where* ν_1, ν_2 *are given by the feedback laws* (1.23) *and where* $\mathcal{F}_1$ *and* $\mathcal{F}_2$ *are defined by* (1.25) *and* (1.26), *respectively.*

2. Existence and uniqueness of solutions

2.1. Function spaces and variational formulation.
We introduce $H = L^2(\Omega)$,

$$W = \{\varphi | \varphi \in H^2(\Omega), \ \varphi = \frac{\partial\varphi}{\partial\nu} = 0 \text{ on } \Gamma_0\}, \quad V = \{\varphi | \varphi \in H^1(\Omega), \ \varphi = 0 \text{ on } \Gamma_0\},$$

where $H^k(\Omega)$ is the standard Sobolev space of order k. Let $b(\varphi; \psi)$ and $c(\varphi; \psi)$ be bilinear forms defined on W and V, respectively, by

$$b(\varphi; \psi) = \int_{\Gamma_1} F\{\varphi_x, \varphi_y, \varphi\} \cdot \{\psi_x, \psi_y, \psi\} d\Gamma,$$

$$c(\varphi; \psi) = (\varphi, \psi) + \gamma(\nabla\varphi, \nabla\psi).$$

These forms are continuous, symmetric and nonnegative, and $c(\cdot; \cdot)$ is strictly coercive on V. If $\Gamma_0 \neq \emptyset$, we may (and do) define new norms on W and V by setting

$$\|\varphi\|_W = a(\varphi)^{\frac{1}{2}}, \quad \|\varphi\|_V = c(\varphi)^{\frac{1}{2}},$$

where $a(\varphi) = a(\varphi; \varphi)$, $c(\varphi) = c(\varphi; \varphi)$. These are equivalent to the usual Sobolev norms on the spaces W and V, respectively. (The equivalence of $a(\varphi)^{\frac{1}{2}}$ and $\|\varphi\|_{H^2(\Omega)}$ on W when $\Gamma_0 \neq \emptyset$ is a version of *Korn's Lemma*; see [4, p.110].)

We identify H with its dual H'. We then have the dense and continuous embeddings $W \subset V \subset H \subset V' \subset W'$. We introduce operators $A \in \mathcal{L}(W, W')$, $B \in \mathcal{L}(W, W')$, $C \in \mathcal{L}(V, V')$ by setting

$$\langle A\varphi, \psi \rangle = a(\varphi; \psi), \quad \langle B\varphi, \psi \rangle = b(\varphi; \psi), \quad \forall \varphi, \psi \in W,$$

$$\langle C\varphi, \psi \rangle = c(\varphi; \psi), \quad \forall \varphi, \psi \in V.$$

A (resp. C) is the canonical isomorphism of W (resp. V) onto W' (resp. V').
   We define also

$$(2.1) \qquad \mathcal{W} = L^2(0, \infty; |D'(\cdot)|; W),$$

the Hilbert space of W-valued functions φ for which $a(\varphi(s))$ is measurable on $(0, \infty)$ and

$$\|\varphi\|_\mathcal{W} = \Big(\int_0^\infty a(\varphi(s))|D'(s)|ds\Big)^{\frac{1}{2}} < +\infty.$$

Let $\mathcal{H}$ denote the product space

$$(2.2) \qquad \mathcal{H} = W \times V \times \mathcal{W},$$

normed by

$$(2.3) \qquad \|\{w, v, u(\cdot)\}\|_\mathcal{H} = \big(D_\infty a(w) + c(v) + \|w - u(\cdot)\|_\mathcal{W}^2\big)^{\frac{1}{2}}.$$

The space $\mathcal{H}$ is a "finite energy space" of the type first introduced for viscoelastic problems by Dafermos [1].
   We define an operator $L \in \mathcal{L}(\mathcal{W}, W)$ by

$$(2.4) \qquad Lu = \int_0^\infty u(s)D'(s)ds$$

where the integral is a Bochner integral in W. For a sufficiently regular function $t \to w(t) :$ $(-\infty, \infty) \to W$ of, say, compact support, we can define a function $t \to w_t : (0, \infty) \to W$ by setting

$$w_t(s) = w(t - s), \quad 0 < s < \infty.$$

With the above definitions, we may now give a variational formulation of the initial-boundary value problem (1.1)–(1.4), (1.15) as follows: Find a W-valued function w defined on $(-\infty, \infty)$ such that

$$(2.5) \qquad \{w(t), w'(t), w_t(\cdot)\} \in \mathcal{H} \quad \text{for} \quad t \geq 0;$$

$$(2.6) \qquad t \to \{w(t), w'(t), w_t(\cdot)\} : [0, \infty) \to \mathcal{H} \quad \text{is continuous;}$$

$$(2.7) \qquad \frac{d}{dt}[c(w'(t); \hat{w}) + b(w(t); \hat{w})] + a(w(t); \hat{w}) + a(Lw_t; \hat{w}) = 0, \ t > 0, \ \forall \hat{w} \in W;$$

$$(2.8) \qquad \{w(0+), w'(0+), w_0(\cdot)\} = \{w^0, w^1, \vartheta^0\} \quad \text{given in } \mathcal{H}.$$

Equivalently, we may replace (2.7) by

$$(2.9) \qquad \frac{d}{dt}[c(w'(t); \hat{w}) + b(w(t); \hat{w})] + D_\infty a(w(t); \hat{w}) + a(L(w_t - w(t)); \hat{w}) = 0,$$

which amounts to writing (1.1), (1.3) in the form (1.7), (1.8). Equation (2.9) is in turn equivalent to

$$(2.10) \qquad \frac{d}{dt}[Cw'(t) + Bw(t)] + D_\infty Aw(t) + AL(w_t - w(t)) = 0 \quad \text{in } W', \ t > 0.$$

2.2. Well-posedness of (2.5) – (2.8).

The problem (2.5) – (2.8) may be solved by a semigroup appoach. First we note the $\mathcal{H}'$, the dual space of $\mathcal{H}$, is

$$\mathcal{H}' = W' \times V' \times W' = W' \times V' \times L^2(0, \infty; |D'(\cdot)|; W').$$

Let A_W denote the canonical isomorphism of W onto W'. It is defined by $(A_W u)(s) = A(u(s))$, $u \in W$. We introduce

$$U(t) = \{w(t), w'(t), w_t(\cdot)\}$$

and define the linear operators

$$C = \begin{bmatrix} A & 0 & 0 \\ 0 & C & 0 \\ 0 & 0 & A_W \end{bmatrix} \in \mathcal{L}(\mathcal{H}, \mathcal{H}'),$$

$$(2.11) \qquad \mathcal{A} = \begin{bmatrix} 0 & -A & 0 \\ A & B & AL \\ 0 & 0 & A_W \frac{\partial}{\partial s} \end{bmatrix},$$

with

$$D(\mathcal{A}) = \{\{w, v, u(\cdot)\} | w \in W, v \in W, u \in W, A(w + Lu) + Bv \in V', \frac{du}{ds} \in W, u(0) = w\}.$$

C is the canonical isomorphism of $\mathcal{H}$ onto $\mathcal{H}'$ and $C^{-1}\mathcal{A} : D(\mathcal{A}) \to \mathcal{H}$.

We write (2.5) – (2.8) as

$$U(t) \in \mathcal{H}, \quad t \geq 0,$$
$$t \to U(t) : [0, \infty) \to \mathcal{H} \quad \text{is continuous,}$$
$$U'(t) + C^{-1}\mathcal{A}U(t) = 0, \quad t > 0,$$
$$U(0) = U^0 \doteq \{w^0, w^1, \vartheta^0(\cdot)\} \in \mathcal{H}.$$

The existence of a unique solution of this last problem is guaranteed by

Theorem 2.1. $-C^{-1}A$ *is the infinitesimal generator of a C_0-semigroup of contractions on $\mathcal{H}$.*

A proof of Theorem 2.1 may be given along the lines of [1, Section 6], and will be omitted.

Remark 2.1: The following inclusion may be shown to hold:

(2.12)
$$D(\mathcal{A}) \supset \{\{w, v, u(\cdot)\} \in \mathcal{H} | w \in H^3(\Omega) \cap W, \; v \in W,$$
$$u \in L^2(0, \infty; |D'(\cdot), |; H^3(\Omega) \cap W), \frac{du}{ds} \in W, \; u(0) = w,$$
$$\mathcal{B}_1 w + \mathcal{B}_1 \int_0^\infty D'(s)u(s)ds = -\mathcal{F}_1(v) \text{ on } \Gamma_1\}.$$

If the initial data $\{w^0, w^1, \vartheta^0(\cdot)\} \in D(\mathcal{A})$, then the mapping $t \to \{w(t), w'(t), w_t(\cdot)\}$ is continuous from $[0, \infty)$ into $D(\mathcal{A})$. From (2.12) it follows that such solutions are *not* classical. In fact, the inclusion (2.12) will be *strict*, in general. However, if the initial data is sufficient by smooth and satisfy appropriate compatibility relations with respect to the boundary conditions, if Γ is smooth *and if $\bar{\Gamma}_0 \cap \bar{\Gamma}_1 = \emptyset$*, it may be proved that the solution is classical, that is, $w \in C([0, \infty); H^4(\Omega) \cap W), w_t \in L^2(0, \infty; |D'(\cdot)|; H^4(\Omega) \cap W), w'' \in C([0, \infty); W)$.

3. Asymptotic energy estimates

We assume that Γ is smooth, $\bar{\Gamma}_0 \cap \bar{\Gamma}_1 = \emptyset$ and that the initial data $\{w^0, w^1, \vartheta^0(\cdot)\}$ are such that the solution of (1.5) – (1.8) is *classical*. The set of such data is dense in $\mathcal{H}$. It is also assumed that $\Gamma_0 \neq \emptyset$ and that the triple $\{\Omega, \Gamma_0, \Gamma_1\}$ satisfies the following geometric conditions: there is a point $X_0 = \{x_0, y_0\} \in \mathbf{R}^2$ such that

(3.1)
$$(X - X_0) \cdot \nu(X) \leq 0, \quad \forall X \in \Gamma_0,$$
$$(X - X_0) \cdot \nu(X) \geq 0, \quad \forall X \in \Gamma_1.$$

We set $m(X) = X - X_0$, and choose the elements f_{ij} of the matrix F according to

(3.2)
$$f_{ij} = (m \cdot \nu)g_{ij}, \quad i, j = 1, 2, 3,$$

where $g_{ij} \in C^1(\Gamma_1)$ and where $G = [g_{ij}]$ satisfies

(3.3)
$$g_0|\xi|^2 \leq G(X)\xi \cdot \xi \leq G_0|\xi|^2, \quad \forall \xi \in \mathbf{R}^3, \; \forall X \in \Gamma_1.$$

for some *positive* constants g_0, G_0.

3.1. An energy identity.

In this subsection we will derive a certain energy identity which will be the basis for our asymptotic energy estimates. In order to do so, the following result will be needed.

Lemma 3.1. *For every $\varphi \in H^4(\Omega)$ and for every $\mu \in \mathbf{R}$,*

(3.4)
$$\int_\Omega (m \cdot \nabla\varphi)\Delta^2\varphi dX = a(\varphi) + \frac{1}{2}\int_\Gamma m \cdot \nu[\varphi_{xx}^2 + \varphi_{yy}^2 + 2\mu\varphi_{xx}\varphi_{yy} + 2(1-\mu)\varphi_{xy}^2]d\Gamma$$
$$+ \int_\Gamma [(\mathcal{B}_2\varphi)m \cdot \nabla\varphi - (\mathcal{B}_1\varphi)\frac{\partial(m \cdot \nabla\varphi)}{\partial\nu}]d\Gamma.$$

Proof: We apply Green's formula (1.15) with $u = \varphi$ and $v = m \cdot \nabla\varphi$. We obtain

(3.5)
$$\int_\Omega (m \cdot \nabla\varphi)\Delta^2\varphi dX = a(\varphi; m \cdot \nabla\varphi) + \int_\Gamma [(\mathcal{B}_2\varphi)m \cdot \nabla\varphi - (\mathcal{B}_1\varphi)\frac{\partial(m \cdot \nabla\varphi)}{\partial\nu}]d\Gamma.$$

Let us calculate $a(\varphi; m \cdot \nabla\varphi)$. We have

$$a(\varphi; m \cdot \nabla\varphi) = \int_\Omega \{\varphi_{xx}(m \cdot \nabla\varphi)_{xx} + \varphi_{yy}(m \cdot \nabla\varphi)_{yy}$$
$$+ \mu[\varphi_{xx}(m \cdot \nabla\varphi)_{yy} + \varphi_{yy}(m \cdot \nabla\varphi)_{xx}] + 2(1-\mu)\varphi_{xy}(m \cdot \nabla\varphi)_{xy}\}dX.$$

When the differentiations of $m \cdot \nabla\varphi$ beneath the integral are carried out, one finds that $a(\varphi; m \cdot \nabla\varphi)$ may be written

(3.6)
$$a(\varphi; m \cdot \nabla\varphi) = 2a(\varphi) + \frac{1}{2}\int_\Omega m \cdot \nabla[\varphi_{xx}^2 + \varphi_{yy}^2 + 2\mu\varphi_{xx}\varphi_{yy} + 2(1-\mu)\varphi_{xy}^2]dX$$
$$= a(\varphi) + \frac{1}{2}\int_\Omega \text{div}\{m[\varphi_{xx}^2 + \varphi_{yy}^2 + 2\mu\varphi_{xx}\varphi_{yy} + 2(1-\mu)\varphi_{xy}^2]\}dX$$
$$= a(\varphi) + \frac{1}{2}\int_\Gamma m \cdot \nu[\varphi_{xx}^2 + \varphi_{yy}^2 + 2\mu\varphi_{xx}\varphi_{yy} + 2(1-\mu)\varphi_{xy}^2]d\Gamma.$$

Inserting (3.6) into (3.5) yields (3.4). ∎

Let us introduce $w^* = L(w_t - w)$, that is

$$w^*(t) = \int_0^\infty D'(s)[w_t(s) - w(t)]ds, \quad t \geq 0.$$

We also introduce the notation

$$a_\Gamma(w) = \int_\Gamma m \cdot \nu[w_{xx}^2 + w_{yy}^2 + 2\mu w_{xx}w_{yy} + 2(1-\mu)w_{xy}^2]d\Gamma.$$

We apply Lemma 3.1 to the function $D_\infty w + w^*$, and integrate the resulting identity in t from 0 to T. In view of (1.11), (1.12) and (1.23) we obtain

(3.7)
$$-\int_0^T \int_\Omega m \cdot \nabla(D_\infty w + w^*)w'' dX dt - \gamma \int_0^T \int_\Omega \nabla w'' \cdot \nabla[m \cdot \nabla(D_\infty w + w^*)]dX dt$$
$$= \int_0^T a(D_\infty w + w^*)dt + \frac{1}{2}\int_0^T a_\Gamma(D_\infty w + w^*)dt$$
$$- \int_0^T \int_{\Gamma_0} \frac{\partial}{\partial\nu}[m \cdot \nabla(D_\infty w + w^*)]\mathcal{B}_1(D_\infty w + w^*)d\Gamma dt$$
$$- \int_0^T \int_{\Gamma_1} \{m \cdot \nabla(D_\infty w + w^*)\mathcal{F}_2(w') - \frac{\partial}{\partial\nu}[m \cdot \nabla(D_\infty w + w^*)]\mathcal{F}_1(w')\}d\Gamma dt.$$

The first term in (3.7) is calculated as follows:

$$(3.8) \quad -\int_0^T \int_\Omega m \cdot \nabla(D_\infty w + w^*) w'' dX \, dt = -Y_1 + \int_0^T \int_\Omega m \cdot \nabla(D_\infty w' + w^{*'}) w' dX \, dt$$

where

$$(3.9) \qquad\qquad Y_1 = (m \cdot \nabla(D_\infty w + w^*), w')|_0^T.$$

The second term on the right side of (3.8) may be written

$$\int_0^T \int_\Omega m \cdot \nabla(D_\infty w + w^{*'}) w' dX \, dt$$

$$(3.10) \qquad \begin{aligned} &= \frac{1}{2} D_\infty \int_0^T \int_\Omega \operatorname{div}(m w'^2) dX \, dt - D_\infty \int_0^T \int_\Omega w'^2 dX \, dt \\ &\quad + \int_0^T \int_\Omega w' m \cdot \nabla w^{*'} dX \, dt \\ &= \frac{1}{2} D_\infty \int_0^T \int_\Omega (m \cdot \nu) w'^2) dX \, dt - D_\infty \int_0^T \int_\Omega w'^2 dX \, dt \\ &\quad + \int_0^T \int_\Omega w' m \cdot \nabla w^{*'} dX \, dt. \end{aligned}$$

We have

$$(3.11) \qquad \begin{aligned} w^{*'}(t) &= \int_0^\infty D'(s)[w'(t-s) - w'(t)] ds \\ &= -\int_0^\infty D'(s) \frac{\partial}{\partial s}[w(t-s) - w(t)] ds + (D(0) - D_\infty) w'(t) \\ &= \int_0^\infty D''(s)[w'(s) - w(t)] ds + (D(0) - D_\infty) w'(t). \end{aligned}$$

Therefore (recall that $D(0) = 1$)

$$(3.12) \qquad \begin{aligned} \int_0^T \int_\Omega w' m \cdot \nabla w^{*'} dX \, dt &= (1 - D_\infty) \int_0^T \int_\Omega w' m \cdot \nabla w' dX \, dt \\ &\quad + \int_0^T \int_\Omega w' \int_0^\infty D''(s) m \cdot \nabla[w_t(s) - w(t)] ds dX \, dt \\ &= \frac{1}{2}(1 - D_\infty) \int_0^T \int_{\Gamma_1} (m \cdot \nu) w'^2 d\Gamma \, dt - (1 - D_\infty) \int_0^T \int_\Omega w'^2 dX \, dt \\ &\quad + \int_0^T \int_\Omega w' \int_0^\infty D''(s) m \cdot \nabla[w_t(s) - w(t)] ds dX \, dt. \end{aligned}$$

Substitution of (3.12) into (3.10) yields

(3.13)
$$\int_0^T \int_\Omega m \cdot \nabla(D_\infty w' + w^{*\prime})w' dX dt$$
$$= -\int_0^T \int_\Omega w'^2 dX dt + \frac{1}{2}\int_0^T \int_{\Gamma_1} (m \cdot \nu)w'^2 d\Gamma dt$$
$$+ \int_0^T \int_\Omega w' \int_0^\infty D''(s)m \cdot \nabla[w_t(s) - w(t)]ds dX dt.$$

Use of (3.8) and (3.13) in (3.7) allows (3.7) to be written

(3.14)
$$Y_1 + \int_0^T \int_\Omega w'^2 dX dt - \int_0^T \int_\Omega w' \int_0^\infty D''(s)m \cdot \nabla[w_t(s) - w(t)]ds dX dt$$
$$+ \gamma \int_0^T \int_\Omega \nabla w'' \cdot \nabla[m \cdot \nabla(D_\infty w + w^*)]dX dt + \int_0^T a(D_\infty w + w^*)dt$$
$$= \frac{1}{2}\int_0^T \int_{\Gamma_1} (m \cdot \nu)w'^2 d\Gamma dt - \frac{1}{2}\int_0^T a_\Gamma(D_\infty w + w^*)dt$$
$$+ \int_0^T \int_{\Gamma_0} \frac{\partial}{\partial \nu}[m \cdot \nabla(D_\infty w + w^*)]B_1(D_\infty w + w^*)d\Gamma dt$$
$$- \int_0^T b(m \cdot \nabla(D_\infty w + w^*); w')dt.$$

We next transform the fourth term on the left side of (3.14) as follows:

(3.15)
$$\int_0^T \int_\Omega \nabla w'' \cdot \nabla[m \cdot \nabla(D_\infty w + w^*)]dX dt$$
$$= Y_2 - \int_0^T \int_\Omega \nabla w' \cdot \nabla[m \cdot \nabla(D_\infty w' + w^{*\prime})]dX dt$$

where

$$Y_2 = (\nabla w', \nabla[m \cdot \nabla(D_\infty w + w^*)])|_0^T.$$

We have

(3.16)
$$\int_0^T \int_\Omega \nabla w' \cdot \nabla(m \cdot \nabla w')dX dt = \frac{1}{2}\int_0^T \int_{\Gamma_1} m \cdot \nu|\nabla w'|^2 d\Gamma dt.$$

From (3.11)

$$\int_\Omega \nabla w' \cdot \nabla (m \cdot \nabla w^{*\prime}) dX \, dt$$

$$= \int_\Omega \nabla w' \cdot \int_0^\infty D''(s) \nabla [m \cdot \nabla (w_t(s) - w(t))] ds \, dX$$

(3.17)
$$+ (1 - D_\infty) \int_\Omega \nabla w' \cdot \nabla (m \cdot \nabla w') dX$$

$$= \int_\Omega \nabla w' \cdot \int_0^\infty D''(s) \nabla [m \cdot \nabla (w_t(s) - w(t))] ds \, dX$$

$$+ \frac{1}{2}(1 - D_\infty) \int_{\Gamma_1} m \cdot \nu |\nabla w'|^2 d\Gamma.$$

It follows from (3.16), (3.17) that

$$\int_0^T \int_\Omega \nabla w' \cdot \nabla [m \cdot \nabla (D_\infty w' + w^{*\prime})] dX \, dt$$

(3.18)
$$= \int_0^T \int_\Omega \nabla w' \cdot \int_0^\infty D''(s) \nabla [m \cdot \nabla (w_t(s) - w(t))] ds \, dX \, dt$$

$$+ \frac{1}{2} \int_0^T \int_{\Gamma_1} m \cdot \nu |\nabla w'|^2 d\Gamma \, dt.$$

Use of (3.15) and (3.18) in (3.14) yields

$$Y_1 + \gamma Y_2 + \int_0^T \int_\Omega w'^2 dX \, dt$$

$$- \int_0^T \int_\Omega w' \int_0^\infty D''(s) m \cdot \nabla [w_t(s) - w(t)] ds \, dX \, dt$$

$$- \gamma \int_0^T \int_\Omega \nabla w' \cdot \int_0^\infty D''(s) \nabla [m \cdot \nabla (w_t(s) - w(t))] ds \, dX \, dt$$

(3.19)
$$+ \int_0^T a(D_\infty w + w^*) dt$$

$$= \frac{1}{2} \int_0^T \int_{\Gamma_1} m \cdot \nu (w'^2 + \gamma |\nabla w'|^2) d\Gamma \, dt - \frac{1}{2} \int_0^T a_\Gamma (D_\infty w + w^*) dt$$

$$+ \int_0^T \int_{\Gamma_0} \frac{\partial}{\partial \nu} [m \cdot \nabla (D_\infty w + w^*)] \mathcal{B}_1 (D_\infty w + w^*) d\Gamma \, dt$$

$$- \int_0^T b(m \cdot \nabla (D_\infty w + w^*); w') dt.$$

Let us next examine the boundary integrals over Γ_0. Write $a_\Gamma = a_{\Gamma_0} + a_{\Gamma_1}$. Since $w = \partial w / \partial \nu = 0$ on Γ_0, we have

$$a_{\Gamma_0}(w + D_\infty w^*) = \int_{\Gamma_0} m \cdot \nu (D_\infty \Delta w + \Delta w^*) d\Gamma.$$

Furthermore,

$$\mathcal{B}_1(D_\infty w + w^*) = \Delta(D_\infty w + w^*),$$

$$\frac{\partial}{\partial\nu}m \cdot \nabla(D_\infty w + w^*) = (m \cdot \nu)\Delta(D_\infty w + w^*) \quad \text{on } \Gamma_0.$$

Therefore the integrals over Γ_0 in (3.19) combine to give

$$\frac{1}{2}\int_{\Gamma_0} m \cdot \nu(D_\infty \Delta w + \Delta w^*)^2 d\Gamma,$$

so that (3.19) may be written

$$\begin{aligned}
(3.20) \quad & Y_1 + \gamma Y_2 + \int_0^T \int_\Omega w'^2 dX dt \\
& - \int_0^T \int_\Omega w' \int_0^\infty D''(s)m \cdot \nabla[w_t(s) - w(t)]ds dX dt \\
& - \gamma \int_0^T \int_\Omega \nabla w' \cdot \int_0^\infty D''(s)\nabla[m \cdot \nabla(w_t(s) - w(t))]ds dX dt \\
& + \int_0^T a(D_\infty w + w^*)dt \\
& = \frac{1}{2}\int_0^T \int_{\Gamma_0} m \cdot \nu(D_\infty \Delta w + \Delta w^*)^2 d\Gamma dt \\
& + \frac{1}{2}\int_0^T \int_{\Gamma_1} m \cdot \nu(w'^2 + \gamma|\nabla w'|^2)d\Gamma dt \\
& - \frac{1}{2}\int_0^T a_{\Gamma_1}(D_\infty w + w^*)dt - \int_0^T b(m \cdot \nabla(D_\infty w + w^*); w')dt.
\end{aligned}$$

Next, in (3.20) we consider the term

$$\begin{aligned}
(3.21) \quad & \int_0^T a(D_\infty w + w^*)dt = D_\infty^2 \int_0^T a(w)dt \\
& + D_\infty \int_0^T a(w; w^*)dt + \int_0^T a(D_\infty w + w^*; w^*)dt.
\end{aligned}$$

We use Green's formula (1.15) with u and v replaced by $D_\infty w + w^*$ and w^*, respectively. The resulting expression may be written

$$\begin{aligned}
(3.22) \quad & a(D_\infty w + w^*; w^*) \\
& = \int_\Omega w^*\Delta^2(D_\infty w + w^*)dX \\
& + \int_{\Gamma_1}[\frac{\partial w^*}{\partial\nu}\mathcal{B}_1(D_\infty w + w^*) - w^*\mathcal{B}_2(D_\infty w + w^*)]d\Gamma.
\end{aligned}$$

Use of (1.11) and (1.12) allows (3.22) to be rewritten (after an integration in t)

$$\int_0^T a(D_\infty w + w^*; w^*)dt = -\int_0^T \int_\Omega w^*(w'' - \gamma \Delta w'')dX dt$$

(3.23)
$$- \int_0^T \int_{\Gamma_1} \{\frac{\partial w^*}{\partial \nu} \mathcal{F}_1(w') - w^*[\mathcal{F}_2(w') + \gamma \frac{\partial w''}{\partial \nu}]\}d\Gamma dt$$

$$= Y_3 + \int_0^T \int_\Omega w' w^{*'} dX dt + \gamma \int_0^T \int_\Omega \nabla w' \cdot \nabla w^{*'} dX dt - \int_0^T b(w^*; w')dt$$

where

(3.24)
$$Y_3 = -[(w^*, w') + \gamma(\nabla w^*, \nabla w')]_0^T.$$

Substitution of (3.23) into (3.21) yields

(3.25)
$$\int_0^T a(D_\infty w + w^*)dt = D_\infty^2 \int_0^T a(w)dt + D_\infty \int_0^T a(w; w^*)dt + Y_3$$

$$+ \int_0^T \int_\Omega (w' w^{*'} + \gamma \nabla w' \cdot \nabla w^{*'})dX dt - \int_0^T b(w^*; w')dt.$$

The second term on the right is calculated as follows:

(3.26)
$$\int_0^T a(w; w^*)dt = \int_0^T \int_0^\infty D'(s)a(w_t(s) - w(t); w(t))dsdt$$

$$= -\frac{1}{2}\int_0^T \int_0^\infty D'(s)a(w_t(s) - w(t))dsdt$$

$$+ \frac{1}{2}\int_0^T \int_0^\infty D'(s)[a(w_t(s)) - a(w(t))]dsdt$$

$$= -\frac{1}{2}\int_0^T \int_0^\infty D'(s)a(w_t(s) - w(t))dsdt$$

$$- \frac{1}{2}\int_0^T \int_0^\infty [D(s) - D_\infty]\frac{\partial}{\partial s}a(w_t(s))dsdt$$

$$+ \frac{1}{2}\int_0^T [D(s) - D_\infty][a(w_t(s)) - a(w(t))]\big|_{s=0}^\infty dt.$$

That is,

(3.27)
$$\int_0^T a(w; w^*)dt = -\frac{1}{2}\int_0^T \int_0^\infty D'(s)a(w_t(s) - w(t))dsdt$$

$$+ \frac{1}{2}\int_0^T \frac{d}{dt}\int_0^\infty [D(s) - D_\infty]a(w_t(s))dsdt$$

$$= -\frac{1}{2}\int_0^T \int_0^\infty D'(s)a(w_t(s) - w(t))dsdt$$

$$+ \frac{1}{2}\int_0^\infty [D(s) - D_\infty]a(w_T(s))ds - \frac{1}{2}\int_0^\infty [D(s) - D_\infty]a(\vartheta^0(s))ds$$

provided that we assume

(3.28) $\vartheta \in L^2(0, \infty; D(\cdot) - D_\infty; W).$

If we now insert (3.27) into (3.25) we obtain

(3.29)
$$
\int_0^T a(D_\infty w + w^*)dt = Y_3 + \int_0^T \int_\Omega (w'w^{*'} + \gamma \nabla w' \cdot \nabla w^{*'})dX \, dt
$$
$$
+ D_\infty^2 \int_0^T a(w)dt - \frac{D_\infty}{2} \int_0^T \int_0^\infty D'(s)a(w_t(s) - w(t))ds \, dt
$$
$$
+ \frac{D_\infty}{2} \int_0^\infty [D(s) - D_\infty]a(w_T(s))ds
$$
$$
- \frac{D_\infty}{2} \int_0^\infty [D(s) - D_\infty]a(\vartheta^0(s))ds - \int_0^T b(w^*; w')dt.
$$

Using (3.11), the second term on the right side of (3.29) can be written

(3.30)
$$
\int_0^T \int_\Omega (w'w^{*'} + \gamma \nabla w' \cdot \nabla w^*)dX \, dt = (1 - D_\infty) \int_0^T \int_\Omega (w'^2 + \gamma |\nabla w'|^2)dX \, dt
$$
$$
+ \int_0^T \int_\Omega w' \int_0^\infty D''(s)[w_t(s) - w(t)]ds \, dX \, dt
$$
$$
+ \gamma \int_0^T \int_\Omega \nabla w' \cdot \int_0^\infty D''(s)\nabla[w_t(s) - w(t)]ds \, dX \, dt.
$$

Use of (3.30) in (3.29) gives

(3.31)
$$
\int_0^T a(D_\infty w + w^*)dt = Y_3 + (1 - D_\infty) \int_0^T \int_\Omega (w'^2 + \gamma |\nabla w'|^2)dX \, dt
$$
$$
+ D_\infty^2 \int_0^T a(w)dt - \frac{D_\infty}{2} \int_0^T \int_0^\infty D'(s)a(w_t(s) - w)ds \, dt
$$
$$
+ \int_0^T \int_\Omega w' \int_0^\infty D''(s)[w_t(s) - w(t)]ds \, dX \, dt
$$
$$
+ \gamma \int_0^T \int_\Omega \nabla w' \cdot \int_0^\infty D''(s)\nabla[w_t(s) - w(t)]ds \, dX \, dt - \int_0^T b(w^*; w')dt
$$
$$
+ \frac{D_\infty}{2} \int_0^\infty [D(s) - D_\infty]a(w_T(s))ds - \frac{D_\infty}{2} \int_0^\infty [D(s) - D_\infty]a(\vartheta^0(s))ds.
$$

We use (3.31) to replace the corresponding term in (3.20). The result is the following identity:

$$(Y_1 + \gamma Y_2 + Y_3) + (1 - D_\infty) \int_0^T \int_\Omega (w'^2 + \gamma |\nabla w'|^2) dX dt + \int_0^T \int_\Omega w'^2 dX dt$$

$$+ D_\infty^2 \int_0^T a(w) dt - \frac{D_\infty}{2} \int_0^T \int_0^\infty D'(s) a(w_t(s) - w) ds dt$$

$$+ \int_0^T \int_\Omega w' \int_0^\infty D''(s)(1 - m \cdot \nabla)(w_t(s) - w) ds dX dt$$

(3.32)
$$+ \gamma \int_0^T \int_\Omega \nabla w' \cdot \int_0^\infty D''(s) \nabla[(1 - m \cdot \nabla)(w_t(s) - w)] ds dX dt$$

$$+ \frac{D_\infty}{2} \int_0^\infty [D(s) - D_\infty] a(w_T(s)) ds - \frac{D_\infty}{2} \int_0^\infty [D(s) - D_\infty] a(\vartheta^0(s)) ds$$

$$= \frac{1}{2} \int_0^T \int_{\Gamma_0} m \cdot \nu (D_\infty \Delta w + \Delta w^*)^2 d\Gamma dt + \frac{1}{2} \int_0^T \int_{\Gamma_1} m \cdot \nu (w'^2 + \gamma |\nabla w'|^2) d\Gamma dt$$

$$- \frac{1}{2} \int_0^T a_{\Gamma_1}(D_\infty w + w^*) dt + \int_0^T b(w^* - m \cdot \nabla(D_\infty w + w^*); w') dt.$$

The identity (3.32) is the basis for the energy estimates to be derived in the next section.

3.2. Energy estimates.
We define

$$\rho(t) = (m \cdot \nabla(D_\infty w(t) + w^*(t)) - w^*(t), w'(t))$$
$$+ \gamma(\nabla[m \cdot \nabla(D_\infty w(t) + w^*(t))] - \nabla w^*(t), \nabla w'(t)).$$

Then

(3.33)
$$\rho(T) - \rho(0) = Y_1 + \gamma Y_2 + Y_3.$$

We note that

(3.34)
$$|\rho(t)| \leq C E(t)$$

for some constant C, where $E(t)$ is given by (1.21). Indeed

(3.35)
$$|\rho(t)| \leq \frac{1}{2}[\|w'(t)\|^2 + \gamma \|\nabla w'(t)\|^2 + \|D_\infty m \cdot \nabla w(t) + m \cdot \nabla w^*(t) - w^*(t)\|^2$$
$$+ \|D_\infty \nabla(m \cdot \nabla w(t)) + \nabla(m \cdot \nabla w^*(t)) - \nabla w^*(t)\|^2]$$
$$\leq \frac{C_1}{2}[\|w'(t)\|^2 + \gamma \|\nabla w'(t)\|^2 + \|\nabla w(t)\|^2 + \|\nabla(m \cdot \nabla w(t))\|^2$$
$$+ \|w^*(t)\|^2 + \|\nabla w^*(t)\|^2 + \|\nabla(m \cdot \nabla w^*(t))\|^2].$$

Since the form $a(\cdot; \cdot)$ is strictly coercive on W, we have

(3.36)
$$\|\nabla w(t)\|^2 + \|\nabla(m \cdot \nabla w(t))\|^2 \leq \alpha_2 a(w(t))$$

for some constant α_2,

$$
\begin{aligned}
\|w^*(t)\|^2 &\leq \int_\Omega [\int_0^\infty |D'(s)| |w_t(s) - w(t)| ds]^2 dX \\
&\leq \int_\Omega \int_0^\infty |D'(s)| ds \int_0^\infty |D'(s)| |w_t(s) - w(t)|^2 ds dX \\
&\leq (1 - D_\infty) \int_0^\infty |D'(s)| \|w_t(s) - w(t)\|^2 ds \\
&\leq \alpha_0 (1 - D_\infty) \int_0^\infty |D'(s)| a(w_t(s) - w(t)) ds.
\end{aligned}
$$

(3.37)

Estimates like (3.37) hold also for $\|\nabla w^*(t)\|^2$ and $\|\nabla(m \cdot \nabla w^*(t))\|^2$. It follows that

(3.38)
$$
\begin{aligned}
|\rho(t)| &\leq \frac{C}{2} [\|w'(t)\|^2 + \gamma \|\nabla w'(t)\|^2 + D_\infty a(w(t)) + \int_0^\infty |D'(s)| a(w_t(s) - w(t)) ds] \\
&= CE(t).
\end{aligned}
$$

We introduce the function

(3.39)
$$
F_\epsilon(t) = E(t) + \epsilon \rho(t).
$$

We are going to prove that for $\epsilon > 0$ and sufficiently small,

(3.40) $\displaystyle F'_\epsilon(t) \leq -\frac{\epsilon}{2} \min(1 - D_\infty, D_\infty) \{ E(t) + \frac{d}{dt} \int_0^\infty [D(s) - D_\infty] a(w_t(s)) ds \} - \frac{\epsilon}{2} E_\Gamma(t),$

where

$$
E_\Gamma(t) = \frac{1}{2} \int_\Gamma m \cdot \nu (w'^2 + \gamma |\nabla w'|^2) d\Gamma + \frac{1}{2} a_{\Gamma_1}(D_\infty w + w^*) - \frac{1}{2} a_{\Gamma_0}(D_\infty w + w^*).
$$

Let us suppose, for the moment, that (3.40) has been established. We then obtain

(3.41)
$$
\begin{aligned}
&\frac{\epsilon D_0}{2} \int_0^t E(\tau) d\tau + \frac{\epsilon}{2} \int_0^t E_\Gamma(\tau) d\tau \leq F_\epsilon(0) - F_\epsilon(t) \\
&\quad - \frac{\epsilon D_0}{2} \int_0^\infty [D(s) - D_\infty] a(w_t(s)) ds + \frac{\epsilon D_0}{2} \int_0^\infty [D(s) - D_\infty] a(\vartheta^0(s)) ds
\end{aligned}
$$

where

$$
D_0 = \min(1 - D_\infty, D_\infty).
$$

From (3.38) and (3.39) we obtain

(3.42) $\qquad\qquad (1 - \epsilon C) E(t) \leq F_\epsilon(t) \leq (1 + \epsilon C) E(t).$

If, therefore, $\epsilon C \leq 1$ we obtain from (3.41)

(3.43)
$$
\begin{aligned}
&\int_0^t E(\tau) d\tau + \frac{1}{D_0} \int_0^t E_\Gamma(\tau) d\tau + \int_0^\infty [D(s) - D_\infty] a(w_t(s)) ds \\
&\quad \leq \frac{2}{\epsilon D_0} (1 + \epsilon C) E(0) + \int_0^\infty [D(s) - D_\infty] a(\vartheta^0(s)) ds, \quad t \geq 0.
\end{aligned}
$$

(3.43) is valid for all initial data for which the right side of (3.38) is finite. We therefore have proved, modulo (3.40), the following energy estimate.

Theorem 3.1. *Assume that the kernal $D(s)$ satisfies (1.9) and (1.10), and the the initial data $\{w^0, w^1, \vartheta^0(\cdot)\}$ satisfy*

(3.44)
$$w^0 \in W, \quad w^1 \in V,$$
$$\vartheta^0 \in L^2(0, \infty; |D'(\cdot)|; W) \cap L^2(0, \infty; D(\cdot) - D_\infty; W).$$

Suppose further that $\Gamma_0 \neq \emptyset$, that Γ_0, Γ_1 satisfy (3.1) and that the gain matrix satisfies (3.2), (3.3). Then the solution of (1.5) - (1.8), (1.23) satisfies

(3.45)
$$\int_0^\infty E(\tau)d\tau + \frac{1}{D_0} \int_0^\infty E_\Gamma(\tau)d\tau + \overline{\lim}_{t\to\infty} \int_0^\infty [D(s) - D_\infty]a(w_t(s))ds$$
$$\leq \frac{1}{\omega}E(0) + \int_0^\infty [D(s) - D_\infty]a(\vartheta^0(s))ds$$

for some $\omega > 0$.

Remark 3.1: ω *is given by*

(3.46)
$$\omega = \frac{\epsilon}{2(1 + \epsilon C)} \min(1 - D_\infty, D_\infty)$$

where C is defined by (3.38) and where $0 < \epsilon < \epsilon_0$ with ϵ_0 depending on elastic and geometric parameters and on the upper and lower bounds on the gain matrix. The value of ϵ_0 will be made explicit in the derivation of (3.34).

We would like to deduce a stronger decay rate for $E(t)$ then (3.45). If it could be shown that

(3.47)
$$\int_0^\infty E(\tau)d\tau < +\infty$$

for *all* initial data $\{w^0, w^1, \vartheta^0(\cdot)\} \in \mathcal{H}$, then a *uniform* exponential decay rate for $E(t)$ would follow from a result of Datko [2]. However, (3.47) *cannot* be expected to hold for all $\{w^0, w^1, \vartheta^0(\cdot)\} \in \mathcal{H}$ without further restrictions on $D(\cdot)$. The reason for this lack of uniform asymptotic stability is not at all obvious, but may be inferred from the work of Prüss [7], for example, at least when $\gamma = 0$. Consider first the uncontrolled system $v_1 = v_2 = 0$. Prüss has obtained an elegant representation formula for the "resolvent" of a class of hyperbolic Volterra equations which seems to include the present model (if $\gamma = 0$). This representation shows that the resolvent decomposes into the sum of a purely elastic part and a viscous part. The viscous part will be exponentially damped only if the kernal has appropriate asymptotic properties such as exponential decay, i.e.,

(3.48)
$$e^{\eta s}[D(s) - D_\infty] \in L^1(0, \infty) \quad \text{for some } \eta > 0$$

(see, e.g. Desch and Miller [3]). Moreover, feedbacks such as (1.23) or, in fact, *any* feedback which does not effect the memory, can influence *only* the elastic part of the resolvent and, therefore, cannot induce exponential decay of the resolvent unless such decay is already present in the uncontrolled system. Therefore, under the hypotheses (1.9), (1.10), one should not expect more than is given by Theorem 3.1.

One may, however, derive certain *nonuniform, algebraic* decay rates for the viscoelastic energy under hypotheses on the asymptotic behavior of the kernal that are much weaker than (3.48). More precisely, in addition to (1.9), (1.10), let us assume

$(3.49)_m$
$$\int_0^\infty s^m [D(s) - D_\infty] ds < \infty \quad \text{for some } m \geq 0.$$

One may then prove the following.

Theorem 3.2. *Assume that the kernal $D(s)$ satisfies (1.9), (1.10) and $(3.49)_m$, and that the initial data $\{w^0, w^1, \vartheta^0(\cdot)\}$ satisfy*

(3.50)
$$w^0 \in W, \quad w^1 \in V,$$
$$\vartheta^0 \in L^2(0, \infty, W) \quad \text{and has compact support in } [0, \infty).$$

Suppose further that $\Gamma_0 \neq \emptyset$, that Γ_0, Γ_1 satisfy (3.1) and that the gain matrix satisfies (3.2), (3.3). Then the solution of (1.5) - (1.8), (1.23) satisfies

$(3.51)_m$
$$\int_0^\infty \tau^{m+1} E(\tau) d\tau + \frac{1}{D_0} \int_0^\infty \tau^{m+1} E_\Gamma(\tau) d\tau$$
$$+ \overline{\lim}_{t \to \infty} t^{m+1} \int_0^\infty [D(s) - D_\infty] a(w_t(s)) ds < +\infty.$$

Proof: The proof uses the key inequality (3.40) together with a nice observation of Leugering [6] (introduced in the course of a derivation of $(3.46)_m$ in the context of a viscoelastic membrane with boundary damping). The proof is by induction on m. Assume $(3.49)_0$, multiply (3.40) by t and integrate the product from 0 to t to obtain

$$tF_\epsilon(t) + \frac{\epsilon D_0}{2} \int_0^t \tau E(\tau) d\tau + \frac{\epsilon}{2} \int_0^t \tau E_\Gamma(\tau) d\tau$$
$$\leq \int_0^t F_\epsilon(\tau) d\tau - \frac{\epsilon D_0}{2} \int_0^t \tau \frac{d}{d\tau} \int_0^\infty [D(s) - D_\infty] a(w_\tau(s)) ds d\tau.$$

Using (3.42) and integrating by parts in the last term on the right we obtain

(3.52)
$$t(1 - \epsilon C) E(t) + \frac{\epsilon D_0}{2} \int_0^t \tau E(\tau) d\tau + \frac{\epsilon}{2} \int_0^t \tau E_\Gamma(\tau) d\tau$$
$$\leq t(1 - \epsilon C) \int_0^t E(\tau) d\tau$$
$$- \frac{\epsilon D_0}{2} \{ t \int_0^\infty [D(s) - D_\infty] a(w_t(s)) ds$$
$$- \int_0^t d\tau \int_0^\infty [D(s) - D_\infty] a(w(\tau - s)) ds \}.$$

We rewrite the last integral on the right as follows:

$$\int_0^t d\tau \int_0^\infty [D(s) - D_\infty] a(w(\tau - s)) ds = \int_0^\infty [D(s) - D_\infty] \int_0^t a(w(\tau - s)) d\tau ds$$

$$= \int_0^\infty [D(s) - D_\infty] \int_{-s}^{t-s} a(w(\tau)) d\tau ds$$

$$(3.53) \quad = \int_{-\infty}^0 a(w(\tau)) \int_{-\tau}^{t-\tau} [D(s) - D_\infty] ds d\tau + \int_0^t a(w(\tau)) \int_0^{t-\tau} [D(s) - D_\infty] ds d\tau$$

$$\leq \int_0^\infty a(\vartheta^0(\tau)) dt \int_\tau^{t+\tau} [D(s) - D_\infty] ds d\tau + \int_0^t E(\tau) d\tau \int_0^{t-\tau} [D(s) - D_\infty] ds d\tau$$

$$\leq \int_0^\infty [a(\vartheta^0(\tau)) + E(\tau)] d\tau + \int_0^\infty [D(s) - D_\infty] ds < +\infty$$

in view of (3.45), (3.49)$_0$ and the compact support of ϑ^0. The estimate (3.51)$_0$ follows from (3.52) and (3.53).

Now assume that (3.51)$_{m-1}$ holds, multiply (3.40) by t^{m+1} and integrate from 0 to t. Proceeding as above we get

$$(1 - \epsilon C) t^{m+1} E(t) + \frac{\epsilon D_0}{2} \int_0^t \tau^{m+1} E(\tau) d\tau + \frac{\epsilon}{2} \int_0^t \tau^{m+1} E_\Gamma(\tau) d\tau$$

$$\leq (m+1)(1 + \epsilon C) \int_0^t \tau^m E(\tau) d\tau$$

$$(3.54)$$

$$- \frac{\epsilon D_0}{2} \{ t^{m+1} \int_0^\infty [D(s) - D_\infty] a(w_t(s)) ds E(\tau) d\tau$$

$$- (m+1) \int_0^t \tau^m d\tau \int_0^\infty [D(s) - D_\infty] a(w(\tau - s)) ds \}.$$

The last integral equals

$$\int_0^\infty [D(s) - D_\infty] \int_{-s}^{t-s} (\tau + s)^m a(w(\tau)) d\tau ds$$

$$\leq 2^{m-1} \int_0^\infty s^m [D(s) - D_\infty] \int_{-s}^{t-s} |\tau|^m a(w(\tau)) d\tau ds$$

$$\leq 2^{m-1} \{ \int_0^\infty \tau^m a(\vartheta^0(\tau)) \int_\tau^{t+\tau} s^m [D(s) - D_\infty] ds d\tau$$

$$+ \int_0^t \tau^m a(w(\tau)) \int_\tau^{t-\tau} s^m [D(s) - D_\infty] ds d\tau \}$$

$$\leq 2^{m-1} \int_0^\infty \tau^m [a(\vartheta^0(\tau)) + E(\tau)] d\tau \int_0^\infty s^m [D(s) - D_\infty] ds.$$

(3.51)$_m$ now follows from (3.54), (3.49)$_m$ and the induction hypothesis (3.51)$_{m-1}$.

Remark 3.2: The derivation of (3.40) and, consequently, (3.45) and (3.51)$_m$ uses the positive definiteness of the matrix G in an essential way. On the other hand, the resolvent already has certain integrability properties wiht respect to the half-line $(0, \infty)$ even when $F = 0$ [7, Theorem 11]. It would be of interest of determine *precisely* what effect the feedback laws (1.23) have on stability, in particular, whether the estimates $(3.51)_m$ remain true if $F = 0$, i.e., assuming only (1.9), (1.10) and $(3.49)_m$.

3.2.1. Proof of (3.40).

We have

$$(3.55) \qquad F'_\epsilon(t) = E'(t) + \epsilon \rho'(t) = -\frac{1}{2} \int_0^\infty D''(s) a(w_t(s) - w) ds - b(w') + \epsilon \rho'(t).$$

The derivative $\rho'(t)$ is calculated using (3.33) and the identity (in T) (3.32). The result is (writing t in place of T)

$$
\begin{aligned}
(3.56) \quad \rho'(t) = & -\int_\Omega w'^2 dX - (1 - D_\infty) \int_\Omega (w'^2 + \gamma |\nabla w'|^2) dX - D_\infty^2 a(w) \\
& + \frac{D_\infty}{2} \int_0^\infty D'(s) a(w_t(s) - w) ds \\
& - \int_\Omega w' \int_0^\infty D''(s)(1 - m \cdot \nabla)(w_t(s) - w) ds dX \\
& - \gamma \int_\Omega \nabla w' \cdot \int_0^\infty D''(s) \nabla [(1 - m \cdot \nabla)(w_t(s) - w)] ds dX \\
& - \frac{D_\infty}{2} \frac{d}{dt} \int_0^\infty [D(s) - D_\infty] a(w_t(s)) ds \\
& + \frac{1}{2} \int_{\Gamma_0} m \cdot \nu (D_\infty \Delta w + \Delta w^*)^2 d\Gamma \\
& + \frac{1}{2} \int_{\Gamma_1} m \cdot \nu (w'^2 + \gamma |\nabla w'|^2) d\Gamma - \frac{1}{2} a_{\Gamma_1}(D_\infty w + w^*) \\
& + b(w^* - m \cdot \nabla(D_\infty w + w^*); w').
\end{aligned}
$$

We proceed to obtain an upper bound for $\rho'(t)$. First, we estimate

$$
\begin{aligned}
(3.57) \quad & \left| \int_\Omega w' \int_0^\infty D''(s)(1 - m \cdot \nabla)(w_t(s) - w) ds dX \right| \\
& \leq \left\{ \int_\Omega w'^2 dX \right\}^{\frac{1}{2}} \left\{ \int_\Omega \left[\int_0^\infty D''(s)(1 - m \cdot \nabla)(w_t(s) - w) ds \right]^2 dX \right\}^{\frac{1}{2}} \\
& \leq \left\{ \int_\Omega w'^2 dX \right\}^{\frac{1}{2}} \left\{ \int_\Omega \int_0^\infty D''(s) ds \int_0^\infty D''(s) |(1 - m \cdot \nabla)(w_t(s) - w)|^2 ds dX \right\}^{\frac{1}{2}} \\
& \leq \frac{\eta}{2} \int_\Omega w'^2 dX + \frac{|D'(0)|}{2\eta} \int_0^\infty D''(s) \| (1 - m \cdot \nabla)(w_t(s) - w) \|^2 ds \\
& \leq \frac{\eta}{2} \int_\Omega w'^2 dX + \alpha_1 \frac{|D'(0)|}{2\eta} \int_0^\infty D''(s) a(w_t(s) - w) ds
\end{aligned}
$$

where α_1 is the smallest constant such that

$$\|(1 - m \cdot \nabla)\varphi\| \leq \alpha_1 a(\varphi), \quad \forall \varphi \in H^2_{\Gamma_0}(\Omega).$$

The next estimate is obtained in a similar fashion:

(3.58)
$$\left| \int_\Omega \nabla w' \cdot \int_0^\infty D''(s)\nabla[(1 - m \cdot \nabla)(w_t(s) - w)dsdX \right|$$
$$\leq \frac{\eta}{2} \int_\Omega |\nabla w'|^2 dX + \alpha_2 \frac{|D'(0)|}{2\eta} \int_0^\infty D''(s)a(w_t(s) - w)ds$$

where α_2 is the smallest constant such that

(3.59)
$$\|\nabla(1 - m \cdot \nabla)\varphi\| \leq \alpha_2 a(\varphi), \quad \forall \varphi \in H^2_{\Gamma_0}(\Omega).$$

Use of the estimates (3.57) and (3.58) in (3.56) yields the upper bound

(3.60)
$$\rho'(t) \leq -(1 - D_\infty - \frac{\eta}{2}) \int_\Omega (w'^2 + \gamma|\nabla w'|^2)dX - D_\infty^2 a(w)$$
$$+ \frac{D_\infty}{2} \int_0^\infty D'(s)a(w_t(s) - w)ds + C_{1,\eta} \int_0^\infty D''(s)a(w_t(s) - w)ds$$
$$- \frac{D_\infty}{2} \frac{d}{dt} \int_0^\infty [D(s) - D_\infty]a(w_t(s))ds + \frac{1}{2} \int_{\Gamma_1} m \cdot \nu(w'^2 + \gamma|\nabla w'|^2)d\Gamma$$
$$+ \frac{1}{2} \int_{\Gamma_0} m \cdot \nu(D_\infty\Delta w + \Delta w^*)^2 d\Gamma - \frac{1}{2}a_{\Gamma_1}(D_\infty w + w^*)$$
$$+ b(w^* - m \cdot \nabla(D_\infty w + w^*); w'),$$

where

(3.61)
$$C_{1,\eta} = \frac{|D'(0)|}{2\eta}(\alpha_1 + \gamma\alpha_2).$$

Let us estimate the last term on the right side of (3.60).

(3.62)
$$|b(w^* - m \cdot \nabla(D_\infty w + w^*); w')| \leq b^{\frac{1}{2}}(w')[b^{\frac{1}{2}}(w^*) + b^{\frac{1}{2}}(m \cdot (D_\infty w + w^*))]$$
$$\leq \frac{1}{\eta}b(w') + \frac{\eta}{2}b(w^*) + \frac{\eta}{2}b(m \cdot \nabla(D_\infty w + w^*)).$$

Recall that $b(\varphi)$ has the form

$$b(\varphi) = \int_{\Gamma_1} (m \cdot \nu)G\{\frac{\partial\varphi}{\partial x}, \frac{\partial\varphi}{\partial y}, \varphi\} \cdot \{\frac{\partial\varphi}{\partial x}, \frac{\partial\varphi}{\partial y}, \varphi\}d\Gamma$$

where G satisfies (3.2). Therefore, since $m \cdot \nu \geq 0$ on Γ_1

(3.63)
$$|b(m \cdot \nabla(D_\infty w + w^*))| \leq G_0 \int_{\Gamma_1} m \cdot \nu[|m \cdot \nabla(D_\infty w + w^*)|^2$$
$$+ |\nabla(m \cdot \nabla(D_\infty w + w^*))|^2]d\Gamma.$$

The terms in (3.63) containing derivatives of $D_\infty w + w^*$ up to order one can be bounded above by $(\text{const.}) a(D_\infty w + w^*)$, while the terms containing second order derivatives can be estimated by $(\text{const.}) a_{\Gamma_1}(D_\infty w + w^*)$. Therefore, we have an estimate of the form

$$
(3.64) \qquad
\begin{aligned}
|b(m \cdot \nabla(D_\infty w + w^*))| &\leq \frac{\beta_1}{2} a(D_\infty w + w^*) + \beta_2 a_{\Gamma_1}(D_\infty w + w^*) \\
&\leq \beta_1 D_\infty^2 a(w) + \beta_1 a(w^*) + \beta_2 a_{\Gamma_1}(D_\infty w + w^*)
\end{aligned}
$$

for suitable constants β_1 and β_2.

To estimate $a(w^*)$ in (3.64), let $\partial^n \varphi$, $n = (n_1, n_2)$ denote the derivative of φ of order n_1 in x and n_2 in y, and similarly for $\partial^m \varphi$. Assume $|n| = n_1 + n_2 \leq 2$ and $|m| \leq 2$. Then

$$
\begin{aligned}
|\partial^n w^*(t) \partial^m w^*(t)| &\leq \int_0^\infty |D'(s)||\partial^n(w_t(s) - w)|ds \int_0^\infty |D'(s)||\partial^m(w_t(s) - w)|ds \\
&\leq (1 - D_\infty)\Big[\int_0^\infty |D'(s)||\partial^n(w_t(s) - w)|^2 ds\Big]^{\frac{1}{2}} \\
&\quad \cdot \Big[\int_0^\infty |D'(s)||\partial^m(w_t(s) - w)|^2 ds\Big]^{\frac{1}{2}} \\
&\leq \frac{1}{2}(1 - D_\infty)\Big[\int_0^\infty |D'(s)||\partial^n(w_t(s) - w)|^2 ds \\
&\quad + \int_0^\infty |D'(s)||\partial^m(w_t(s) - w)|^2 ds\Big].
\end{aligned}
$$

Therefore

$$
\int_\Omega |\partial^n w^*(t) \partial^m w^*(t)| dX \leq (\text{const.}) \int_0^\infty |D'(s)| a(w_t(s) - w) ds,
$$

from which follows that

$$
(3.65) \qquad a(w^*) \leq \beta_3 \int_0^\infty |D'(s)| a(w_t(s) - w) ds
$$

for a suitable constant β_3.

Inserting (3.65) back into (3.64) gives

$$
(3.66) \qquad
\begin{aligned}
|b(m \cdot \nabla(D_\infty w + w^*))| &\leq \beta_1 D_\infty^2 a(w) \\
&+ \beta_1 \beta_3 \int_0^\infty |D'(s)| a(w_t(s) - w) ds + \beta_2 a_{\Gamma_1}(D_\infty w + w^*).
\end{aligned}
$$

We also have

$$
(3.67) \qquad b(w^*) \leq \beta_4 a(w^*) \leq \beta_3 \beta_4 \int_0^\infty |D'(s)| a(w_t(s) - w) ds.
$$

We may now use (3.66), (3.67) in (3.62) to obtain

$$|b(w * -m \cdot \nabla((D_\infty w + w^*); w')| \leq \frac{1}{\eta} b(w') + \frac{\eta}{2} \beta_1 D_\infty^2 a(w)$$

(3.68)

$$+ \frac{\eta}{2} \beta_3 (\beta_1 + \beta_4) \int_0^\infty |D'(s)| a(w_t(s) - w) ds + \frac{\eta}{2} \beta_2 a_{\Gamma_1}(D_\infty w + w^*).$$

Substituting (3.68) for the corresponding term in (3.60) yields

$$\rho'(t) \leq -(1 - D_\infty - \frac{\eta}{2}) \int_\Omega (w'^2 + \gamma |\nabla w'|^2) dX - (1 - \frac{\eta \beta_1}{2}) D_\infty^2 a(w)$$

(3.69)

$$- C_{2,\eta} \int_0^\infty |D'(s)| a(w_t(s) - w) ds + C_{1,\eta} \int_0^\infty D''(s) a(w_t(s) - w) ds$$

$$- \frac{D_\infty}{2} \frac{d}{dt} \int_0^\infty [D(s) - D_\infty] a(w_t(s)) ds + \frac{1}{2} \int_{\Gamma_1} m \cdot \nu (w'^2 + \gamma |\nabla w'|^2) d\Gamma$$

$$- \frac{1}{2}(1 - \eta \beta_2) a_{\Gamma_1}(D_\infty w + w^*) + \frac{1}{2} \int_{\Gamma_0} m \cdot \nu (D_\infty \Delta w + \Delta w^*)^2 d\Gamma,$$

where

$$C_{2,\eta} = \frac{1}{2}[D_\infty - \eta \beta_3 (\beta_1 + \beta_4)].$$

Choose $\eta > 0$ so small that

$$1 - D_\infty - \frac{\eta}{2} \geq \frac{1}{4}(1 - D_\infty), \quad 1 - \eta \beta_1 \geq \frac{1}{4}, \quad 1 - \eta \beta_2 \geq \frac{1}{2}, \quad C_{2,\eta} \geq \frac{1}{4} D_\infty.$$

We then obtain from (3.69) the estimate

$$\rho'(t) \leq -\frac{1}{2} \min(1 - D_\infty, D_\infty)\{E(t) + \frac{d}{dt} \int_0^\infty [D(s) - D_\infty] a(w_t(s)) ds\}$$

(3.70)

$$- \frac{1}{4} a_\Gamma(D_\infty w + w^*) + C_{1,\eta} \int_0^\infty D''(s) a(w_t(s) - w) ds$$

$$+ \frac{1}{2} \int_{\Gamma_1} m \cdot \nu (w'^2 + \gamma |\nabla w'|^2) d\Gamma,$$

where

$$a_\Gamma(D_\infty w + w^*) = a_{\Gamma_1}(D_\infty w + w^*) - \int_{\Gamma_0} m \cdot \nu (D_\infty \Delta w + \Delta w^*)^2 d\Gamma$$

$$= a_{\Gamma_1}(D_\infty w + w^*) - a_{\Gamma_0}(D_\infty w + w^*).$$

Therefore, from (3.55) and (3.70)

$$F'_\epsilon(t) \leq -\frac{\epsilon D_0}{2} \{E(t) + \frac{d}{dt} \int_0^\infty [D(s) - D_\infty] a(w_t(s)) ds\} - b(w') - \frac{\epsilon}{4} a_\Gamma(D_\infty w + w^*)$$

$$- (\frac{1}{2} - \epsilon C_{1,\eta}) \int_0^\infty D''(s) a(w_t(s) - w) ds + \frac{\epsilon}{2} \int_{\Gamma_1} m \cdot \nu (w'^2 + \gamma |\nabla w'|^2) d\Gamma$$

where $D_0 = \min(1 - D_\infty, D_\infty)$. From (3.1) it may be seen that

$$b(w') \geq g_0 \int_{\Gamma_1} m \cdot \nu(w'^2 + |\nabla w'|^2) d\Gamma \geq g_0 \int_{\Gamma_1} m \cdot \nu(w'^2 + \gamma |\nabla w'|^2) d\Gamma$$

provided $\gamma = h^2/12 \leq 1$, as we may assume. Consequently

$$F'_\epsilon(t) \leq -\frac{\epsilon D_0}{2} \{ E(t) + \frac{d}{dt} \int_0^\infty [D(s) - D_\infty] a(w_t(s)) ds \} - \frac{\epsilon}{4} a_\Gamma (D_\infty w + w^*)$$

$$- (\frac{1}{2} - \epsilon C_{1,\eta}) \int_0^\infty D''(s) a(w_t(s) - w) ds$$

$$- (g_0 - \frac{\epsilon}{2}) \int_{\Gamma_1} m \cdot \nu(w'^2 + \gamma |\nabla w'|^2) d\Gamma.$$

The proof of (3.40) is now completed by choosing $\epsilon > 0$ so small that

$$\frac{1}{2} - \epsilon C_{1,\eta} \geq 0, \quad g_0 - \frac{\epsilon}{2} \geq \frac{\epsilon}{4}.$$

REFERENCES

[1] Dafermos, C.M., *Contraction semigroups and trend to equilibrium in continuum mechanics*, in "Lecture Notes in Mathematics, Vol. 503," Springer-Verlag, Berlin, 1976, pp. 295–306.

[2] Datko, R., *Extending a theorem of Liapunov to Hilbert spaces*, J. Math. Anal. Appl. **32** (1970), 610–616.

[3] Desch, W., and R.K. Miller, *Exponential stabilization of Volterra integrodifferential equations in Hilbert space*, J. Differential Eqs. **70** (1987), 366–389.

[4] Duvaut, G., and. J.L. Lions, "Inequalities in Mechanics and Physics," Springer-Verlag, Berlin, 1976.

[5] Lagnese, J.E., and J.L. Lions, *Modelling, Analysis and Control of Thin Plates*, in "Collection Recherches en Mathématiques Appliquées, Vol. 6," Masson, Paris, 1988.

[6] Leugering, G., *On boundary feedback stabilization of a viscoelastic membrane*, Proc. Royal Soc. Edinburgh (to appear).

[7] Prüss, J., *Positivity and regularity of hyperbolic Volterra equations in Banach spaces*, Math. Ann. **279** (1987), 317–344.

John E. Lagnese
Department of Mathematics
Georgetown University
Washington, DC 20057
USA

International Series of
Numerical Mathematics, Vol. 91
© 1989 Birkhäuser Verlag Basel

Controllability of a viscoelastic Kirchhoff plate

I. Lasiecka

Applied Mathematics Department
University of Virginia

Abstract. Exact controllability with boundary controls for a Kirchhoff plate and viscoelastic
Kirchhoff plate is established. The results are formulated in "optimal" spaces, i.e.; the space
of exact controllability coincides with the space of regularity.

Keywords. Exact controllability, Kirchhoff plate, semigroups.

1980 *Mathematics subject classifications*: 35L20

1. Introduction

In this paper we shall study by using a "direct method" a question of exact controllability
of a Kirchhoff plate with memory. What we call here a "direct method" is a natural
generalization to infinite dimensional spaces of a classical approach to controllability used
in the finite dimensional theory and revolving around the construction of the "Gramian"
or "controllability matrix".

This approach has been used a number of times in the past see e.g. [D-R], [S.2], [S-W]
and more recently in the case of wave or plate problems see [K-L-S], [L.1], [L-T.1], [L-
T.2], [L-T.3], [T.1], [T.2]. This method is conceptually different from the HUM-Method
(or RHUM) recently introduced by J.L. Lions (also called (F, F', Λ)-method), see [L.3],
which seeks to provide at the start an explicit procedure to construct a control driving the
initial data to the rest, by performing certain integrations by parts on the original p.d.e.
model and on its adjoint and then by using certain estimates obtained via multiplier's
method as to impose the structure of the control law.

Instead the "direct method" simply rephrases in its first step the notion of exact controllability (say, for a time reversible problem) by stating that this notion means that the
solution operator (starting from zero initial data of the time reversed problem) is surjective
or onto, or equivalently that its adjoint has a continuous inverse and hence has a bound
from below. The crux of the controllability issue becomes precisely this: to establish that
the appropriate dynamics satisfies this characterizing bound from below. Once this technical point is solved, exact controllability is established, albeit in a non constructive manner
at least at this stage. However, this deficiency is readily remedied: once exact controllability is known, a straightforward minimization argument – which can be performed for an
'abstract' model and then specialized to the particular mixed problem, see Appendices in
[L-T.1] – [L-T.3], [T.1] – provides say the desired minimal norm steering control. In many
situations (wave or plate equations with $L_2(\Sigma)$ controls) both procedures are equivalent.
In some cases, it appears to us that the "direct method" offers a more natural and simpler

Part of this research was carried out while the author was a visiting scientist at the Interdisciplinary Center
for Applied Mathematics, Virginia Polytechnic Institute and State University, which is partially funded
under DARPA Contract No. F49620-87-0116.

solution to the problem. This is the case when the class of controls is irregular in time, for example in $(H^1(0,T))'$ (see [L-T.1], [L-T.2], [L-T.3]), or when a 'small' perturbation (say, damping) is added to the dynamics [T.2], or when the dynamics involved is rather complicated, like for example the viscoelastic model considered in this paper. It is technically easier to esablish the property of the "ontoness" or surjectivity of the solution map, rather than constructing directly from the p.d.e. model the control driving the system to the rest: (this can always be done later, via operator techniques, once "ontoness" is established as mentioned above).

Consider the following model of a viscoelastic Kirchhoff plate.

$$\text{(i)}\qquad y_{tt} - \gamma\Delta y_{tt} + \Delta^2 y + \beta\Delta^2 Q(t)y = 0 \text{ in } Q \equiv (\Omega \times 0T),$$

(1.1) $$\text{(ii)}\qquad y|_\Gamma = 0; \ \frac{\partial u}{\partial \eta}|_\Gamma = u \in L_2(\Sigma) \text{ in } \Sigma \equiv (\Gamma \times 0T);$$

$$\text{(iii)}\quad y(t=0) = y_0 \in H_0^1(\Omega), \ y_t(t=0) = y_1 \in \tilde{L}_2(\Omega).$$

Here $\Omega \subset \mathbf{R}^n$ is an open bounded set with a boundary Γ, γ and β are real numbers, $Q \in \mathcal{L}(L_2(Q))$ is given by $Q(t)y \equiv \int_0^t \hat{Q}(t-s)y(s)ds$, where $\hat{Q} = \hat{Q}(t,x) \in C^1(\bar{Q})$.

$\tilde{L}_2(\Omega)$ is the dual to $H_0^2(\Omega)$ with respect to $H_0^1(\Omega)$ inner product[1]. Since $L_2(\Omega)$ is isomorphic to the dual of $H_0^1(\Omega) \cap H^2(\Omega)$ with respect to $H_0^1(\Omega)$ inner product and $H_0^2(\Omega) \subset H_0^1(\Omega) \cap H^2(\Omega)$, $L_2(\Omega) \subset \tilde{L}_2(\Omega)$.

We shall prove that the system (1.1) with β "small" is exactly null controllable for some $T > 0$, i.e.: there exists $T > 0$ and $u \in L_2(\Sigma)$ such that for all initial data

(1.2) $$y_0 \in H_0^1(\Omega); \ y_1 \in \tilde{L}_2(\Omega)$$

the solution of (y, y_t) of (1.1) satisfies $y(T) = y_t(T) = 0$.

The problem of the exact controllability of (1.1) has been considered in [L-L]. In [L-L], the authors, using HUM method have established : (i) exact controllability on the space $H_0^1(\Omega) \times L_2(\Omega)$ for the model (1.1) with $\beta = 0$ (without the memory term) and (ii) exact controllability for the model (1). (i), (iii) with β "small" and with the boundary conditions:

(1.3) $$\frac{\partial}{\partial \eta}(y + BQy) = u$$

instead of (1.1ii).

Our exact controllability results with $\beta = 0$, obtained by "direct method" hold on the space $H_0^1 \times \tilde{L}_2(\Omega)$ which is larger than $H_0^1(\Omega) \times L_2(\Omega)$. This plays a crucial role in studying the case with the memory, (i.e. when $\beta \neq 0$) and it allows us to treat the natural boundary conditions (1.1ii) instead of (1.3). In fact, since the case $\beta \neq 0$ is treated via perturbation technique from the case $\beta = 0$, it is critically important that the unperturbed result hold on "the right space" i.e. on $\tilde{L}_2(\Omega)$ not on $L_2(\Omega)$ (otherwise perturbation becomes "unbounded"). As the result, exact controllability results for $\beta \neq 0$ also hold on the larger

---

[1]If $V \subset H$ with a dense injection, the dual V' to V with respect to the inner product of H is defined as following: $y \in V'$ iff $(y,v)_H < \infty$ for all $x \in V$ and $|y|_{V'} \equiv \sup_{x \in V} \frac{(y,x)_H}{|x|_V}$.

space $H_0^1(\Omega) \times \tilde{L}_2(\Omega)$. The problem of exact null controllability result, but with $\gamma = 0$ – has been also considered in [L.1], [L.2]. In [L.2], [L.3] however, the effect of the memory acts on the system as a compact perturbation, (thus smallness of β is not needed). In our case, instead, due to the presence of $\Delta^2 Qy$, the memory term is not a compact perturbation and the main difficulty consists precisely of showing that this term can be appropriately absorbed for the perturbations techniques to apply. In fact, if one would take $H_0^1(\Omega) \times L_2(\Omega)$ as the state space, this absorption would not be possible. The key element of our approach is to provide a correct semigroup model for the dynamics (1.1) in the correct functional spaces.

The paper is organized as follows: In section 2, for the reader's convenience, we shall provide a brief outline of the "direct method". In section 3, by using the "direct method" we establish the exact null controllability of the plate without memory, i.e., $\beta = 0$, and in section 4 we shall apply perturbation techniques to treat the case $\beta \neq 0$.

2. Exact controllability by "direct method"

Consider the following "abstract" control system:

(2.1)
$$y_t = Ay + Bu;$$
$$y(0) = y_0 \in H.$$

Here A is a generator of C_0-semigroup e^{At} on a Hilbert space H with the domain denoted by $D(A)$ and we assume that $R(e^{At})$ is dense. Let U be another Hilbert space. We assume that $B \in \mathcal{L}(U; C[0T; D(A)'])$ where $(D(A))'$ denotes the dual to $D(A)$ (equipped with a natural graph norm) with respect to the inner product in H.

We say that the system (1.1) is *exactly null-controllable* $(T; U; H)$ iff for any $y_0 \in H$ there exists $u \in U$ such that

(2.2)
$$y(T) = 0.$$

In this section we shall collect some results on exact controllability. To study the exact controllability, it is natural to introduce the solution map $L_T : U \to H$ given by

(2.3)
$$(L_T u)(t) \equiv A \int_0^T e^{A(t-\tau)} A^{-1} Bu(\tau) d\tau.$$

By the virtue of the assumption on B,

(2.4)
$$L_T \in \mathcal{L}(U; (D(A)')).$$

Let L_T^* denote the adjoint of L_T i.e., $(L_T u, v)_{H_1} = \langle u, L_T^* v \rangle_U$; $u \in U$; $v \in D(A^*)$, where H_1 is another Hilbert space such that $L_T : U \to H_1$ is closed and dense. We shall assume that the system (2.1) is approximately controllable i.e.:

(2.5)
$$L_T^* v = 0 \Longrightarrow v = 0.$$

As it is well-known, the approximate controllability implies exact controllability in "some space". More precisely, by (2.5), $|L_T^* v|_U$ defines a norm on $D(A^*)$. Let $D(L_T^*)$ be a completion of functions $v \in D(A^*)$ under the norm $|L_T^* v|_U$. Hence, $D(L_T^*) \equiv \{v; |L_T^* v|_U < \infty\}$ and $D(L_T^*)$ equipped with a norm $|L_T^* v|_U$ is a Hilbert space. Let $(D(L_T^*))'$ denotes dual to $D(L_T^*)$ with respect to H_1-inner product.

Lemma 2.1. *Assume (2.5). Then for any initial data y_0 such that*

$$(2.6) \qquad\qquad y_0 \in D(\tilde{L}_T^*)'; \quad \text{where} \quad \tilde{L}_T \equiv e^{-AT} L_T$$

there exists $u \in U$ such that $y(T) = 0$. Moreover, there exists a minimal norm control $u^0 \in U$ which is given by

$$(2.7) \qquad\qquad u^0 = -L_T^*(L_T L_T^*)^{-1} e^{AT} y_0 = \tilde{L}_T^*(\tilde{L}_T \tilde{L}_T^*)^{-1} y_0.$$

Proof: Since, $|L_T^* v|_U$ is a norm on $D(L_T^*)$, by the Lax-Millgram Theorem

$$(2.8) \qquad\qquad L_T L_T^* \quad \text{is an isomorphism:} \quad D(L_T^*) \to D(L_T^*)'.$$

Therefore

$$(L_T L_T^*)^{-1} \in \mathcal{L}(D(L_T^*)') \to D(L_T^*))$$

and the expression on the RHS of (2.7) defines an element $u \in U$. The representation (2.7) follows now from the standard Lagrange multipliers technique.

Remark 2.1. It is straightforward to show that $D(\tilde{L}_T^*)'$ is independent on the choice of H_1, (it depends only on the choice of U and of the structure of the operators L_T).

Remark 2.2. Notice that by (2.8) $D(\tilde{L}_T^*)'$ is the optimal space of null controllability for a given control space U.

The main issue in the exact controllability of infinite dimensional systems is the characterization of $D(\tilde{L}_T^*)$. Thus we are looking for the estimate of the form

$$(2.9) \qquad\qquad |\tilde{L}_T^* v|_U \geq C|v|_V$$

where V is a "known" space.

In fact, once we have (2.9), then

$$D(\tilde{L}_T^*) \subset V \quad \text{and} \quad V' \subset (D\tilde{L}_T^*)'$$

where V' is the dual to V with respect to H_1 inner product. If moreover we have

$$(2.10) \qquad\qquad |\tilde{L}_T^* v|_U \leq C|v|_V$$

then $V' = (D(\tilde{L}_T^*))'$ and V' is the optimal space of exact controllability. Thus from Lemma 2.1 we deduce

Lemma 2.2. *Assume (2.9). Then for any $y_0 \in V'$ there exists $u \in U$ such that $y(T) = 0$. The minimal norm control u^0 is given by the formula (2.7).*

Remark 2.3. The reader familiar with HUM method would recognize that in some canonical situations (Petrowsky type of systems with $L_2(\Sigma)$ controls), the operator $\tilde{L}_T \tilde{L}_T^*$ would coincide with Λ operator of HUM method, and the approximate controllability condition (2.5) is nothing else but the Hilbert uniqueness requirement of HUM approach.

To conclude this section, we notice that in concrete applications, the main technical issue is to establish the inequality (2.9). Once this is done the construction of the minimal norm control is straightforward and it is given by the formula (2.7).

3. Exact controllability with $\beta = 0$

In this section we shall apply the "direct method" to prove the exact controllability of (1.1) with $\beta = 0$.

Lemma 3.1. *Assume $\beta = 0$ in (1.1). Then there exists $T > 0$ such that for any $(y_0, y_1) \in H_0^1(\Omega) \times \tilde{L}_2(\Omega)$, there exists $u \in L_2(\Sigma)$ such that $y(T) = y_t(T) = 0$. The minimal norm control u^0 is given by the formula (3.16) below and the space $H_0^1(\Omega) \times \tilde{L}_2(\Omega)$ is the optimal space of exact null-controllability.*

Proof: We begin by constructing the operator L_T. To this end we need to represent the solution of (1.1) in the semigroup form. We introduce the following operators:

$$(3.1) \qquad A_1 y \equiv \Delta^2 y; \ y \in D(A_1) = \{y \in H^{-1}(\Omega); \Delta^2 y \in H^{-1}(\Omega); y|\Gamma = \frac{\partial y}{\partial \eta}|_\Gamma = 0\}.$$

It is well known [G-I] that $D(A_1)$ is isomorphic with $H^3(\Omega) \cap H_0^2(\Omega)$

$$(3.2) \qquad A_2 y \equiv -\Delta y; \ y \in D(A_2) = H_0^1(\Omega) \cap H^2(\Omega).$$

Since the operator A_2 has its representation also on $H_0^1(\Omega)$ (denoted by the same symbol), A_2 is an isomorphism between $H_0^1(\Omega)$ and $H^{-1}(\Omega)$. Thus $(1 + \gamma A_2)^{-1} \in \mathcal{L}(H^{-1}(\Omega) \to H_0^1(\Omega))$, and we can define the operator $A : H_0^1(\Omega) \to H_0^1(\Omega)$ by the formula:

$$(3.3) \qquad A \equiv (1 + \gamma A_2)^{-1} A_1.$$

Let $G \in \mathcal{L}(L_2(\Gamma) \to H_0^{3/2}(\Omega))$ be defined as

$$(3.4) \qquad Gu \equiv v \text{ iff } \Delta^2 v = 0 \text{ in } \Omega; \ v|_\Gamma = 0, \ \frac{\partial v}{\partial \eta}|_\Gamma = g.$$

By the energy methods and direct calculations one can show that

(i) A is positive and selfadjoint on $H_{0,\gamma}^1(\Omega)$ when $|y|_{H_{0,\gamma}^1} \equiv |(1 + \gamma A_2)^{1/2} y|_{L_2(\Omega)}$

(ii) $D(A^{1/2}) = H_0^2(\Omega)$ where $A : H_{0,\gamma}^1(\Omega) \to H_{0,\gamma}^1(\Omega)$

$(3.5) \qquad$ and $|A^{1/2} y|_{H_{0,\gamma}^1(\Omega)} = |\Delta y|_{L_2(\Omega)}$

(iii) $(D(A^{1/2}))' = \tilde{L}_2(\Omega)$ (duality with respect to $H_{0,\gamma}^1(\Omega)$ – inner product) and $|y|_{\tilde{L}_2(\Omega)} = |A^{-1/2} y|_{H_{0,\gamma}^1(\Omega)}$.

The operator A generates cosine $C(t)$ and sine $S(t)$ operators with the following properties:

$$(3.6) \qquad \begin{array}{lll} C(t) \in \mathcal{L}(H_0^\alpha(\Omega) \to H_0^\alpha(\Omega)) & \alpha = 0, 1; 2 & \text{uniformly in } t \in [0, T] \\ S(t) \in \mathcal{L}(H_0^\alpha(\Omega) \to H_0^{\alpha+1}(\Omega)) & \alpha = 0, 1 & \text{uniformly in } t \in [0, T] \\ AS(t) \in \mathcal{L}(H_0^{\alpha+1}(\Omega) \to H_0^\alpha(\Omega)) & \alpha = 0, 1 & \text{uniformly in } t \in [0, T] \end{array}$$

where we denote $H_0^0(\Omega) \equiv \tilde{L}_2(\Omega)$. With the above notation our solution operator L_T : $L_2(\Sigma) \to H_{0,\gamma}^1(\Omega) \times \tilde{L}_2(\Omega)$ is given by:

$$(3.7) \qquad L_T u \equiv \begin{pmatrix} \int_0^T AS(T-\tau)Gu(\tau)d\tau \\ \int_0^T AC(T-\tau)Gu(\tau)d\tau \end{pmatrix} = \int_0^T e^{\mathcal{A}(T-\tau)} \begin{pmatrix} 0 \\ A\,Gu(\tau) \end{pmatrix} d\tau$$

where $\mathcal{A} = \begin{pmatrix} y_1 \\ y_2 \end{pmatrix} = \begin{pmatrix} y_2 \\ Ay_1 \end{pmatrix}$.

We shall prove that (2.9) holds with

$$(3.8) \qquad U \equiv L_2(\Sigma); \; V = H = H_1 = H_{0,\gamma}^1(\Omega) \times \tilde{L}_2(\Omega).$$

Notice first that since $e^{\mathcal{A}t}$ is a group on $H_{0,\gamma}^1(\Omega) \times L_2(\Omega)$, it is enough to prove (2.9) with $\tilde{L}_T$ replaced by L_T. To this end we compute L_T^* i.e.

$$(L_T u, v)_{H_{0,\gamma}^1 \times \tilde{L}_2(\Omega)} = \langle u, L_T^* v \rangle_{L_2(\Sigma)}.$$

Rather straightforward computations give

$$(3.9) \qquad L_T^* \begin{pmatrix} v_0 \\ v_1 \end{pmatrix} (\tau) = G^\sharp A_1 [S(T-\tau)v_0 + G(T-\tau)A^{-1}v_1]$$

where $G^\sharp$ is the L_2 adjoint of the Green may G. Since $G^\sharp \in \mathcal{L}(H^{-1}(\Omega) \to L_2(\Gamma))$, on application of Green's Theorem yields

$$(3.10) \qquad G^\sharp A_1 \psi = \Delta\psi|_\Gamma; \; \psi \in D(A_1).$$

Thus, (2.9) in our notation becomes

$$(3.11) \qquad |\Delta\psi|_{L_2(\Sigma)} \geq C[|\psi_1|_{H_{0,\gamma}^1(\Omega)} + |A\psi_0|_{\tilde{L}_2(\Omega)}]$$

where

$$(3.12) \; \psi(t) \equiv C(T-t)A^{-1}v_1 + S(T-t)v_0 \quad \text{with} \quad \psi_0 = \psi(T) = A^{-1}v_1; \; \psi_1 = \psi_t(T) = v_0.$$

Since the norms $|A\psi_0|_{\tilde{L}_2(\Omega)}$ and $|\psi_0|_{H_0^2(\Omega)}$ are equivalent, (3.11) amounts to showing that

$$(3.13) \qquad |\Delta\psi|_{L_2(\Sigma)} \geq C[|\psi_1|_{H_0^1(\Omega)} + |\psi_0|_{H_0^2(\Omega)}]$$

for ψ solution of

$$(3.14) \qquad \begin{aligned} &\psi_{tt} - \gamma\Delta\psi_{tt} + \Delta^2\psi = 0; \; \psi(T) = \psi_0; \; \psi_t(T) = \psi_1 \\ &\psi|_\gamma = \frac{\partial\psi}{\partial\eta}|_\Gamma = 0. \end{aligned}$$

On the other hand the validity of (3.13) was established, by multiplier's method, in [L-L]. Since in (3.13) the inequality in opposite direction also holds (see [L-L], one has

(3.15)
$$D(L_T^*) = H_0^1(\Omega) \times \tilde{L}_2(\Omega), \quad \text{hence}$$
$$D(L_T^*)' = H_0^1(\Omega) \times \tilde{L}_2(\Omega).$$

Thus the result of Lemma 3.1 follows now from Lemma 2.1. To determine the minimal norm control we follow the procedure in (2.7). Let

$$(\psi_0, \psi_1) \equiv (\tilde{L}_T \tilde{L}_T^*)^{-1}(y_0, y_1) \in H_0^1(\Omega) \times \tilde{L}_2(\Omega)$$

with L_T, L_T^* given by (3.7), (3.9). Then from (2.7) and (3.10)

(3.16)
$$u^0 = -\tilde{L}_T^*(\psi_0, \psi_1) = \Delta\phi(t)|_\Gamma$$

where

$$\phi_{tt} - \gamma\Delta\phi_{tt} + \Delta^2\phi(t)|_\Gamma = 0$$
$$\phi(0) = A^{-1}\psi_1 \subset H_0^2(\Omega)$$
$$\phi_t(0) = \psi_0 \in H_0^1(\Omega)$$
$$\phi|_\Gamma = \frac{\partial\phi}{\partial\eta}|_\Gamma = 0. \quad \blacksquare$$

4. Exact controllability of the plate with a memory–case $\beta \neq 0$

In this section we consider the case when $\beta \neq 0$; but β is 'small'. We shall show that the problem with $\beta \neq 0$ can be reduced via perturbation argument to the problem when $\beta = 0$. Our main result is:

Theorem 4.1. There exists a constant c_0 such that if $|\beta| < c_0$ then the exact null controllability $(T, L_2(\Sigma); H_0^1(\Omega) \times \tilde{L}_2(\Omega))$ of (1.1) is equivalent to the exact null controllability $(T, L_2(\Sigma); H_0^1(\Omega) \times \tilde{L}_2(\Omega))$ of (1.1) with $\beta = 0$.

Corollary 4.1. There exists $T > 0$ and $u \in L_2(\Sigma)$ such that for β sufficiently small and for all initial data $(y_0, y_1) \in H_0^1(\Omega) \times \tilde{L}_2(\Omega)$ the solution (y, y_t) of (1.1) satisfies $y(T) = y_t(T) = 0$.

Proof of Theorem 4.1: Using the notation of section 3, we rewrite (1.1) as:

(4.1)
$$y_{tt} - (1 + \gamma A_2)^{-1}\Delta^2 y + \beta(1 + \gamma A_2)^{-1}\Delta^2 Qy = 0$$
$$\frac{\partial y}{\partial \eta} = u \quad \text{on} \quad \Gamma.$$

Since $(1 + \gamma A_2)^{-1} \Delta^2 Q y = A[Q y - G(Q|_\Gamma u)]$, the solution (y, y_t) of (4.1) can be written in the semigroup form as

$$y(t) = C(t)y_0 + S(t)y_1 - A \int_0^t S(t - \tau)G[1 - \beta Q|_\Gamma] \cdot u(\tau)d\tau$$

(4.2a)
$$+ \beta A \int_0^t S(t - \tau)Q y(\tau)d\tau;$$

$$y_t(t) = AS(t)y_0 + C(t)y_1 + A \int_0^t C(t - \tau)G[1 - \beta Q|_\Gamma] \cdot u(\tau)d\tau$$

(4.2b)
$$+ \beta A \int_0^t C(t - \tau)Q y(\tau)d\tau.$$

Integrating by parts the last term on the RHS of (4.2b) gives

(4.3) $$A \int_0^t C(t - \tau)Q y(\tau)d\tau = A \int_0^t S(t - \tau)Q y_t(\tau)d\tau + A \int_0^t S(t - \tau)\hat{Q}(\tau)y_0 d\tau.$$

Denoting

$$(Py)(t) \equiv A \int_0^t S(t - \tau)Q y(\tau)d\tau,$$

$$\mathcal{L}_t u \equiv \int_0^t AS(t - \tau)G u(\tau)d\tau,$$

$$\mathcal{L}_{t,\beta} u \equiv \mathcal{L}_t(1 - \beta Q|_\Gamma)u,$$

we rewrite (4.2) in the operator form as

(4.4)
$$(I - \beta P)y(t) = C(t)y_0 + S(t)y_1 + \mathcal{L}_{t,\beta} u$$
$$(I - \beta P)y_t(t) = AS(t)y_0 + C(t)y_1 + \frac{d}{dt}\mathcal{L}_{t,\beta} u + \beta A \int_0^t S(t - \tau)\hat{Q}(\tau)y_0 d\tau.$$

If we set $\gamma_T y \equiv y(T)$, (4.4) becomes

(4.5)
$$y(T) = \gamma_T(I - \beta P)^{-1}\mathcal{L}_{t,\beta} u + \gamma_T(I - \beta P)^{-1}[C(\cdot)y_0 + S()y_1]$$
$$y_t(T) = \gamma_T(I - \beta P)^{-1}\frac{d}{dt}\mathcal{L}_{t,\beta} u + \gamma_T(I - \beta P)^{-1}[AS()y_0 + C()y_1$$
$$+ \beta A \int_0^{\cdot} S(\cdot - \tau)\hat{Q}(\tau)y_0 d\tau].$$

Using the notation

$$\hat{L}_T \equiv \begin{pmatrix} \gamma_T(I - \beta P)^{-1}\mathcal{L}_{t,\beta} \\ \gamma_T(I - \beta P)^{-1}\frac{d}{dt}\mathcal{L}_{t,\beta} \end{pmatrix}$$

$$C_1 \equiv \gamma_T(I - \beta P)^{-1}[C(\cdot)y_0 + S()y_1]$$

$$C_2 \equiv \gamma_T(I - \beta P)^{-1}[AS()y_0 + C()y_1 + \beta A \int_0^t S(t - \tau)\hat{Q}(\tau)y_0 d\tau]$$

we rewrite (4.5) as

$$(4.6) \qquad \begin{pmatrix} y(T) \\ y_t(T) \end{pmatrix} = \hat{L}_T u + \begin{pmatrix} C_1 \\ C_2 \end{pmatrix}.$$

Thus our null controllability problem amounts to showing that

$(A-1)$ $\qquad\qquad C_1 \in H_0^1(\Omega);\ C_2 \in \tilde{L}_2(\Omega),$

$(A-2)$ $\qquad\qquad \hat{L}_T$ is from $L_2(\Sigma)$ onto $H_0^1(\Omega) \times \tilde{L}_2(\Omega).$

To assert $(A-2)$ it is enough to establish (2.9) with $U \equiv L_2(\Sigma);\ H_1 = H \equiv H_0^1(\Omega) \times \tilde{L}_2(\Omega).$

$(A-2^1)$ $\qquad\qquad |\hat{L}_T^* v|_{L_2(\Sigma)} \geq C[|v_0|_{H_0^1(\Omega)} + |v_1|_{\tilde{L}_2(\Omega)}]$

where the adjoint $\hat{L}_T^*$ is taken with respect to $L_2(\Sigma)$ and $H_0^1(\Omega) \times \tilde{L}_2(\Omega)$ topologies.
On the other hand, writing

$$(I - \beta P)^{-1} = I - \beta P(I - \beta P)^{-1},$$

using the definition of $\mathcal{L}_t^\beta$ and noting that

$$L_T u \equiv \gamma_T \begin{pmatrix} \mathcal{L}_t u \\ \frac{d}{dt}\mathcal{L}_t u \end{pmatrix}$$

where L_T is given by (3.7), we obtain the following perturbation formula for $\hat{L}_T$.

$$(4.7) \qquad \hat{L}_T u = L_T u - \beta L_T(Q|_\Gamma u) - \beta \gamma_T P(I - \beta P)^{-1} \begin{pmatrix} \mathcal{L}_{t,\beta} u \\ \frac{d}{dt}\mathcal{L}_{t,\beta} u \end{pmatrix}.$$

Lemma 4.1.

$$(4.8) \qquad L_T \in \mathcal{L}(L_2(\Sigma) \rightarrow H_0^1(\Omega) \times \tilde{L}_2(\Omega)).$$

For β sufficiently small we have

$$(4.9) \qquad \gamma_T P(I - \beta P)^{-1} \begin{pmatrix} \mathcal{L}_{t,\beta} \\ \frac{d}{dt}\mathcal{L}_{t,\beta} \end{pmatrix} \in \mathcal{L}(L_2(\Sigma) \rightarrow H_0^1(\Omega) \times \tilde{L}_2(\Omega)).$$

Assuming for a moment validity of Lemma 4.1 we see the $(A-2^1)$ with β sufficiently small is equivalent to the same estimate with $\hat{L}_T$ replaced by L_T. On the other hand, this is equivalent to the null controllability of the plate equation with $\beta = 0$. Thus, to complete the proof of the Theorem 4.1, it remains to prove $(A-1)$ and Lemma 4.1. To accomplish this, we shall first formulate and prove the following propositions:

Proposition 4.1.

$$\mathcal{L}_t \in \mathcal{L}(L_2(\Sigma) \to H_0^1(\Omega))$$

$$\frac{d}{dt}\mathcal{L}_t \in \mathcal{L}(L_2(\Sigma) \to \tilde{L}_2(\Omega))$$

uniformly in $t \in [0, T]$.

The statement of Proposition 4.1 follows, via duality, from the opposite to the inequality (2.23) on p. 123 of [L-L] combined with the techniques of [L-L-T].

Proposition 4.2.

(4.10) $\qquad\qquad P \in \mathcal{L}(L_2(0, T; H_0^s(\Omega)) \to C(0, T; H_0^s(\Omega)) \quad$ for $s = 0, 1$.

Proof of Proposition 4.2: We write

$$Py(t) = A \int_0^t \frac{d}{dt} C(t - \tau) A^{-1} \int_0^t \hat{Q}(\tau - s)y(s)ds d\tau$$

$$= \int_0^t \hat{Q}(t - s)y(s)ds + \int_0^t C(t - \tau)[\hat{Q}(0)y(\tau) + \int_0^\tau \hat{Q}_\tau(\tau - s)y(s)ds]d\tau.$$

Now the conclusion follows from the regularity of the kernel $\hat{Q}(t)$ and from the regularity properties of sine and cosine operators listed in (3.6).

Proof of Lemma 4.1: (4.8) follows directly from Proposition 4.1. As for (4.9), we notice first that for β small enough, by (4.10) we have

(4.11) $\qquad\qquad (I - \beta P)^{-1} \in \mathcal{L}(C(0T; H_0^s(\Omega)); \ s = 0, 1.$

Thus (4.9) follows now from the results of Propositions 4.1, 4.2 combined with (4.11).

Proof of $(A - 1)$: With $y_0 \in H_0^1(\Omega); y_1 \in \tilde{L}_2(\Omega)$, (3.6) gives

$$C(t)y_0 + S(t)y_1 \in C[0T; H_0^1(\Omega)]$$

$$AS(t)y_0 + C(t)y_1 \in C[0T; \tilde{L}_2(\Omega)].$$

Thus by (4.11), $C_1 \in H_0^1(\Omega)$. Since by the regularity of $\hat{Q}; A \int_0^{\cdot} S(\cdot - \tau)\hat{Q}(\tau)y_0 d\tau \in \tilde{L}_2(\Omega)$, (4.11) also implies that $C_2 \in \tilde{L}_2(\Omega)$, as desired.

To complete the proof of Theorem 4.1 it is enough to establish that $(A - 2')$ holds for β small iff (2.9) is satisfied with L_T given by (3.7).

The above assessment follows from (4.7) and the result of Lemma 4.1. The proof of Theorem 4.1 is thus completed. Corollary 4.1 is a direct consequence of Theorem 4.1 and Lemma 3.1.

Remark 4.1. The fact that the exact null controllability in the case of $\beta = 0$ holds on $H_0^1(\Omega) \times \tilde{L}_2(\Omega)$ instead of $H_0^1(\Omega) \times L_2(\Omega)$ is of crucial importance. Indeed, otherwise the perturbation argument used to prove $(A - 2')$ will not be valid as the operators P is not bounded from $C(0T; L_2(\Omega))$ into itself (compare with the result of Proposition 4.2).

Remark 4.2. Having established the exact null controllability for the system (1.1), one may construct minimal norm control by applying formula (2.7) with L_T replaced by $\hat{L}_T$ and $e^{AT}y_0$ by C_1, C_2.

Acknowledgment. Personal thanks go to Profs. J. Burns, M. Gunzburger, K. Hannsgen, T. Swobodny and R. Wheeler for very helpful discussions during my visit at ICAM.

REFERENCES

[D-R] D. Dolecki and L. Russell, *A general theory of observation and control*, SIAM J. Control & Opt. **15** (1977), 185–220.

[F.1] H.O. Fattorini, *Reachable states in boundary control of the head equation are independent on time*, Proc. Royal Soc. **81** (1976), 71–77.

[G-R] K.D. Graham and D.L. Russell, *Boundary value control of the wave equation in a spherical region*, SIAM J. Control & Opt. **13** (1975), 174–196.

[G.1] P. Grisvard, *Caracterization de quelques espaces d'interpolation*, Arch. Rat. Mech. Anal. **25** (1967), 40–63.

[K-L-S] W. Krabs, G. Leugering and T. Seidman, *On boundary controllability of a vibrating plate*, A.M.O. **13** (1985), 205-229.

[L-L-T] I. Lasiecka, J. Lions and R. Triggiani, *Non homogeneous boundary value problems for second order hyperbolic operators*, J.M.P.A. **65** (1986), 149–192.

[L-T.1] I. Lasiecka and R. Triggiani, *Exact boundary controllability for the wave equation with Neumann Boundary control*, Applied Math. Optimiz. **19** (1989), 243–290.

[L-T.2] I. Lasiecka and R. Triggiani, *Exact Controllability of the Euler-Bernoulli Equation with boundary controls for displacement and moment*, J. of Mathematical Anal. and Applications (to appear).

[L-T.3] I. Lasiecka and R. Triggiani, *Exact controllability of the Euler-Bernoulli Equations with control in the Dirichlet and Neumann Boundary Conditions – A nonconservative case*, SIAM J. Control & Opt. **27** (1989), 330–373.

[L.1] G.Leuering, *Exact boundary controllability of an integro-differential equation*, Applied Math. and Opt. **15** (1987), 223–250.

[L.2] G. Leugering, *On the reachability problem of a viscoelastic beam during a slewing maneuver.* in this proceedings

[L.3] J.L. Lions, *Exact Controllability, stabilization and perturbations for distributed systems*, in "John Von Neumann Lectures," Boston, July 1986.

[L.4] W. Littman, *Near Optimal time boundary controllability for a class of hyperbolic equations*, in "LNCIS Vol. 97," Springer-Verlag, 1987, pp. 307–312.

[L-L] J. Lagnese and J.L. Lions, "Modelling, Analysis and Control of Thin Plates," Masson, 1988.

[R.1] D.L. Russell, *Controllability and stabilizability theory for linear partial differential equations: recent progress and open questions*, SIAM Rev. **20** (1978), 639–739.

[S-W] G. Schmidt and N. Weck, *On the boundary behavior of solutions to elliptic and parabolic equations – with applications to boundary control for parabolic equations*, SIAM J. Control & Opt. (1978), 593–598.

[S.2] T.I. Seidman, *Observation and prediction for the heat equation II*, J.M.A.A. **38** (1972), 149–166.

[T.1] R. Triggiani, *Exact boundary controllability on $L_2(\Omega) \times H^{-1}(\Omega)$ of the wave equation with Dirichlet boundary controls acting on a portion of the boundary control*, Appl. Math. & Optim. **18** (1988), 241–277.

[T.2] R. Triggiani, *Exact controllability in the presence of damping*, in "Proceedings of International Conference in Differential Equations," Columbus, Ohio, March 1988; and Proceedings of 30 years of modern optimal, University of Rhode Island, 1988.

I. Lasiecka
Applied Mathematics Department
University of Virginia
Charlottesville, VA 22903
USA

International Series of
Numerical Mathematics, Vol. 91
© 1989 Birkhäuser Verlag Basel

On the reachability problem of a viscoelastic beam during a slewing maneuver

G. Leugering

Fachbereich Mathematik
Technische Hochschule Darmstadt

Abstract. The reachability problem for a hybrid integro-partial differential equation of the Volterra-type is considered. A result of J. Prüß [10] is used to show a uniqueness result which leads to approximate or exact reachability via the HUM-method, applied to a reference problem, and a Fredholm-property of the control-to-state map. Energy estimates are given with the aid of energy-multipliers.

Keywords. Convolution equations, Laplace-transform, asymptotic behavior, energy multipliers, HUM (RHUM).

1980 *Mathematics subject classifications:* 93B05, 45K05, 73F15, 49F15

We consider the problem of point controllability (reachability) of a rotating beam, which is supposed to represent a robot arm during a slewing maneuver. Those robot-structures are, in reality, rather complex, a global mathematical modelling is far from being complete, and is presumably very difficult to handle with respect to a particular control problem. Nevertheless, significant for all those structures is a more or less weak damping due to the action of internal (hidden) parameters. Therefore, the variables which are chosen to represent the "state" of the system are necessarily determined not only at an actual time, t, but also through their whole history. In continuum mechanics this type of dependence has traditionally been framed within the theory of viscoelasticity.

Therefore, even though we do not know about a particular constitutive material equation, which in fact appears to be a problem of its own, we treat the beam configuration within the context of history-value problems, revealing a very subtle dissipation of energy, which, as a phenomenon, can be observed in any experiment.

Now, conservation of energy is an important property, which is vitally used in most of the controllability results for vibrating systems. In reality, the energy simply not being conserved, the question remains, whether null-controllability or reachability, which are then different concepts, hold for the model under consideration, where one usually is not able to take into account all state variables. In order to tackle this problem it is not adequate to look for the null-controllability problem in the classical sense: bring the system to rest and stay there, without further control action applied to the system; simply because of the hidden variables.

After this pre-justification we proceed to formulate the mathematical problem.

We simply adopt the model from the recent paper of W. Desch and R. Miller [3], see also M. Delfour et. al. [1]. There a beam of length ℓ is assumed attached to a cylinder of radius r. The (plane (!)) system is slewed by an angle Θ round the z-axis. $y(x)$ is the vertical deflection of a point x of the centerline with respect to the rotated reference frame. Denote by ρ, S, I, the density, cross-section area and its moment w.r.t. the z-axis. If we assume a purely elastic material, we have to consider a material function E, called Young's modulus.

Suppose we have the option to control the rotation – by applying moments, $M = -c\Theta + f$, to the cylinder – and the deflection, y, of the beam – by regulating, with φ, the angle (the bending moment would be another reasonable choice) – at the point where the beam is attached to the cylinder. Then we have, as a reference system with perfect memory

$$(EIy'')'' + \rho S\ddot{y} + \rho S(x+r)\ddot{\Theta} = 0$$

(1)
$$\int_0^\ell \rho S(x+r)\ddot{y}\,dx + \int_0^\ell \rho\{S(x+r)^2 + I\}dx\ddot{\Theta} = M = -c\Theta + f$$

$$\text{for} \quad (t,x) \in (0,T) \times (0,\ell),$$

$$
\begin{aligned}
y(0) &= 0 & (EIy'')(\ell) &= 0 \\
y'(0) &= \varphi & (EIy'')'(\ell) &= 0
\end{aligned}
$$
(2)
$$\text{for} \quad t \in (0,T)$$

(3)
$$y(0,\cdot) = y_0, \quad \dot{y}(0,\cdot) = y_1, \quad \Theta(0) = \Theta_0, \quad \dot{\Theta}(0) = \Theta_1$$
$$\text{for} \quad x \in (0,\ell),$$

where we have denoted the derivative with respect to x and t by a prime and a superdot, respectively. We have suppressed the explicit notation of the variable x, t where it did not seem misleading. As already explained, and because of the limited space available, we do not want to discuss the model neither from the mechanical point of view – this can be seen from the references in [3] – nor from an operator-theoretic point of view, stating, however, that on introducing the following notation

$$w := y + (x+r)\Theta,$$
$$z := \begin{pmatrix} w \\ \Theta \end{pmatrix},$$
$$H := L_2(0,\ell,\rho S) \times \mathbb{R}_b,$$

where $L_2(0,\ell\rho S), \mathbb{R}_b$ (b is the coefficient of $\ddot{\Theta}$) are weighted spaces with inner product

$$(u,v) = \int_0^\ell uv\rho S\,dx, \quad (\Theta \cdot \rho) = b\theta \cdot \rho,$$

$$Lz := L\begin{pmatrix} w \\ \Theta \end{pmatrix} = \begin{pmatrix} \frac{1}{\rho S}(EIw'')'' \\ \frac{1}{b}(c\Theta + r(EIw'')'(0) - EIw''(0)) \end{pmatrix}$$
$$D(L) := \{z \in H \,|\, Lz \in H\}$$
$$U := \{z \in H \,|\, w, w' \ \text{abs. cont.} \ (EI)^{1/2}w'' \in L_2(0,\ell)\},$$
$$V := \{z \in U \,|\, w(0) = r\Theta, \ w'(0) = \Theta\},$$
$$Az := Lz,$$
$$D(A) := \{z \in V \cap D(L) \,|\, (EIw'')(\ell) = (EIw'')'(\ell) = 0\},$$
$$Bf := \begin{pmatrix} 0 \\ f \end{pmatrix}, \quad \phi := x\begin{pmatrix} \varphi \\ 0 \end{pmatrix},$$
$$Nz := \begin{pmatrix} EIw''(\ell) \\ (EIw'')'(\ell) \end{pmatrix},$$

the system (1) – (3) is equivalent to the abstract system

$$\ddot{z} + Lz = Bf$$

(4)
$$z - \phi \in V, \quad Nz = 0$$

$$z(0) = z_0, \quad \dot{z}(0) = z_1$$

which can be handled for instance, in its variational form, in $L_2(0, T, U^*)$.

Note, that A, defined above, is a positive definite selfadjoint operator in H (because of the term $c\Theta$ (!)). According to the remarks above, we introduce a kernel, $a(t)$, (see J. Prüß [10])

$$(A_1) \quad \begin{cases} a(t) = 0 \quad\quad t < 0 \\ a_0 + ta_\infty + \int_0^t a_1(s)ds \quad t \geq 0 \\ \text{with} \\ a_0 \geq 0, \quad a_\infty \geq 0 \quad a_1(0+) < \infty, \\ a_1(t) \geq 0, \quad a_1(t) \quad \text{log-convex}; \end{cases}$$

in fact, in the mechanical context, $\dot{a}(t)$ is the relaxation modulus of an isotropic visco-elastic body, and (A_1) constitutes a very weak set of assumptions, compatible with the laws of thermodynamics. (One may think of a completely monotone kernel: $(-1)^j a^j(t) \geq 0$ $\forall j$ (!).) Denoting the Stieltjes convolution with the measure $da(\cdot)$ by L_a $(L_a z = z * da)$ we have

$$\ddot{z} + \dot{L}_a(Lz) = Bf$$

(5)
$$z - \phi \in V, \quad \dot{L}_a(Nz) = 0$$

$$z(0) = z_0, \quad \dot{z}(0) = z_1.$$

Note that $\dot{L}_a(Lz) = a_0 Lz + (a_1(0+) + a_\infty)Lz + \dot{a}_1 * Lz$ (* denotes the usual convolution: $f * g := \int_0^t f(t-s)g(s)ds$.) Note also that with $(a_0 > 0, a_\infty = 0, a_1 \equiv 0)\hat{=}(a_0, 0, 0)$, (5) is essentially a parabolic "PDE" while the choice $(0, a_\infty, 0)$ reduces (5) to (4) $(a_\infty$ normalized to 1), which is a hyperbolic type "PDE". The case $(a_0, a_\infty, 0)$ is an example of Kelvin-Voigt-type visco-elastic material, with a short memory, while $(0, a_\infty, a_1)$ represents a visco-elastic solid of Boltzmann-type, if $\dot{a}_1(0+) > -\infty$, or of "spring-pot"-type else. The most general case (a_0, a_∞, a_1) can be viewed as describing an "interpolation" of the other cases and is physically relevant in homogenization theory, see [12], see also [11]. Since we are interested in the configurations, z, which we can reach from the starting position $z_0 = z_1 = 0$, only, we set the initial values equal to zero (which is no restriction). Now, $\dot{L}_a(Lz) = \frac{d^2}{dt^2} a * Lz$ and, therefore, (5) can then be integrated to

$$z + a * Lz = F$$

(6)
$$z - \phi \in V, \quad a * Nz = 0$$

with $F(t) = \int_0^t \int_0^\tau Bf(\delta)d\delta d\tau$. Now, obviously $L\phi = 0$, and $N\phi = 0$, hence, defining: $u = z - \phi$ we have $u \in D(A)$ and

(7)
$$u + a * Au = F - \phi$$

which is exactly the problem treated by J. Prüß [10]. As a matter of fact, it is shown there that there exists a Volterra-resolvent operator family, $R(t)$, such that for $f, \varphi \in L_2(0, T)$, the solution $u(\cdot)$, to (7) is given by

$$(8) \qquad\qquad u = \frac{d}{dt} R * (F - \phi)$$

hence, $z = \phi + \frac{d}{dt} R * (F - \phi)$. Since ϕ is not necessarily differentiable, it is plain that the substitution $u = z - \phi$ is not the most appealing one, if one is mainly interested in regularity. One may take any other "reference problem", like

$$(9) \qquad\qquad \begin{aligned} v + b * Lv + d * v &= 0 \\ v - \phi \in V, \quad Nv &= 0 \end{aligned}$$

with a different kernel b (to be chosen conveniently), and conclude, with $z = u - v, c = a - b$, that u has to satisfy

$$(10) \qquad\qquad u + a * Au = c * Lv - d * v + F$$

(For example, if $a_0 = 0$ one may take $b = (a_\infty + a_1(0))t$ and $d = \omega$, then (5) is a damped wave equation with damping rate $\omega/2$ which can be chosen and that the right-hand side of (8) is a convolution: $e * v$.) It is then a matter of applying well-known results on well-posedness of Volterra-(integrodifferential-)equations (see [2]) to see that our system (6) is well-posed in $H \times V^*$, using $L_2(0, T)$-controls φ, f, in the sense that

$$z \in C^1(0, T, V^*) \cap C(0, T, H)$$

The problem we are looking at is to describe the reachable set

Definition.

(i) *The system (6) is called approximately reachable at T, if the set*

$$\Omega = \{(z(T, \varphi, f), \dot{z}(T, \varphi, f) | \varphi, f \in L_2(0, T)\}$$

is dense in $H \times V^*$.

(ii) *The system (6) is called exactly reachable in*

$$X \times Y \subseteq H \times V^* \quad \text{at} \quad T, \quad \text{if} \quad \Omega = X \times Y, \quad X, Y \quad \text{Hilbert-spaces.}$$

We consider the weaker property first. It can be shown as in [8], that in the case $(a_0, a_\infty, 0)$ the system (6) satisfies (i) but not (ii). The same holds true for (a_0, a_∞, a_1) if $a_1(\cdot)$ is completely monotone, see [7]. The key to these results is a uniqueness argument for analytic functions, which can not be applied if a_1 is not analytic.

Suppose we are given a pair $(h^0, h^1) \in H \times V^*$ satisfying $< (h^0, h^1), \Omega_s > = 0$ with

$$\Omega_s = \{(z(T, \varphi, f), \dot{z}(T, \varphi, f) | \varphi \in H_0^2(0, T), f \in L_2(0, T))\}$$

which is obviously a dense subset of Ω. We would like to conclude that this already implies $(h^0, h^1) = 0$. We assume regular kernels

(A_2) $\qquad\qquad\qquad\qquad\qquad\qquad \dot{a}_1(0) > -\infty$

Thanks to the regularity met in Ω_s we can directly apply (8)

$$z = \phi + \frac{d}{dt} R * (F - \phi)$$
$$= \phi + (F - \phi)(0)R + R * (\dot{F} - \dot{\phi})$$
$$= \phi + R * (\dot{F} - \dot{\phi})$$

and

$$\dot{z} = \dot{\phi} + (\dot{F} - \dot{\phi})(0)R + R * (\ddot{F} - \ddot{\phi})$$
$$= \dot{\phi} + R * (\ddot{F} - \ddot{\phi})$$

At $t = T$ this gives $(\varphi \in H_0^2(0, T))$

$$z(T) = \int_0^T R(\tau)d\tau\, F(T) + \int_0^T \int_0^{T-s} R(\tau)d\tau (\ddot{F} - \ddot{\phi})(s)ds$$
$$\dot{z}(T) = \int_0^T R(T - s)(\ddot{F} - \ddot{\phi})(s)ds$$

Hence

$$0 = \;<(h^0, h^1), (z(T), \dot{z}(T))>_{H\times V^\bullet} = \int_0^T \{(\int_0^T R(\tau)d\tau\, h^0, e_2)$$

$$+ \int_0^{T-s} (R(\tau)h^0, e_2)_H d\tau + (R(T - s)h^1, e_2)_{V^\bullet}\} f(s)ds$$

$$+ \int_0^T \{\int_0^{T-s} (R(\tau)h^0, xe_1)_H d\tau + (R(T - s)h^1, xe_1)_{V^\bullet}\} \ddot{\varphi}(s)ds$$

with

$$e_1 = \begin{pmatrix} I \\ 0 \end{pmatrix}, \qquad e_2 = \begin{pmatrix} 0 \\ I \end{pmatrix} \quad \text{in} \quad H, \quad \text{for all} \quad \varphi \in H_0^2, f \in L_2.$$

This implies

(11) $\qquad (\int_0^T R(\tau)d\tau\, h^0, e_2) + \int_0^t (R(\tau)h^0, e_2)d\tau + (R(t)h^1, e_2) = 0 \quad \text{on} \quad (0, T),$

(12) $\qquad \int_0^t (R(\tau)h^0, xe_1)d\tau + (R(t)h^1, xe_1) = 0 \quad \text{on} \quad (0, T).$

It is this point at which we need more information about the resolvent operator R.

Taking the Laplace-transform of equation (12), we find

(13)
$$\frac{1}{s}(\hat{R}(s)h^0, xe_1) + (\hat{R}(s)h^1, xe_1) = o(e^{-TRe(s)})$$

as $s \to \infty$. In order to proceed further, we have to investigate two different cases: $a_0 > 0$ and $a_0 = 0$. It has been shown by J. Prüß [10] that $\hat{R}(s)$ is given by

$$\hat{R}(s) = \frac{1}{s\sqrt{\hat{a}(s)}} \int_0^\infty e^{-\frac{1}{\sqrt{\hat{a}(s)}}t} C(t)dt,$$

where

$$\frac{1}{\sqrt{\hat{a}(s)}} = \frac{s}{\sqrt{sa_0 + a_\infty + s\hat{a}_1(s)}} = \begin{cases} s\hat{k}_1(s), & a_0 > 0 \\ \frac{1}{\mu}s + s\hat{k}_1(s), & a_0 = 0 \end{cases} \quad (\mu^2 := a_\infty + a_1(0)),$$

with k_1 absolutely continuous and concave, and where $C(t)$ denotes the cosine family generated by $-A$. In the mechanical context, k_1 is the creep-compliance associated with $a(\cdot)$.

If $a_0 > 0$

$$s\hat{k}_1(s) = (\frac{s}{a_0})^{1/2}(1 + \frac{a_\infty + s\hat{a}_1(s)}{a_0 s})^{-1/2}$$

and since we have assumed $a_1(0+) < \infty$

$$\frac{a_\infty + s\hat{a}_1(s)}{a_0 s} \sim \frac{1}{2}\frac{a_\infty + a_1(0)}{a_0}, \quad s \to \infty$$

therefore

$$r(s) := \frac{a_\infty + a_1(0)}{2a_0^{1/2}}\frac{1}{s^{1/2}}(1 + \frac{s\hat{a}_1(s) - a_1(0)}{a_\infty + a_1(0)})$$

$$\sim \frac{\mu^2}{2a_0^{3/2}}\frac{1}{s^{1/2}}, \quad s \to \infty,$$

and

$$\frac{1}{\sqrt{\hat{a}(s)}} = s\hat{k}_1(s) = (\frac{s}{a_0})^{1/2} + r(s)(1 + o(1))$$

$$=: (\frac{s}{a_0})^{1/2} + z(s).$$

If $a_0 = 0$

$$\frac{1}{\sqrt{\hat{a}(s)}} = \frac{1}{\mu}s + s\hat{k}_1(s) = \frac{s}{\mu} + s\hat{k}_1(s) - \dot{k}_1(0) + \dot{k}_1(0)$$

$$=: \frac{s}{\mu} + r(s) + \dot{k}_1(0)$$

with

$$r(s) := s\hat{k}_1(0) = \hat{\dot{k}}_1(s) \sim \frac{\ddot{k}_1(0)}{s}$$

if $\dot{a}_1$ is absolutely continuous, or $r(s) = o(1)$ else. We therefore have

$$\hat{R}(s) = \frac{1}{s\sqrt{\hat{a}(s)}} \cdot \begin{cases} \int_0^\infty e^{-(\frac{s}{a_0})^{1/2}t} e^{-z(s)t} C(t)dt, & a_0 > 0 \\ \int_0^\infty e^{-\frac{s}{\mu}t} e^{-r(s)t} e^{-k_1(0)t} C(t)dt, & a_0 = 0. \end{cases}$$

I. $a_0 > 0, \quad a_1 \neq 0, \quad a_1(0) < \infty, \quad \dot{a}_1(0) < -\infty$

$$(14) \qquad \hat{R}(s) = \frac{1}{s\sqrt{\hat{a}(s)}} \sum_{j=0}^\infty (-1)^j \frac{1}{j!} \{ \int_0^\infty e^{-(\frac{s}{a_0})^{1/2}t} t^j C(t)dt \} z(s)^j$$

II. $a_0 = 0, \quad a_1 \quad$ as above

$$(15) \qquad \hat{R}(s) = \frac{1}{s\sqrt{\hat{a}(s)}} \sum_{j=0}^\infty (-1)^j \frac{1}{j!} \{ \int_0^\infty e^{-\frac{s}{\mu}t} t^j e^{-k_1(0)t} C(t)dt \} r(s)^j.$$

Now,

$$\int_0^\infty e^{-(\frac{s}{a_0})^{1/2}t} t^j C(t)dt = (-1)^j (\frac{a_0^{3/2}}{s^{1/2}})^j \frac{d^j}{ds^j} \int_0^\infty e^{-(\frac{s}{a_0})t^{1/2}} C(t)dt$$

$$= (-1)^j (\frac{a_0^{3/2}}{s^{1/2}})^j \frac{d^j}{ds^j} (\frac{s}{a_0})^{1/2} (\frac{s}{a_0} + A)^{-1}$$

$$= \frac{(-1)^j}{a_0^{3/2}} (\frac{a_0^{3/2}}{s^{1/2}})^j \frac{d^j}{ds^j} s^{1/2} \int_0^\infty e^{-st} T(t)dt$$

where, now, $T(t)$ is the analytic semigroup generated by $-a_0 A$. Hence,

$$(16) \qquad \int_0^\infty e^{-(\frac{s}{a_0})^{1/2}t} t^j C(t)dt = \frac{1}{a_0^{3/2}} (\frac{a_0^{3/2}}{s^{1/2}})^j \sum_{k=0}^j \binom{j}{k} (s^{1/2})^{(j-k)} \int_0^\infty e^{-st} t^k T(t)dt$$

Define

$$p(t) := \begin{cases} (T(t)h^0, xe_1)_H, & a_0 > 0 \\ e^{-k_1(0)t}(C(t)h^0, xe_1)_H, & a_0 = 0, \end{cases}$$

$$q(t) := \begin{cases} (T(t)h^1, \quad xe_1)_{V^*}, & a_0 > 0 \\ e^{-k_1(0)t}(C(t)h^1, \quad xe_1)_{V^*}, & a_0 = 0. \end{cases}$$

We proceed to discuss the asymptotic equation (13). We consider case II (15) first, the I. can then be deduced from II.

Here (12) reads like

(17)
$$
(\frac{1}{s}\hat{p}(s) + \hat{q}(s))(1 + \sum_{j=1}^{\infty}(-1)^j\frac{1}{j!}\frac{1}{s}\int_0^{\infty}e^{-\frac{s}{\mu}t}t^jp(t)dt
$$

$$
+ \int_0^{\infty}e^{-\frac{s}{\mu}t}t^jq(t)dtr(s)^j/(\frac{1}{s}\hat{p}(s) + \hat{q}(s))
$$

$$
= o(e^{-(T-\varepsilon)Re(s)}), \quad s \to 0, \quad \varepsilon > 0 \quad \text{arbitrary small.}
$$

Assume that $\hat{p}(s)$ and $\hat{q}(s)$ do not have the asymptotic property of (17),seperately. Then $\hat{q}(s) \sim -\frac{1}{s}\hat{p}(s)$, since otherwise $\frac{1}{s}\hat{q}(s) + \hat{p}(s)$ is the lowest order term in (17) contradicting (17).Thus

$$
\hat{q}(s) = -\frac{1}{s}\hat{p}(s) - \frac{\varepsilon(s)}{s}\hat{p}(s), \quad \varepsilon(s) \to 0, \quad (s \to \infty).
$$

Hence

$$
\int_0^{\infty}e^{-\frac{1}{\mu}st}t^jq(t)dt = -\mu^j(-1)^j\sum_{k=0}^{j}\binom{j}{k}(\frac{1}{s})^{(j-k)}\hat{p}^k(s) - \mu^j(-1)^j\frac{d^j}{ds^j}\varepsilon(s)\frac{1}{s}\hat{p}(s)
$$

Inserted into (17), this gives

$$
- \frac{\varepsilon(s)}{s}\hat{p}(s)(1 + \sum_{j=1}^{\infty}\frac{\mu^j}{j!}\{\sum_{k=0}^{j-1}\binom{j}{k}(\frac{1}{s})^{(j-k)}\hat{p}^k(s)
$$

$$
+ (\varepsilon(\cdot)\frac{1}{(\cdot)}\hat{p})^j(s)\}r(s)^j/(\frac{\varepsilon(s)}{s}\hat{p}(s)))
$$

$$
= - \frac{\varepsilon(s)}{s}\hat{p}(s)(1 + \sum_{j=1}^{\infty}(-1)^j\mu^j\{\frac{1}{s^{j+1}}\sum_{k=0}^{j-1}\frac{(-1)^k}{k!}s^k\hat{p}^k(s)
$$

$$
+ \frac{1}{j!}(\varepsilon(\cdot)\frac{1}{(\cdot)}\hat{p})^j(s)\}r(s)^j/\frac{\varepsilon(s)}{s}\hat{p}(s))
$$

$$
=: - \frac{\varepsilon(s)}{s}\hat{p}(s)(1 + \sum Q_j(s)r(s)^j/\frac{\varepsilon(s)}{s}\hat{p}(s)) = o(e^{-(T-\varepsilon)Re(s)}).
$$

Suppose $\varepsilon(s) = o(e^{-T_0Re(s)})$, and assume that $\frac{\varepsilon(s)}{s}\hat{p}(s) = o(e^{-(T-\varepsilon)Re(s)})$, $(T_0 < T)$. Then (17) gives

$$
\sum_{j=1}^{\infty}(-1)^j\mu^j\frac{1}{s^{j+1}}\{\sum_{k=0}^{j-1}\frac{(-1)^k}{k!}s^k\hat{p}(s)^k + \frac{1}{j!}s^{j+1}(\varepsilon(\cdot)\frac{1}{(\cdot)}\hat{p})^j(s)\}r(s)^j
$$

$$
= o(e^{-(T-\varepsilon)Re(s)}) \quad s \to \infty
$$

which in turn requires

$$
\sum_{k=0}^{j-1}\frac{(-1)^k}{k!}s^k\hat{p}^k(s) + \frac{1}{j!}s^{j+1}(\varepsilon(\cdot)\frac{1}{(\cdot)}\hat{p}(\cdot))^j(s)
$$

$$
= o(e^{-(T-\varepsilon)Re(s)}), \quad s \to \infty,
$$

for all $j \geq 1$, a contradiction. In the remaining cases, the expression in parenthesis in position (17) has to contribute to the asymptotic equation; if $\frac{\varepsilon(s)}{s}\hat{p}(s) = o(e^{-T_1 Re(s)})$ $(T_1 < T)$ then by (17)

$$(19) \qquad (1 + \sum_{j=1}^{\infty} Q_j(s)r(s)^j / \frac{\varepsilon(s)}{s}\hat{p}(s)) = o(e^{-(T-T_1-\varepsilon)Re(s)})$$

(ε arbitrarily small!). This implies that the lowest order term $(j = 1)$ of the series has to compensate $r(s)$ in order to have $Q_1(s)r(s) \sim -1$, that is

$$\varepsilon(s) \sim \frac{r(s)}{s} \quad (\to \hat{p}(s) = o(e^{-(T_1-\varepsilon)Re(s)}))$$

But then we have the same problem as in the first case (here $\varepsilon(s)$ is even of zero type). If, finally, $(\frac{\varepsilon(s)}{s}\hat{p}(s))^{-1}$ is of arbitrarily small exponential type type, or of finite polynomial order at infinity, (17) leads again to (19) but with $e^{-(T-\varepsilon)Re(s)}$ on the right-hand side rather than merely $e^{-(T-T_1-\varepsilon)Re(s)}$). Therefore, the only possible case left is the one where $\hat{p}(s)$, $\hat{q}(s) = o(e^{-TRe(s)})$, $s \to \infty$. By (16) the same arguments apply to the case I.

We conclude: $p(t) = q(t) \equiv 0$ on $(0,T)$ if $a_1 \neq 0$, or $p(t) + \int_0^t q(\tau)d\tau \equiv 0$ on $(0,T)$ if $a_1 \equiv 0$, which is then the classical case. This is a uniqueness problem! Since our system operator A has a compact resolvent $(\lambda - A)^{-1}$, we have a family of orthoprojections onto the eigenspaces H_k associated with an eigenvalue λ_k, $\lim_{k\to\infty} \lambda_k = +\infty$, $\lambda_k < \lambda_{k+1}$. Hence, we have

$$(20) \qquad \sum_{k=0}^{\infty} e^{-a_0\lambda_k t}(P_k h^0, xe_1)_H = 0 \quad \text{on} \quad (0,T),$$

if $a_0 > 0$, or

$$(21) \qquad \sum_{k=0}^{\infty} \cos\lambda_k^{1/2}t(P_k h^0, xe_1)_H = 0 \quad t \in (0,T)$$

if $a_0 = 0$ and analogous equations for h^1.

While (20) implies $(P_k h^0, xe_1) = 0, \forall k((P_k h^1, xe_1)_{V^*} = 0, \forall k)$ without further assumptions, the same conclusion is not necessarily true for (21), and is in fact very strongly related to the asymptotic behavior of λ_k as $k \to \infty$. It is obvious, from the above that all arguments apply to the equation (10) as well. Now suppose for a moment

$$(P_k h^0, e_2) = (P_k h^0, xe_1) = 0, \quad (P_k h^1 e_2) = (P_k h^1 xe_1) = 0$$

for all $k \in \mathbb{N}$.

But $P_k h^0$ satisfies

$$AP_k h^0 = \lambda_k P_k h^0,$$

and

$$0 = (P_k h^0, x e_1) = (x P_k h^0, e_1)$$
$$= \frac{1}{\lambda_k} \int_0^e x (EI(P_k h^0)_1'')'')dx$$
$$= \frac{1}{\lambda_k} (P_k h^0)_1''(0),$$
$$0 = (P_k h^0, e_2) = \frac{1}{\lambda_k} (P_k h^0)_2.$$

Therefore,

$$P_k h^0 = \begin{pmatrix} w_{0,k} \\ 0 \end{pmatrix}$$

with $w_{0,k}$ satisfying

$$(EIw_{0,k}'')'' = \lambda w_{0,k}, \quad r(EIw_{0,k}'')' = EIw_{0,k}''(0)$$
$$w_{0,k}(0) = 0, \quad (EIw_{0,k})''(\ell) = 0$$
$$w_{0,k}'(0) = 0, \quad (EIw_{0,k}'')'(\ell) = 0$$

and from above

$$w_{0,k}''(0) = 0.$$

and therefore even

$$(EIw_{0,k}'')'(0) = 0.$$

Hence, the eigenelement $w_{0,k}$ has zero Cauchy-data at the boundary $x = 0$. This implies $w_{0,k} \equiv 0$. It is interesting to observe that the presence of the boundary control φ causes the additional boundary condition $w_{0,k}''(0) = 0$, which alone is sufficient for the same conclusion, independent of the control f, which, in effect, contributes the equation $r(EIw_{0,k}'')' = EIw_{0,k}''(0)$. But even this condition at the boundary $x = 0$ is in addition to the others sufficient for $w_{0,k} \equiv 0$. The argument is the same as in the second case: If both are zero, we are in the first case of Cauchy-data. If not, $EIw_{0,k}''(0) = r(EIw_{0,k}'')'(0)$ have a common sign, say positive. Then $w_{0,k}''$ (which is zero at $x = 0$) increases and is positive on an interval $(0, \delta)$ and so is $w_{0,k}$ and a forteriori $(EIw_{0,k}'')''$ is increasing, by the same procedure and so forth. But, after a finite number of steps, we arrive at $x = \ell$, concluding that the boundary conditions at $x = \ell$ cannot be zero, as required.

We finally conclude that approximate reachability (or controllability) of system (6) is just a matter of the uniqueness property of the almost periodic functions $\sum_{k=\infty}^{\infty} e^{i\lambda_k^{1/2} t} c_k$ (for $a_0 = 0$) and is for free if $a_0 > 0$.

Theorem 1. *Let $a(\cdot)$ satisfy A_1, A_2. Then the system (6),(or (5), (1)) is approximately reachable in the sense above, with one or two controls φ and f, in case*

I):

$$a_0 > 0, \quad \text{with} \quad T > 0 \quad \text{arbitrary};$$

II):

$$a_0 = 0, \quad \text{with a finite } \; T > 0, \quad \text{if } \sum_{k=-\infty}^{\infty} e^{-i\lambda_k^{1/2}t} c_k \equiv 0 \quad \text{on} \quad (0,t) \Rightarrow c_k = 0.$$

Note that the eigenvalue problem connected with A is not entirely trivial, because of the appearance of the eigenvalue parameter in the boundary conditions. It is reasonable, however, to expect the eigenvalues to behave asymptotically as the corresponding values for the cantilivered beam, so that the condition in case II. would be met. Since, in this case, we are much more interested in exact reachability, we have to look at the exact-reachability (controllability)-property of the system (6) with $a_1 \equiv 0$ first.

Theorem 2. *Let ρ, S, E, I be constant (with $\rho S\ell < 1 \leq 12EI/\ell^2$). Then the system (5), ((6)) is exactly reachable in $H \times V^*$ within a suitable large time $T > 0$ by $L_2(0,T)$-controls φ and f.*

Remark 1:

   (i) Here we need both controls!

  (ii) It is likely to be able to relax the assumption on T to arbitrary times $T > 0$, using a recent result of Komornik [4]

 (iii) The relations between the material functions seem awkward but they indicate that thin and stiff beams are controllable this way. The method of proof allows even for non-constant coefficients with some limitations.

Sketch of Proof:: The proof is done using (HUM) (see [9]) and an energy multiplier, in order to get the estimates. Because of limites space we are very brief.
   Introduce:

$$\psi = \begin{pmatrix} \zeta \\ \omega \end{pmatrix}, \qquad \ddot{\psi} + A\psi = 0$$
$$\psi(T) = \psi_0 \qquad \dot{\psi}(T) = \psi_1$$

then by: (HUM) (in fact (RHUM))

$$< (\dot{z}(T), -z(T)), (\psi_0, \psi_1) >_{H \times H} = \int_0^T \varphi EI\zeta''(0)dt + \int_0^T f\omega \, dt$$

hence, with

$$\varphi = \zeta''(0), \quad f = \omega, \quad \Lambda(\psi_0, \psi_1) := (\dot{z}(T), -z(T))$$

we have

$$< \Lambda(\psi_0, \psi_1), (\psi_0, \psi_1) >_{H \times H} = \int_0^T EI(\zeta''(0))^2 + \omega^2 \, dr$$

We already know, when the right side defines a norm. But the information needed is not readily availlable: we do not know the eigenvalues analytically,and we do not have an asymptotic expansion yet. Nevertheless, we introduce the multiplier:

$$M(x) = \begin{pmatrix} m\zeta' \\ \frac{m(0)}{r}\omega \end{pmatrix}$$

where $m(x)$ may be chosen as $m(x) = \frac{x-\ell}{\ell}$.

Then, on utilizing the equation

$$0 = \int_0^T < \ddot{\psi} + A\psi, M(x) >_H dt$$

one can show, with some estimation effort (!) including the estimation of the coefficients, that there are constants $T, C_T, c_T > 0$ such that

$$c_T E(0) \le \int_0^T EI(\zeta''(0))^2 + \omega^2 dt \le C_T E(0)$$

with $E(0) = \|(\psi_0, \psi_1)\|_{V \times H}$. This gives the result. ∎

Remark: This result has been proved by G. Schmidt (Montreal) independently (not using HUM) ((at the same afternoon)), when he was visiting Darmstadt.

In order to answer the question of whether exact reachability holds for system (5) or (6), when $a_0 = 0$, we rely on results of W. Desch and R. Grimmer [2] and, by different means of J. Prüß [10].

In [10] it is shown that

$$R(t) = \int_0^{\mu t} w_t(t, \tau)C(\tau)d\tau + e^{-\frac{\mu}{2k}t}C(\mu t) = R_0(t) + e^{-\frac{\mu}{2k}t}C(\mu t),$$

with $\hat{dw}_t(s) := \frac{1}{s\sqrt{\hat{a}(s)}}e^{-\frac{t}{\sqrt{a(s)}}}$ such that $R_0(t)$ is norm-continuous. As has been shown in [6,5] this representation implies that the control-to-state map,$(\varphi, f) \to (z, \dot{z})$ at $t = T$ is a semi-Fredholm-operator. Since by Thm 1 its range is dense, we conclude:

Corollary. *The system (5), ((6)) is exactly reachable within $H \times V^*$ and $(0, T)$, T as in Theorem 2.*

REFERENCES

[1] M.C. Delfour, M. Kern, L. Passeron and B. Sevenne, *Modelling of a rotating flexible beam*, in "Control of Distributed Parameter Systems 1986," H.E. Rauch, editor, 4th IFAC Symposium, Los Angeles, Pergamon Press, 1986, pages 383–387.

[2] W. Desch and R.C. Grimmer, *Initial boundary value problems for integro-differential equations*, J. Int. Equs **10(1-3)** (1985), 73–97.

[3] W. Desch and R.K. Miller, *Exponential stabilization of volterra integrodifferential equations in Hilbert spaces*, Journal of Differential Equations **70(3)** (1987), 366–389.

[4] V. Komornik, *Controlabilte exacte en un temps minimal in minimal time*, Comptes Rend. Acad. Sci. Paris Ser. I Math. **304** (1987), 223–225.

[5] G. Leugering, *Boundary controllability of volterra integrodifferential equations in Hilbert spaces*, in "Distributed Parameter Systems," F. Kappel, K. Kunisch, W. Schappacher, editors, Springer-Verlag, 1987,, pp. 234–252.

[6] G. Leugering, *Exact boundary controllability of an integrodifferential equation*, Applied Mathem. and Optimization **15** (1987), 223–250.

[7] G. Leugering, *On boundary controllability of viscoelastic systems*, in "Distributed Parameter Systems," Bermundez, editor, IFIP TC7-workshop, July 1987 Santiago Compostela, Spain 1987, 190–201.

[8] G. Leugering and E.J.P.G. Schmidt, *Boundary control of a vibrating plate with internal damping*, Math. Meth. in the Appl. Sciences (to appear).

[9] J.L. Lions, *Controllabilite exacte des systemes distribues*, Comptes R.A.S. Paris **302** (1986), 471–475.

[10] J. Prüß, *Positivity and regularity of hyperbolic volterra equations in Banach spaces*, Math. Annalen **279** (1987), 317–344.

[11] M. Renardy, *Initial value problems for viscoelastic liquids*, in "Trends and Applications of Pure Mathematics to Mechanics," P.G. Ciarlet and M. Roseau, editors, Ecole Polytechnique, Palaiseau, Springer-Verlag, 1983, pages 333–345.

[12] E. Sanchez-Palencia, "Non-Homogeneous Media and Vibration Theory," Springer-Verlag, 1980.

G. Leugering
Fachbereich Mathematik
Technische Hochschule Darmstadt
Schloßgartenstraße 10
D-6100 Darmstadt
West Germany

International Series of
Numerical Mathematics, Vol. 91
© 1989 Birkhäuser Verlag Basel

A note on source term identification
for parabolic equations

Tao Lin and Richard E. Ewing

Departments of Mathematics
University of Wyoming

Abstract. We discuss a source term identification problem for the one-dimensional parabolic
equation on the quarter plane. The source term is supposed to be in the form $f(x)g(t)$, in
which the function $f(x)$ is to be identified. The extra information about the solution to the
parabolic equation is given at the boundary where both the solution itself and its flux are
supposed to be known. Under some conditions, existence and uniqueness for this problem are
proved. We also give a direct numerical scheme for computing $f(x)$ locally. Several numerical
examples are supplied to show some properties of this scheme.

Keywords. Parameter identification, inverse problem, PDE, numerical method.

1980 *Mathematics subject classifications*: Primary 35R30, 93C20; Secondary 93B30.

1. Introduction

In this paper, we discuss the source term identification problem for a one-dimensional
parabolic equation. Specifically, the problem is to identify $f(x)$ in the following equations:

$$u_t(x,t) = (a(x)u_x(x,t))_x + f(x)g(t) \quad x > 0, t > 0,$$

(1) $$u(x,0) = \phi(x) \quad x > 0,$$

$$a(0)u_x(0,t) = \psi(t) \quad t > 0,$$

with the overspecified boundary data

(2) $$u(0,t) = r(t) \quad t > 0.$$

Here, we assume that functions $g(t)$, $\phi(x)$, $\psi(t)$, and $r(t)$ are given.

The source term identification problem for parabolic equations has been discussed in
many papers [1–5]. The main results now available are based on the assumption that the
source term has a special form; for example, the source term depends on only one variable,
or the source is of the form of separation of variables, or the component functions have
compact support. In [2], the component of the source term which depends only upon the
spatial variable is given, and the identification problem for the component of the source
term depending on temporal variables is discussed. Also, these papers mainly consider
the determination of uniqueness or the restoration of the continuous dependence of the
solutions upon data.

In this paper, we also discuss the source term identification problem for a parabolic
equation with the assumption that the source term is of the form of separation of variables
$f(x)g(t)$ where the spatial component f is independent of t and the temporal component
g is independent of x. The unknown component is the spatially variable function f, and

g is assumed to be known. We do not assume that either the temporal component or the spatial component has any special form. We discuss both constant and variable coefficient cases.

For the constant coefficient case, i.e., $a(x)$ is constant, we prove the existence and uniqueness of the solution to the identification problem in Section 2 under certain sufficient conditions. In the proof of existence, we reduce the problem to two integral equation problems, and then solve these integral equations in a closed form under some sufficient conditions. Uniqueness for the identification problem follows also from the uniqueness of the solution to the integral equations. Even though the closed-form solution of the identification problem is only valid for the constant coefficient case, it can help us understand the ill-posedness of the identification problem: not only can we obtain existence and uniqueness from this solution procedure, but we can also use it to construct examples to show that the solution does not depend continuously on data. To our knowledge, this is the first time that a closed-form solution formula has been presented for this kind of parameter identification problem on the quarter plane.

For the variable coefficient case, the previous closed form solution procedure does not work. Instead of proving the existence and uniqueness of the solution to the problem, we try to develop a numerical scheme to estimate the unknown, in Section 3. The usual approach is to set the estimation problem as a nonlinear least squares problem. However, this approach is very time consuming because solving a least squares problem numerically involves an iteration procedure. Usually, in each iteration, one has to solve a PDE at least once. The number of iterations needed to solve the estimation problem is usually unknown; this means that the number of PDE solutions required to solve the estimation problem satisfactorily is unknown, and may be in the hundreds or even thousands. Another difficulty in the least squares approach is that the unknowns usually number too many. Even for a simple, one-dimensional parabolic equation, the unknowns often number more than one hundred for even a very moderate estimation. These are the two main reasons that the least squares approach is too time consuming.

In this paper, we develop an approach to estimate the unknown in a much faster way. The idea is based on perturbing the parabolic equation into a hyperbolic equation with a small parameter ϵ, and then solving the estimation problem via the perturbed partial differential equation. Specifically, the numerical scheme we develop is based on solving for $f(x)$ numerically in the following equations:

$$
\begin{aligned}
\epsilon u_{tt}(x,t) + u_t(x,t) &= (a(x)u_x(x,t))_x + f(x)g(t), \\
u(x,0) &= \phi(x), \\
a(0)u_x(0,t) &= \psi(t), \\
u(0,t) &= r(t).
\end{aligned}
$$

(3)

This hyperbolic perturbation approach has been used [6,7] to solve the noncharacteristic Cauchy problem for a parabolic equation. It has also been used to estimate the coefficient $a(x)$ in a one-dimensional parabolic equation [8].

Because of the perturbation, the numerical scheme we develop is a local one. By 'local,' we mean that it works satisfactorily only in some finite interval $[0, \ell]$, with $\ell < \infty$. Since all the data we have comes only from the boundary $x = 0$, a global estimation scheme is very difficult to obtain. On the other hand, a local scheme is usually good enough in practice. Also, since the scheme we develop is a linear explicit one, its works very fast. Roughly speaking, it is as fast as solving the PDE once to estimate the parameter in $[0, \ell]$. Several numerical examples are carried out below to show the effectiveness of this approach.

2. Existence and uniqueness

In this section, we develop a procedure to solve the identification problem in a closed form for the one-dimensional parabolic equation from (1) with constant coefficient $a(x)$. Without loss of generality, we can suppose that the coefficient in the PDE is 1; i.e., in this section we will try to solve for $f(x)$ in the following equations:

(4)
$$
\begin{aligned}
u_t(x,t) &= u_{xx}(x,t) + f(x)g(t), \\
u(x,0) &= \phi(x), \\
u_x(0,t) &= \psi(t), \\
u(0,t) &= r(t),
\end{aligned}
$$

with g, ϕ, ψ, and r given. The idea here is to reduce this identification problem to two integral equations which are first type Fredholm and Volterra equations, respectively. Then, under certain sufficient conditions, we can solve these two integral equations in closed forms via the Laplace transform. First, we consider the solution to the following Fredholm equation:

(5)
$$
\frac{1}{\sqrt{\pi t}} \int_0^\infty \exp\left(-\frac{\xi^2}{4t}\right) f(\xi) d\xi = F(t).
$$

Lemma 1. *If the function*

(6)
$$
\frac{1}{\sqrt{t}} F\left(\frac{1}{4t}\right)
$$

has a Laplace inverse transformation, then the integral equation (5) has a unique solution which can be represented as follows:

(7)
$$
f(x) = x\sqrt{\pi} \mathcal{L}^{-1}\left(\frac{1}{\sqrt{t}} F\left(\frac{1}{4t}\right)\right)(s)\Big|_{s=x^2}.
$$

Proof: Let $\xi^2 = x$ and $t = (4\tau)^{-1}$. Then (5) becomes

(8)
$$
\int_0^\infty \exp(-\tau x) f(\sqrt{x}) \frac{1}{\sqrt{x}} dx = \frac{\sqrt{\pi}}{\sqrt{\tau}} F\left(\frac{1}{4\tau}\right).
$$

Hence, solving (5) is equivalent to finding the function $F_1(x) = \frac{1}{\sqrt{x}} f(\sqrt{x})$ whose Laplace transformation is $\frac{\sqrt{\pi}}{\sqrt{\tau}} F\left(\frac{1}{4\tau}\right)$. Since $\frac{1}{\sqrt{\tau}} F\left(\frac{1}{4\tau}\right)$ has an inverse Laplace transformation, we obtain

$$
\frac{1}{\sqrt{x}} f(\sqrt{x}) = \sqrt{\pi} \mathcal{L}^{-1}\left(\frac{1}{\sqrt{\tau}} F\left(\frac{1}{4\tau}\right)\right)(x),
$$

and therefore

$$(9) \qquad f(x) = x\sqrt{\pi}\mathcal{L}^{-1}\left(\frac{1}{\sqrt{\tau}}F\left(\frac{1}{4\tau}\right)\right)(x^2).$$

The uniqueness of the solution to this integral equation is due to the fact that the Laplace transform is a 1–1 mapping.

To simplify the notation, we use the following well-known functions [11]:

$$(10) \qquad K(x,t) = \frac{1}{\sqrt{4\pi t}}\exp\left(-\frac{x^2}{4t}\right),$$

$$(11) \qquad N(x,\xi,t) = K(x - \xi,t) + K(x + \xi,t).$$

Before giving our existence and uniqueness theorem, we make the following assumptions about the given functions $g(t)$, $\phi(x)$, $\psi(t)$, and $r(t)$:

Assumption 1: The function

$$(12) \qquad F_1(t) = r(t) + \int_0^t \frac{1}{\sqrt{\pi(t-\tau)}}\psi(\tau)d\tau - \frac{1}{\sqrt{\pi t}}\int_0^\infty \exp\left(-\frac{\xi^2}{4t}\right)\phi(\xi)d\xi$$

has a Laplace transformation.

Assumption 2: $g(t)$ has a Laplace transformation, and

$$(13) \qquad \mathcal{L}(g(t))(s) \neq 0.$$

On the basis of Lemma 1 and Assumptions 1 and 2, we can prove the following existence and uniqueness theorem:

Theorem 2. *Suppose* $g(t)$, $\phi(x)$, $\psi(t)$, *and* $r(t)$ *are smooth enough and satisfy Assumptions 1 and 2. If*

$$\frac{\mathcal{L}(F_1(t))(s)}{\mathcal{L}(g(t))(s)}$$

and

$$\frac{1}{\sqrt{t}}\mathcal{L}^{-1}\left(\frac{\mathcal{L}(F_1(t))(s)}{\mathcal{L}(g(t))(s)}\right)$$

have inverse Laplace transformations, then the identification problem (4) for $f(x)$ *has a unique solution, which can be represented as*

$$(14) \qquad f(x) = x\sqrt{\pi}\mathcal{L}^{-1}\left(\frac{1}{\sqrt{t}}\mathcal{L}^{-1}\left(\frac{\mathcal{L}(F_1)(s)}{\mathcal{L}(g)(s)}\right)\left(\frac{1}{4t}\right)\right)(x^2).$$

Proof: It is well known that functions $u(x,t)$ and $f(x)$ in (4) satisfy the following relationship [11]:

(15)
$$u(x,t) = -2 \int_0^t K(x, t-\tau)\psi(\tau)d\tau + \int_0^\infty N(x, \xi, t)\phi(\xi)d\xi$$
$$+ \int_0^t \int_0^\infty N(x, \xi, t-\tau)f(\xi)g(\tau)d\tau d\xi.$$

Letting $x = 0$ in (15), we have

(16)
$$r(t) = -2 \int_0^t K(0, t-\tau)\psi(\tau)d\tau + \int_0^\infty N(0, \xi, t)\phi(\xi)d\xi$$
$$+ \int_0^t \left(\int_0^\infty N(0, \xi, t-\tau)f(\xi)g(\tau)d\xi \right) d\tau.$$

Plugging (10) and (11) into (16), we see that

(17)
$$r(t) = - \int_0^t \frac{1}{\sqrt{\pi(t-\tau)}}\psi(\tau)d\tau + \frac{1}{\sqrt{\pi t}} \int_0^\infty \exp\left(-\frac{\xi^2}{4t}\right)\phi(\xi)d\xi$$
$$+ \int_0^t \left(\frac{1}{\sqrt{\pi(t-\tau)}} \int_0^\infty \exp\left(-\frac{\xi^2}{4(t-\tau)}\right)f(\xi)d\xi \right) g(\tau)d\tau.$$

Using the definition of $F_1(t)$, (12), we then obtain an integral equation which relates the unknown $f(x)$ and the other given functions:

(18)
$$\int_0^t \left(\frac{1}{\sqrt{\pi(t-\tau)}} \int_0^\infty \exp\left(-\frac{\xi^2}{4(t-\tau)}\right)f(\xi)d\xi \right) g(\tau)d\tau = F_1(t).$$

This integral equation is obviously equivalent to the following two integral equations:

(19)
$$\int_0^t y(\tau)g(t-\tau)d\tau = F_1(t),$$

(20)
$$\frac{1}{\sqrt{\pi t}} \int_0^\infty \exp\left(-\frac{\xi^2}{4t}\right)f(\xi)d\xi = y(t).$$

Using Assumptions 1 and 2 about $F_1(t)$ and $g(t)$, we obtain

(21)
$$y(t) = \mathcal{L}^{-1}\left(\frac{\mathcal{L}(F_1(t))(s)}{\mathcal{L}(g(t))(s)} \right).$$

Hence,

(22)
$$\frac{1}{\sqrt{\pi t}} \int_0^\infty \exp\left(-\frac{\xi^2}{4t}\right)f(\xi)d\xi = y(t) = \mathcal{L}^{-1}\left(\frac{\mathcal{L}(F_1(t))(s)}{\mathcal{L}(g(t))(s)} \right)(t).$$

Now, applying Lemma 1 to this first type Fredholm equation, we obtain

$$(23) \qquad f(x) = x\sqrt{\pi}\mathcal{L}^{-1}\left(\frac{1}{\sqrt{t}}\mathcal{L}^{-1}\left(\frac{\mathcal{L}(F_1(t))(s)}{\mathcal{L}(g(t))(s)}\right)\left(\frac{1}{4t}\right)\right)(x^2).$$

From this existence proof, we can easily see that the uniqueness is based on Lemma 1 and the 1-1 mapping property of the Laplace transform.

The proof of existence in the theorem above is a constructive one: it relates the unknown $f(x)$ directly to the given functions $g(t)$, $\phi(x)$, $\psi(t)$, and $r(t)$ via (12) and (14). Note that (12) and (14) do not depend on $u(x,t)$ at all; hence, we can use the procedure developed in the proof to solve the identification problem in a closed form if all the conditions required by the theorem are satisfied. The following example shows how this is done.

Example 1: Let $u(x,t)$ and $f(x)$ satisfy

$$u_t(x,t) = u_{xx}(x,t) + f(x)e^{-2t},$$
$$u(x,0) = x^2,$$
$$u_x(0,t) = 0,$$
$$u(0,t) = 0.$$

The true solution to this problem is

$$u(x,t) = e^{-2t}x^2,$$
$$f(x) = -2(x^2 + 1).$$

To solve this problem by the procedure developed above, we let

$$g(t) = e^{-2t}, \qquad \phi(x) = x^2,$$
$$\psi(x) = 0, \qquad r(t) = 0.$$

Then by (12),

$$F_1(t) = -\frac{1}{\sqrt{\pi t}}\int_0^\infty e^{-\frac{\xi^2}{4t}}\xi\,d\xi = -2t,$$

and

$$\mathcal{L}(g(t))(s) = \frac{1}{s+2},$$
$$\mathcal{L}(F_1(t))(s) = -\frac{2}{s^2},$$
$$\mathcal{L}^{-1}\left(\frac{\mathcal{L}(F_1(t))(s)}{\mathcal{L}(g(t))(s)}\right)(t) = \mathcal{L}^{-1}\left(-2\left(\frac{1}{s}+\frac{2}{s^2}\right)\right)(t)$$
$$= -2(1+2t),$$
$$\mathcal{L}^{-1}\left(\frac{\mathcal{L}(F_1(t))(s)}{\mathcal{L}(g(t))(s)}\right)\left(\frac{1}{4t}\right) = -2\left(1+\frac{1}{2t}\right).$$

Then, applying (14),

$$f(x) = x\sqrt{\pi}\mathcal{L}^{-1}\left(\frac{1}{\sqrt{t}}\mathcal{L}^{-1}\left(\frac{\mathcal{L}(F_1(t))(s)}{\mathcal{L}(g(t))(s)}\right)\left(\frac{1}{4t}\right)\right)(x^2)$$

$$= x\sqrt{\pi}\mathcal{L}^{-1}\left(-2\left(\frac{1}{\sqrt{t}} + \frac{1}{2t^{\frac{3}{2}}}\right)\right)(x^2)$$

$$= x\sqrt{\pi}\left(-2\left(\frac{1}{\sqrt{\pi s}} + \sqrt{\frac{s}{\pi}}\right)\right)\bigg|_{s=x^2}$$

$$= -2x\left(\frac{1}{x} + x\right) = -2(x^2 + 1).$$

This is the true solution to the source term identification problem.

Even though the existence and uniqueness of the identification problem are usually assured under certain conditions, the identification problem is essentially ill-posed, since solving the identification problem is equivalent to solving two integral equations of the first type according to Theorem 1. Generally speaking, Fredholm and Volterra integral equations of the first type are ill-posed since they usually do not exhibit continuous dependence on data. Hence, the ill-posedness of the identification problem comes only from the violation of the continuous dependence on data. We can easily see this from the following example.

Example 2: Let $u^n(x,t)$ and $f_n(x)$ satisfy

(24)
$$\begin{aligned}
u_t^n(x,t) &= u_{xx}^n(x,t) + f_n(x)g_n(t), \\
u^n(x,0) &= \phi_n(x), \\
u_x^n(0,t) &= \psi_n(t), \\
u^n(0,t) &= r_n(t),
\end{aligned}$$

with

$$\begin{aligned}
g_n(t) &= \cos(t) + \frac{1}{n^2}\sin(t), \\
\phi_n(x) &= 0, \\
\psi_n(t) &= \frac{1}{n}\sin(t), \\
r_n(t) &= 0.
\end{aligned}$$

By applying (14), we can easily obtain the true solution to (24):

$$f_n(x) = \sin\left(\frac{1}{n}x\right).$$

We can see immediately that:

(1) $(g_n(t), \phi_n(x), \psi_n(t), r_n(t))$ converge uniformly to

$$(g_0(t), \phi_0(x), \psi_0(t), r_0(t)) = (\cos(t), 0, 0, 0).$$

(2) The solution to the same problem with data $(g_0(t), \phi_0(x), \psi_0(t), r_0(t))$ is

$$f_0(x) = 0.$$

(3) $\{f_n(x)\}_{n=1}^{\infty}$ does not converge to $f_0(x)$ in any of the usual topologies, e.g., uniform or L^2.

Hence, the solution of this problem does not depend continuously on the data, and the problem is ill posed. We can possibly restore the continuous dependence on data of the problem by restricting the solution space or by some other technique [2,9]. We do not discuss this aspect of the identification problem further in this paper.

3. A numerical scheme for local estimation

In the last section, we proved the existence and uniqueness of the solution to the source term identification problem for the constant coefficient case under certain sufficient conditions. In the proof, we developed a procedure by which we can solve the problem in a closed form if all the conditions required by Theorem 1 are satisfied. In many applications, we cannot assume that the coefficient in the PDE is constant. A closed form solution for the source term identification problem for parabolic equations with non-constant coefficients is rather difficult even for the one-space-dimension case. Hence, practically, we have to use numerical methods to solve the problem in this case.

A frequently used approach for solving the identification problem is to formulate it as a non-linear least squares problem. If we want to get higher accuracy from one estimation in this approach, we encounter a very large, highly ill-conditioned, non-linear optimization problem [10]. Moreover, this type of optimization problem is very time-consuming if we use a gradient-type optimization procedure to solve it, since gradient-type procedures are iteration procedures in which we have to solve a PDE in every iteration in order to evaluate the functional, or the gradient, or both. Needless to say, we also have to solve a large linear system, usually with a full matrix, in each iteration. In some situations, we need more efficient estimations, even if they are rough. These observations show us the need to develop faster methods which may not require iteration.

We also observe that the numerical scheme that we wish to derive has to work locally, i.e., in a finite interval, since:
(1) Even though the problem is to identify $f(x)$ for all x in an infinite interval $[0, \infty)$, the computer can only deal with a finite problem.
(2) The problem is essentially ill-posed: it is very difficult to extract useful information about $f(x)$ at a point x far away from the left, data bearing boundary.

Hence, we will derive a local direct scheme to identify $f(x)$ for $x \in [0, 1]$, without loss of generality.

First, we present a scheme based on discretizing the parabolic equation by a finite difference method. Although this scheme is very unstable, it serves as a basis for our final scheme, which will be derived from a new PDE obtained from the original parabolic equation via a perturbation.

Let τ, h be the time and space variable steps with

$$h = \frac{1}{M}$$

and let u_j^k, $a_{j+\frac{1}{2}}$, g_k, and f_j be the approximations of $u(jh, k\tau)$, $a((j + \frac{1}{2})h)$, $g(k\tau)$, and $f(jh)$, respectively. Then, the following finite difference formula for a one-dimensional parabolic equation is standard:

$$(25) \qquad \frac{u_j^{k+1} - u_j^k}{\tau} = \frac{1}{h^2} \left[a_{j+\frac{1}{2}} \left(u_{j+1}^k - u_j^k \right) - a_{j-\frac{1}{2}} \left(u_j^k - u_{j-1}^k \right) \right] + f_j g_k.$$

The numerical schemes we develop require the following reasonable assumption:

Assumption 3: $f(0)$ is given, $g(0) > 0$, and $a(x) \geq a_0 > 0$ for some a_0.

Using Assumption 3, (25), and the extra boundary data, we obtain the first scheme as follows:

Scheme I:

(1) Compute

$$(26) \qquad u_1^k = r(k\tau) + \frac{h\psi(k\tau)}{a(0)}$$

for $k = 1, \cdots, M - 1$.

(2) Compute

$$(27) \qquad f_1 = \frac{1}{g_0} \left[\frac{u_1^1 - \phi(h)}{\tau} - \frac{1}{h^2} \left[a_{1+\frac{1}{2}}(\phi(2h) - \phi(h)) - a_{\frac{1}{2}}(\phi(h) - \phi(0)) \right] \right].$$

Then for $j = 2, \cdots, M - 1$:

3. Compute for $k = 1, \cdots, M - j$,

$$(28) \qquad u_j^k = \frac{1}{a_{j-\frac{1}{2}}} \left[h^2 \left(\frac{u_{j-1}^{k+1} - u_{j-1}^k}{\tau} - f_{j-1} g_k \right) + a_{j-1-\frac{1}{2}} \left(u_{j-1}^k - u_{j-2}^k \right) \right] + u_{j-1}^k.$$

4. Compute

$$(29) \qquad \begin{aligned} f_j = \frac{1}{g_0} \Bigg[& \frac{u_j^1 - \phi(jh)}{\tau} \\ & - \frac{1}{h^2} \left[a_{j+\frac{1}{2}}(\phi((j+1)h) - \phi(jh)) - a_{j-\frac{1}{2}}(\phi(jh) - \phi((j-1)h)) \right] \Bigg]. \end{aligned}$$

5. If $j < M - 1$, go to (3) for next j; if $j = M - 1$, stop.

Scheme I is a complete, direct scheme which allows us to estimate $f(x)$ for $x \in [0,1]$. Unfortunately, this scheme is very unstable; thus, it cannot give us a reasonable estimate of $f(x)$ outside a very small neighborhood of $x = 0$. The instability is mainly because we are computing the finite difference equation (25) for the parabolic equation in the direction of the spatial variable instead of the time direction. This motivates the use of a hyperbolic equation in deriving the numerical scheme, since the time variable and space variable play symmetric roles in a hyperbolic equation. Hence, we try to stabilize Scheme I by introducing a small second-order time derivative term in the parabolic equation; the second scheme is based on discretizing the hyperbolic equation

$$\text{(30)} \qquad \epsilon u_{tt}^{\epsilon}(x,t) + u_t^{\epsilon}(x,t) = (a(x)u_x^{\epsilon}(x,t))_x + f(x)g(t),$$

with the same initial and boundary conditions:

$$\text{(31)} \qquad \begin{aligned} u^{\epsilon}(x,0) &= \phi(x), \\ u_x^{\epsilon}(0,t) &= \psi(t), \\ u^{\epsilon}(0,t) &= r(t). \end{aligned}$$

The first question we have to answer is, what is the difference between this hyperbolic equation and the original parabolic equation in (1)? For the constant coefficient case, Eldén [7] estimates this difference via a Fourier transform. For the variable coefficient case, energy-like quantities have been used [8] to measure the difference and to estimate this difference. We present the estimates here; for details, please refer to [8].

First, we assume that the coefficient $a(x)$ satisfies

$$\text{(32)} \qquad a(x) \geq a_0 > 0,$$

and let

$$\text{(33)} \qquad U^{\epsilon}(x,t) = u^{\epsilon}(x,t) - u(x,t).$$

The difference $U^{\epsilon}(x,t)$ then satisfies the following PDE:

$$\text{(34)} \qquad \begin{aligned} \epsilon U_{tt}^{\epsilon}(x,t) + U_t^{\epsilon}(x,t) &= a(x)U_{xx}^{\epsilon}(x,t) + a_x(x)U_x^{\epsilon}(x,t) - \epsilon u_{tt}(x,t), \\ U^{\epsilon}(x,0) &= 0, \\ U^{\epsilon}(0,t) &= 0, \\ U_x^{\epsilon}(0,t) &= 0. \end{aligned}$$

The quantity used to measure the difference is defined by

$$\text{(35)} \qquad I_0^{\epsilon} = \frac{1}{2\sqrt{\epsilon} \int_x^1 \frac{1}{\sqrt{a(\xi)}} d\xi} \int_{D^{\epsilon}(x,t)}^{E^{\epsilon}(x,t)} (U^{\epsilon}(x,\tau))^2 \, d\tau,$$

where

$$E^\epsilon(x,t) = t - \int_1^x \sqrt{\frac{\epsilon}{a(\xi)}} d\xi, \quad \text{and}$$

(36)

$$D^\epsilon(x,t) = \begin{cases} t + \int_1^x \sqrt{\frac{\epsilon}{a(\xi)}} d\xi, & \text{if } t + \int_1^2 \sqrt{\frac{\epsilon}{a(\xi)}} d\xi \geq 0, \\ 0, & \text{otherwise.} \end{cases}$$

Since the characteristic equation for the perturbed equation is

(37) $$-a(x)(dt)^2 + \epsilon(dx)^2 = 0,$$

we know that $(x, E^\epsilon(x,t))$ is on one of the characteristic lines,

(38) $$C_-^\epsilon(t) : \left(x, t - \int_1^x \sqrt{\frac{\epsilon}{a(\xi)}} d\xi \right),$$

and that $(x, D^\epsilon(x,t))$ is on another characteristic line,

(39) $$C_+^\epsilon(t) : \left(x, t + \int_1^x \sqrt{\frac{\epsilon}{a(\xi)}} d\xi \right),$$

or just on the x-axis. Hence, $I_0^\epsilon(x,t)$ is a kind of average $(U^\epsilon(x,t))^2$ on the line connecting $(x, D^\epsilon(x,t))$ and $(x, E^\epsilon(x,t))$. The basic result from [8] is as follows:

Theorem 3. *If $a(x)$ and $u(x,t)$ are smooth enough and satisfy the conditions*

(40) $$a(x) \geq a_0 > 0,$$
$$\max_{0 \leq x \leq 1, t > 0} |u_{tt}(x,t)| \leq M$$

for some constants a_0 and M, then

(41) $$I_0^\epsilon \leq \left[\frac{\epsilon^2}{2} \left(\frac{x^4}{3} + \epsilon x^3 + 2\epsilon^2 x^2 \right) + \epsilon^4 x e^{x\left(\frac{1}{\epsilon}+1\right)} \right] \cdot \frac{M^2}{a_0^{2+\frac{1}{2}} \int_x^1 \frac{1}{\sqrt{a(\xi)}} d\xi},$$

for ϵ satisfying

(42) $$\frac{1}{\epsilon} \geq \frac{\max_{0 \leq x \leq 1}(|a_x(x)| + 1)}{a_0}.$$

The estimate (41) is by no means optimal, but it tells us that for any given δ, we can find a neighborhood of $x = 0$ in which I_0^ϵ is less than the δ by choosing a suitable ϵ.

The second scheme is based on using Assumption 3, the data in (31), and the following well-known finite difference formula for the one-dimensional hyperbolic equation:

(43)
$$\frac{\epsilon \left(u_j^{\epsilon k+1} - 2u_j^{\epsilon k} + u_j^{\epsilon k-1} \right)}{\tau^2} + \frac{u_j^{\epsilon k+1} - u_j^{\epsilon k-1}}{2\tau}$$
$$= \frac{1}{h^2} \left(a_{j+\frac{1}{2}} \left(u_{j+1}^{\epsilon k} - u_j^{\epsilon k} \right) - a_{j-\frac{1}{2}} \left(u_j^{\epsilon k} - u_{j-1}^{\epsilon k} \right) \right) + f_j g_k.$$

Scheme II:.

(1) Compute, for $k = 1, \cdots, M - 1$,

$$(44) \qquad u_1^{\epsilon k} = r(k\tau) + \frac{h\psi(k\tau)}{a(0)}.$$

(2) Compute

$$(45) \qquad f_1 = \frac{1}{g_0} \left[\frac{u_1^{\epsilon 1} - \phi(h)}{\tau} - \frac{1}{h^2} \left[a_{1+\frac{1}{2}}(\phi(2h) - \phi(h)) - a_{\frac{1}{2}}(\phi(h) - \phi(0)) \right] \right].$$

Then for $j = 2, \cdots, M - 1$:

3. Compute $k = 1, \cdots, M - j$,

$$(46) \qquad \begin{aligned} u_j^{\epsilon k} = \frac{1}{a_{j-\frac{1}{2}}} & \left[h^2 \left(\frac{\epsilon \left(u_{j-1}^{\epsilon k+1} - 2u_{j-1}^{\epsilon k} + u_{j-1}^{\epsilon k-1} \right)}{\tau^2} + \frac{u_{j-1}^{\epsilon k+1} - u_{j-1}^{\epsilon k-1}}{2\tau} - f_{j-1}g_k \right) \right. \\ & \left. + a_{j-1-\frac{1}{2}} \left(u_{j-1}^{\epsilon k} - u_{j-2}^{\epsilon k} \right) \right] + u_{j-1}^{\epsilon k}. \end{aligned}$$

4. Compute

$$(47) \qquad \begin{aligned} f_j = \frac{1}{g_0} & \left[\frac{u_j^{\epsilon 1} - \phi(jh)}{\tau} \right. \\ & \left. - \frac{1}{h^2} \left[a_{j+\frac{1}{2}}(\phi((j+1)h) - \phi(jh)) - a_{j-\frac{1}{2}}(\phi(jh) - \phi((j-1)h)) \right] \right]. \end{aligned}$$

5. If $j < M - 1$, go to (3) for next j; if $j = M - 1$, stop.

Note that the main computation for $u_j^{\epsilon k}$ via (46) in Scheme II is of second order in local truncation error, but that the scheme starts with (44) which is only of first order. The accuracy of Scheme II is therefore lowered because (46) is an approximation of the hyperbolic equation (30) which has an error propogation property. In order to increase accuracy, we use a new difference equation instead of (44) to start Scheme II. The new starting computation can be derived as follows. Suppose the PDE in (1) holds up to the line $x = 0$; then, by a Taylor expansion, we obtain

$$\begin{aligned} \frac{u(h, k\tau) - u(0, k\tau)}{h} &= u_x(0, k\tau) + \frac{h}{2}u_{xx}(0, k\tau) + \frac{h^2}{6}u_{xxx}(\xi, k\tau) \\ &= \frac{\psi(k\tau)}{a(0)} + \frac{h}{2a(0)}(u_t(0, k\tau) - a_x(0)u_x(0, k\tau) - f(0)g(k\tau)) \\ &\quad + \frac{h^2}{6}u_{xxx}(\xi, k\tau) \\ &= \frac{\psi(k\tau)}{a(0)} + \frac{h}{2a(0)}(r_t(k\tau) - \frac{a_x(0)}{a(0)}\psi(k\tau) - f(0)g(k\tau)) + \frac{h^2}{6}u_{xxx}(\xi, k\tau), \end{aligned}$$

and we have the more accurate difference equation

$$(48) \qquad u_1^{\epsilon k} = r(k\tau) + h \left[\frac{\psi(k\tau)}{a(0)} + \frac{h}{2a(0)} (r_t(k\tau) - a_x(0)\psi(k\tau) - f(0)g(k\tau)) \right].$$

Now, we discuss some properties of Scheme I and Scheme II through the following examples.

Example 3: Let $u(x,t)$ and $f(x)$ satisfy

$$u_t(x,t) = u_{xx}(x,t) + (25\pi^2 \sin(t) + \cos(t))f(x),$$
$$u(x,0) = 0,$$
$$u(0,t) = 0,$$
$$u_x(0,t) = 5\pi \sin(t).$$

This problem can be solved in a closed form by the procedure developed in Section 2; the true solution is

$$f(x) = \sin(5\pi x).$$

The numerical results from Scheme I and Scheme II are shown in Figure 1. Note that the computational result from Scheme I diverges from the true solution on the right half of $[0,1]$, while the result from Scheme II with $\epsilon = 0.01$ is very accurate on the whole interval.

Example 4: This example has a variable coefficient and cannot be solved by the closed form procedure. Let $u(x,t)$ and $f(x)$ satisfy

$$u_t(x,t) = ((1+x^3)u_x)_x + e^{-t}f(x),$$
$$u(x,0) = e^{-x},$$
$$u(0,t) = e^{-t},$$
$$u_x(0,t) = -e^{-t}.$$

The true solution is $f(x) = 3x^2 - 2 - x^3$. The results from both schemes are plotted in Figure 2. We see that the result from Scheme I blows up outside a small neighborhood of $x = 0$, but Scheme II with $\epsilon = 0.009$ works well up to $x = 1$.

Example 5: Even though Scheme II is derived only for $x \in [0,1]$, it is very easy to extend it to a larger interval. Here, we use Scheme II to solve the same problem as in Example 3, but on a larger interval: $[0,3]$. The result is plotted in Fig. 3. We see that the computed $f(x)$ is still very satisfactory up to $x = 3$.

Example 6: From the previous section, we know that the solution to the identification problem discussed here does depend continuously on the data. In applications, the data always contain some noise, so it would be interesting to see how random noise in the data affects the computational results. Let $u(x,t)$, $f(x)$ satisfy:

$$u_t = u_{xx} + f(x)(-2e^{-t} + \text{magn3} \cdot \text{ran3}),$$
$$u(x,0) = e^{-x} + \text{magn4} \cdot \text{ran4},$$
$$u(0,t) = e^{-t} + \text{magn1} \cdot \text{ran1},$$
$$u_x(0,t) = -e^{-t} + \text{magn2} \cdot \text{ran2},$$

where $\text{ran}i$ for $i = 1,2,3,4$ denote the random error and $\text{magn}i$ for $i = 1,2,3,4$ denote the magnitudes of the random error. It is easy to check that the true solution to this identification problem is $f(x) = e^{-x}$ if $\text{magn}i = 0$ for $i = 1,2,3,4$. In our computation, $\text{ran}i$'s are random numbers between 0 and 1 generated by a Vax-Fortran random function. In Figure 4, $\text{magn}i = 0.00001$, and in Figure 5 $\text{magn}i = 0.000001$, for $i = 1,2,3,4$. In Figures 4 and 5, we observe that the smaller the magnitudes of the random errors, the smoother the computational results from both schemes, but that the results from Scheme II are obviously better than those from Scheme I. We also observe that the error in the initial condition data $u(x,0)$ has much more influence on the numerical results than the errors in any other data. To see this, we use Scheme II to compute $f(x)$ with $\text{magn}i = 0$, for $i = 1,2,3$ and $\text{magn4} = 0.00001$. The result is plotted in Figure 6. Then we use Scheme II to compute $f(x)$ with $\text{magn}i = 0.00001$ for $i = 1,2,3$, and $\text{magn4} = 0$, and plot the result in Figure 7. The result in Figure 7 is much better than that in Figure 6.

Acknowledgments. This research was supported in part by U.S. Army Research Office Contract No. DAAG29–84–K–0002, by U.S. Air Force Office of Scientific Research Contract No. AFOSR–85–0117, by National Science Foundation Grant No. DMS–8504360, by Office of Naval Research Contract No. N00014–88–K–0370, and by the Institute for Scientific Computation through NSF Grant No. RII–8610680.

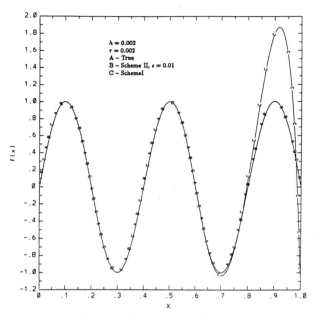

Figure 1. Results from Example 3

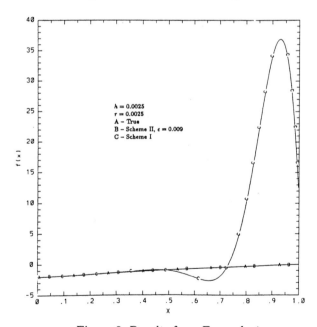

Figure 2. Results from Example 4

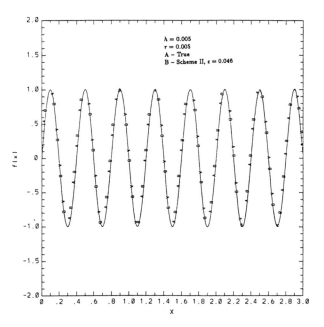

Figure 3. Results from Example 5

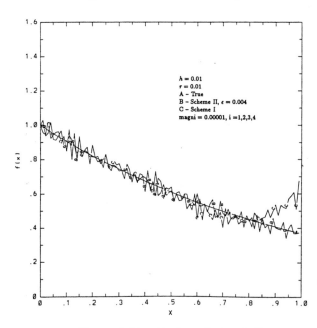

Figure 4. Results from Example 6

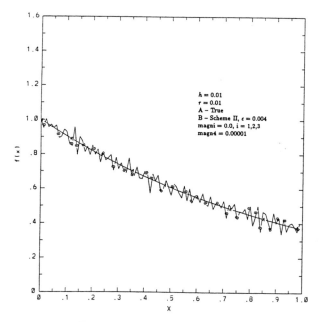

Figure 5. Results from Example 6

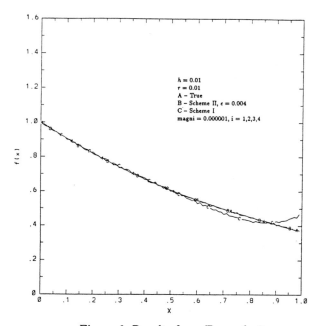

Figure 6. Results from Example 6

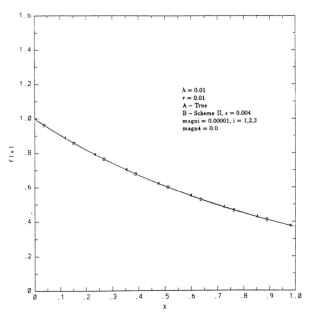

Figure 7. Results from Example 6

REFERENCES

[1] J.R. Cannon, *Some remarks on locating radiation sources,*, in "Inverse and Ill-Posed Problems," Academic Press, 1986, 315–325.

[2] J.R. Cannon and S.P. Esetva, *An inverse problem for the heat equation*, Inverse Problems **2** (1986), 395–403.

[3] J.R. Cannon and Y. Lin, *Determination of a source term in a linear parabolic differential equation with mixed boundary conditions*, in "Inverse Problems, International Series of Numerical Mathematics, vol. 77," Birkhäuser Verlag, 1986, 31–49.

[4] J.R. Cannon and R.E. Ewing, *Determination of a source term in a linear parabolic partial differential equation*, Journal of Applied Mathematics and Physics **27** (1976), 393–401.

[5] J.R. Cannon, *Determination of an unknown heat source from overspecified boundary data*, SIAM J. Numer. Anal. **5** (1968), 275–286.

[6] C.F. Weber, Int. J. Heat Mass Transfer **24** (1981), 1783–1792.

[7] L. Eldén, *Modified equation for approximating the solution of Cauchy problem for the heat equation*, in "Inverse and Ill-Posed Problems," Academic Press, 1987, 345–350.

[8] T. Lin and R.E. Ewing, *Hyperbolic perturbation method for the parameter estimation problem of parabolic equations*, (in preparation).

[9] J. Baumeister, "Stable Solution of Inverse Problems," Friedr. Vieweg Sohn, Braunschweig, 1987.

[10] R.E. Ewing, T. Lin and R. Falk, *Inverse and ill-posed problems in reservoir simulation*, in "Inverse and Ill-Posed Problems," Academic Press, 1987, 483–497.

[11] J.R. Cannon, *The one-dimensional heat equation*, in "The Encyclopedia of Mathematics and Its Applications," Addison-Wesley, 1984.

Tao Lin and Richard E. Ewing
Departments of Mathematics
Petroleum Engineering, and Chemical Engineering
University of Wyoming
Laramie, Wyoming 82071
USA

International Series of
Numerical Mathematics, Vol. 91
© 1989 Birkhäuser Verlag Basel

On numerical methods for state constrained optimal shape design problems

R. Mäkinen

Department of Mathematics
University of Jyväskylä

Abstract. We consider the numerical solution of optimal shape design problems where the state problem is linear elliptic boundary value problem. Existence of solution and convergence of approximation is studied. The theory is then applied to model problems of elasticity. Numerical examples are given.

Keywords. Constrained distributed control problem, optimal shape design problem.

1980 *Mathematics subject classifications*: 49A22, 65K10

1. Abstract setting

Let V and H denote two Hilbert spaces such that $V \subset H \subset V^*$ with compact imbeddings when H is identified with its dual.

Let $\mathcal{U}^{ad}$ be a subset of a Banach space $\mathcal{U}$. To each $\alpha \in \mathcal{U}^{ad}$ we associate a symmetric, continuous and coercive bilinear form $a(\alpha; \cdot, \cdot) : V \times V \to R$. We suppose that

(1.1) $\exists C_1 > 0$ such that $\forall \alpha \in \mathcal{U}^{ad}$ $|a(\alpha; u, v)| \leq C_1 \|u\| \|v\|$ $\forall u, v \in V$
(1.2) $\exists C_2 > 0$ such that $\forall \alpha \in \mathcal{U}^{ad}$ $a(\alpha; v, v) \geq C_2 \|v\|^2$ $\forall v \in V$.

This form defines a continuous linear operator $A(\alpha) : V \to V^*$ by

$$\big(A(\alpha)u, v\big) = a(\alpha; u, v) \quad \forall u, v \in V.$$

Similarly, we associate to each $\alpha \in \mathcal{U}^{ad}$ a linear form $L(\alpha; \cdot) : V \to R$. We suppose that

(1.3) $\exists C_3 \geq 0$ such that $\forall \alpha \in \mathcal{U}^{ad}$ $|L(\alpha; v)| \leq C_3 \|v\|$ $\forall v \in V$.

This form can also be expressed as an element of V^*

$$\big(F(\alpha), v\big) = L(\alpha; v) \quad \forall v \in V.$$

Then the variational problem

$(\mathcal{P}(\alpha))$
$$\begin{cases} \text{Find } u(\alpha) \in V \text{ such that} \\ a(\alpha; u(\alpha), v) = L(\alpha; u(\alpha)) \quad \forall v \in V \end{cases}$$

has a unique solution $u(\alpha)$ for every $\alpha \in \mathcal{U}^{ad}$ when the assumptions (1.1)-(1.2) are valid. We also assume that

(1.4) the mappings $\alpha \mapsto A(\alpha) : \mathcal{U}^{ad} \to \mathcal{L}(V, V^*)$ and $\alpha \mapsto F(\alpha) : \mathcal{U}^{ad} \to V^*$ are continuous.

This research was supported by the Academy of Finland

Let S be a non-empty subset of $\mathcal{U}^{ad} \times V$ and define the subset $\mathcal{E}^{ad} \subset \mathcal{U}^{ad}$ by

$$\mathcal{E}^{ad} = \{\alpha \in \mathcal{U}^{ad} \mid (\alpha, u(\alpha)) \in S\},$$

(assumed non-empty) where $u(\alpha)$ is the solution of $(\mathcal{P}(\alpha))$.

Now we can state our abstract optimal shape design problem:

$$(\mathbf{P}) \qquad \begin{cases} \text{Find } \alpha^* \in \mathcal{E}^{ad} \text{ such that} \\ J(\alpha^*) \leq J(\alpha) \quad \forall \alpha \in \mathcal{E}^{ad}, \end{cases}$$

where $J(\alpha) = I(\alpha, u(\alpha))$ for some real valued functional I which is defined on $\mathcal{U}^{ad} \times V$.

Theorem 1.1. *Suppose that assumptions (1.1)-(1.4) are valid. Moreover suppose that*

(1.5) $\mathcal{U}^{ad}$ *is compact in* $\mathcal{U}$,
(1.6) $I(\alpha, u)$ *is lower semicontinuous in* $\mathcal{U}^{ad} \times V$,
(1.7) $S \subset \mathcal{U}^{ad} \times V$ *is closed.*

Then the problem $(\mathbf{P})$ *has at least one solution.*

Proof: Let $\{\alpha_n\} \subset \mathcal{U}^{ad}$ be a minimizing sequence and $\alpha_n \to \alpha$ in $\mathcal{U}$. As in [**Haslinger– Neittaanmäki–Tiba,1987**] it can be proved that $u(\alpha_n) \to u$ in V for a subsequence, where u is the solution of $(\mathcal{P}(\alpha))$. As S is closed we obtain $(\alpha, u) \in S$. Then $(\alpha, u(\alpha))$ is an optimal pair for $(\mathbf{P})$.

2. Relaxing state constraints

Next the problem $(\mathbf{P})$ is approximated by a sequence of problems without state constraints.

We define real valued functionals $\Psi : \mathcal{U} \times V \to R$ and $\psi : \mathcal{U}^{ad} \to R$, $\psi(\alpha) = \Psi(\alpha, u(\alpha))$ with the following properties

(2.1) Ψ is lower semicontinuous,
(2.1) $\psi(\alpha) \geq 0 \quad \forall \alpha \in \mathcal{U}^{ad}$ and $\psi(\alpha) = 0 \iff \alpha \in \mathcal{E}^{ad}$.

We associate with the problem $(\mathbf{P})$ the relaxed problem

$$(\mathbf{P}_\varepsilon) \qquad \begin{cases} \text{Find } \alpha_\varepsilon^* \in \mathcal{U}^{ad} \text{ such that} \\ J_\varepsilon(\alpha_\varepsilon^*) \leq J_\varepsilon(\alpha) \quad \forall \alpha \in \mathcal{U}^{ad}, \end{cases}$$

where

$$J_\varepsilon(\alpha) = J(\alpha) + \frac{1}{\varepsilon}\psi(\alpha)$$

$$= I(\alpha, u(\alpha)) + \frac{1}{\varepsilon}\Psi(\alpha, u(\alpha))$$

and where $u(\alpha)$ is the solution of $(\mathcal{P}(\alpha))$.

Theorem 2.1. *Let the assumptions of Theorem 1.1 and (2.1)-(2.2) be valid. Then there exists a solution* α_ε^* *of* $(\mathbf{P}_\varepsilon)$ *for any* $\varepsilon > 0$.

Proof: is evident.

Let $\varepsilon_j \to 0$, $j \to \infty$ be a sequence of penalty parameters.

Theorem 2.2. *Let $\{\alpha_j^*\}$ be a sequence of solutions of problems $(\mathbf{P}_{\varepsilon_j})$ and $u_j := u(\alpha_j^*)$ corresponding solutions of $(\mathcal{P}(\alpha_j^*))$. Then there exist: subsequences (denoted with the same symbol) $\{\alpha_j^*\}$, $\{u_j\}$ and elements $\alpha^* \in \mathcal{E}^{ad}$ and $u \in V$ such that*

$$\alpha_j^* \to \alpha^* \text{ in } \mathcal{U}$$
$$u_j \to u \text{ in } V.$$

Moreover α^* is the solution of $(\mathbf{P})$ and $u = u(\alpha^*)$ solves $(\mathcal{P}(\alpha))$.

Proof: One can easily show the existence of a subsequence of $\{\alpha_j^*\}$ (still denoted by $\{\alpha_j^*\}$) and $\alpha^* \in \mathcal{U}^{ad}$ such that $\alpha_j^* \to \alpha^*$ in $\mathcal{U}$ and $u(\alpha_j) \to u(\alpha^*)$ in V. Next we prove that α^* solves $(\mathbf{P})$.

First we show that $\alpha^* \in \mathcal{E}^{ad}$. From the definition of $(\mathbf{P}_{\varepsilon_j})$ it follows that

(2.2)
$$J(\alpha_j^*) + \frac{1}{\varepsilon_j}\psi(\alpha_j^*) \le J(\alpha) + \frac{1}{\varepsilon_j}\psi(\alpha) \quad \forall \alpha \in \mathcal{U}^{ad}.$$

Substituting an element α of $\mathcal{E}^{ad}$ $(\mathcal{E}^{ad} \ne \emptyset)$ into the righthand side of (2.2) we see that

$$J(\alpha_j^*) + \frac{1}{\varepsilon_j}\psi(\alpha_j^*) \le J(\alpha).$$

As α is fixed,
$$0 \le \psi(\alpha_j^*) \le \varepsilon_j(J(\alpha) - J(\alpha_j^*)) \to 0 \text{ as } j \to \infty.$$

Thus $\psi(\alpha^*) = 0$ and consequently $\alpha^* \in \mathcal{E}^{ad}$.

Finally
$$J(\alpha_j^*) \le J(\alpha_j^*) + \frac{1}{\varepsilon_j}\psi(\alpha_j^*) \le J(\alpha) \quad \forall \alpha \in \mathcal{E}^{ad}.$$

Letting $j \to \infty$ we see that $J(\alpha^*) \le J(\alpha)$.

3. Finite dimensional approximation of the relaxed problem

Let us associate for each $h \in (0, 1)$

- a (finite dimensional) Hilbert space $V_h \subset V$
- a (finite dimensional) set of admissible controls $\mathcal{U}_h^{ad} \subset \mathcal{U}$
- a functional $I_h : \mathcal{U}_h^{ad} \times V_h \to R$ (approximation of I)
- a functional $\Psi_h : \mathcal{U}_h^{ad} \times V_h \to R$ (approximation of Ψ).

Consider the following problem

$(\mathbf{P}_{\varepsilon h})$
$$\begin{cases} \text{Find } \alpha_{\varepsilon h}^* \in \mathcal{U}_h^{ad} \text{ such that} \\ J_{\varepsilon h}(\alpha_{\varepsilon h}^*) \le J_{\varepsilon h}(\alpha_h) \quad \forall \alpha_h \in \mathcal{U}_h^{ad}. \end{cases}$$

Here $J_{\varepsilon h}(\alpha_h) = I_h(\alpha_h, u_h(\alpha_h)) + \frac{1}{\varepsilon}\Psi_h(\alpha_h, u_h(\alpha_h))$ and $u_h(\alpha_h)$ is the unique solution of the approximate state problem

$$(\mathcal{P}_h(\alpha_h)) \qquad\qquad a(\alpha_h; u_h(\alpha_h), v_h) = L(\alpha_h; v_h) \quad \forall v_h \in V_h.$$

Remark. In $(\mathcal{P}_h(\alpha_h))$ no approximation is used for the bilinear and linear form $a(\alpha; \cdot, \cdot)$ and $L(\alpha; \cdot, \cdot)$ (exact numerical integration).

Theorem 3.1. *Let the following hypotheses be satisfied:*

(3.1) $\mathcal{U}_h^{ad}$ *is compact in $\mathcal{U}$ $\forall h \in (0,1)$*

(3.2) *For every sequence $\{\alpha_h\}$, $\alpha_h \in \mathcal{U}_h^{ad}$ $\forall h$ there exists a converging subsequence and the limit belongs to $\mathcal{U}^{ad}$. Conversely, for every $\alpha \in \mathcal{U}^{ad}$ there is a sequence $\alpha_h \to \alpha$, $\alpha_h \in \mathcal{U}_h^{ad}$.*

(3.3) I_h *and Ψ_h are lower semicontinuous in $\mathcal{U}_h^{ad} \times V_h$ $\forall h \in (0,1)$ and for all sequences $\alpha_h \to \alpha$ in $\mathcal{U}$ $u_h \to u$ in V*

$$\lim_{h \to 0} I_h(\alpha_h, u_h) = I(\alpha, u)$$

$$\lim_{h \to 0} \Psi_h(\alpha, u_h) = \Psi(\alpha, u)$$

(3.4) *If $\alpha_h \to \alpha$, $u_h \rightharpoonup u$ (weakly) and $v_h \to v$ then $a(\alpha_h; u_h, v_h) \to a(\alpha; u, v)$*

(3.5) *If $\alpha_h \to \alpha$ and $v_h \rightharpoonup v$ then $\lim_{h \to 0} L(\alpha_h; v_h) = L(\alpha; v)$.*

Then for each h fixed, there exists an optimal pair $(\alpha_{\varepsilon h}^*, u_h(\alpha_{\varepsilon h}^*))$ of $(\mathbf{P}_{\varepsilon h})$. Moreover there exists a subsequence of $\{\alpha_{\varepsilon h}^*\}_{h>0}$ denoted with the same symbol such that

$$\alpha_{\varepsilon h}^* \to \alpha_\varepsilon^* \text{ in } \mathcal{U}$$

$$u_h(\alpha_{\varepsilon h}^*) \to u(\alpha_\varepsilon^*) \text{ in } V,$$

where $(\alpha_\varepsilon^*, u(\alpha_\varepsilon^*))$ is an optimal pair of $(\mathbf{P}_\varepsilon)$.

Proof: For each $h \in (0,1)$ fixed, the assumptions of Theorem 1.1 are satisfied and, consequently there exists an optimal pair $(\alpha_{\varepsilon h}^*, u_h(\alpha_{\varepsilon h}^*))$. By (3.2) there exists a subsequence $\{\alpha_{\varepsilon h}^*\}_{h>0}$ such that $\alpha_{\varepsilon h}^* \to \alpha_\varepsilon \in \mathcal{U}^{ad}$. The sequence $u_h(\alpha_{\varepsilon h}^*)$ is bounded in V by virtue of (1.2)–(1.3). Thus we can choose a subsequence of $u_h(\alpha_{\varepsilon h}^*)$ that converges weakly to some $u \in V$. Let $v \in V$ be fixed. Because V_h is an internal approximation of V one can find $v_h \in V_h$ such that $v_h \to v$ in V. Now

$$a(\alpha_{\varepsilon h}^*; u_h(\alpha_{\varepsilon h}^*), v_h) = L(\alpha_{\varepsilon h}^*; v_h).$$

Using (3.4)–(3.5) and passing to the limit we obtain

$$a(\alpha_\varepsilon; u, v) = L(\alpha_\varepsilon; v).$$

As v was arbitrarily chosen it follows that $u = u(\alpha_\varepsilon)$.

Next we show that $u_h(\alpha^*_{\varepsilon h}) \to u(\alpha^*_\varepsilon)$ (strongly):

$$C_2\|u_h - z_h\|^2 \le a(\alpha_h; u_h - z_h, u_h - z_h)$$
$$= L(\alpha_h; u_h - z_h) - a(\alpha_h; z_h, u_h - z_h) \to 0 \text{ as } h \to 0,$$

making use of (3.4)-(3.5). Here $z_h \in V_h$ was chosen such that $z_h \to u$ in V.

It remains to show that α^*_ε is a minimizer of J_ε. Let $\bar{\alpha} \in \mathcal{U}^{ad}$ be arbitrary and let $\bar{\alpha}_h \to \bar{\alpha}$ be the (sub)sequence given by (3.2). Using the same technique as above we can obtain that $u_h(\bar{\alpha}_h) \to u(\bar{\alpha})$. Thus we have from (3.3) that

$$I(\alpha^*_\varepsilon, u(\alpha^*_\varepsilon)) + \frac{1}{\varepsilon}\Psi(\alpha^*_\varepsilon, u(\alpha^*_\varepsilon))$$
$$= \lim_{h \to 0}\left(I_h(\hat{\alpha}_h, u_h(\hat{\alpha}_h)) + \frac{1}{\varepsilon}\Psi_h(\hat{\alpha}_h, u_h(\hat{\alpha}_h))\right)$$
$$\le \lim_{h \to 0}\left(I_h(\bar{\alpha}_h, u_h(\bar{\alpha}_h)) + \frac{1}{\varepsilon}\Psi_h(\bar{\alpha}_h, u_h(\bar{\alpha}_h))\right)$$
$$= I(\bar{\alpha}, u(\bar{\alpha})) + \frac{1}{\varepsilon}\Psi(\bar{\alpha}, u(\bar{\alpha})) \quad \forall \bar{\alpha} \in \mathcal{U}^{ad}.$$

4. On the numerical realization of the problem ($\mathbf{P}_{\varepsilon h}$)

As the space V_h and the set $\mathcal{U}^{ad}_h$ are finite dimensional, functions α_h and u_h are completely determined by vectors $\mathbf{a} = (a_1, ..., a_M)$ and $\mathbf{u} = (u_1, ..., u_N)$. Here $M = M(h) = \dim \mathcal{U}^{ad}_h$ and $N = N(h) = \dim V_h$.

The problem ($\mathbf{P}_{\varepsilon h}$) in matrix form reads now

(4.1) $\qquad \begin{cases} \text{Find } \mathbf{a}^* \in U \text{ (the set of admissible design vectors) such that} \\ \mathcal{J}_\varepsilon(\mathbf{a}^*) \le \mathcal{J}_\varepsilon(\mathbf{a}) \quad \forall \mathbf{a} \in U. \end{cases}$

Here $\mathcal{J}_\varepsilon(\mathbf{a}) = \mathcal{I}(\mathbf{a}, \mathbf{u}) + \frac{1}{\varepsilon}\tilde{\Psi}(\mathbf{a}, \mathbf{u})$ and $\mathbf{u} = \mathbf{u}(\mathbf{a})$ is obtained as the solutions of the linear system of equations

(4.2) $$\mathbf{K}(\mathbf{a})\mathbf{u} = \mathbf{f}(\mathbf{a}).$$

The problem (4.1) can now be solved using standard nonlinear programming algorithms.

Efficient nonlinear programming algorithms need the gradient of $\mathcal{J}_\varepsilon$. Partial derivatives with respect to design variable a_k can be approximated using finite differences i.e.

(4.3) $$\frac{\partial \mathcal{J}_\varepsilon(\mathbf{a})}{\partial a_k} \approx \frac{\mathcal{J}_\varepsilon(\mathbf{a} + \delta \mathbf{e}_k) - \mathcal{J}_\varepsilon(\mathbf{a})}{\delta}, \quad \delta > 0$$

or computed exactly, calculating analytical formulas for them using the classical adjoint state technique:

Theorem 4.1. *If the mappings* $\mathbf{a} \mapsto \mathbf{K}(\mathbf{a})$ *and* $\mathbf{a} \mapsto \mathbf{f}(\mathbf{a})$ *are smooth enough, then*

$$(4.4) \qquad \frac{\partial \mathcal{J}_\varepsilon(\mathbf{a})}{\partial a_k} = \frac{\partial \mathcal{J}_\varepsilon(\mathbf{a}, \mathbf{u})}{\partial a_k} + \mathbf{p}^T \left(\frac{\partial \mathbf{f}(\mathbf{a})}{\partial a_k} - \frac{\partial \mathbf{K}(\mathbf{a})}{\partial a_k} \mathbf{u} \right)$$

where $\mathbf{p}$ *is the solution of the adjoint equation*

$$(4.5) \qquad \qquad \mathbf{K}(\mathbf{a})\mathbf{p} = \nabla_{\mathbf{u}} \mathcal{J}_\varepsilon(\mathbf{a}, \mathbf{u})$$

Proof: Using the implicit function theorem we can differentiate the state equation (4.2) to obtain

$$(4.6) \qquad \qquad \mathbf{K}(\mathbf{a}) \frac{\partial \mathbf{u}}{\partial a_k} = \frac{\partial \mathbf{f}(\mathbf{a})}{\partial a_k} - \frac{\partial \mathbf{K}(\mathbf{a})}{\partial a_k} \mathbf{u}.$$

Then

$$\begin{aligned} \frac{\partial \mathcal{J}_\varepsilon(\mathbf{a})}{\partial a_k} &= \frac{\partial \mathcal{J}_\varepsilon(\mathbf{a}, \mathbf{u})}{\partial a_k} + (\nabla_{\mathbf{u}} \mathcal{J}_\varepsilon)^T \frac{\partial \mathbf{u}}{\partial a_k} \\ &= \frac{\partial \mathcal{J}_\varepsilon(\mathbf{a}, \mathbf{u})}{\partial a_k} + (\mathbf{K}(\mathbf{a})^{-1} \nabla_{\mathbf{u}} \mathcal{J}_\varepsilon)^T \left(\frac{\partial \mathbf{f}(\mathbf{a})}{\partial a_k} - \frac{\partial \mathbf{K}(\mathbf{a})}{\partial a_k} \mathbf{u} \right) \\ &= \frac{\partial \mathcal{J}_\varepsilon(\mathbf{a}, \mathbf{u})}{\partial a_k} + \mathbf{p}^T \left(\frac{\partial \mathbf{f}(\mathbf{a})}{\partial a_k} - \frac{\partial \mathbf{K}(\mathbf{a})}{\partial a_k} \mathbf{u} \right) \end{aligned}$$

making use of (4.5) and (4.6).

Remark. It can be seen that evaluation of the gradient of the cost functional at a given point requires M additional solutions of the state problem when (4.3) is used for gradient calculations. On the other hand (4.4) requires only solution of the adjoint problem. Therefore (4.4)–(4.5) are superior when efficiency of the algorithm is concerned. On the other hand programming of (4.4)–(4-5) may be very tedious. For further study of sensitivity analysis of optimal shape design problems we refer to monograph [**Haslinger–Neittaanmäki**,1988].

5. Applications

5.1. Weight minimization of a beam.

5.1.1. Setting of the problem. Let

$$\begin{aligned} &\Omega = (0, L) \subset R, \quad \mathcal{U} = C^{0,1}(\Omega), \quad V = H_0^2(\Omega), \\ &\mathcal{U}^{ad} = \{ \alpha \in \mathcal{U} \mid \alpha_0 \le \alpha(x) \le \alpha_1, \quad |\alpha'(x)| \le c_0 \text{ a.e. in } \Omega \}, \\ &A(\alpha)u = (b\alpha^3 u'')'' \quad \text{and} \\ &F(\alpha) = \rho \alpha + f, \quad \rho \in R. \end{aligned}$$

This is the model of a clamped Euler-Bernoulli beam loaded by the force f and body force $\rho \alpha$. $\alpha(x)$ is the thickness of the beam, $b > 0$ is a constant depending on the shape of the cross section of the beam and material constants.

Let the cost function be $J(\alpha) = \int_\Omega \alpha(x)\, dx$, i.e. we have the classical weight minimization problem.

Utilizing the theorem of Arzela-Ascoli we see that $\mathcal{U}^{ad} \subset \mathcal{U}$ is compact. It is easy to see that the assumptions (1.1)–(1.4) and (1.6) are also satisfied.

As the first example we consider the case where the maximum displacement of the beam is required to stay in certain limits i.e. $S = \mathcal{U}^{ad} \times K$ where K is the convex, closed set

$$K = \{v \in V \mid \|v\|_{\infty,\Omega} \le r\}, \quad r > 0.$$

As K is closed, the assumption (1.7) follows.

As the second example we consider the case where the bending stress in the beam may not exeed a given value σ i.e.

$$S = \{(\alpha, v) \in \mathcal{U}^{ad} \times V \mid \|\alpha v''\|_{\infty,\Omega} \le \sigma\}.$$

Let us check that assumption (1.7) is satisfied: Let $\alpha_n \in \mathcal{E}^{ad}$, $\alpha_n \to \alpha$ in $\mathcal{U}$. Then it follows from the proof of Theorem 1.1 that $u(\alpha_n) \to u(\alpha)$ in V. We denote $\langle \cdot \rangle = \max(\cdot, 0)$. Now

$$
\begin{aligned}
0 \le 2\int_\Omega \langle \alpha u(\alpha)'' - \sigma \rangle \, dx &= \int_\Omega \left(|\alpha u(\alpha)'' - \sigma| + (\alpha u(\alpha)'' - \sigma) \right) dx \\
&\le \int_\Omega |\alpha u(\alpha)'' - \alpha_n u(\alpha_n)''| \, dx + \int_\Omega (\alpha u(\alpha)'' - \alpha_n u(\alpha_n)'') \, dx + 2\int_\Omega \langle \alpha_n u(\alpha_n)'' - \sigma \rangle \, dx.
\end{aligned}
$$

As $\alpha_n \in \mathcal{E}^{ad}$ it follows that $\int_\Omega \langle \alpha_n u(\alpha_n)'' - \sigma \rangle \, dx = 0$. On the other hand

$$
\int_\Omega |\alpha u(\alpha)'' - \alpha_n u(\alpha_n)''| \, dx \le c_1 \|\alpha - \alpha_n\|_{\infty,\Omega} \|u(\alpha)''\|_{0,\Omega}
$$
$$
+ c_2 \|\alpha_n\|_{\infty,\Omega} \|u(\alpha)'' - u(\alpha_n)''\|_{0,\Omega} \to 0 \text{ as } n \to \infty.
$$

Therefore $\alpha u(\alpha)'' \le \sigma$ a.e. in $\bar\Omega$. Similarly we can show that $-\alpha u(\alpha)'' \le \sigma$ a.e. in $\bar\Omega$.

For another type of shape optimization problems for beams see [**Haug–Arora**,1979], [**Hlavaček–Bock–Lovisek**,1984] and [**Mäkinen**,1989].

5.1.2. Penalization. In the first case we can choose

$$\Psi(\alpha, u(\alpha)) = \int_\Omega \langle u(\alpha) - r \rangle^2 \, dx + \int_\Omega \langle -u(\alpha) - r \rangle^2 \, dx$$

and in the second case

$$\Psi(\alpha, u(\alpha)) = \int_\Omega \langle \alpha u(\alpha)'' - \sigma \rangle^2 \, dx + \int_\Omega \langle -\alpha u(\alpha)'' - \sigma \rangle^2 \, dx.$$

5.1.3. Discretization. Let $0 = a_0 < a_1 < ... < a_{N(h)+1} = L$ be a partition of the interval $[0, L]$. The points a_i are called nodes and the intervals $T_i = [a_{i-1}, a_i]$, $i = 1, ..., N+1$ are called elements. We denote $h_i = a_i - a_{i-1}$ and we suppose that $\max h_i / \min h_i \le \beta$, where

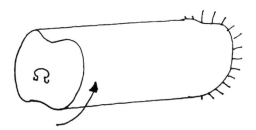

Figure 5.1

$\beta > 0$ is a given absolute constant. We use the notations of Section 5.1.1 and we consider the following approximations

$$V_h = \{v_h \in C^1 \mid v_h|_{T_i} \in P_3\} \subset V$$
$$U_h = \{u_h \in C^0 \mid u_h|_{T_i} \in P_1\} \subset \mathcal{U}, \quad \mathcal{U}_h^{ad} = U_h \cap \mathcal{U}^{ad},$$

where P_3 and P_1 are the sets of cubic and linear polynomials respectively. The functionals I_h and Ψ_h are evaluated using some suitable quadrature rules.

The hypotheses (3.1)–(3.2) are satisfied (see [**Begis-Glowinski**,1975], [**Tiihonen**,1987]). To obtain (3.3) we write

$$\Psi_h(\alpha_h, u_h) - \Psi(\alpha, u) = \Psi_h(\alpha_h, u_h) - \Psi(\alpha_h, u_h) + \Psi(\alpha_h, u_h) - \Psi(\alpha, u).$$

In the first case $\Psi_h(\alpha_h, u_h) = \Psi_h(u_h)$. As $u, u_h \in C^1$, it is not difficult to find a quadrature rule such that $|\Psi_h(u_h) - \Psi(u_h)| \le ch\|u'\|_\infty$. In the second case the integrand is still a piecewise polynomial of low order. Therefore $\Psi_h(\alpha_h, u_h)$ can be exactly evaluated (i.e. $\Psi_h = \Psi$). Moreover in both cases $I_h = I$.

Finally

$$a(\alpha_h; u_h, v_h) =$$
$$= \int_\Omega b(\alpha_h^3 - \alpha^3) u_h'' v_h'' \, dx + \int_\Omega b\alpha^3 u_h''(v_h'' - v) \, dx + \int_\Omega b\alpha^3 u_h'' v'' \, dx$$
$$\rightarrow a(\alpha; u, v).$$

Similarly $L(\alpha; v_h) \rightarrow L(\alpha; v)$. Thus (3.4)-(3.5) hold.

5.2. Weight minimization of a bar.

5.2.1. Setting of the problem. Consider the problem of torsion of an elastic cylindrical bar $\Omega \times (0, L) \subset R^3$ (see Figure 5.1). The material of the bar is assumed homogeneous and isotropic.

From St. Venant theory of torsion it follows that the deformation of the system is governed by the boundary value problem

$$\begin{cases} -\Delta u = 2 & \text{in } \Omega \\ \quad u = 0 & \text{on } \partial\Omega, \end{cases}$$

where u is the Prandtl stress potential. The stress field $\{\sigma_{ij}\}_{i,j=1}^3$ satisfies

$$\sigma_{ij} = 0 \quad \text{exept for } \sigma_{31} = \sigma_{13} = \mu\Theta\frac{\partial u}{\partial x_2}, \quad \sigma_{32} = \sigma_{23} = -\mu\Theta\frac{\partial u}{\partial x_1}.$$

Here Θ is the angular deflection per unit length and $\mu > 0$ is a material constant.

The angular deflection torque relation is given by $\Theta = \frac{T}{\mu R}$, where R is the torsional rigidy, given by $R = 2\int_\Omega u\, dx$. At each $x \in \Omega$ the stress field $\{\sigma_{ij}\}$ can be characterized by the von Mises equivalent stress which reduces to

$$\sigma_e = \tilde{\mu}\Theta|\nabla u|, \quad \tilde{\mu} = \text{const} > 0.$$

The classical optimization problem related to this model is to find the shape of the cross section with given area such that R is maximized (see [**Banichuk**,1976]). Instead of this we state the following optimal shape design problem: Find the shape of the cross section Ω such that the weight of the bar is minimized subject to the constraint that the equivalent stress does not exeed a given value.

We restrict ourselves to the following special geometry. Let the cross section of the bar be symmetric with respect to the x_2-axis (see Figure 5.2). Then only a half of the cross section of the bar is analyzed.

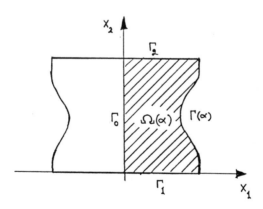

Figure 5.2

Let α be a Lipschitz continuous and strictly postive function on $[0, s]$, $s > 0$. The α defines a domain $\Omega(\alpha)$ by

$$\Omega(\alpha) = \{x \in R^2 \mid 0 < x_1 < \alpha(x_2), \ 0 < x_2 < s\}.$$

The boundary $\partial\Omega(\alpha)$ consists of four parts

$$\begin{aligned}
\Gamma(\alpha) &= \{x \in R^2 \mid x_1 = \alpha(x_2), \ x_2 \in [0, s]\}, \\
\Gamma_0 &= \{0\} \times (0, s), \\
\Gamma_1 &= [0, \alpha(0)] \times \{0\}, \\
\Gamma_2 &= [0, \alpha(0)] \times \{s\}.
\end{aligned}$$

Let $\mathcal{U} = W^{1,\infty}([0, s])$ and

$$\mathcal{U}^{ad} = \{\alpha \in W^{2,\infty}([0, s]) \mid 0 < \alpha_0 \leq \alpha \leq \alpha_1, \ |\alpha'| \leq c_0, \ |\alpha''| \leq c_1 \text{ a.e. in } (0, s)\}.$$

For each $\alpha \in \mathcal{U}^{ad}$ we define the state problem

$(\mathcal{P}(\alpha))$
$$\begin{cases}
-\Delta u(\alpha) = 2 \text{ in } \Omega(\alpha) \\
\ u(\alpha) = 0 \text{ on } \Gamma_1 \cup \Gamma_2 \cup \Gamma(\alpha) \\
\ \dfrac{\partial u(\alpha)}{\partial n} = 0 \text{ on } \Gamma_0
\end{cases}$$

The problem $(\mathcal{P}(\alpha))$ has an unique (weak) solution $u(\alpha)$ for each α. The shape optimization problem reads now

$(\mathbf{P}(\alpha))$
$$\begin{cases}
\text{Find } \alpha^* \in \mathcal{E}^{ad} \text{ such that} \\
J(\alpha^*) \leq J(\alpha) \quad \forall \alpha \in \mathcal{E}^{ad},
\end{cases}$$

where

$$J(\alpha) = \int_{\Omega(\alpha)} dx$$

and

$$\begin{aligned}
\mathcal{E}^{ad} &= \{\alpha \in \mathcal{U}^{ad} \mid (\alpha, u(\alpha)) \in S\}, \\
S &= \{(\alpha, v) \in \mathcal{U}^{ad} \times V(\alpha) \mid \|\nabla v\|_{\infty, \Omega(\alpha)} \leq cR(\alpha)\}, \\
V(\alpha) &= \{v \in H^1(\Omega(\alpha)) \mid v|_{\Gamma_1 \cup \Gamma_2 \cup \Gamma(\alpha)} = 0\}, \\
c &= \text{constant independent of } \alpha \text{ and } v.
\end{aligned}$$

In order to be able to utilize the theory of Chapter 1, following [**Begis–Glowinski,1975**], we pose the problem in the fixed domain $\hat{\Omega} = (0, 1) \times (0, s)$. Let $\hat{V} = \{\hat{v} \in H^1(\hat{\Omega}) \mid \hat{v} = 0 \text{ on } \partial\hat{\Omega} \setminus \hat{\Gamma}_0\}$, where

$$\begin{aligned}
\hat{\Gamma} &= \{1\} \times (0, s), \\
\hat{\Gamma}_0 &= \{0\} \times (0, s), \\
\hat{\Gamma}_1 &= [0, 1] \times \{0\}, \\
\hat{\Gamma}_2 &= [0, 1] \times \{s\}.
\end{aligned}$$

For each $\alpha \in \mathcal{U}^{ad}$ there exists a bijection $T = T(\alpha) : \hat{\Omega} \to \Omega(\alpha)$ defined by

$$T(\alpha)(\hat{x}_1, \hat{x}_2) = (\alpha(\hat{x}_2)\hat{x}_1, \hat{x}_2)$$

and an associated isomorphism between $V(\alpha)$ and $\hat{V}$

$$\hat{v}(\hat{x}) = v(T(\alpha)\hat{x}).$$

The Jacobian of the transformation is given by

$$J(\alpha)(\hat{x}_1, \hat{x}_2) = \begin{pmatrix} \alpha(\hat{x}_2) & \hat{x}_1\alpha'(\hat{x}_2) \\ 0 & 1 \end{pmatrix}$$

and $|\det J(\alpha)| = \alpha$.

The shape optimization problem reads now

$$\begin{cases} \text{Find } \alpha^* \in \hat{\mathcal{E}}^{ad} \text{ such that} \\ \hat{J}(\alpha^*) \le \hat{J}(\alpha) \quad \forall \alpha \in \hat{\mathcal{E}}^{ad}, \end{cases}$$

where $\hat{J}(\alpha) = \int_0^s \alpha(\hat{x}_2)\, d\hat{x}_2$. The set of admissible controls is given by $\hat{\mathcal{E}}^{ad} = \{\alpha \in \mathcal{U}^{ad} \mid (\alpha, \hat{u}(\alpha)) \in \hat{S}\}$, where $\hat{u}(\alpha)$ is the solution of the state problem

$$\begin{cases} \text{Find } \hat{u}(\alpha) \in \hat{V} \text{ such that} \\ a(\alpha; \hat{u}(\alpha), v) = L(\alpha; v) \quad \forall v \in \hat{V} \end{cases}$$

and $\hat{S} = \{(\alpha, \hat{v}) \in \mathcal{U}^{ad} \times \hat{V} \mid \|J(\alpha)^{-T}\nabla\hat{v}\|_{\infty,\hat{\Omega}} \le c\hat{R}(\alpha)\}$.

Here we have denoted

$$a(\alpha; u, v) = \int_{\hat{\Omega}} \left(J(\alpha)^{-T}\nabla u\right) \cdot \left(J(\alpha)^{-T}\nabla v\right) \alpha\, d\hat{x},$$

$$L(\alpha; v) = \int_{\hat{\Omega}} 2v\alpha\, d\hat{x} \quad \text{and} \quad \hat{R}(\alpha) = \int_{\hat{\Omega}} \hat{u}(\alpha)\alpha\, d\hat{x}.$$

The assumptions (1.1)–(1.5) of Theorem 1.1 are satisfied (see [**Tiihonen**,1987]). Also (1.6) is satisfied. Assumption (1.7) follows as in Section 5.1 using the facts

$$\alpha_n \to \alpha \text{ in } \mathcal{U},$$
$$J(\alpha)^{-T} \to J(\alpha)^{-T} \text{ in } \left(L^\infty(\hat{\Omega})\right)^4,$$
$$\hat{u}(\alpha_n) \to \hat{u}(\alpha) \text{ in } \hat{V}$$

and

$$\hat{R}(\alpha_n) \to \hat{R}(\alpha).$$

Then the existence of at least one solution follows.

5.2.2. Penalization. The state constraint can be relaxed by introducing the following penalty functional

$$\Psi(\alpha, \hat{u}(\alpha)) = \int_{\hat{\Omega}} \left\langle |J(\alpha)^{-T} \nabla \hat{u}(\alpha)|^2 - c^2 \hat{R}(\alpha)^2 \right\rangle^2 \, d\hat{x}.$$

5.2.3. Discretization. Let $\hat{\Omega}$ be divided into $2N^2$ triangles of equal size and form. By S_h, $h = 1/N$ we denote the set of functions which are continuous in $\hat{\Omega}$ and linear polynomials within each triangle. We denote $\hat{V}_h = \{v_h \in S_h \mid v_h = 0 \text{ on } \partial\hat{\Omega} \setminus \hat{\Gamma}_0\}$. It is easy to see that $\hat{V}_h \subset \hat{V}$, $\forall h > 0$. A finite dimensional approximation of the control set $\mathcal{U}^{ad}$ is given by

$$\mathcal{U}_h^{ad} = \{\alpha_h \in W^{1,\infty}([0,1]) \mid \alpha_h|_{[\frac{i-1}{N}, \frac{i}{N}]} \in P_1, \ i = 1, ..., N,$$

$$\alpha_h(\frac{i}{N}) = \alpha(\frac{i}{N}), \ i = 0, ..., N \quad \text{for some } \alpha \in \mathcal{U}^{ad}\}.$$

The hypothesis (3.1), the second part of (3.2) and (3.3)–(3.5) can be verified in the same way as in the Section 5.1. Let us have a sequence $\{\alpha_{h_j}\}$, $\alpha_{h_j} \in \mathcal{U}_{h_j}^{ad}$. By the definition of $\mathcal{U}_h^{ad}$, we can find a sequence $\{\alpha^j\}$, $\alpha^j \in \mathcal{U}^{ad}$ such that

$$\alpha^j(\frac{i}{N(h_j)}) = \alpha_{h_j}(\frac{i}{N(h_j)}), \quad i = 0, ..., N(h_j).$$

The sequence $\{\alpha^j\}$ has a converging subsequence with a limit $\alpha \in \mathcal{U}^{ad}$. Now by interpolation results and by the definition of $\mathcal{U}^{ad}$ we have

$$\|\alpha_{h_j} - \alpha\|_{1,\infty} \leq \|\alpha_{h_j} - \alpha^j\|_{1,\infty} + \|\alpha^j - \alpha\|_{1,\infty}$$

$$\leq h_j \|(\alpha^j)''\|_\infty + \|\alpha^j - \alpha\|_{1,\infty} \leq c_1 h_j + \|\alpha^j - \alpha\|_{1,\infty},$$

so that $\alpha_{h_j} \to \alpha$ in $\mathcal{U}$ for a subsequence. This gives the first part of (3.2).

6. Numerical results

6.1. The beam bending problem. Let $\mathcal{U}^{ad} = \{\alpha \in C^{0,1} \mid \frac{1}{10} \leq \alpha \leq 1, |\alpha'| \leq c_0\}$. The problem is discretized using 20 elements. Therefore we have 21 design variables.

In Figure 6.1 we see the final designs for weight minimization problem with displacement constraint $\|u(\alpha)\|_\infty \leq \frac{1}{100}$ when the Lipschitz constant is varied.

$$\mathcal{U}^{ad} = \{\alpha \in C^{0,1}([0,1]) \mid 0.1 \leq \alpha \leq 1, \quad |\alpha'| \leq c_0\}$$
$$+ \text{displacement constraint} \quad \|u(\alpha)\|_\infty \leq 0.01$$

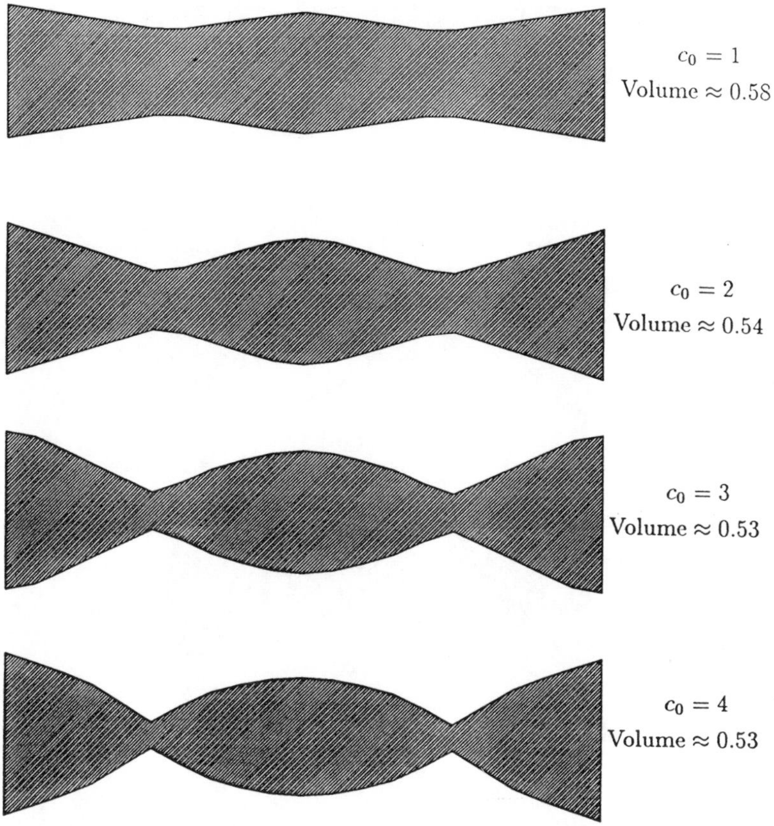

$c_0 = 1$
Volume ≈ 0.58

$c_0 = 2$
Volume ≈ 0.54

$c_0 = 3$
Volume ≈ 0.53

$c_0 = 4$
Volume ≈ 0.53

Figure 6.1

As the second example we consider the problem with $c_0 = 5$ and with stress constraint $\|\alpha u(\alpha)''\|_\infty \leq \frac{1}{10}$. In Figure 6.2 we see the final designs with different discretization parameters.

$$\mathcal{U}^{ad} = \{\alpha \in C^{0,1}([0,1]) \mid 0.1 \le \alpha \le 1, \quad |\alpha'| \le 5\}$$
$$+ \text{ stress constraint } \|\alpha u(\alpha)''\|_\infty \le 0.1$$

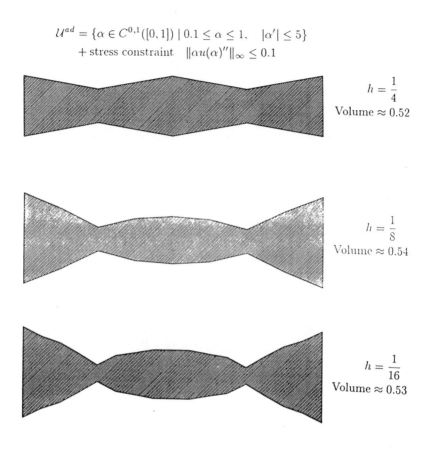

$$h = \frac{1}{4}$$
Volume ≈ 0.52

$$h = \frac{1}{8}$$
Volume ≈ 0.54

$$h = \frac{1}{16}$$
Volume ≈ 0.53

Figure 6.2

As the last example of this Section we consider simply supported (i.e. $u(0) = u(1) = u''(0) = u''(1) = 0$) beam with a point load $-\delta(x - \frac{1}{2})$ and the stress constraint $\|\alpha u(\alpha)''\|_\infty \le \frac{1}{2}$ with $\mathcal{U}^{ad} = \{\alpha \in C^{0,1} \mid \frac{1}{100} \le \alpha \le 1, |\alpha'| \le 5\}$ and $h = \frac{1}{20}$.

In Figure 6.3 we see a plot of stresses in the uniform beam $\alpha \equiv 0.69$ (marked with $\circ$) and in the optimal beam (marked with $\bullet$) with volume 0.47 shown in Figure 6.4.

In all cases the penalty parameter $\varepsilon = 10^{-4}$ was used.

6.2. The torsion problem. Let $\mathcal{U}^{ad} = \{\alpha \in C^{0,1} \mid \frac{1}{2} \le \alpha \le \frac{3}{2}, |\alpha'| \le 1, |\alpha''| \le 2\}$ and let the state constraint be $\|\nabla u(\alpha)\|_\infty \le 5R(\alpha)$.

Assuming symmetry also in the x_1-direction, only a quarter of the original problem needs to be analyzed. The problem is discretized using 200 triangular elements resulting 11 degrees of freedom in optimization of the domain. The penalty parameter was chosen to be $\varepsilon = 10^{-3}$.

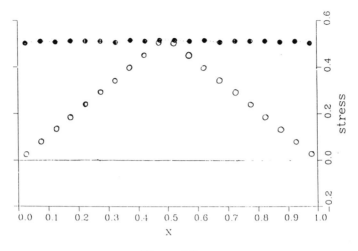

Figure 6.3

As an initial quess square cross section with area equal to 4.0 was chosen. In Figure 6.5 a quarter of the initial and optimal cross sections with equivalent stress contours are plotted. The area of the optimal cross section is 3.31 .

Figure 6.4

6.3. Remarks. All computations for the numerical results were performed in VAX 8600-computer with double precision arithmetic. The resulting nonlinear mathematical programming problems were solved using E04VCF-routine from NAG-subroutine library. This routine is essentially the NPSOL-routine due to Gill et al. (see [**Gill–Murray–Saunders–Wright**,1984]). The state problems were solved using Cholesky's method (beam problem) and preconditioned conjugate gradient method (torsion problem).

As the problems are generally nonconvex, the optimization algorithm may give only a local minima.

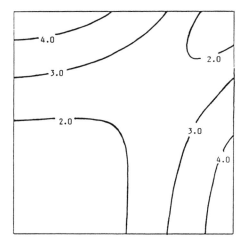 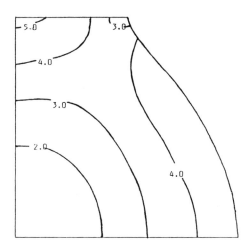

Figure 6.5

The parameters that bound α' and α'' do not have physical background. They are used only to quarantee the compactness of $\mathcal{U}^{ad}$. On the other hand, bounding the derivatives makes the iterative optimization process more stable. The choice of proper discretization is also a problem. Therefore in practise a sequence of optimal shape design problems with different parameters must be solved to find the 'true' optimal solution.

The computing time required to solve optimal design problems with fixed discretization parameter depends also on implementation details: choice of optimization and analysis algorithms, gradient calculations (analytic formulas vs. finite difference approximations), tolerance within which a design is considered optimal, etc.

For the example problems approximate execution times in VAX 8600-computer varied between 10 seconds (beam problems, analytic gradients) to 4.5 minutes (torsion problem, finite difference approximation for gradient).

REFERENCES

[Banichuk,1976] N.V. Banichuk, *Optimization of Elastic Bars in Torsion*, Int. J. Solids and Structures 12 (1976), 275–286.

[Begis–Glowinski,1975] D. Begis and R. Glowinski, *Application de la méthode des éléments finis à l'approximation d'un problème de domaine optimal. Méthodes de résolution des problèmes approchés*, Applied Mathematics and Optimization 2 (1975), 130–169.

[Gill–Murray–Saunders–Wright,1984] P.E. Gill, W. Murray, M.A. Saunders and M.H. Wright, "User's Guide for NPSOL: A Fortran Package for Nonlinear Programming," Technical report SOL 84-7, Stanford University, 1984.

[Haslinger–Neittaanmäki,1988] J. Haslinger and P. Neittaanmäki, "Finite Element Approximation for Optimal Design: Theory and Applications," J. Wiley & Sons, Chichester, 1988.

[Haslinger-Neittaanmäki-Tiba,1987] J. Haslinger, P. Neittaanmäki and T. Tiba, *On State Constrained Control Problems*, "Optimal control of partial differential equations," K.-H. Hoffmann and W. Krabs (ed.), International series of numerical mathematics, vol 78, 1987.

[Haug–Arora,1979] E.J. Haug and J.S. Arora, "Applied Optimal Design," John Wiley & Sons, New York Chichester Brisbane Toronto, 1979.

[Hlaváček–Bock–Lovíšek,1984] I. Hlaváček, I. Bock and J. Lovíšek, *Optimal Control of a Variational Inequality with Applications to Structural Analysis. I. Optimal Design of a Beam with Unilateral Supports*, Applied Mathematics and Optimization 11 (1984), 111-143.

[Mäkinen,1989] R. Mäkinen, Theses, University of Jyväskylä (1989) (to appear).

[Tiihonen,1987] T. Tiihonen, *Abstract Approach to a Shape Design Problem for Variational Inequalities*, University of Jyväskylä, Department of mathematics, Preprint 62 (1987).

R. Mäkinen
Department of mathematics
University of Jyväskylä
Seminaarinkatu 15
SF-40100 Jyväskylä
Finland

International Series of
Numerical Mathematics, Vol. 91
© 1989 Birkhäuser Verlag Basel

Sensitivity analysis of convex optimization problems

Kazimierz Malanowski

Systems Research Institute
Polish Academy of Sciences

Abstract. A family of convex optimization problems in an abstract Hilbert space, that depend
on a parameter is considered. The idea of differentiability of projection onto the cone inducing
the order in the space of constraints is used to find the form of the weak directional derivative,
with respect to the parameter, of the solutions to the optimization problems. The abstract
results are applied to linear-quadratic optimal control problems subject to control and state
constraints.

Keywords. Convex optimization problems, sensitivity analysis, projection onto convex cone,
optimal control subject to control and state constraints.

1980 *Mathematics subject classifications*: 49B50, 49B27, 49B10

1. Introduction

Sensitivity analysis of mathematical programming problems in finite dimension is fairly
well developed (see [3]). Sensitivity results for optimization problems in infinite dimension
are less complete. They concern mostly sensitivity of the optimal value function (see, e.g.
[9,15]). As far as the sensitivity of the solutions to changes of the parameters are concerned
one should mention the results of Haraux [7] and Mignot [16] on differentiability of the
projection mappings onto convex sets in a Hilbert space with applications to variational
inequalities. These results were used by Sokołowski to sensitivity analysis of different optimal control problems for distributed parameter systems (see, e.g., [18,19]). In [10–12,14]
the structure of pointwisely constrained convex optimal control problems was used to find
the form of directional derivatives of the solutions with respect to the parameters.

In this paper a family of convex optimization problems in a Hilbert space that depend on
a vector parameter is considered. It is assumed that the primal and dual optimal variables
are Lipschitz continuous functions of the parameter.

Optimality conditions, involving projection on a closed convex cone are analysed. Using
differentiability properties of projection onto a cone, obtained in [13], it is shown that under
suitable assumptions, the solutions of the optimization problems are weakly directionally
differentiable functions of the parameter. The directional derivative is characterized as the
solution of an auxiliary linear-quadratic optimization problem.

Sufficient conditions of weak Gâteaux differentiability are formulated and the second
order sensitivity formula for the optimal value function is derived.

The abstract results are applied to quadratic optimal control problems subject to state
and control constraints.

This research was supported by the Research Program CPBP/01.01

2. Convex optimization problems in Hilbert space

Let $H \subset \mathbf{R}^q$ be an open and convex set of admissible parameters. Z and Y are two Hilbert spaces (space of arguments and of constraints respectively).

$K \subset Y$ is a closed, convex cone having its vertex at 0, which induces a partial order in Y.

$$K^+ = \{y \in Y | (x, y)_Y \geq 0, \, \forall \, x \in K\}$$

denotes the polar cone to K.

By $P_K : Y \to K$ we denote the operator of metric projection onto K, i.e.,

$$\|P_K(y) - y\|_Y = \inf_{x \in K} \|x - y\|_Y.$$

There are given two functions

$$F(\cdot, \cdot) : Z \times H \to \mathbf{R}^1,$$
$$\phi(\cdot, \cdot) : Z \times H \to Y.$$

For each $h \in H$ we define the following problem of optimization:

(0_h) find $z_h \in \Phi_h$ such that

$$F(z_h, h) = \min_{z \in \Phi_h} F(z, h),$$

where

$$\Phi_h = \{z \in Z | - \phi(z, h) \in K\}.$$

It is assume that the following conditions hold:

(A1) $F(\cdot, \cdot)$ and $\phi(\cdot, \cdot)$ are twice Fréchet differentiable on $Z \times H$,

(A2) $F(\cdot, h)$ are strongly convex, uniformly on compact subsets $G \subset H$, i.e. for any compact $G \subset H$ there exists $\beta > 0$ such that

$$(v, D_{zz}^2 F(z, h)v)_Z \geq \beta \|v\|_Z^2 \, \forall \, v, z \in Z, \, \forall \, h \in G,$$

(A3) for each $h \in H$, $\phi(\cdot, h)$ is convex in sense of the order induced by K,

(A4) for each $h \in H$, the feasible set Φ_h is nonempty:

$$\Phi_h \neq 0.$$

It is well known that under conditions (A1) - (A4), for each $h \in H$, there exists a unique solution z_h of (0_h).

Let us introduce the following Lagrangian associated with (0_h)

(2.1)
$$\mathcal{L}(\cdot, \cdot, \cdot) : Z \times Y \times H \to \mathbf{R}^1,$$
$$\mathcal{L}(z, \lambda, h) = F(z, h) + (\lambda, \phi(z, h))_Y.$$

In addition to (A1) - (A4) it is assumed:

(A5) for each $h \in H$ there exists a unique normal Lagrange multiplier, i.e. the element $\lambda_h \in K^+$ such that z_h is characterized by the following Kuhn-Tucker conditions:

(2.2) $$D_z \mathcal{L}(z_h, \lambda_h, h) = D_z F(z_h, h) + [D_z \phi(z_h, h)]^* \lambda_h = 0,$$

(2.2a) $$(\lambda_h, \phi(z_h, h))_Y = 0, \quad \lambda_h \in K^+.$$

Since $-\phi(z_h, h) \in K$, (2.2a) implies

(2.3a) $$P_K(-(\lambda_h + \phi(z_h, h))) = -\phi(z_h, h),$$

(2.3b) $$P_{K^+}(\lambda_h + \phi(z_h, h)) = \lambda_h.$$

(A6) for any compact $G \subset H$ there exists $c > 0$ such that

$$\|z_{h_1} - z_{h_2}\|_Z, \|\lambda_{h_1} - \lambda_{h_2}\|_Y \leq c|h_1 - h_2| \quad \text{for all} \quad h_1, h_2 \in G.$$

We intend to analyse differential properties of z_h treated as a function of the parameter h. To this end we will use the characterization of z_h given by (2.2), (2.3a), (2.3b).

We will need some differential properties of the projection onto a closed convex cone, which are given below.

For any $y \in Y$, let us define the following closed and convex cone

(2.4) $$\Sigma(y) = \overline{(K + [P_K(y)])} \cap [y - P_K(y)]^\perp,$$

where $[y]$ denotes one dimensional subspace generated by y and $[y]^\perp$ is the annihilator of $[y]$.

Following [7] we call cone K polyhedric if

(2.5)
$$\Sigma(y) := \overline{(K + [P_K(y)])} \cap [y - P_K(y)]^\perp$$
$$= (K + [P_K(y)]) \cap [y - P_K(y)]^\perp \quad \text{for all} \quad y \in Y.$$

The following theorem is proved in [13]:

Theorem 2.1. *Let $f : H \to Y$ be Lipschitz continuous, and $h \in H$, $g \in \mathbf{R}^q$ be arbitrary but fixed. Suppose that for some $\{\alpha_h\} \downarrow 0$ we have*

$$\frac{1}{\alpha_n}[(f(h + \alpha_n g) - g(h)] \rightharpoonup d \quad \text{weakly in} \quad Y.$$

If K is polyhedric and one of the following two conditions is satisfied:

  (i) $\Sigma(f(h))$ — *is a subspace,*

  (ii) *(a) K is proper, i.e. $y \in K$ and $-y \in K$ implies $y = 0$,*

      (b) P_K is continuous in weak topology, i.e., $y_n \rightharpoonup y$ weakly in Y, implies $P_K(y_n) \rightharpoonup$

$P_K(y)$ weakly in Y,

(c) P_K is convex in the sense of the order induced in Y by K, i.e.

$$[P_K(y_1) + P_K(y_2)] - P_K(y_1 + y_2) \in K \quad \text{for all} \quad y_1, y_2 \in Y,$$

then

$$\frac{1}{\alpha_n}[P_K(f(h + \alpha_n g)) - P_K(f(h))] \rightharpoonup P_{\Sigma(f(h))}(d) \quad \text{weakly in} \quad Y$$

and

$$\frac{1}{\alpha_n}[P_{-K}(f(h + \alpha_n g)) - P_{-K^+}(f(h))] \rightharpoonup P_{-\Sigma^+(f(h))}(d) \quad \text{weakly in} \quad Y.$$

In order to be able to apply Theorem 2.1 to (0_h) we have to add the following assumptions:

(A7) K is polyhedric, i.e. (2.5) holds:

(A8) one of the following two conditions holds:

    (A8') $\Sigma_h := \overline{(K + \phi(z_h, h))} \cap [\lambda]^\perp$ — is a subspace,

    (A8") (a) K is proper,

          (b) P_K is continuous in weak topology,

          (c) P_K is convex.

Now we can prove our principal results:

Theorem 2.2. *If assumptions (A1) – (A8) hold, then the solution z_h of (0_h) is a weakly directionally differentiable function of h at any $h \in H$, i.e., for any $g \in \mathbf{R}^q$ and $\alpha > 0$*

$$z_{h+\alpha g} = z_h + \alpha \delta_{h,g} z_h + o(\alpha),$$

where $\frac{o(\alpha)}{\alpha} \rightharpoonup 0$ weakly in Z, and the directional derivative $\delta_{h,g} z_h$ is given by the unique solution of the following quadratic problem of optimization:

$(Q0_{h,g})$ find $w_{h,g} \in M_h$ such that

$$(2.6) \qquad R(w_{h,g}, h, g) = \min_{w \in M_h} \{R(w, h, g) := \frac{1}{2}(w, T_1(h)w)_Z + (w, T_2(h)g)_Z\}$$

where

$$(2.6a) \qquad T_1(h) = D_{zz}^2 \mathcal{L}(z_h, \lambda_h, h),$$

$$(2.6b) \qquad T_2(h) = D_{zh}^2 \mathcal{L}(z_h, \lambda_h, h),$$

$$(2.6c) \qquad M_h = \{w \in Z | - [D_z\phi(z_h, h)w + D_h\phi(z_h, h)g] \in \Sigma_h\}.$$

Proof: Let $g \in \mathbf{R}^q$ be an arbitrary but fixed direction and let $\{\alpha_n\} \downarrow 0$ be an arbitrary sequence. By (A6) there exists a subsequence $\{\alpha'_n\} \subset \{\alpha_n\}$ such that

$$(2.7a) \qquad \frac{1}{\alpha'_n}(z_{h+\alpha'_n g} - z_h) \rightharpoonup w \quad \text{weakly in} \quad Z,$$

$$(2.7b) \qquad \frac{1}{\alpha'_n}(\lambda_{h+\alpha'_n g} - \lambda_h) \rightharpoonup w \quad \text{weakly in} \quad Y.$$

Taking the difference quotient of (2.2) at $h + \alpha'_n g$ and h, passing to the limit and using (A1) and (2.7) we obtain

$$(2.8) \qquad \begin{aligned} &D^2_{zz}\mathcal{L}(z_h, \lambda_h, h)w + D^2_{z\lambda}\mathcal{L}(z_h, \lambda_h, h)\mu + D^2_{zh}\mathcal{L}(z_h, \lambda_h, h)g \\ &= D^2_{zz}\mathcal{L}(z_h, \lambda_h, h)w + D^2_{zh}\mathcal{L}(z_h, \lambda_h, h)g + [D_z\phi(z_h, \lambda_h)]^*\mu = 0. \end{aligned}$$

Similarly applying Theorem 2.1 to (2.3) we find that

$$(2.9a) \qquad \begin{aligned} P_{\Sigma_h}(-(\mu + D_z\phi(z_h, h)w + D_h\phi(z_h, h)g)) \\ = -(D_z\phi(z_h, h)w + D_h\phi(z_h, h)g), \end{aligned}$$

$$(2.9b) \qquad P_{\Sigma_h^+}(\mu + D_z\phi(z_h, h)w + D_h\phi(z_h, h)g) = \mu.$$

It is easy to see that (2.9) is equivalent to

$$(2.10) \qquad \begin{aligned} &-(D_z\phi(z_h, h)w + D_h\phi(z_h, h)g) \in \Sigma_h, \\ &(-\mu, D_z\phi(z_h, h)w + D_h\phi(z_h, h)g)_Y = 0, \quad \mu \in \Sigma_h^+. \end{aligned}$$

Obviously (2.8) and (2.10) constitute the Kuhn-Tucker conditions for $(Q0_{h,g})$. But by (A2) and (A3), $T_1(h)$ given by (2.6a) is positive definite, hence $(Q0_{h,g})$ has a unique solution, which is characterized by the Kuhn-Tucker conditions. It shows that w is independent of the choice of the sequences $\{\alpha_n\} \downarrow 0$ and $\{\alpha'_n\} \subset \{\alpha_n\}$, i.e., it is the directional derivative of z_h. ∎

Corollary 2.1. *It follows from the proof of Theorem 2.1 that any cluster point of $\{\frac{1}{\alpha_n}(\lambda_{h+\alpha_n g} - \lambda_h)\}$ is a Lagrange multiplier associated with $(Q0_{h,g})$. Hence if this multiplier is defined uniquely, then it is equal to the directional derivative of the Lagrange multiplier λ_h associated with the original problem (0_h).*

It is easy to see that in case where (A8') holds the solution $w_{h,g}$ of $(Q0_{h,g})$ is a linear continuous function of g. Hence we obtain:

Corollary 2.2. *If (A1) – (A7) and (A8') hold, then z_h is weakly Gâteaux differentiable at h.*

Remark 2.1: In case of finite-dimensional convex programming problems (A8') reduces to well known strict complementarity condition, which assures continuous differentiability of the solutions with respect to the parameter (see [3]). Hence it is natural to call (A8') *strict complementarity condition*.

In sensitivity analysis of optimization problems an important role is played by the so-called optimal value function

$$F^0(\cdot) : H \to \mathbf{R}^1,$$
$$F^0(h) = F(z_h, h).$$

which to every value of the parameter assigns the optimal value of the cost functional.

Using exactly the same argument as in [13] we obtain the following second order expansion of F^0:

Corollary 2.3. *If assumptions of Theorem 2.2 hold, then for any $h \in H$, $g \in \mathbf{R}^q$ and $\alpha > 0$ we have*

$$F^0(h + \alpha g) = F^0(h) + \alpha \langle D_h \mathcal{L}(z_h, \lambda_h, h), g \rangle$$
$$+ \frac{\alpha^2}{2} \left([w_{h,g}, g], \begin{bmatrix} D_{zz}^2 \mathcal{L}(z_h, \lambda_h, h), D_{zh}^2 \mathcal{L}(z_h, \lambda_h, h) \\ D_{hz}^2 \mathcal{L}(z_h, \lambda_h, h), D_{hh}^2 \mathcal{L}(z_h, \lambda_h, h) \end{bmatrix} [w_{h,g}, g] \right) + o(\alpha^2)$$

where $w_{h,g}$ is the solution of $(Q0_{h,g})$ and $\frac{0(\alpha^2)}{\alpha^2} \to_{\alpha \downarrow 0} 0$.

Note that the result of the form of Corollary 2.3, for mathematical programming problems in finite dimension, was first obtained by Shapiro [17], who used a different technique of the proof.

3. Linear-quadratic optimal control problem

In this section we apply Theorem 2.2 to obtain the form of the directional derivative with respect to the parameter of the solutions to linear-quadratic optimal control problem subject to state and control constraints.

The results of this type have already been obtained in [12] using different technique.

As before by $H \subset \mathbf{R}^q$ we denote an open and convex set of admissible parameters. For each $h \in H$ we consider the following problem of optimal control:

(OC_h) find $(u_h, x_h) \in L^2(0, T; \mathbf{R}^m) \times W^{1,2}(0, T; \mathbf{R}^n)$ such that

$$(3.1) \qquad F(u_h, x_h, h) = \min\{F(u, x, h) := \int_0^T f(u(t), x(t), h) dt$$

subject to

$$(3.2) \qquad \dot{x}(t) = A(h)x(t) + B(h)u(t),$$
$$(3.2a) \qquad x(0) = x^0(h),$$
$$(3.3) \qquad E(h)u(t) + e(h) \le 0 \quad \text{for almost all} \quad t \in [0, T],$$
$$(3.4) \qquad L(h)x(t) + \ell(h) \le 0 \quad \text{for all} \quad t \in [0, T],$$

where

(3.5)

$$u(t) \in \mathbf{R}^m, x(t) \in \mathbf{R}^n$$

$$f(u, x, h) = \frac{1}{2}[u^T, x^T]Q(h)\begin{bmatrix} u \\ x \end{bmatrix} + \langle q^1(h), u \rangle + \langle q^2(h), x \rangle$$

$$Q(\cdot) := \begin{bmatrix} Q^{11}(\cdot), & Q^{12}(\cdot) \\ Q^{21}(\cdot), & Q^{22}(\cdot) \end{bmatrix} : H \to \mathbf{R}^{(m+n)\times(m+n)},$$

$$q^1(\cdot) : H \to \mathbf{R}^m, \quad q^2(\cdot) : H \to \mathbf{R}^n,$$

$$A(\cdot) : H \to \mathbf{R}^{n\times n}, \quad B(\cdot) : H \to \mathbf{R}^{n\times m},$$

$$E(\cdot) : H \to \mathbf{R}^{r\times m}, \quad e(\cdot) : H \to \mathbf{R}^r,$$

$$L(\cdot) : H \to \mathbf{R}^{s\times m}, \quad \ell(\cdot) : H \to \mathbf{R}^s.$$

Assume that the following conditions hold:

(D1) $Q(\cdot), q^1(\cdot), q^2(\cdot), A(\cdot), B(\cdot), E(\cdot), e(\cdot), L(\cdot), \ell(\cdot)$ are twice Fréchet differentiable on H,

(D2) for each $h \in H$ there exists $\beta(h) > 0$ such that

$$\langle v, Q(h)v \rangle \geq \beta(h)|v|^2 \quad \text{for all} \quad v \in \mathbf{R}^{n+m},$$

(D3) for each $h \in H$ there exists a pair $(\hat{u}_h, \hat{x}_h) \in L^\infty(0, T; \mathbf{R}^m) \times W^{1,\infty}(0, T; \mathbf{R}^n)$ which satisfies (3.2) and a constant $\rho(h) < 0$ such that

$$\langle E^i(h), \hat{u}_h(t) \rangle + e^i(h) \leq \rho(h) \quad \text{for all} \quad i \in I := \{1, \dots, r\}$$
$$\text{and for a.a.} \quad t \in [0, T]$$

$$\langle L^j(h), \hat{x}_h(t) \rangle + \ell^j(h) \leq \rho(h) \quad \text{for all} \quad j \in J := \{1, \dots, s\}$$
$$\text{and for all} \quad t \in [0, T],$$

(D4) for each $h \in H$

$$L(h)x^0(h) + \ell(h) < 0.$$

It is well known (see e.g. [6,8]) that under the above conditions (OC_h) has a unique solution.

For each $h \in H$ and $t \in [0, T]$ we introduce the sets of indices of binding constraints:

$$I_h(t) = \{i \in I | E^i(h)\hat{u}_h(t) + e^i(h) = 0\},$$
$$J_h(t) = \{j \in J | L^j(h)\hat{x}_h(t) + \ell^j(h) = 0\}.$$

In addition to (D1) – (D4) we assume that the following constraints regularity conditions hold:

(D5) for each $h \in H$ there exists $\gamma(h) > 0$ such that

$$|[E^*_{I_h(t)}(h), B^*(h)L^*_{J_h(t)}(h)]v| \geq \gamma(h)|v|,$$

for almost all $t \in [0, T]$ and for all v of appropriate dimension, where $E^*_{I_h(t)}(h)$ and $L^*_{J_h(t)}(h)$ denote the matrices, whose columns are all columns of $E^*(h)$ and $L^*(h)$ respectively corresponding to the sets $I_h(t)$ and $J_h(t)$.

To express problem (OC_h) in the form of (O_h) we put

$$
\begin{aligned}
Z &:= L^2(0,T;\mathbf{R}^m) \times W^{1,2}(0,T;\mathbf{R}^n),\\
(3.6)\qquad Y &:= \mathbf{R}^n \times L^2(0,T;\mathbf{R}^n) \times L^2(0,T;\mathbf{R}^r) \times W^{1,2}(0,T;\mathbf{R}^s),\\
K &:= K^1 \times K^2 \times K^3 \times K^4
\end{aligned}
$$

where

(3.6a)
$$
K^1 := \{0\}, \; K^2 := \{0\},
$$
(3.6b) $\quad K^3 := \{v \in L^2(0,T;\mathbf{R}^n)|v(t) \geq 0 \text{ for a.a. } t \in [0,T]\},$
(3.6c) $\quad K^4 := \{y \in W^{1,2}(0,T;\mathbf{R}^s)|v(t) \geq 0 \text{ for all } t \in [0,T]\},$
$$
F(z,h) := F(u,x,h),
$$
$$
\phi(z,h) := \{x(0) - x^0(h), \dot{x} - A(h)x - B(h)u, E(h)u - e(h), L(h)x - \ell(h)\}.
$$

Let us introduce the following Lagrangian associated with (OC_h)

$$
\begin{aligned}
\mathcal{L}(\cdot,\cdot,\cdot,\cdot,\cdot,\cdot,\cdot) &\\
: L^2(0,T;\mathbf{R}^m) &\times W^{1,2}(0,T;\mathbf{R}^n) \times \mathbf{R}^n \times L^2(0,T;\mathbf{R}^n) \times L^2(0,T;\mathbf{R}^r)\\
&\times W^{1,2}(0,T;\mathbf{R}^s) \times H \to \mathbf{R}^1
\end{aligned}
$$

$$
\begin{aligned}
\mathcal{L}(u,x,\rho,p,\kappa,\nu,h) &\\
:= F(u,x,h) &+ \langle \rho, x(0) - x^0(h)\rangle + (p, \dot{x} - A(h)x - B(h)u) + (\kappa, E(h)u + e(h))\\
&+ \langle \nu(0), L(h)x(0) - \ell(h)\rangle + (\dot{\nu}, L(h)\dot{x}).
\end{aligned}
$$

It is known (see [4 – 6]) that by (D2) - (D5) for each $h \in H$ there exist unique normal Lagrange multipliers $(\rho_h, p_h, \kappa_h, \nu_h) \in \mathbf{R}^n \times L^2(0;T;\mathbf{R}^n) \times (K^3)^+ \times (K^4)^+$ associated with (OC_h), such that the solution (u_h, x_h) is characterized by the following Kuhn-Tucker conditions:

$$
\begin{aligned}
D_u\mathcal{L}(u_h, x_h, \rho_h, p_h, \kappa_h, \nu_h, h) &= 0,\\
D_x\mathcal{L}(u_h, x_h, \rho_h, p_h, \kappa_h, \nu_h, h) &= 0,\\
(\kappa_h, E(h)u_h + e(h)) &= 0, \quad \kappa_h \geq 0,\\
\langle \nu_h(0), L(h)x_h(0) + \ell(h)\rangle + (\dot{\nu}_h, L(h)\dot{x}_h) &= 0\\
\nu_h \geq 0, \; \dot{\nu}_h &- \text{is non-increasing and } 0 \leq \dot{\nu}_h(t) \leq \nu_h(0).
\end{aligned}
$$

The following regularity result is proved in [4]:

Lemma 3.1. *If (D2) - (D5) hold, then* $u_h(\cdot), \dot{x}_h(\cdot), p_h(\cdot), \kappa_h(\cdot), \dot{\nu}_h(\cdot)$ *are Lipschitz continuous on* $[0,T]$.

Moreover it was shown in [1,2] that:

Lemma 3.2. *If (D1) – (D5) hold, then for any compact $G \in H$ there exists $c > 0$ such that*

$$\|u_{h_1} - u_{h_2}\|, \|x_{h_1} - x_{h_2}\|_{1,2}, |\rho_{h_1} - \rho_{h_2}|, \|p_{h_1} - p_{h_2}\|, \|\kappa_{h_1} - \kappa_{h_2}\|,$$
$$\|\nu_{h_1} - \nu_{h_2}\|_{1,2} \le c|h_1 - h_2| \quad \text{for all} \quad h_1, h_2 \in G.$$

To apply Theorem 2.2 we need the form of the set Σ_h defined in (A8'). It is easy to see that:

$$\Sigma_h = \Sigma_h^1 \times \Sigma_h^2 \times \Sigma_h^3 \times \Sigma_h^4,$$
$$\Sigma_h^1 = \{0\}, \quad \Sigma_h^2 = \{0\},$$
$$\Sigma_h^3 = \{v \in L^2(0, T; \mathbf{R}^r)| \quad v^i(t) = 0 \quad \text{for a.a.} \quad t \in \hat{\Xi}_h^i,$$
$$v^i(t) \ge 0 \quad \text{for a.a.} \quad t \in \Xi_h^i \backslash \hat{\Xi}_h^i, i \in I\},$$

where

$$\Xi_h^i = \{t \in [0, T]|\langle E^i(h), u_h(t)\rangle + e^i(h) = 0\}$$
$$\hat{\Xi}_h^i = \{t \in \Xi_h^i|\kappa_h^i(t) > 0\},$$
$$\Sigma_h^4 = \{y \in W^{1,2}(0, T; \mathbf{R}^s)| \quad y^i(t) = 0 \quad \text{for all} \quad t \in \hat{N}_h^j,$$
$$y^i(t) \ge 0 \quad \text{for all} \quad t \in N_h^j \backslash \hat{N}_h^j, j \in J\},$$

where

$$N_h^j = \{t \in [0, T]|\langle L^j(h), u_h(t)\rangle + \ell^j(h) = 0\}$$
$$\hat{N}_h^j = \overline{\{t \in N_h^j|\nu_h^j(\cdot) - \text{is decreasing in a neighbourhood of } t\}}.$$

Certainly Σ_h is a subspace if and only if

(3.7a) $$\text{meas}\,\{\Xi_h^i \backslash \hat{\Xi}_h^i\} = 0 \quad \text{for all} \quad i \in I,$$
(3.7b) $$\text{meas}\,\{N_h^j \backslash \hat{N}_h^j\} = 0 \quad \text{for all} \quad j \in J.$$

We have to verify assumptions (A1) – (A8). It is easy to see that (D1) – (D4) imply (A1) – (A4). Conditions (A5) – (A6) hold by Lemmas 3.1 and 3.2. By Corollary 2 in [7] the cones K^3 and K^4 in (3.6) are polyhedric, i.e., (A7) holds. Hence it remains to check (A8).

Conditions (A8")are not satisfied by K^3. Namely (A8")(b) is violated. Therefore, in order to apply Theorem 2.2 we have to assume (3.7a).

On the other hand it is proved in [13] that (A8") is satisfied by K^4.

Hence we can apply Theorem 2.2 to (OC_h) and by this theorem we obtain:

Theorem 3.1. *If (D1) – (D5) and (3.7a) are satisfied, then for any $h \in H$, $g \in \mathbf{R}^q$ and any $\alpha > 0$*

$$u_{h+\alpha g} = u_h + \alpha \delta_{h,g} u_h + o_u(\alpha), \quad \text{where} \quad \frac{o_h(\alpha)}{\alpha} \rightharpoonup 0 \quad \text{weakly in} \quad L^2(0,T;\mathbf{R}^m),$$

$$x_{h+\alpha g} = x_h + \alpha \delta_{h,g} x_h + o_x(\alpha), \quad \text{where} \quad \frac{o_x(\alpha)}{\alpha} \rightharpoonup 0 \quad \text{weakly in} \quad W^{1,2}(0,T;\mathbf{R}^n)$$

and the directional derivatives $\delta_{h,g} u_h$, $\delta_{h,g} x_h$ are given by the unique solution $(v_{h,g}, y_{h,g})$ of the following quadratic optimal control problem:

$(QC_{h,g})$ find $(v_{h,g}, y_{h,g}) \in L^2(0,T;\mathbf{R}^m) \times W^{1,2}(0,T;\mathbf{R}^n)$ such that

$$Q(v_{h,g}, y_{h,g}, h, g) = \min\{Q(v,y,h,g)$$

$$:= \int_0^T \left(\frac{1}{2}[v^T(t), y^T(t)]Q(h) \begin{bmatrix} v(t) \\ y(t) \end{bmatrix} \right.$$

$$\left. + [v^T(t), y^T(t)] \begin{bmatrix} S_h^1(t) \\ S_h^2(t) \end{bmatrix} g \right) dt + y^T(T)W_h g\}$$

subject to

$$\dot{y}(t) = A(h)y(t) + B(h)v(t) + (D_h A(h)g)x_h(t) + (D_h B(h)g)u_h(t),$$

$$y(0) = D_h x^0(h)g$$

and

$$\langle E^i(h), v(t) \rangle + \langle D_h[E^i(h)u_h(t) + e^i(h)], g \rangle = 0 \quad \text{if} \quad t \in \Xi_h^i = \hat{\Xi}_h^i,$$

$$\langle L^j(h), y(t) \rangle + \langle D_h[L^j(h)u_h(t) + \ell^j(h)], g \rangle \begin{cases} = 0 & \text{if} \quad t \in \hat{N}_h^j \\ \leq 0 & \text{if} \quad t \in N_h^j \backslash \hat{N}_h^j, \end{cases}$$

where

$$S_h^1(t) = D_h Q^{11}(h)u_h(t) + D_h Q^{12}(h)x_h(t) + \sum_{i=1}^r \kappa_h^i(t)D_h E^i(h) - D_h B^*(h)p_h(t),$$

$$S_h^2(t) = D_h Q^{21}(h)u_h(t) + D_h Q^{22}(h)x_h(t) + \sum_{j=1}^s \check{\nu}_h^j(t)D_h L^j(h) - D_h A^*(h)p_h(t),$$

$$W_h = \sum_{j=1}^s \check{\nu}_h^j(T)D_h L^j(h).$$

References

[1] Dontchev, A.L., *Perturbations, Approximations and Sensitivity Analysis of Optimal Control Systems*, in "Lecture Notes in Control and Information Sciences, Vol. 52," Springer-Verlag, Berlin, 1983.

[2] Dontchev, A.L., *The constrained linear-quadratic optimal control problem*, in "Mathematics and Education in Mathematics, 1987," Proceedings of the Sixteenth Spring Conference of the Union of Bulgarian Mathematicians, Publishing House of Bulgarian Academy of Sciences, Sofia, 1987, pp. 83–90.

[3] Fiacco, A.V., "Introduction to Sensitivity and Stability Analysis in Nonlinear Programming," Academic Press, New York, 1987.

[4] Hager, W.W., *Lipschitz continuity for constrained processes*, SIAM J. Control Optim. **17** (1979), 321–337.

[5] Hager, W.W., *Convex control and dual approximations, Pt.I*, Control Cybern. **8** (1979), 5–22.

[6] Hager, W.W. and Mitter, S.K., *Lagrange duality theory for convex control problems*, SIAM J. Control Optim. **14** (1976), 843–856.

[7] Haraux, A., *How to differentiate the projection on a convex set in Hilbert space: Some applications to variational inequalities*, J. Math. Soc. Japan **29** (1977), 615–631.

[8] Lee, E.B. and Marcus L., "Foundations of Optimal Control Theory," J. Wiley, New York, 1967.

[9] Lempio, F. and Maurer H., *Differential stability in infinite dimensional nonlinear programming*, Appl. Math. Optim. **6** (1980), 139–152.

[10] Malanowski, K., *Differential stability of solutions to convex, control constrained optimal control problems*, Appl. Math. Optim. **12** (1984), 1–14.

[11] Malanowski, K., *Stability and sensitivity of solutions to optimal control problems for systems with control appearing linearly*, Appl. Math. Optim. **16** (1987), 73–91.

[12] Malanowski, K., *Stability of Solutions to Convex Problems of Optimization*, in "Lecture Notes in Control and Information Sciences, Vol. 93," Springer-Verlag, Berlin, 1987.

[13] Malanowski, K., *Sensitivity analysis of optimization problems in Hilbert space with application to optimal control*, (to appear).

[14] Malanowski, K. and Sokołowski, J., *Sensitivity of solutions to convex, constrained optimal control problems for distributed parameter systems*, J. Math. Anal. Appl. **120** (1986), 240–263.

[15] Maurer, H., *Differential stability of optimal control problems*, Appl. Math. Optim. **5** (1979), 63–82.

[16] Mignot, F., *Côntrol dans les inequations variationelles*, J. Funct. Anal. **22** (1976), 130–185.

[17] Shapiro, A., *Second order sensitivity analysis and asymptotic theory of parametrized nonlinear programs*, Math. Program. **33** (1984), 280–299.

[18] Sokołowski, J., *Differential stability of control constrained optimal control problems for distributed parameter systems*, in "Distributed Parameter Systems"," Kappel, F., Kunisch, K., Schappacher, W. (eds.), Lecture Notes in Control and Information Sciences, Vol. 75, Springer-Verlag, Berlin, 1985.

[19] Sokołowski, J., *Sensitivity analysis of control constrained optimal control problems for distributed parameter sytems*, SIAM J. Control Optim. **25** (1987), 1542–1556.

Kazimierz Malanowski
Systems Research Institute
of the Polish Academy of Sciences
ul. Newelska 6
01-447 Warszawa
Poland

International Series of
Numerical Mathematics, Vol. 91
© 1989 Birkhäuser Verlag Basel

A geometric structure of the Ljapunov equation related to stabilization of parabolic systems

Takao Nambu

Department of Mathematics
Kumamoto University

Abstract. We are going to study in this paper a Ljapunov equation described by $XL - BX = C$. Here, L, B, and C are given linear operators acting in separable Hilbert spaces, and are derived from a specific feedback control system. An important geometric structure of the operator solution X to the Ljapunov equation has been obtained so far for a general elliptic differential operator L of order 2 in a bounded domain of an Euclidean space. When L is especially a self-adjoint operator, we will derive a stronger geometric character of X by introducing a new approach. The result is then applied to stabilization of a linear system with a self-adjoint L as a coefficient operator.

Keywords. the Ljapunov equation, stabilization by feedback, evolution equations of parabolic type.

1980 *Mathematics subject classifications:* 47A62, 93D15 (primary), 35K22, 35B37 (secondary)

1. Introduction

Dynamic compensators often appears in stabilization studies of linear parabolic control systems. A geometric structure of the related operator equation, the so called Ljapunov equation, plays a central role in the stabilization. The equation is described as $XL - BX = C$, where L, B, and C are linear operators derived from a specific feedback control system, and will be defined below. The operator B stands for the coefficient of the compensator, and contains several parameters to be designed. Most important is the parameter γ indicating the growth rate of the spectrum $\sigma(B)$. The allowable range of γ for stabilization has been limited to the open interval $(0, 2)$ [4]. It is important to extend the allowable range of γ for the following reason: The extension guarantees much wider gaps between the eigenvalues of B as well as a more freedom of design of feedback control schemes. In fact, the algebraic conditions satisfied by the actuators of the compensator (Theorem 2.2 below) are likely to be easily broken if the gap between the eigenvalues of B at infinity becomes very small. Thus, the extension of the allowable range of γ leads to a higher degree of safety for the stabilization scheme against unexpected perturbations on B. The extension is also interesting from the mathematical viewpoint since it requires a new development of analysis. In [5], an extension to the interval $(0, 2]$ has been achieved, where L is restricted within a class of operators derived from spatial self-adjoint operators in a bounded interval of $\mathbb{R}^1$. The purpose of this paper is to present a further extension of our previous results to the case where L is a more general self-adjoint operator in a Hilbert space. A new approach is employed to the proof of the extension although the admissible interval $(0, 2)$ of γ is merely extended to $(0, 2]$.

The setting of the problem is as follows: Let H be a separable real Hilbert space, and consider the following differential equation in H:

$$(1.1) \qquad \frac{d}{dt}u + Lu = \sum_{k=1}^{M} f_k(t)h_k, \quad t > 0, \quad u(0) = u_0.$$

Here, L indicates a linear self-adjoint operator acting in H with dense domain $\mathcal{D}(L)$, $f_k(t)$, $1 \le k \le M$, inputs to be designed in feedback form, and h_k, $1 \le k \le M$, actuators belonging to H. It is assumed that L has a compact resolvent and is bounded from below. Output of (1.1) is given by a set of bounded linear functionals;

$$(1.2) \qquad \langle u(t), w_k \rangle, \quad 1 \le k \le N,$$

where w_k, $1 \le k \le N$, are observation weighting vectors belonging to H. Hereafter, the inner product and the norm in H are denoted by $\langle \cdot, \cdot \rangle$ and $\|\cdot\|$ respectively. The assumption on the operator L guarantees the existence of the eigenpairs for L having the following properties [1]:

(i) $\sigma(L) = \{\lambda_1, \lambda_2, \dots\}$, $\quad -\infty < \lambda_1 < \lambda_2 < \cdots < \lambda_i < \cdots \to \infty$;

(ii) $L\varphi_{ij} = \lambda_i \varphi_{ij}$, $\quad i \ge 1$, $\quad 1 \le j \le m_i (< \infty)$; and

(iii) the set $\{\varphi_{ij}; \ i \ge 1, \ 1 \le j \le m_i\}$ forms a complete orthonormal system in H.

A good example of L arises from the well known elliptic theory [1]. It is given as follows: Let $\mathcal{L}$ be a formally self-adjoint and uniformly elliptic differential operator of order 2 in a bounded domain Ω of $\mathbb{R}^m$ with a smooth boundary Γ of $(m-1)$-dimension. Associated with $\mathcal{L}$ is a boundary operator τ of Dirichlet or generalized Neumann type. Then, the operator L is defined in $H = L^2(\Omega)$ by $\mathcal{L}$ and τ as

$$Lu = \mathcal{L}u, \quad u \in \mathcal{D}(L) = \{u \in H^2(\Omega); \ \tau u = 0 \text{ on } \Gamma\}.$$

A "compensator" is an abstract differential equation in a separable Hilbert space $\mathcal{H}$, and is described by

$$(1.3) \qquad \frac{d}{dt}v + Bv = \sum_{k=1}^{N} \langle u(t), w_k \rangle \xi_k + \sum_{k=1}^{M} f_k(t)\alpha_k,$$

$$v(0) = v_0,$$

where ξ_k's and α_k's indicate actuators in $\mathcal{H}$ to be designed. Let us choose $f_k(t)$ in (1.1) and (1.3) as

$$(1.4) \qquad f_k(t) = \langle v(t), \rho_k \rangle_{\mathcal{H}}, \quad 1 \le k \le M,$$

$\langle \cdot, \cdot \rangle_{\mathcal{H}}$ indicating the inner product in $\mathcal{H}$, and ρ_k's belonging to $\mathcal{H}$. Then, we obtain a closed loop system, (1.1), (1.3), and (1.4). Eqns. (1.1), (1.3), and (1.4) are written as an equation in $H \times \mathcal{H}$;

$$(1.5) \qquad \left(\frac{d}{dt} + \mathcal{M}\right)\begin{bmatrix} u \\ v \end{bmatrix} = \begin{bmatrix} 0 \\ 0 \end{bmatrix}, \quad \begin{bmatrix} u(0) \\ v(0) \end{bmatrix} = \begin{bmatrix} u_0 \\ v_0 \end{bmatrix}.$$

The operator B in (1.3) is specified in the following lemma. Here, the structure of B is determined independently of L.

Lemma 1.1 [4,5]. *Let A be a positive-definite self-adjoint operator in a separable Hilbert space H_0 with a compact resolvent. Let $\{\mu_i^2, \zeta_{ij}; i \geq 1, 1 \leq j \leq n_i(< \infty)\}$ denote the eigenpairs of A (μ_i^2 are labelled according to increasing order, and ζ_{ij} normalized). Define $\mathcal{H}$ and B as*

$$\mathcal{H} = \mathcal{D}(A^{1/2}) \times H_0,$$

and

$$B = \begin{bmatrix} 0 & -1 \\ A & 2aA^{1/2} \end{bmatrix}, \quad \mathcal{D}(B) = \mathcal{D}(A) \times \mathcal{D}(A^{1/2}), \quad a \in (0,1)$$

respectively. Furthermore, set

$$\eta_{ij}^{\pm} = \frac{1}{\sqrt{2}\mu_i} \begin{bmatrix} \zeta_{ij} \\ -\mu_i \omega^{\pm} \zeta_{ij} \end{bmatrix}, \quad i \geq 1, \quad 1 \leq j \leq n_i, \quad \omega^{\pm} = a \pm \sqrt{1-a^2}\, i.$$

Then

(i) $\sigma(B) = \{\mu_i \omega^{\pm}; i \geq 1\}$, $0 < \mu_1 < \mu_2 < \cdots \to \infty$;

(ii) $B\eta_{ij}^{\pm} = \mu_i \omega^{\pm} \eta_{ij}^{\pm}$, $i \geq 1$, $1 \leq j \leq n_i$;

(iii) *the set $\{\eta_{ij}^{\pm}; i \geq 1, 1 \leq j \leq n_i\}$ forms a normalized Riesz basis for $\mathcal{H}$; and*

(iv) $-B$ *generates an analytic semigroup such that*

$$\|e^{-tB}\|_{\mathcal{L}(\mathcal{H})} \leq \sqrt{\frac{2}{1-a}}\, e^{-a\mu_1 t}, \quad t \geq 0.$$

Remark 1. It is clear by this definition of B that $-\mathcal{M}$ in (1.5) generates an analytic semigroup $e^{-t\mathcal{M}}$, $t > 0$.

Remark 2. Define a real Hilbert space $\hat{\mathcal{H}}$ by

$$\hat{\mathcal{H}} = \left\{ h \in \mathcal{H}; \quad h = \sum_{i=1}^{\infty} \sum_{j=1}^{n_i} (h_{ij}\eta_{ij}^{+} + \overline{h_{ij}}\eta_{ij}^{-}), \quad \sum_{i,j} |h_{ij}|^2 < \infty \right\}.$$

Then, it is easy to see that

$$B; \quad \mathcal{D}(B) \cap \hat{\mathcal{H}} \xrightarrow{\text{onto}} \hat{\mathcal{H}}, \quad \text{and} \quad e^{-tB} \in \mathcal{L}(\hat{\mathcal{H}}).$$

The operator C is defined by

$$Cu = -\sum_{k=1}^{N} \langle u, w_k \rangle \xi_k, \quad u \in H.$$

Based on these definitions of L, B, and C, we are going to study the Ljapunov equation $XL - BX = C$ in §2, and establish the extension stated above. In §3, the result is then applied to stabilization of eqn.(1.1) under the feedback scheme described above. We close this section with the following remark: It is possible to design another type of compensators [2,4,8], where $\mathcal{H} = H$, and the operator B is determined via L. In this case, the construction of the compensator becomes easier. On the other hand, we do not know precisely how $\sigma(B)$ behaves at infinity.

2. The Ljapunov equation

We begin with defining matrices W_i and Ξ_i by

$$W_i = \left[\langle w_k, \varphi_{ij} \rangle ; \begin{array}{l} k \downarrow 1, \ldots, N \\ j \to 1, \ldots, m_i \end{array} \right], \quad \text{and} \quad \Xi_i = \left[\xi_{ij}^k ; \begin{array}{l} k \downarrow 1, \ldots, N \\ j \to 1, \ldots, n_i \end{array} \right], \quad i \geq 1,$$

where we have set $\xi_k = \sum_{i,j} (\xi_{ij}^k \eta_{ij}^+ + \overline{\xi_{ij}^k} \eta_{ij}^-) \in \hat{\mathcal{H}}$, $1 \leq k \leq N$. Let us assume that the operator B satisfies a spectrum growth condition:

$$(2.1) \qquad\qquad\qquad \exists c_1 > 0; \quad \mu_n \leq c_1 n^2, \quad n \geq 1,$$

that is, $\gamma = 2$. Note that the condition (2.1) is weaker than in [4, Theorem 4.2] and [5, Theorem 3.1]. It will be shown that Theorem 2.2 below holds for a general self-adjoint L under (2.1). A new approach will be employed to the proof. Let P_K be the projection operator in H corresponding to the first K eigenvalues of L, $\lambda_1, \ldots, \lambda_K$. First, we have the following

Theorem 2.1 [4]. *The Ljapunov equation $XL - BX = C$ on $\mathcal{D}(L)$ has a unique operator solution $X \in \mathcal{L}(H; \hat{\mathcal{H}})$. The solution X is expressed by*

$$(2.2) \qquad Xu = \sum_{i,j} \sum_{k=1}^N f_k(\mu_i \omega^+ ; u) \xi_{ij}^k \eta_{ij}^+ + \sum_{i,j} \sum_{k=1}^N f_k(\mu_i \omega^- ; u) \overline{\xi_{ij}^k} \eta_{ij}^-,$$

$$f_k(\lambda ; u) = \langle (\lambda - L)^{-1} u, w_k \rangle, \quad 1 \leq k \leq N.$$

The following theorem is our main result in this section:

Theorem 2.2. *Suppose that the spectrum growth condition (2.1) is satisfied. Suppose further that the algebraic conditions*

$$(2.3) \qquad \text{rank } W_i = m_i, \quad 1 \leq i \leq K, \quad \text{and} \quad \text{rank } \Xi_i = N, \quad i \geq 1$$

are satisfied. Then, we have an inclusion relation

$$(2.4) \qquad \overline{X^* \hat{\mathcal{H}}} \supset P_K H = \text{span}\{\varphi_{ij} ; 1 \leq i \leq K, 1 \leq j \leq m_i\}.$$

Proof of Theorem 2.2. It will suffice to show the following relation equivalent to (2.4);

$$\text{Ker } X \subset \{P_K H\}^\perp = \{u \in H; \langle u, \varphi_{ij} \rangle = 0, 1 \leq i \leq K, 1 \leq j \leq m_i\}.$$

Let u be any element of Ker X. Since the set $\{\eta_{ij}^\pm\}$ forms a basis for $\hat{\mathcal{H}}$ (Lemma 1.1), we see from Theorem 2.1 that

$$\sum_{k=1}^N f_k(\mu_i \omega^+ ; u) \xi_{ij}^k = 0, \quad i \geq 1, \quad 1 \leq j \leq n_i.$$

Since rank $\Xi_i = N$, $i \geq 1$, the above relation implies that

(2.5) $\qquad f_k(\mu_i \omega^+; u) = \langle (\mu_i \omega^\pm - L)^{-1} u, w_k \rangle = 0, \quad i \geq 1, \quad 1 \leq k \leq N.$

It will be shown below that (2.5) implies that $f_k(\lambda; u) \equiv 0$, $1 \leq k \leq N$ under the condition (2.1) weaker than those in [4,5]. We are going to employ a new approach in order to establish this. Set $L_c = L + c$, c being a positive number in order that $\sigma(L_c)$ is entirely contained in the right half-plane. First, let us note that

(2.6) $\qquad 2\lambda(\lambda^2 - L_c)^{-1} = (\lambda + L_c^{1/2})^{-1} + (\lambda - L_c^{1/2})^{-1}$ on H, $\quad \lambda \in \rho(\pm L_c^{1/2}).$

We are going to consider a function $h(\lambda)$ defined by

$$h(\lambda) = -2\lambda f(-\lambda^2),$$

where

$$f(\lambda) = f_k(\lambda - c; u) = \sum_{i,j} \frac{1}{\lambda - (\lambda_i + c)} \langle u, \varphi_{ij} \rangle \langle w_k, \varphi_{ij} \rangle$$

and the subscript k is fixed for a while. According to (2.6), $h(\lambda)$ is expressed as

(2.7) $\qquad h(\lambda) = \langle (\lambda - iL_c^{1/2})^{-1} u, w_k \rangle + \langle (\lambda + iL_c^{1/2})^{-1} u, w_k \rangle.$

Since $f(\mu_n \omega^\pm + c) = 0$, $n \geq 1$, the zeros of $h(\lambda)$ must be at least $\pm i(\mu_n \omega^+ + c)^{1/2}$ and $\pm(\mu_n \omega^- + c)^{1/2}$, $n \geq 1$. Here, $\arg(\mu_n \omega^+ + c)^{1/2}$ is monotone increasing, and asymptotically approaches $\frac{1}{2} \mathrm{Tan}^{-1}(\sqrt{1-a^2}/a)$. Similarly, $\arg(\mu_n \omega^- + c)^{1/2}$ is monotone decreasing, and approaches $-\frac{1}{2} \mathrm{Tan}^{-1}(\sqrt{1-a^2}/a)$.

Expression (2.7) leads to a consideration of the strongly continuous groups $\exp(it L_c^{1/2})$ and $\exp(-it L_c^{1/2})$ generated by $iL_c^{1/2}$ and $-iL_c^{1/2}$ respectively. They satisfy estimates

$$\|\exp(\pm it L_c^{1/2})\|_{\mathcal{L}(H)} = 1 \quad \text{for any} \quad t \in \mathbb{R}^1.$$

We remark that self-adjointness of L is essential to ensure the above estimates. If $\mathrm{Re}\,\lambda > 0$, we see that

$$\int_0^\infty e^{-\lambda t} e^{\pm it L_c^{1/2}} u \, dt = (\lambda \mp iL_c^{1/2})^{-1} u, \quad u \in H.$$

Thus, $h(\lambda)$ has another expression written by

$$h(\lambda) = \int_0^\infty e^{-\lambda t} \langle (e^{it L_c^{1/2}} + e^{-it L_c^{1/2}}) u, w_k \rangle dt, \quad \mathrm{Re}\,\lambda > 0.$$

Let us recall that $h(i(\mu_n \omega^- + c)^{1/2}) = 0$, $n \geq 1$, and that $\arg i(\mu_n \omega^- + c)^{1/2}$ is monotone decreasing and approaches $\frac{\pi}{2} - \frac{1}{2} \mathrm{Tan}^{-1}(\sqrt{1-a^2}/a)$. Choosing $\delta > 0$ small enough, we find that

(2.8) $\qquad \int_0^\infty e^{\{-i(\mu_n \omega^- + c)^{1/2} + \delta\}t} e^{-\delta t} \langle (e^{it L_c^{1/2}} + e^{-it L_c^{1/2}}) u, w_k \rangle dt = 0, \quad n \geq 1.$

Now, let us recall Szász's theorem in the classical Fourier analysis. It is stated as follows [7]:

Let $\operatorname{Re}\lambda_n > -\frac{1}{2}$, $n \geq 1$. The set of functions $\{\exp(-(\lambda_n + \frac{1}{2})t); \ n \geq 1\}$ is closed on $L^2(0,\infty)$ if and only if

$$\sum_{n=1}^{\infty} \frac{1 + 2\operatorname{Re}\lambda_n}{1 + |\lambda_n|^2} = \infty.$$

It is not difficult to show that the above condition is satisfied in our problem where $i(\mu_n\omega^- + c)^{1/2} - \delta = \lambda_n + \frac{1}{2}$. In fact, set $\mu_n\omega^- + c = r_n\exp(-i\theta_n)$ with $\theta_n \nearrow \theta_0 = \operatorname{Tan}^{-1}(\sqrt{1-a^2}/a)$. If c_2 is small enough, e.g., $0 < c_2 < 2\cos\left(\frac{\pi-\theta_0}{2}\right)$, then the inequality

$$\frac{1 + 2\operatorname{Re}\lambda_n}{1 + |\lambda_n|^2} > \frac{c_2}{r_n^{1/2}}$$

holds for n large enough. We also note that $r_n - \mu_n$ tends to $c\cos\theta_0$ as $n \to \infty$. Thus, according to the assumption (2.1), we find that

$$\sum_{n\geq N} \frac{1 + 2\operatorname{Re}\lambda_n}{1 + |\lambda_n|^2} \geq \operatorname{const}\sum_{n\geq N} \frac{1}{n} = \infty$$

for a large integer N. Therefore, we conclude from (2.8) that

$$e^{-\delta t}\langle(e^{itL_c^{1/2}} + e^{-itL_c^{1/2}})u, w_k\rangle = 0, \quad t \geq 0,$$

and that $h(\lambda) = 0$ for $\operatorname{Re}\lambda > 0$. Since $f(\lambda)$ is a meromorphic function, it immediately follows that $f(\lambda)$ is identically equal to zero except at its possible poles $\lambda_i + c$, $i \geq 1$. Calculating the residue of $f(\lambda)$ at $\lambda_i + c$, we find that

$$\sum_{j=1}^{m_i} \langle u, \varphi_{ij}\rangle\langle w_k, \varphi_{ij}\rangle = 0, \quad i \geq 1, \quad 1 \leq k \leq N.$$

Since rank $W_i = m_i$, $1 \leq i \leq K$, we conclude that $\langle u, \varphi_{ij}\rangle = 0$, $1 \leq i \leq K$, $1 \leq j \leq m_i$, that is $P_K u = 0$. The proof of Theorem 2.2 is thereby completed. Q.E.D.

As we have seen, self-adjointness of L acts essentially in the proof of Theorem 2.2. Thus, the same method never applies to the case where L is a non self-adjoint operator. In this case, the proof is instead based on Carleman's theorem [10]. The admissible range of γ is, however, limited to the open interval $(0, 2)$ [4]. We remark that a further extension of the admissible range of γ in Theorem 2.2 to a non self-adjoint L is possible if a more number of w_k's are available. This extension is to be reported in the author's forthcoming paper.

3. Application to stabilization

In this section, we will see how Theorem 2.2 is applied to stabilization of eqn. (1.1). The problem is stated as follows: *Given w_k's and h_k's, design a compensator described by (1.3) such that the state $(u(t), v(t))$ in (1.5) decays exponentially as $t \to \infty$ with a given decay rate.* Thus, we will design the vectors, ξ_k's, α_k's, and especially ρ_k's. Since the procedure to establish stabilization is the same as in [4,5], only its outline is given here.

Set $\alpha_k = X h_k \in \hat{\mathcal{H}}$, $1 \le k \le M$ in (1.3). Then, it is easily seen that $Xu(t) - v(t)$ satisfies

$$\left(\frac{d}{dt} + B \right)(Xu(t) - v(t)) = 0, \quad Xu(0) - v(0) = Xu_0 - v_0 .$$

Thus, $Xu(t) - v(t) = e^{-tB}(Xu_0 - v_0)$, $t \ge 0$. Substituting this into (1.1), we see that $u(t)$ satisfies

$$(3.1) \qquad \frac{d}{dt}u(t) + Mu(t) = \sum_{k=1}^{M} \langle e^{-tB}(v_0 - Xu_0), \rho_k \rangle_{\mathcal{H}} \, h_k .$$

Here, M indicates the operator defined by

$$Mu = Lu - \sum_{k=1}^{M} \langle u, X^*\rho_k \rangle h_k , \quad u \in \mathcal{D}(M) = \mathcal{D}(L).$$

The operator M generates an analytic semigroup e^{-tM}, $t > 0$. Our aim is to choose ρ_k's so that $\sigma(M)$ entirely lies in the right half-plane. Let us define matrices H_i by

$$H_i = \left[\langle h_k, \varphi_{ij} \rangle \, ; \, \begin{matrix} k \downarrow 1, \dots, M \\ j \to 1, \dots, m_i \end{matrix} \right], \quad i \ge 1.$$

The following result is what is now well known as the most fundamental stabilization theorem: If the algebraic conditions

$$(3.2) \qquad \text{rank } H_i = m_i, \quad 1 \le i \le K, \quad \text{where} \quad \lambda_{K+1} > a\mu_1 ,$$

are satisfied, there exist suitable vectors $y_1, \dots, y_M \in P_K H$ such that

$$\left\| \exp t\left\{ -L + \sum_{k=1}^{M} \langle \cdot, y_k \rangle h_k \right\} \right\|_{\mathcal{L}(H)} \le \text{const } e^{-\lambda_{K+1}t}, \quad t \ge 0 .$$

Under the assumptions of Theorem 2.2, we can choose $\rho_1, \dots, \rho_M \in \hat{\mathcal{H}}$ such that $X^*\rho_k$ is arbitrarily close to y_k for each k. Thus, for a given κ; $a\mu_1 < \kappa < \lambda_{K+1}$, we can find $\rho_1, \dots, \rho_M \in \hat{\mathcal{H}}$ such that

$$\left\| e^{-tM} \right\|_{\mathcal{L}(H)} \le \text{const } e^{-\kappa t}, \quad t \ge 0 .$$

Here, $\rho_1, \ldots, \rho_M$ are expressed as linear combinations of a finite number of $\eta_{ij}^{\pm}$, say, $1 \leq i \leq I$, $1 \leq j \leq n_i$. In view of Lemma 1.1, the right-hand side of (3.1) decays exponentially with exponent $a\mu_1$, and so does $u(t)$. Therefore,

$$\left\| e^{-tM} \right\|_{\mathcal{L}(H \times \hat{\mathcal{H}})} \leq \text{const } e^{-a\mu_1 t}, \quad t \geq 0. \tag{3.3}$$

Stabilization of eqn. (1.1) has been just established.

Reduction of the compensator (1.3) to a finite-dimensional equation is easily carried out. Let $\tilde{P}_I$ be the projection operator in $\mathcal{H}$ corresponding to the eigenvalues $\mu_i \omega^{\pm}$, $1 \leq i \leq I$; $\tilde{P}_I = (2\pi i)^{-1} \oint (\lambda - B)^{-1} d\lambda$, where the integral is taken along a contour encircling $\mu_i \omega^{\pm}$, $1 \leq i \leq I$. Let us set $v_1(t) = \tilde{P}_I v(t)$. Then, the pair $(u(t), v_1(t))$ satisfies an equation in $H \times \tilde{P}_I \hat{\mathcal{H}}$, described by

$$\left(\frac{d}{dt} + \mathcal{N} \right) \begin{bmatrix} u \\ v_1 \end{bmatrix} = \begin{bmatrix} 0 \\ 0 \end{bmatrix}, \quad \begin{bmatrix} u(0) \\ v_1(0) \end{bmatrix} = \begin{bmatrix} u_0 \\ v_{10} \end{bmatrix} \in H \times \tilde{P}_I \hat{\mathcal{H}}. \tag{3.4}$$

The semigroup $e^{-t\mathcal{N}}$ satisfies the same estimate as that for e^{-tM}. Note that the equation in $\tilde{P}_I \hat{\mathcal{H}}$ satisfied by $v_1(t)$ is equivalent to a finite-dimensional compensator with the dependent variable $y(t)$ in $\mathbb{R}^S$, $S = 2(n_1 + \cdots + n_I)$. Let us summerize the result obtained so far.

Theorem 3.1. *Suppose that the assumptions in Theorem 2.2 as well as (3.2) are satisfied. Then, we can find α_k and $\rho_k \in \hat{\mathcal{H}}$, $1 \leq k \leq M$ which ensure (3.3). Eqn. (1.5) is reduced to eqn. (3.4) satisfying an estimate*

$$\left\| e^{-t\mathcal{N}} \right\|_{\mathcal{L}(H \times \tilde{P}_I \hat{\mathcal{H}})} \leq \text{const } e^{-a\mu_1 t}, \quad t \geq 0. \tag{3.5}$$

Remark 1. The equation equivalent to (3.4) is described in concrete form as follows:

$$\frac{d}{dt} u(t) + L u(t) = \sum_{k=1}^{M} \langle y(t), \tilde{\rho}_k \rangle_{\mathbb{R}^S} h_k, \quad u(0) = u_0,$$

$$\frac{d}{dt} y(t) + \mathcal{F} y(t) = \sum_{k=1}^{N} \langle u(t), w_k \rangle \tilde{\xi}_k, \quad y(0) = y_0,$$

where $y(t)$, $\tilde{\rho}_k$, and $\tilde{\xi}_k$ belong to $\mathbb{R}^S$, and $\langle \cdot, \cdot \rangle_{\mathbb{R}^S}$ denotes the inner product in $\mathbb{R}^S$. Derivation of these vectors and the matrix $\mathcal{F}$ is simple, but seems tedious. Thus, it is not stated here.

Remark 2. The operator B in Lemma 1.1 is arbitrary as far as it satisfies (2.3): It is chosen to satisfy, as a necessary condition, $n_i \geq N$, $i \geq 1$, whereas $N \geq \max_{1 \leq i \leq K} m_i$. Thus, the operator A constituting B may be given, for example, as

$$H_0 = \ell^2,$$

$$A = \text{diag} \left[\underbrace{\mu_1^2 \cdots \mu_1^2}_{n_1} \, \mu_2^2 \cdots \underbrace{\mu_i^2 \cdots \mu_i^2}_{n_i} \cdots \right].$$

REFERENCES

[1] S. Agmon, "Lectures on Elliptic Boundary Value Problems," Van Nostrand, Princeton, 1965.

[2] R.F. Curtain, *Finite dimensional compensators for parabolic distributed systems with unbounded control and observation*, SIAM J. Control Optim. **22** (1984), 255–276.

[3] T. Kato, "Perturbation Theory for Linear Operators," Springer-Verlag, New York, 1976.

[4] T. Nambu, *On stabilization of partial differential equations of parabolic type: boundary observation and feedback*, Funkcial. Ekvac. **28** (1985), 267–298.

[5] T. Nambu, *The Ljapunov equation and an application to stabilisation of one-dimensional diffusion equations*, Proc. Roy. Soc. Edinburgh **104A** (1986), 39–52.

[6] T. Nambu, *Degenerate self-adjoint perturbation in Hilbert space*, Proc. Japan Acad. Ser. **A 63** (1987), 379–381.

[7] R.E.A.C. Paley and N. Wiener, "Fourier Transforms in the Complex Domain," Amer. Math. Soc. Colloq. Publ., New York, 1934.

[8] Y. Sakawa, *Feedback stabilization of linear diffusion systems*, SIAM J. Control Optim. **21** (1983), 667–676.

[9] O. Szász, *Über die Approximation stetiger Funktionen durch lineare Aggregate von Potenzen*, Math. Ann. **77** (1916), 482–496.

[10] E.C. Titchmarsh, "The Theory of Functions," Oxford Univ. (Clarendon) Press, Oxford, 1939.

[11] K. Yosida, "Functional Analysis," 6th Ed., Springer-Verlag, Berlin Heidelberg, 1980.

Takao Nambu
Department of Mathematics
Faculty of Engineering
Kumamoto University
Kumamoto 860
Japan

International Series of
Numerical Mathematics, Vol. 91
© 1989 Birkhäuser Verlag Basel

Optimal solutions for a
free boundary problem for crystal growth

Pekka Neittaanmäki and Thomas I. Seidman

Department of Mathematics
University of Jyväskylä

Department of Mathematics and Statistics
University of Maryland Baltimore County

Abstract. We consider a free boundary problem modeling the growth/dissolution of a crystal in a radially symmetric setting. Existence of an optimal boundary control, minimizing a cost functional of a standard 'integral-quadratic' form, is already known and we here consider the characterization and computation of such an optimal control.

Keywords. Free boundary problem, nonlinear, parabolic, partial differential equation, boundary control, optimal control, optimality conditions, computation.

1980 *Mathematics subject classifications*: 49B25

1. Introduction

In [7] an existence proof was given for an optimal boundary control minimizing a fairly standard sort of cost functional $\mathcal{J}$ with dynamics given by a free boundary problem corresponding to a model of growth (dissolution) of a radially symmetric crystal grain. Our present concerns are: the development of the corresponding optimality conditions, with some consequent additional regularity, and also the computational aspects of the optimization.

Consider a solution of some substance surrounding a 'crystal grain' of the pure material. Letting $\Omega = \Omega(t)$ be the (bounded) spatial region in $\mathbb{R}^d$ occupied by the solution, we assume that the concentration C will satisfy in $\mathcal{Q} := \{(t, x) : x \in \Omega(t), \, 0 < t < T\}$ a (nonlinear) diffusion-reaction equation of standard form:

$$(1.1) \qquad C_t = \Delta C + F(C).$$

We consider the problem only in a context of radial symmetry: $0 \le r := |x| \le 1$ with the crystal grain occupying $0 \le r \le R(t)$ so

$$\Omega(t) = \{x : R(t) < r := |x| < 1\}.$$

Note that by a suitable choice of units, we have taken the diffusion coefficient and the outer radius each to be 1; we will also use units for which the concentration for the crystal, itself, is 1. We then consider the data $\alpha(\cdot)$ for the Dirichlet boundary condition

$$(1.2) \qquad C(t, x) = \alpha(t) \qquad \text{for } |x| = 1, \, 0 < t \le T$$

The research by TIS was partially supported by the Air Force Office of Scientific Research under grants #AFOSR-87-0190 and #AFOSR-87-0350. Thanks also are due to the University of Jyväskylä for its hospitality and support.

as our *boundary control*.

The motion of the crystal boundary is given by $\dot{R} = h$ for a suitable function $h(\cdot)$: growth for $h > 0$, dissolution for $h < 0$. If h were given *a priori*, then we could simply adjoin the 'conservation of mass' boundary condition (2.3) at $r = R(t)$ to have a *moving boundary problem*, specified by the pair $[\alpha, h]$ and the initial conditions. Actually, the crystal growth is determined by a *constitutive relation* of the form:[1]

$$(1.3) \qquad\qquad h = H(\omega, R) \qquad \text{with } \omega := C|_{r=R+}$$

and, coupling with this, we have the free boundary problem of the title. We note that the most commonly used special form of this constitutive relation is

$$(1.4) \qquad H(\omega, R) = K(\omega - C_*(R)) \qquad \text{with } C_*(R) = \gamma_0 e^{\gamma/R},$$

following Gibbs-Nernst and Ostwald-Freundlich, cf. [2]. The problem will then be radially symmetric and we can reduce to one space dimension, replacing ΔC by $[C_{rr} + \frac{d-1}{r} C_r]$ in taking C as $C(t,r)$. Indeed, we later introduce the variable $y := \frac{1-|x|}{1-R}$ to work with the *fixed* spatial domain $(0,1)$.

We note at this point that (under suitable conditions) well-posedness of the free boundary problem is demonstrated in [3] for smooth $\alpha(\cdot)$ while an existence and closure result (needed for the existence of optimal boundary controls) is given in [7] for less smooth α; more detail later. Our theoretical goal in this paper is to obtain the optimality condition (first order necessary conditions) characterizing minimizers of the cost functional $\mathcal{J}$. Since we do not even have uniqueness of the solution map: $\alpha \mapsto C$, much less differentiability of $\mathcal{J}$ as a function of α, for α with the regularity known for the optimal control, there are some technical problems involved in the justification and computation of these conditions. Our approach will be to note differentiability of $\mathcal{J}$ as a function of the pair $[\alpha, h]$ for the decoupled problem and then to use the Implicit Function Theorem, along with the existence result noted above, for the coupled problem.

We then also present some corresponding numerical computation for the optimal boundary control. For this computation we will take Ω to be 2-dimensional $(d = 2)$ and will take $K = 1$, $\gamma_0 = .01$, and $\gamma = 1$ in (1.4) so, written in terms of ω and $\rho = 1/(1 - R)$, we will then be considering the specific relation

$$(1.5) \qquad\qquad H = \omega - .01\, e^{1/(1-1/\rho)}.$$

2. Statement of the Problem

We consider somewhat more precisely the problem under consideration. We begin with the *decoupled problem* which, in its first formulation, consists of (1.1) on $Q = Q_T$ given by $Q := \{(t,x) : R(t) < r := |x| < 1,\ 0 < t < T\}$, with $R(\cdot) > 0$ given by the ordinary differential equation

$$(2.1) \qquad\qquad\qquad \dot{R} = h(\cdot),$$

---

[1]If it were not for the assumption of radial symmetry, we would have H as a function of the local concentration ω and of the *curvature* κ; see, e.g., [5]. Here, that dependence is on $R = 1/\kappa$.

together with the initial conditions

$$(2.2) \qquad R(0) = R_0, \qquad C(0, \cdot) = C_0(\cdot) \quad \text{on } (R_0, 1),$$

the Dirichlet boundary condition (1.2) at $r = 1$, and the flux boundary condition

$$(2.3) \qquad C_r = (1 - C)h \qquad \text{at } r = R(t),$$

corresponding to conservation of mass at the moving crystal boundary, with the concentration normalized to 1 for the crystal itself. We assume throughout that:

$$(\mathbf{H}) \qquad \begin{cases} (i) & F \text{ is nonincreasing with } F(0) \geq 0 \geq F(1); \\ (ii) & 0 < R_0 < 1 \text{ and } 0 \leq C_0 \leq 1 \text{ on } (R_0, 1); \\ (iii) & 0 \leq \alpha \leq 1 \text{ on } (0, T); \\ (iv) & h \text{ is bounded and gives } 0 \leq R \leq 1 \text{ on } (0, T) \end{cases}$$

with F smooth and C_0, α, h measurable. Fixing the initial data, we expect a map $[\alpha, h] \longmapsto [C, \omega]$ so we will have $\mathcal{J}$ well-defined as a function of $[\alpha, h]$. The free boundary problem couples this with (1.3), and we assume that

$$(2.4) \qquad H \text{ is smooth with } H(\cdot, 0) \leq 0.$$

Note that this is certainly satisfied (for $R > 0$) for H as in (1.4).

As indicated earlier, it is convenient to reformulate the problem in order to work on a fixed, one-dimensional spatial domain. Thus we make a change of variables: $x \mapsto y$, setting

$$(2.5) \qquad u(t, y) = C(t, x) \qquad \text{with } y := \frac{1 - |x|}{1 - R(t)}.$$

Note that $y = \rho(1 - r)$ where $\rho(\cdot) := 1/(1 - R(\cdot)) > 1$ so $R = 1 - 1/\rho$ and $|x| = r = 1 - y/\rho$. The problem $[(2.1), (1.1), (1.2), (2.3), (2.2)]$ now just becomes:

$$(2.6) \qquad \begin{cases} \dot{\rho} = \rho^2 h & \text{on } (0, T) \\ u_t = \rho^2 u_{yy} - \psi u_y + F(u) & \text{on } (0, T) \times (0, 1) \\ -\rho u_y = (1 - \omega)h & \text{at } y = 1 \\ u = \alpha \; [= \text{control}] & \text{at } y = 0 \\ \rho = \rho_0 \text{ and } u = u_0(\cdot) & \text{at } t = 0 \end{cases}$$

where $\psi := \rho \left[\dfrac{(d - 1)\rho}{\rho - y} + yh \right]$ and $\omega := u(\cdot, 1)$ with $\rho_0 = 1/(1 - R_0) > 1$. Note that $y = 0, 1$ correspond to $r = 1, 0$. Of course we obtain $\rho_0 > 1$ and $u_0(\cdot)$ on $(0, 1)$ from (2.2). The free boundary problem just couples this with (1.3), no longer taking h as given; we assume H is now formulated as $H(\omega, \rho)$.

Theorem 1 [[7], [3]]. *Assume* (H)-(*i*, *ii*) *and* (2.4). *Then there is a (weak) solution* $u \in L^\infty(\to [0,1])$ *of the coupled problem* (2.6), (1.3) *for each* $\alpha \in L^\infty(\to [0,1])$; *the set of such* $[\alpha, u, h]$ *(with h given by* (1.3) *and necessarily satisfying* (H)-(*iv*)) *is closed (in, say, the weak* L^2 *product topology). If* $\alpha \in H^1(0,T)$, *then the solution u is unique.*

Some caution is needed here. Since the existence results of [7] and [3] only give existence on some interval $(0, T_\alpha)$, it is then an assumption of the control problem that we can consider controls for which $T_\alpha \geq T$. We remark that existence may fail on the full interval $[0, T]$ only by having $R \to 0$ or $R \to 1$. We also note that [7] does not assert uniqueness of the solution but, under the stronger regularity assumptions, this is given by the different argument of [3]. Each of these proves existence first for a modified problem and then uses a maximum principle argument to show the modification has no effect. Finally, and most important for our subsequent considerations, we note that the existence argument of [7], using interior regularity results for parabolic equations, gives (local) $L^2(\to H^1)$ regularity for u away from the outer boundary $y = 0$ — where trace theory tells us that we *cannot* achieve this without requiring $\alpha \in H^{1/2}[0,T]$; we write $\mathcal{H}$ for this space of functions "H^1 away from $y = 0$" so $u \in L^2(\to \mathcal{H})$. On the other hand, the resulting function h will be as regular as the smoothness of F and H will permit.

We will be somewhat specific and consider here an integral cost functional of the fairly standard quadratic type:

$$(2.7) \qquad \mathcal{J} = \frac{1}{2} \int_0^T (\rho - \rho_*)^2 + \frac{\mu}{2} \int_0^T (\alpha - \alpha_*)^2$$

with ρ_*, α_* given. This is not quite the form considered in [7] but, as the argument is essentially identical, we take as proven there for *this* functional the following existence result:

Theorem 2 [[7]]. *Assume* (H)-(*i*, *ii*) *and* (2.4); *assume there is at least one admissible control* $\alpha \in L^\infty(\to [0,1])$ *for which one has existence on* $[0, T]$, *making* $\mathcal{J}$ *finite. Then there is an optimal boundary control* $\hat{\alpha} \in L^\infty(\to [0,1])$, *minimizing the cost functional* $\mathcal{J}$ *of* (2.7), *subject to* (2.6), (1.3) *as constraints.*

Our objective is the characterization and computation of this $\hat{\alpha}$. We will obtain the (first order) optimality condition for $\hat{\alpha}$ and will also use this to obtain some improved regularity, showing that the minimizer is actually in $H^1(0,1)$ (with greater regularity on intervals for which the constraints $0 \leq \hat{\alpha} \leq 1$ are not active. (We do this for $\mathcal{J}$ given by (2.7) but note that the arguments would be essentially the same for a variety of cost functionals of roughly similar form.)

3. Differentiability

In this section and the next we present our main theoretical results, regarding the optimality condition characterizing the optimal boundary control whose existence was shown in [7]. In order to obtain such conditions one needs some form of differentiability of the cost functional with respect to variation of the control α and, unfortunately, we do not even know uniqueness, much less continuity or differentiability of the dependence of the solution u on α. We approach this by observing that u depends nicely on the pair $[\alpha, h]$ if one were to consider the *uncoupled problem* and that one can then apply the Implicit Function Theorem to obtain the effect of the exogenous constraint (1.3).

Theorem 3. *Assume* **(H)**. *Then there is a unique (weak) solution u of (2.6) in $L^2(\to \mathcal{H}) \cap L^\infty(\to L^2)$ for each admissible pair $[\alpha, h]$. This solution u (and $\omega := u(\cdot, 1)$) depends Frechet differentiably on $[\alpha, h]$.*

Proof: The system (2.6) is of a quite standard form for a quasilinear parabolic PDE once one has obtained $\rho(\cdot)$ from h and then shifted u by the (fixed) solution $\tilde{u}$ of the *linear* equation

$$\tilde{u}_t = \rho^2 \tilde{u}_{yy} - \psi \tilde{u}_y; \qquad \tilde{u} = \alpha \ [y = 0]; \qquad -\rho \tilde{u}_y + h\tilde{u} = h \ [y = 1],$$

There is then no difficulty in obtaining (by entirely standard methods, e.g., treating the nonlinearity by a contraction argument) $u - \tilde{u}$ as the unique solution of the resulting nonlinear equation (now with homogeneous BC). [We note that, at this point, there is no need for the hypotheses that $0 \le \alpha \le 1$, $0 \le u_0 \le 1$.]

Now consider the same system, replacing α by $\hat{\alpha} = \hat{\alpha}^\varepsilon := \alpha + \varepsilon \bar{\alpha}$ and h by $\hat{h} = \hat{h}^\varepsilon := h + \varepsilon \bar{h}$ with $\varepsilon \ne 0$; let the solution be $\hat{u} = \hat{u}^\varepsilon$. Note that $\alpha, h, \hat{\alpha}, \hat{h}$ are all arbitrary functions in, say, $L^2(0, T)$ except that **(H)**-(iv) holds for $h, \hat{h}$.

Now consider $\bar{u}^\varepsilon := (\hat{u}^\varepsilon - u)/\varepsilon$. This (together with the corresponding divided differences corresponding to ρ, ω, etc.) then satisfies a system which we will not bother to write out explicitly — other than to note that there is a term: $[F(u + \varepsilon \bar{u}^\varepsilon) - F(u)]/\varepsilon$ to which we may apply the Mean Value Theorem, etc.

It is not difficult to estimate $\bar{u}^\varepsilon$ with a bound (in the same space as for the existence of u) independent of ε, for $0 \ne \varepsilon$ near 0. Thus, as $\varepsilon \to 0$, one always obtains weakly convergent subsequences.

For any such subsequence one then easily sees, again by a standard argument, that the weak limit $\bar{u}$ satisfies the (formally obtained) limit system:

$$(3.1) \quad \begin{cases} \dot{\bar{\rho}} = (2\rho h)\bar{\rho} + \rho^2 \bar{h} & \text{on } (0, T) \\ \bar{u}_t = \rho^2 \bar{u}_{yy} - \psi \bar{u}_y + F'(u)\bar{u} + a\bar{\rho} + b\bar{h} & \text{on } (0, T) \times (0, 1) \\ -\rho \bar{u}_y = -h\bar{\omega} - [h(1-\omega)/\rho]\bar{\rho} + (1-\omega)\bar{h} & \text{at } y = 1 \\ \bar{u} = \bar{\alpha} & \text{at } y = 0 \\ \bar{\rho} = 0 \text{ and } \bar{u} = 0 & \text{at } t = 0, \end{cases}$$

where $a := 2\rho u_{yy} - \psi_\rho u_y$, $b := -y\rho u_y$ and $\bar{\omega} := \bar{u}(\cdot, 1)$. Since (3.1) is easily seen to have a unique solution, this shows that $\bar{u}$ is, indeed, the derivative of the (nonlinear) solution map $\mathbf{U} : [\alpha, h] \mapsto u$ of (2.6) at $[\alpha, h]$ in the direction of $[\bar{\alpha}, \bar{h}]$, i.e., this solution map is differentiable with a Gateaux derivative $\mathbf{U}'(\alpha, h) : [\bar{\alpha}, \bar{h}] \mapsto \bar{u}$ defined by (3.1); our interest really will be in the induced map: $[\bar{\alpha}, \bar{h}] \mapsto \bar{\omega}$. Some caution is needed as to the space in which to consider $\bar{u}$ in view of our concerns as to the regularity of u, and so of u_{yy} appearing in a, etc. The important consideration for us is that this concern applies only near $y = 0$ and we have no difficulties at all, in this respect, in treating the trace $\bar{\omega}$ on $y = 1$, where the regularity is essentially independent of the regularity of $\hat{\alpha}, \bar{\alpha}$. The observation that this is actually a Frechet derivative follows on verification of the continuous dependence of the operator on $[\alpha, h]$ in L^2.

The system (3.1) is the *variational system* for the decoupled problem given by (2.6). We now consider the relation of this to the free boundary problem, obtained by coupling (2.6) with (1.3). We begin with the obvious differential corollary to the constraint (1.3):

$$\bar{h} = H_\omega \bar{\omega} + H_\rho \bar{\rho}$$

and use this to eliminate $\bar{h}$ in (3.1), obtaining

(3.2)
$$\begin{cases} \dot{\bar{\rho}} = \rho c_1 \bar{\rho} + \rho d_1 \bar{\omega} & \text{on } (0, T) \\ \bar{u}_t = \rho^2 \bar{u}_{yy} - \psi \bar{u}_y + F'(u)\bar{u} + c_2 \bar{\rho} + d_2 \bar{\omega} & \text{on } (0, T) \times (0, 1) \\ -\rho \bar{u}_y = c_3 \bar{\rho} - d_3 \bar{\omega} & \text{at } y = 1 \\ \bar{u} = \bar{\alpha} & \text{at } y = 0 \end{cases}$$

with homogeneous initial condition at $t = 0$ as before and setting:

$$c_1 := 2h + \rho H_\rho \qquad\qquad d_1 := \rho H_\omega$$
$$c_2 := 2\rho u_{yy} - \psi_\rho u_y - \rho y H_\rho \qquad d_2 := -y u_y H_\omega$$
$$c_3 := (1 - \omega)[H_\rho - h/\rho] \qquad d_3 := h - (1 - \omega)H_\omega$$

Theorem 4. *Assume* **(H)**-(i, ii) *and let* $\hat{\alpha}, \hat{h}, \hat{u}$, *etc., satisfy the coupled problem* [(2.6), (1.3)] *with, necessarily,* $\hat{\rho} > 1$ *continuous. Then (locally, for* α *in an* L^2 *neighborhood of this* $\hat{\alpha}$) *there is a (uniquely) well-defined and differentiable map:* $\alpha \mapsto [h, \omega]$ *(giving* h, u *close to* $\hat{h}, \hat{u}$). *The derivative of this is the linear map:* $\bar{\alpha} \mapsto [\bar{h}, \bar{\omega}]$, *given by (3.2). This also gives a bounded linear operator*

(3.3) $$\mathbf{L} = \mathbf{L}(\hat{\alpha}, \hat{u}) : \bar{\alpha} \mapsto \bar{\rho} : L^2(0, T) \to L^2(0, T).$$

Proof: This will follow from the Implicit Function Theorem, applied to the map $\mathbf{D}$: $[\alpha, h] \mapsto [h - H(\omega, \rho)]$, as given by (2.6). It is clear that $\mathbf{D}$ is (continuously) Frechet differentiable and we must verify that $\mathbf{D}_h$ (at $[\hat{\alpha}, \hat{h}]$) is a boundedly invertible linear operator on, say, $L^2(0, T)$.

Note that $\mathbf{D}_h : \bar{h} \mapsto [\bar{h} - H_\omega \bar{\omega} + H_\rho \bar{\rho}]$, with $\bar{\omega}, \bar{\rho}$ obtained from (3.1) with $\hat{\alpha} = 0$. Since it is clear that $\bar{\omega}, \bar{\rho}$ depend compactly on $\bar{h}$, we have $\mathbf{D}_h = [\mathbf{I} + (compact)]$ and spectral theory tells us that the bounded invertibility follows if 0 is not an eigenvalue, i.e., from the observation (e.g., by a Gronwall estimate) that the system (3.2), has only the unique trivial solution when $\bar{\alpha} = 0$. The Implicit Function Theorem then asserts the local existence of a C^1 map: $\alpha \mapsto h$ giving (1.3), as desired, with derivative $[\mathbf{D}_h]^{-1}\mathbf{D}_\alpha$ — just corresponding to (3.2).

Note that this ensures that the optimal boundary control, whose existence is guaranted by Theorem 2, cannot be isolated (among admissible controls for which the free boundary problem has a solution) and that one will have (directional) differentiability of $\mathcal{J}$ with respect to the relevant 'directions' $\bar{\alpha}$ of admissible variation — where we call $\bar{\alpha}$ *admissible* provided $\alpha = \hat{\alpha} + s\bar{\alpha}$ satisfies **(H)**-(iii) for sufficiently small $s \geq 0$.

Corollary. *The solution map:* $\alpha \mapsto u$ *for the free boundary problem — as in Theorem 1 — is well-defined and continuous from* $L^2((0,T) \to [0,1])$.

Proof: We have *local* continuity and, by [3], uniqueness from $H^1((0,T) \to [0,1])$ — whose density then gives global uniqueness.

4. Optimality conditions

From Theorem 4 we may differentiate $\mathcal{J}$ at $\hat{\alpha}$ (using $\hat{u}$, etc., for evaluation) to obtain

(4.1) $\langle \mathcal{J}', \bar{\alpha} \rangle = \langle \hat{\alpha} - \alpha_*, \bar{\alpha} \rangle + \mu \langle \hat{\rho} - \rho_*, \bar{\rho} \rangle$

for every admissible variation $\bar{\alpha}$. Note that we can rewrite this as $\langle \mathcal{J}', \bar{\alpha} \rangle = \langle \hat{\alpha} - \hat{\beta}, \bar{\alpha} \rangle$ with

(4.2) $\hat{\beta} := \alpha_* + \mu \mathbf{L}^*(\hat{\rho} - \rho_*)$

where $\mathbf{L}$ is given, as in (3.3), by (3.2).

To use this we must be able to calculate the adjoint $\mathbf{L}^*$ and we now proceed to do this. We are considering $\bar{\rho}$ related to $\bar{\alpha}$ by (3.2) and wish to find a mapping: $\zeta \mapsto \Upsilon$ such that $\int_0^T \zeta \bar{\rho} = \int_0^T \Upsilon \bar{\alpha}$.

Consider φ, v satisfying

(4.3)
$$\begin{cases} -\dot{\varphi} = c_1 \varphi + f_1 & \text{on } (0,T) \\ -v_t = \rho^2 v_{yy} - \psi v_y + F'(u)v & \text{on } (0,T) \times (0,1) \\ \rho v_y = f_2 & \text{at } y = 1 \\ v = 0 & \text{at } y = 0 \end{cases}$$

where

$$f_1 := [\zeta + (\int_0^T c_2 v) - \rho c_3 v|_{y=1}]$$

$$f_2 := [d_1 \varphi + (\int_0^T d_2 v) + [d_3 - \psi/\rho]v|_{y=1}]$$

and $\Upsilon := \rho^2 v_y|_{y=0}$ with homogeneous 'initial condition' at $t = T$ and using c_1, etc., as in (3.2). Note that the integrals appearing in the definitions of f_1 and f_2 are with respect to y, over $(0,1)$. (In particular, we have (from c_2) such an integral of $2\rho u_{yy} v$ and we use an integration by parts to rewrite this term as $-2[(1-w)hv|_{y=1} + \rho \int_0^T u_y v_y]$; there is some question about the regularity of u_y near $y = 0$ but we have enough regularity for v there to compensate, giving f_1 in $L^1(0,T)$.)

Lemma 4.2. *Let* $\mathbf{L}$ *be given as in (3.3). Then the adjoint is given, through (4.3), by* $\mathbf{L}^* : \zeta \mapsto \Upsilon := \rho^2 v_y|_{y=0}$.

Proof: From the initial conditions we have $\int_0^T [\dot{\bar{\rho}}\varphi + \bar{\rho}\dot{\varphi}] = 0$ and $\int_0^T \int_\Omega [\bar{u}_t v + \bar{u} v_t] = 0$. Now use the PDEs to substitute for $\bar{u}_t, v_t$; then integrate by parts and use the boundary conditions to obtain:

$$\int_0^T \zeta \bar{\rho} = \int_0^T \Upsilon \bar{\alpha} + \int_0^T f_1 \bar{\rho} + \int_0^T \rho[(f_2 - d_1\varphi) - \rho v_y|_{y=1}]\bar{w}.$$

Using the identity $\int_0^T f_1 \bar{\rho} = \int_0^T \rho d_1 \varphi \bar{w}$, obtained from the ODEs, then gives $\mathbf{L}^* \zeta = \Upsilon$.

Theorem 5. *Assume* (H)-(i, ii) *and* (2.4) *and let* $\hat{\alpha}, \hat{u}$ *be such that the cost functional* $\mathcal{J}$ *of* (2.7) *is minimized, subject to* (2.6), (1.3), *over* $\alpha \in L^\infty(\to [0, 1])$. *Then the optimal boundary control* $\hat{\alpha}$ *is given by*

$$(4.4) \qquad \hat{\alpha} = \begin{cases} 1 & \text{when } \alpha_* + \Upsilon \geq 1 \\ 0 & \text{when } \alpha_* + \Upsilon \leq 0 \\ \alpha_* + \Upsilon & \text{when } 0 \leq \alpha_* + \Upsilon \leq 1 \end{cases}$$

where $\Upsilon = \rho^2 v_y|_{y=0}$ *from* (4.3) *with* $\zeta := \mu(\hat{\rho} - \rho_*)$ — *and this is to be further coupled with* [(2.6), (1.3)].

Proof: At the minimizer $\hat{\alpha}$ we have $\langle \mathcal{J}', \bar{\alpha} \rangle \geq 0$ for any admissible $\bar{\alpha}$. Let $\Upsilon := \rho^2 v_y|_{y=0}$ be given by (4.3) with $\zeta := \mu(\hat{\rho} - \rho_*)$ so (4.2) gives $\hat{\beta} := \alpha_* + \Upsilon$. Now, for $\varepsilon > 0$, consider $S = \{t : \hat{\alpha} \leq (1 - \varepsilon), (\hat{\beta} - \varepsilon)\}$ and let $\bar{\alpha} = $ (characteristic function of S); one easily sees that $\bar{\alpha}$ is admissible. However, $(\hat{\alpha} - \hat{\beta}) \leq -\varepsilon$ on S so $\langle \hat{\alpha} - \hat{\beta}, \bar{\alpha} \rangle \leq -\varepsilon |S|$ — which would contradict the non-negativity of $\langle \mathcal{J}', \bar{\alpha} \rangle$ unless the measure $|S|$ were 0. As $\varepsilon > 0$ was arbitrary, this gives $\hat{\alpha} \geq \min\{1, \hat{\beta}\}$ ae. We get $\hat{\alpha} \geq \max\{0, \hat{\beta}\}$ ae similarly and together these give (4.4).

Corollary. *The optimal boundary control* $\hat{\alpha}$ *obtained by Theorem 2 is actually in* $H^1(0, T)$ *if* α_* *is.*

Proof: By (4.4), it is sufficient to show that $\Upsilon \in H^1$ For which we need only see that (4.3) gives sufficient regularity for v near $y = 0$. We can *almost* just cite [6] to get $v \in H^{1,2}$ as we have $F'(u) \in L^2$, etc. We must observe that regularity for parabolic equations localizes so we can ignore the question near $y = 1$ and then note that the only relevant boundary data is that at $y = 0$ — obviously smooth; the leading coefficient $\hat{\rho}^2$ is smooth enough not to interfere with this.

Our arguments do not permit us to treat the case $\mu = 0$ in (2.7) directly but it is interesting to consider this as a limit. If α_μ denotes a minimizer corresponding to $\mu > 0$, then the constraint ensures existence of a subsequence $\alpha_\mu \rightharpoonup \alpha_0$ weakly convergent in $L^2(0, T)$ for which the arguments of [7] give suitable convergence to a minimizer of the limit functional. It is then only necessary to pass to the limit in Theorem 5 to obtain the optimality condition for α_0; the relevant change is that we now have $\zeta = 0$ in f_1 as used in (4.3). Clearly, this argument does not preserve, for α_0, the regularity obtained in the Corollary above.

5. Numerical results

We will consider numerically the problem

$$(5.1) \qquad \min_{\alpha \in U} \left\{ \mathcal{J}(\alpha) \equiv \frac{1}{2} \int_0^1 (\rho(t) - \rho_*(t))^2 \, dt \right\}$$

subject to the 'constraint' that (ρ, u) satisfies the nonlinear parabolic system (2.6) with H

specifically as in (1.4) — i.e.,

(5.2)
$$\begin{cases} \dot{\rho}(t) = \rho(t)^2 \left(u(t,1) - \gamma_0 e^{\delta\rho(t)/(\rho(t)-1)} \right), \quad t \in (0,1) \\ \dot{u}(t,y) = \rho(t)^2 u_{yy}(t,y) - \rho(t) \left[\dfrac{\rho(t)}{\rho(t) - y} + y\left(u(t,1) - \gamma_0 e^{1/(1-1/\rho(t))} \right) \right] u_y, \\ \qquad\qquad\qquad\qquad\qquad\qquad y \in (0,1), \ t \in (0,1) \end{cases}$$

with the boundary conditions

(5.3)
$$\begin{cases} u(t,0) = \alpha(t) \\ u_y(t,1) = -\dfrac{1}{\rho(t)} (1 - u(t,1))\left(u(t,1) - \gamma_0 e^{1/(1-1/\rho(t))}\right), \quad t \in (0,1) \end{cases}$$

and the initial conditions

(5.4)
$$\begin{cases} \rho(0) = \rho_0 \\ u(0,y) = u_0(y), \qquad y \in (0,1). \end{cases}$$

Note that (5.1) corresponds to (2.7) with $\mu = 0$. The set of admissible controls is $U = \{\alpha \mid \alpha_{\min} \le \alpha \le \alpha_{\max}\}$.

One can derive discrete analogues to (5.1)–(5.4) in standard ways: we have employed the trapezoidal rule for computing $J(\alpha)$, the finite element method with linear elements for the spatial discretization of (5.2)–(5.3), and the implicit Euler method for time discretization. The resulting nonlinear state system can then be solved, for example, by a combination of Newton and overrelaxation methods. In order to apply efficient nonlinear programming algorithms for minimizing the discretized cost functional $\mathcal{J}$, one must be able to compute the (discretized) gradient $\nabla_\alpha \mathcal{J}$. This can be done in a well known manner by applying the adjoint state technique for the discretized state system (see [4]).

Example 5.1. In (5.1)–(5.4) let $\rho_*(t) = 2 + 2t$ (i.e., $R_*(t) = 1 - 1/(2 + 2t)$), $\gamma_0 = 0.01$, $\rho_0 = 2$ and $u_0(y) = -y(y-2)$ (i.e.

$$C(0,r) = C_0(r) = \begin{cases} 1, & 0 \le r \le 1/2 \\ -4r(r-1), & 1/2 < r \le 1 \ ). \end{cases}$$

In the numerical results presented here we have taken $\Delta y = 1/20$ for the space discretization and $\Delta t = 1/20$ for the time discretization. As a result one has 21 unknowns in the minimization of $\mathcal{J}$. The conjugate gradient algorithm with box constraints $\alpha_{\min} = 0.05$ and $\alpha_{\max} = 0.85$ was then used for this minimization.

As an initial guess for α we have chosen a constant control, $\alpha(t) \equiv 0.15$ (Figure 5.1). Figure 5.2 shows $R(t)$ (— line) and $R_*(t)$ (--- line). In Figure 5.3 we see the corresponding concentration $C(t,r)$.

In Figure 5.4 we then see the numerical approximation to the optimal control α after 3 iterations of the conjugate gradient algorithm. Figure 5.5 shows corresponding value of $R(t)$ (— line) and $R_*(t)$ (--- line) and Figure 5.6 shows the corresponding values of the evolving concentration.

The authors are indebted to Dr. E. Laitinen for his help in carrying out these numerical tests.

A forthcoming paper of the authors' will provide further details on the numerical realization of algorithms for finding optimal controls in such crystal growth/dissolution problems.

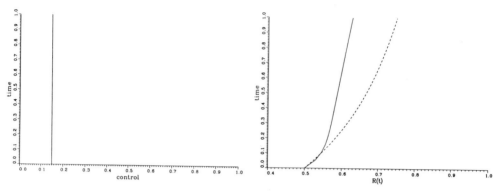

Figure 5.1   Figure 5.2

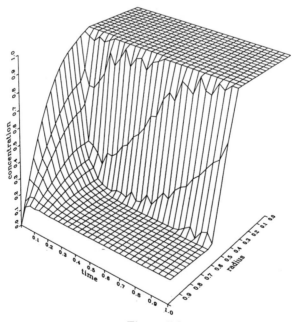

Figure 5.3

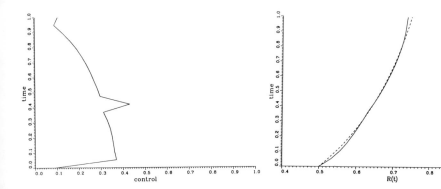

Figure 5.4   Figure 5.5

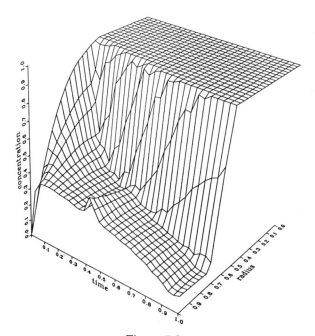

Figure 5.6

REFERENCES

[1] J. P. Aubin, *Un théorème de compacité*, CRAS de Paris **265** (1963), 5042–5045.
[2] F. Conrad and M. Cournil, *Free boundary problems in dissolution-growth processes*, preprint.
[3] F. Conrad, D. Hilhorst, and T.I. Seidman, *Well-posedness of the free boundary problem for a dissolution-growth process*.
[4] E. Laitinen and P. Neittaanmäki, *On numerical solution of the problem connected with the control of the secondary cooling in the continuous casting process*, Control-Theory and Advanced Technology **4** (1988).
[5] J. S. Langer, *Instability and pattern formation in crystal growth*, Rev. Mod. Phys. **52** (1980), 1–28.
[6] J.L. Lions and E. Magenes, "Non-Homogeneous Boundary Value Problems and Applications, vol. II," Springer-Verlag, New York, 1972.
[7] T.I. Seidman, *Some control-theoretic questions for a free boundary problem*, in "Control of Partial Differential Equations," LN in Control and Inf. Sci. #114, edit. A. Bermudez, Springer-Verlag, New York, 1988, pp. 265–276.

Pekka Neittaanmäki
Department of Mathematics
University of Jyväskylä
Jyväskylä
Finland
(Bitnet: neittaanmaki@FINJYU)

Thomas I. Seidman
Department of Mathematics and Statistics
University of Maryland Baltimore County
Baltimore, MD 21228
USA
(Bitnet: seidman@UMBC).

International Series of
Numerical Mathematics, Vol. 91
© 1989 Birkhäuser Verlag Basel

On Hilbert-Schmidt norm convergence of Galerkin approximation for operator Riccati equations

I.G. Rosen[1]

Department of Mathematics
University of Southern California

Abstract. An abstract approximation framework for the solution of operator algebraic Riccati equations is developed. The approach taken is based upon a formulation of the Riccati equation as an abstract nonlinear operator equation on the space of Hilbert-Schmidt operators. Hilbert-Schmidt norm convergence of solutions to generic finite dimensional Galerkin approximations to the Riccati equation to the solution of the original infinite dimensional problem is argued. The application of the general theory is illustrated via an operator Riccati equation arising in the linear-quadratic design of an optimal feedback control law for a one dimensional heat/diffusion equation. Numerical results demonstrating the convergence of the associated Hilbert-Schmidt kernels are included.

Keywords. Operator algebraic Riccati equation, Hilbert-Schmidt operator, nonlinear operator equation, Galerkin approximation, linear quadratic regulator

1980 *Mathematics subject classifications*: 41A65, 47B10, 65J15, 93C25

1. Introduction

In this paper we develop an abstract approximation theory for algebraic Riccati equations on spaces of Hilbert-Schmidt operators. Our approach is based upon Barbu's [1] formulation of a class of Riccati equations as abstract nonlinear operator equations on a space of Hilbert-Schmidt operators. We argue that solutions to generic finite dimensional Galerkin approximations to the Riccati equation converge in Hilbert-Schmidt norm to the solution of the original infinite dimensional equation.

Our effort here is closely related to results in one of our earlier papers [8] wherein we developed an approximation theory for operator Riccati differential equations using techniques similar to those which will be employed below. Our treatments here and in [8] differ from the standard approach to the analysis of operator Riccati equation approximation in that we consider the nonlinear operator equations directly in the space of Hilbert-Schmidt operators rather than integral equation equivalents and their limiting properties as the time horizon tends to infinity (see, for example, [4] and [5]). While we do in fact obtain a stronger convergence result than the ones yielded by the standard approach, the class of problems to which our theory applies is somewhat more restricted. Indeed, the linear part of the equation must be strongly monotone in a space of Hilbert-Schmidt operators, and the nonlinear component must be monotone. Also, the operator on the right-hand side of

[1] This research supported in part by the United States Air Force Office of Scientific Research under grant AFOSR-87-0356. Part of this research was carried out while the author was a visiting scientist and consultant at the Institute for Computer Applications in Science and Engineering at the NASA Langley Research Center, Hampton, VA, which is operated under NASA contract NAS1-18107.

the equation must be Hilbert-Schmidt. In the context of linear-quadratic control problems, for example, in order to apply our theory we must require that the system dynamics be regularly dissipative (i.e. abstract parabolic and uniformly exponentially stable) and that the state penalty operator be Hilbert-Schmidt. There are also some restrictions on the input and control penalty operators as well. In the case of the infinite dimensional filtering problem on the other hand, formulation in a space of Hilbert-Schmidt operators is more natural (see [2] and [4]). In any case, however, our primary intent in this effort and in [8] was to make a first attempt at applying some of the available nonlinear functional analytic approximation techniques to abstract operator Riccati equations. It is hoped and intended that further investigation and research in this general direction will follow.

In section 2 we briefly outline and summarize Barbu's [1] results for operator algebraic Riccati equations on spaces of Hilbert-Schmidt operators. Our approximation results are then given in section 3. The convergence arguments given there depend heavily upon a factorization result for Hilbert-Schmidt operators and its subsequent implications regarding the convergence of Galerkin approximations. In the fourth section we discuss the application of our results to a Riccati equation arising in the linear-quadratic design of an optimal feedback control law for a one dimensional heat/diffusion equation. Numerical results illustrating the convergence of spline-based approximations are also presented.

2. Operator Algebraic Riccati Equations in Spaces of Hilbert-Schmidt Operators

A rather complete existence theory for solutions to operator algebraic Riccati equations can be found in Barbu's book [1]. We provide a brief outline of those results here. Barbu's theory forms the basis for the approximation and convergence results which will be developed in the subsequent section below.

Let H be a real separable Hilbert space with inner product $(\cdot, \cdot)$ and associated induced norm $|\cdot|$. Let V also be a real separable Hilbert space with inner product $\langle \cdot, \cdot \rangle$ and corresponding norm $\|\cdot\|$. We assume that V is densely, continuously, and compactly embedded in H. Indentifying H with its dual, H^*, we have $V \hookrightarrow H = H^* \hookrightarrow V^*$ with the final embedding dense and continuous as well. If we let $\|\cdot\|_*$ denote the standard operator norm on V^*, then the continuity of the above embeddings imply the existence of a constant $\mu > 0$ for which $|\varphi| \leq \mu \|\varphi\|$, $\varphi \in V$, and $\|\varphi\|_* \leq \mu |\varphi|$, $\varphi \in H$.

Let $\gamma \in \mathcal{L}(V, V^*)$ denote the canonical isomorphism (Riesz map) from V onto V^*. That is, for $\varphi, \psi \in V$, $(\gamma \varphi, \psi) = \langle \varphi, \psi \rangle$, where in this case $(\cdot, \cdot)$ denotes the usual extension of the H inner product to the duality pairing between V and V^*. It follows that $\gamma^{-1} \in \mathcal{L}(V^*, V) \cap \mathcal{L}(H, V)$ is self-adjoint, positive, and compact as a mapping from H into H. With inner product $\langle \cdot, \cdot \rangle_*$ given by

$$\langle \varphi, \psi \rangle_* = \langle \gamma^{-1}\varphi, \gamma^{-1}\psi \rangle = (\varphi, \gamma^{-1}\psi), \quad \varphi, \psi \in V^*,$$

V^* is a Hilbert space with $\|\varphi\|_* = \sqrt{\langle \varphi, \varphi \rangle_*}$, for $\varphi \in V^*$. The mapping γ^{-1} being self-adjoint, positive, and compact on H yields the existence of an orthonormal basis $\{e_k\}_{k=1}^{\infty}$ for H such that $\gamma^{-1}e_k = \rho_k^{-2}e_k$, $k = 1, 2, \ldots$ with $\rho_k > 0$, $k = 1, 2, \ldots$. It follows that $\{\rho_k^{-1}e_k\}_{k=1}^{\infty}$ and $\{\rho_k e_k\}$ are orthonormal bases for V and V^* respectively.

Let $HS(X,Y)$ denote the Hilbert space of Hilbert-Schmidt operators defined on the separable Hilbert space X with range in the separable Hilbert space Y. Let $[\cdot,\cdot]_{HS(X,Y)}$ and $|\cdot|_{HS(X,Y)}$ denote the usual Hilbert-Schmidt inner product and corresponding induced norm on $HS(X,Y)$. Let $\mathcal{H} = HS(H,H)$ with $[\cdot,\cdot]_{\mathcal{H}} = |\cdot|_{HS(H,H)}$ and let $\mathcal{V} = HS(V^*,H)\cap HS(H,V)$ with $[\cdot,\cdot]_{\mathcal{V}} = [\cdot,\cdot]_{HS(V^*,H)} + [\cdot,\cdot]_{HS(H,V)}$ and $\|\Phi\|_{\mathcal{V}} = \sqrt{[\Phi,\Phi]_{\mathcal{V}}}$, for $\Phi \in \mathcal{V}$. The space $\mathcal{V}$ together with the innerproduct $[\cdot,\cdot]_{\mathcal{V}}$ is a Hilbert space. Moreover it can be argued that the inclusions $HS(V^*,H) \subset HS(H,H) \subset HS(V,H)$ and $HS(H,V) \subset HS(H,H) \subset HS(H,V^*)$ are dense and continuous and that $HS(V^*,H)$ and HS(V,H), and HS(H,V) and $HS(H,V^*)$ are dual pairs with respect to the duality pairing derived from the $\mathcal{H}$ inner product, $[\cdot,\cdot]_{\mathcal{H}}$. Consequently we have $\mathcal{V}^* = HS(V,H) + HS(H,V^*)$, and identifying $\mathcal{H}$ with its dual, the dense and continuous embeddings $\mathcal{V} \hookrightarrow \mathcal{H} = \mathcal{H}^* \hookrightarrow \mathcal{V}^*$.

Let $a(\cdot,\cdot) : V \times V \to \mathbb{R}$ be a bounded, strongly V-elliptic bilinear form. More precisely, we assume that there exist positive constants α and β for which $a(\varphi,\varphi) \geq \alpha\|\varphi\|^2$, $\varphi \in V$ and $|a(\varphi,\psi)| \leq \beta\|\varphi\|\,\|\psi\|$, $\varphi,\psi \in V$. Let $A \in \mathcal{L}(V,V^*)$ be the operator defined by $(A\varphi,\psi) = a(\varphi,\psi)$, $\varphi,\psi \in V$. Let $a^*(\cdot,\cdot) : V \times V \to \mathbb{R}$ be the form adjoint to $a(\cdot,\cdot)$. That is $a^*(\varphi,\psi) = a(\psi,\varphi)$, $\varphi,\psi \in V$. It follows that $a^*(\varphi,\varphi) \geq \alpha\|\varphi\|^2, \varphi \in V$ and $|a^*(\varphi,\psi)| \leq \beta\|\varphi\|\,\|\psi\|$, $\varphi,\psi \in V$. Let $A^* \in \mathcal{L}(V,V^*)$ be the operator defined by $(A^*\varphi,\psi) = a^*(\varphi\ \psi)$, $\varphi,\psi \in V$. Then, if we define the operators $\tilde{A} : Dom(\tilde{A}) \subset H \to H$ and $\tilde{A}^* : Dom(\tilde{A}^*) \subset H \to H$ to be the restrictions of the operators A and A^* to the sets $Dom(\tilde{A}) = \{\varphi \in V : A\varphi \in H\}$ and $Dom(\tilde{A}^*) = \{\varphi \in V : A^*\varphi \in H\}$ respectively, it can be argued that $\tilde{A}$ and $\tilde{A}^*$ are densely defined and are H-adjoints of one another. That is $(\tilde{A}\varphi,\psi) = (\varphi,\tilde{A}^*\psi), \varphi \in Dom(\tilde{A})$, $\psi \in Dom(\tilde{A}^*)$.

Define the closed convex cone $\mathcal{C}$ in $\mathcal{H}$ by $\mathcal{C} = \{\Phi \in HS(H,H) : \Phi = \Phi^*,\ \Phi \geq 0\}$ and let $\Phi \to \mathcal{B}(\Phi)$ be a single valued mapping defined for every $\Phi \in \mathcal{C}$ with range in $\mathcal{H}$ which is continuous from $\mathcal{H}$ into itself. (Such a mapping $\mathcal{B}$ can be defined via the operational calculus for bounded linear operators by $\mathcal{B}(\Phi) = f(\Phi), \Phi \in \mathcal{C}$, where f is a single valued complex function of a complex variable satisfying $f(0) = 0$ and which is analytic on the nonnegative real axis (see Dunford and Schwartz, Part II [3], Theorem XI. 6.7.7).) We assume that $\mathcal{B}$ is bounded and monotone on $\mathcal{C}$. That is that it maps $\mathcal{H} - bounded$ subsets of $\mathcal{C}$ into $\mathcal{H} - bounded$ subsets and that it has the property

$$(2.1) \qquad\qquad [\mathcal{B}(\Phi) - \mathcal{B}(\Psi), \Phi - \Psi]_{\mathcal{H}} \geq 0$$

for every $\Phi, \Psi \in \mathcal{C}$. We assume further that

$$(2.2) \qquad\qquad (I + \lambda\mathcal{B})\mathcal{C} \supset \mathcal{C}, \quad \lambda > 0.$$

Let $\Theta \in \mathcal{C}$ be given, and we consider the generalized algebraic Riccati equation for $\Pi \in \mathcal{H}$ given by

$$(2.3) \qquad\qquad A^*\Pi + \Pi A + \mathcal{B}(\Pi) = \Theta.$$

We seek a solution $\Pi \in \mathcal{C}$ to (2.3). We note that for n a positive integer, the operator $\mathcal{B}$ given by $\mathcal{B}(\Phi) = \Phi^n$, $\Phi \in \mathcal{C}$ (i.e. $f(z) = z^n$) can be shown to satisfy the conditions above.

Indeed, boundedness follows from the esitmate $|\Phi^n|_{\mathcal{H}} \leq |\Phi|_{\mathcal{H}}^n$, while monotonicity can be established via

$$[\Phi^n - \Psi^n, \Phi - \Psi]_{\mathcal{H}} = \sum_{j=1}^{n} \left[\Phi^{n-j}\{\Phi - \Psi\}\Psi^{j-1}, \Phi - \Psi\right]_{\mathcal{H}} \geq 0,$$

for $\Phi, \Psi \in \mathcal{C}$ (see [11]). Finally, for $\Psi \in \mathcal{C}$ and $\lambda > 0$ let $\{\psi_i\}_{i=1}^{\infty}$ be the orthonormal set of eigenvectors of Ψ with corresponding eigenvalues $\{\alpha_i\}_{i=1}^{\infty}$. (The fact that $\Psi \in \mathcal{C}$ implies $\alpha_i \geq 0$, $i = 1, 2, \dots$). Then, if we let γ_i denote a nonnegative solution to the equation $\gamma_i + \lambda\gamma_i^n - \alpha_i = 0$ (the intermediate value theorem guarantees that such a γ_i exists with $0 \leq \gamma_i \leq \alpha_i$) and set $\Phi\varphi = \sum_{i=1}^{\infty} \gamma_i(\varphi, \psi_i)\psi_i$, $\varphi \in H$, it follows that $\Phi \in \mathcal{C}$ and $\Phi + \lambda\Phi^n = \Psi$. This establishes (2.2). When n = 2, (2.3) becomes the familiar quadratic Riccati equation.

Define the operator $\mathcal{A} \in \mathcal{L}(\mathcal{V}, \mathcal{V}^*)$ by

$$\mathcal{A}\Phi = A^*\Phi + \Phi A, \qquad \Phi \in \mathcal{V}.$$

It can be argued (see [1]) that $\mathcal{A}$ is strongly $\mathcal{V}$−elliptic - that is there exists a constant $\omega > 0$ for which

(2.4) $[\mathcal{A}\Phi, \Phi]_{\mathcal{H}} \geq \omega\|\Phi\|_{\mathcal{V}}^2, \quad \Phi\epsilon\mathcal{V}.$

If we define the subspace $Dom(\mathcal{A}) = \{\Phi \in \mathcal{V} : \mathcal{A}\Phi \in \mathcal{H}\}$ then it follows (see [10]) that the operator $\mathcal{A} : Dom(\mathcal{A}) \subset \mathcal{H} \to \mathcal{H}$ is densely defined and m-accretive in $\mathcal{H}$. With the above definitions the problem of finding a solution to the operator algebraic Riccati equation (2.3) becomes one of finding a solution $\Pi \in Dom(\mathcal{F})$ to the abstract nonlinear operator equation in $\mathcal{H}$ given by

(2.5) $\mathcal{F}(\Pi) = \Theta$

where $\Theta \in \mathcal{C}$ is given and $\mathcal{F} : Dom(\mathcal{F}) \subset \mathcal{H} \to \mathcal{H}$ is the operator defined by $\mathcal{F}(\Phi) = \mathcal{A}\Phi + \mathcal{B}(\Phi)$ for $\Phi \in Dom(\mathcal{F}) = \mathcal{C} \cap Dom(\mathcal{A})$.

Using a standard fixed point argument on the closed convex subset $\mathcal{C}$, Barbu [1] argues that the equation

$$\mathcal{F}_\lambda(\Phi_\lambda) = \Theta$$

has a unique solution $\Phi_\lambda \in Dom(\mathcal{F})$ for each $\lambda > 0$ where $\mathcal{F}_\lambda : Dom(\mathcal{F}) \subset \mathcal{H} \to \mathcal{H}$ is the Yosida -like approximation to $\mathcal{F}$ given by $\mathcal{F}_\lambda = \mathcal{A} + \lambda^{-1}\{I - (I + \lambda\mathcal{B})^{-1}\}$. Then using the boundedness and monotonicity of $\mathcal{B}$, Barbu argues further that the Π_λ converge in $\mathcal{H}$ as $\lambda \to 0$ to an operator $\Pi \in Dom(\mathcal{F})$ which is a solution to (2.5) (or, equivalently, (2.3)). The strong $\mathcal{V}$−ellipticity of $\mathcal{A}$ (i.e. (2.4)), and the monotonicity of $\mathcal{B}$ (i.e. (2.1)) are then used in the usual way to establish the uniqueness of Π. Note that $\Pi \in Dom(\mathcal{F})$ implies that $\Pi \in \mathcal{C}$, i.e. that it is a nonnegative, self-adjoint operator in $\mathcal{H}$, and that $\Pi \in \mathcal{V}$ with $\mathcal{A}\Pi = A^*\Pi + \Pi A \in \mathcal{H}$.

3. Galerkin Approximation and Convergence Theory

For each $n = 1, 2, \ldots$ let H_n be a finite dimensional subspace of H with $H_n \subset V$, $n = 1, 2, \ldots$. Let $P_n : H \to H_n$ be the orthogonal projection of H onto H_n with respect to the $(\cdot, \cdot)$ inner product on H. We assume that

$$(3.1) \qquad \lim_{n \to \infty} \|P_n \varphi - \varphi\| = 0, \qquad \varphi \in V.$$

Note that assumption (3.1) implies that $\lim_{n \to \infty} |P_n \varphi - \varphi| = 0$, $\varphi \in H$, and that the P_n are uniformly bounded in the uniform operator topologies on $\mathcal{L}(H)$ and $\mathcal{L}(V)$.

Lemma 3.1. *The operators P_n admit extensions, which we shall again call P_n, to idempotent, uniformly bounded operators in $\mathcal{L}(V^*)$. Moreover, $\lim_{n \to \infty} \|P_n \varphi - \varphi\|_* = 0, \varphi \in V^*$, the V^*-adjoint (i.e. the operator P_n^* satisfying $\langle P_n \varphi, \psi \rangle_* = \langle \varphi, P_n^* \psi \rangle_*, \varphi, \psi \in V^*$) is given by $P_n^* = \gamma P_n \gamma^{-1}$, and $\lim_{n \to \infty} \|P_n^* \varphi - \varphi\|_* = 0, \varphi \in V^*$.*

Proof: For $\varphi \in V^*$ set $P_n \varphi = \varphi_n$ where φ_n is the representer of the functional on H_n which is the restriction of φ to H_n. That is $(\varphi_n, \psi_n) = (P_n \varphi, \psi_n) = (\varphi, \psi_n), \psi_n \in H_n$. It is clear that P_n as given above is a well defined extension of the orthogonal projection of H onto H_n and that it is idempotent. Moreover, since $H_n \subset V \subset V^*$, for $\varphi \in V^*$ we may consider $P_n \varphi = \varphi_n \in H_n$ an element in V^* via the duality pairing $(P_n \varphi, \psi) = (\varphi_n, \psi), \psi \in V$. Then for $\varphi \in V^*, \psi \in V$, and $\varphi_n = P_n \varphi$ we have $(P_n \varphi, \psi) = (\varphi_n, \psi) = (\varphi_n, P_n \psi) = (\varphi, P_n \psi)$. Consequently for $\varphi \in V^*$ we have

$$\|P_n \varphi\|_* = \sup_{\substack{\psi \in V \\ \|\psi\| \leq 1}} |(P_n \varphi, \psi)| = \sup_{\substack{\psi \in V \\ \|\psi\| \leq 1}} |(\varphi, P_n \psi)| \leq \|\varphi\|_* \|P_n\|,$$

or $\|P_n\|_* \leq \|P_n\|$. Thus assumption (3.1) implies that the P_n are uniformly bounded in $\mathcal{L}(V^*)$. (We note that alternatively the same extension of the projections P_n to operators on V^* could have been obtained by using the standard approach based upon the density of H in V^* and the usual extension construction in terms of limits.) Now for $\varphi \in H$ we have

$$\langle P_n \varphi, \psi \rangle_* = (P_n \varphi, \gamma^{-1} \psi) = (\varphi, P_n \gamma^{-1} \psi) = (\varphi, \gamma^{-1} \gamma P_n \gamma^{-1} \psi)$$
$$= \langle \varphi, \gamma P_n \gamma^{-1} \psi \rangle_*,$$

and consequently that the V^*-adjoint of P_n, P_n^*, is given by $P_n^* = \gamma P_n \gamma^{-1}$. Finally, assumption (3.1) yields

$$\lim_{n \to \infty} \|P_n^* \varphi - \varphi\|_* = \lim_{n \to \infty} \|\gamma^{-1} P_n^* \varphi - \gamma^{-1} \varphi\|$$
$$= \lim_{n \to \infty} \|\gamma^{-1} \gamma P_n \gamma^{-1} \varphi - \gamma^{-1} \varphi\|$$
$$= \lim_{n \to \infty} \|P_n \gamma^{-1} \varphi - \gamma^{-1} \varphi\| = 0$$

for each $\varphi \in V^*$ and the lemma is proved.

Define the sequence of finite dimensional subspaces $\mathcal{H}_n, n = 1, 2, \ldots$ of $\mathcal{H}$ by

$$\mathcal{H}_n = \{\Phi_n P_n : \Phi_n \in \mathcal{L}(H_n)\}.$$

Clearly H_n finite dimensional implies that all operators in $\mathcal{H}_n$ are of finite rank and thus that $\mathcal{H}_n$ is a subspace of both $\mathcal{H}$ and $\mathcal{V}$. For each $n = 1, 2, \ldots$ let $\mathcal{C}_n \subset \mathcal{H}_n$ be the closed convex cone given by $\mathcal{C}_n = \{\Phi_n P_n \in \mathcal{H}_n : \Phi_n = \Phi_n^*, \Phi_n \geq 0\}$. Note that $\mathcal{C}_n \subset \mathcal{C}$, $n = 1, 2, \ldots$. We define Galerkin approximations to the operator $\mathcal{F}, \mathcal{F}_n : Dom(\mathcal{F}_n) \subset \mathcal{H}_n \to \mathcal{H}_n$, as follows:

$$(3.2) \qquad \mathcal{F}_n(\Phi_n P_n) = \{\mathcal{A}\Phi_n P_n + \mathcal{B}(\Phi_n P_n)\}|_{\mathcal{H}_n}, \quad \text{for} \quad \Phi_n P_n \in Dom(\mathcal{F}_n) = \mathcal{C}_n.$$

That is, for $\Phi_n P_n \in \mathcal{C}_n$, $\mathcal{F}_n(\Phi_n P_n) = \Psi_n P_n \in \mathcal{H}_n$ where $\Psi_n P_n$ is that element in $\mathcal{H}_n$ (guaranteed to exist and be unique by the Riesz Representation Theorem applied to the functional $\mathcal{A}\Phi_n P_n + \mathcal{B}(\Phi_n P_n) \in \mathcal{V}^*$ restricted to a functional on the finite dimensional Hilbert space $\mathcal{H}_n$) which satisfies $[\mathcal{A}\Phi_n P_n + \mathcal{B}(\Phi_n P_n), \chi_n P_n]_{\mathcal{H}} = [\Psi_n P_n, \chi_n P_n]_{\mathcal{H}}, \ \chi_n P_n \in \mathcal{H}_n$.

It is of some value to note that the approximations to $\mathcal{F}$ given in (3.2) are the same ones that would be obtained via the standard approach which is based upon the replacement of the operators A and $\mathcal{B}$ in (2.3) by their respective Galerkin approximations on H_n and $\mathcal{H}_n$. Indeed, for each $n = 1, 2, \ldots$ define the operators $A_n \in \mathcal{L}(H_n)$ and $\mathcal{B}_n : \mathcal{C}_n \subset \mathcal{H}_n \to \mathcal{H}_n$ by $A_n \varphi_n = \psi_n$, where for $\varphi_n \in H_n$, ψ_n is that element in H_n which satisfies $(A\varphi_n, \chi_n) = (\psi_n, \chi_n)$, $\chi_n \in H_n$, and $\mathcal{B}_n(\Phi_n P_n) = P_n \mathcal{B}(\Phi_n P_n) P_n$. From the fact that P_n is the orthogonal projection of H onto H_n it follows that $[\Phi P_n, \Psi_n]_{\mathcal{H}} = [\Phi, \Psi_n]_{\mathcal{H}}$ for every $\Phi \in \mathcal{V}^*$ and $\Psi_n \in \mathcal{H}_n$. Then, for $\Phi_n P_n, \Psi_n P_n \in \mathcal{H}_n$ we have

$$
\begin{aligned}
[\mathcal{F}_n(\Phi_n P_n), \Psi_n P_n]_{\mathcal{H}} &= [\mathcal{A}\Phi_n P_n + \mathcal{B}(\Phi_n P_n), \Psi_n P_n]_{\mathcal{H}} \\
&= [\{\mathcal{A}\Phi_n P_n + \mathcal{B}(\Phi_n P_n)\} P_n, \Psi_n P_n]_{\mathcal{H}} \\
&= [A^*\Phi_n P_n + \Phi_n P_n A P_n + \mathcal{B}(\Phi_n P_n) P_n, \Psi_n P_n]_{\mathcal{H}} \\
&= \sum_{k=1}^{\infty} \{(A^*\Phi_n P_n e_k, \Psi_n P_n e_k) + (\Phi_n P_n A P_n e_k, \Psi_n P_n e_k) \\
&\quad + (\mathcal{B}(\Phi_n P_n) P_n e_k, \Psi_n P_n e_k)\} \\
&= \sum_{k=1}^{\infty} \{(A_n^*\Phi_n P_n e_k, \Psi_n P_n e_k) + (A_n P_n e_k, \Phi_n^* \Psi_n P_n e_k) \\
&\quad + (P_n \mathcal{B}(\Phi_n P_n) P_n e_k, \Psi_n P_n e_k)\} \\
&= \sum_{k=1}^{\infty} (\{A_n^*\Phi_n + \Phi_n A_n + \mathcal{B}_n(\Phi_n P_n)\} P_n e_k, \Psi_n P_n e_k),
\end{aligned}
$$

or,

$$\mathcal{F}_n(\Phi_n P_n) = \{A_n^*\Phi_n + \Phi_n A_n + \mathcal{B}_n(\Phi_n P_n)\} P_n.$$

In the particular case when $\mathcal{B}(\Phi) = \Phi^2$, for example, the operators $\mathcal{F}_n$ take the form $\mathcal{F}_n(\Phi_n P_n) = \{A_n^*\Phi_n + \Phi_n A_n + \Phi_n^2\} P_n$.

Set $\Theta_n = P_n \Theta P_n \in C_n$ and consider the problem of finding a solution $\Pi_n \in C_n$ to the nonlinear operator equation

(3.3) $\mathcal{F}_n(\Pi_n) = \Theta_n$

in $\mathcal{H}_n$. Arguments similar to those described in section 2 above yield that for each $n = 1, 2, \ldots$, the equation (3.3) admits a unique solution $\Pi_n \in C_n$. We shall argue that $\lim_{n \to \infty} \|\Pi_n - \Pi\|_{\mathcal{V}} = 0$; that is, that the Π_n converge in the $\mathcal{V}$-Hilbert-Schmidt norm to the solution Π to the equation (2.5) (or equivalently (2.3)). In order to do this we shall require the following lemmas.

Lemma 3.2. *If $\{a_i\}_{i=1}^{\infty}$ is an absolutely summable sequence of real numbers, then there exist sequences $\{b_i\}_{i=1}^{\infty}$ and $\{c_i\}_{i=1}^{\infty}$ for which $\lim_{i \to \infty} b_i = 0$, $\{c_i\}_{i=1}^{\infty}$ is absolutely summable, and $a_i = b_i c_i$, $i = 1, 2, \ldots$.*

Proof: Set $\alpha = \sum_{i=1}^{\infty} |a_i|$ and, for $j = 0, 1, 2, \ldots$ define the nonnegative integers k_j by $k_0 = 0$ and for $j = 1, 2, \ldots$ let k_j be the first index for which

$$\sum_{i=1}^{k_j} |a_i| > \alpha - \frac{1}{j^3}.$$

Then setting $b_i = 1/j$ and $c_i = j a_i$, for $i = k_{j-1} + 1, \ldots, k_j$, $j = 1, 2, \ldots$, we have $b_i c_i = a_i$, $i = 1, 2, \ldots$, $\lim_{i \to \infty} b_i = 0$, and

$$\sum_{i=1}^{\infty} |c_i| = \sum_{j=1}^{\infty} j \sum_{k=k_{j-1}+1}^{k_j} a_k = \sum_{j=1}^{\infty} j \left\{ \sum_{k=1}^{k_j} a_k - \sum_{k=1}^{k_{j-1}} a_k \right\}$$

$$\leq \sum_{k=1}^{k_1} a_k + \sum_{j=2}^{\infty} j \left\{ \alpha - (\alpha - \frac{1}{(j-1)^3}) \right\}$$

$$\alpha + \sum_{j=1}^{\infty} \frac{1}{j^2} + \sum_{j=1}^{\infty} \frac{1}{j^3} < \infty$$

Lemma 3.3. *Let X and Y be real separable Hilbert spaces with inner products denoted by $\langle \cdot, \cdot \rangle_X$ and $\langle \cdot, \cdot \rangle_Y$ respectively. Then every $\Phi \in HS(X, Y)$ can be written in factored form as $\Phi = \Phi^1 \Phi^2$ with $\Phi^1 \in \mathcal{L}(Y)$ compact and $\Phi^2 \in HS(X, Y)$.*

Proof: Let $\{x_i\}_{i=1}^{\infty}$ be an orthonormal basis for X and let $\{y_i\}_{i=1}^{\infty}$ be an orthonormal basis for Y. If $\Phi \in HS(X, Y)$ then it has a representation in the form of an infinite matrix $\Phi \leftrightarrow [\varphi_{ij}]$ where $\varphi_{ij} = \langle y_i, \Phi x_j \rangle_Y$, and $\sum_{i=1}^{\infty} \sum_{j=1}^{\infty} \varphi_{ij}^2 < \infty$. Now for $i = 1, 2, \ldots$ set $a_i = \sum_{j=1}^{\infty} \varphi_{ij}^2$. The sequence $\{a_i\}_{i=1}^{\infty}$ is absolutely summable, so applying Lemma 3.2 we obtain sequences $\{b_i\}_{i=1}^{\infty}$ and $\{c_i\}_{i=1}^{\infty}$ for which $a_i = b_i c_i$, $i = 1, 2, \ldots$, $\lim_{i \to \infty} b_i = 0$, and $\sum_{i=1}^{\infty} |c_i| = \sum_{i=1}^{\infty} c_i < \infty$. Define $\Phi^1 \in \mathcal{L}(Y)$ by $\Phi^1 y = \sum_{i=1}^{\infty} \sqrt{b_i} \langle y, y_i \rangle_Y y_i$, $y \in Y$, and $\Phi^2 \in \mathcal{L}(X, Y)$ by $\Phi^2 x = \sum_{i=1}^{\infty} \sum_{j=1}^{\infty} \frac{\varphi_{ij}}{\sqrt{b_i}} \langle x, x_j \rangle_X y_i$, $x \in X$. Then $\Phi^1 \Phi^2 = \Phi$, and, since $\lim_{i \to \infty} \sqrt{b_i} = 0$ and $\sum_{i=1}^{\infty} \sum_{j=1}^{\infty} \left(\frac{\varphi_{ij}}{\sqrt{b_i}} \right)^2 = \sum_{i=1}^{\infty} \frac{1}{b_i} \sum_{j=1}^{\infty} \varphi_{ij}^2 = \sum_{i=1}^{\infty} \frac{a_i}{b_i} = \sum_{i=1}^{\infty} c_i < \infty$, it follows that Φ^1 is compact and $\Phi^2 \in HS(X, Y)$.

Lemma 3.4.

(a) For every $\Phi \in \mathcal{H}$, $\lim_{n \to \infty} |P_n \Phi P_n - \Phi|_{\mathcal{H}} = 0$.

(b) For every $\Phi \in \mathcal{V}$, $\lim_{n \to \infty} \|P_n \Phi P_n - \Phi\|_{\mathcal{V}} = 0$.

Proof: (a). For $\Phi \in \mathcal{H}$ we have

$$
\begin{aligned}
|P_n \Phi P_n - \Phi|_{\mathcal{H}} &\leq |P_n \Phi P_n - P_n \Phi|_{\mathcal{H}} + |P_n \Phi - \Phi|_{\mathcal{H}} \\
&\leq |\Phi P_n - \Phi|_{\mathcal{H}} + |P_n \Phi - \Phi|_{\mathcal{H}} \\
&= |(\Phi P_n)^* - \Phi^*|_{\mathcal{H}} + |P_n \Phi - \Phi|_{\mathcal{H}} \\
&= |P_n \Phi^* - \Phi^*|_{\mathcal{H}} + |P_n \Phi - \Phi|_{\mathcal{H}}
\end{aligned}
$$

where in the above estimate we have used the fact that $P_n \in \mathcal{L}(H)$ with $|P_n| \leq 1$ and that $|P_n \Psi|_{\mathcal{H}} \leq |P_n| \, |\Psi|_{\mathcal{H}} \leq |\Psi|_{\mathcal{H}}$ for every $\Psi \in \mathcal{H}$. If we apply Lemma 3.3 with $X = Y = H$ to $\Phi, \Phi^* \in \mathcal{H} = HS(H, H)$, then we obtain $\Phi = \Phi^1 \Phi^2$, $\Phi^* = (\Phi^*)^1 (\Phi^*)^2$ with $\Phi^1, (\Phi^*)^1 \in \mathcal{L}(H)$ compact and $(\Phi^*)^2 \in HS(H, H)$. It follows that

$$
|P_n \Phi^* - \Phi^*|_{\mathcal{H}} = |(P_n - I)(\Phi^*)^1 (\Phi^*)^2|_{\mathcal{H}} \leq |(P_n - I)(\Phi^*)^1| \, |(\Phi^*)^2|_{\mathcal{H}},
$$

and

$$
|P_n \Phi - \Phi|_{\mathcal{H}} \leq |(P_n - I)\Phi^1| \, |\Phi^2|_{\mathcal{H}},
$$

which together with assumption (3.1) and the fact that Φ^1 and $(\Phi^*)^1$ are compact yield the desired result.

(b). For $\Phi \in \mathcal{V} = HS(V^*, H) \cap HS(H, V)$ we have

$$
\begin{aligned}
|P_n \Phi P_n - \Phi|_{HS(V^*, H)} &\leq |P_n \Phi P_n - P_n \Phi|_{HS(V^*, H)} + |P_n \Phi - \Phi|_{HS(V^*, H)} \\
&\leq |P_n| \, |\Phi P_n - \Phi|_{HS(V^*, H)} + |P_n \Phi - \Phi|_{HS(V^*, H)} \\
&\leq |\Phi P_n - \Phi|_{HS(V^*, H)} + |P_n \Phi - \Phi|_{HS(V^*, H)}.
\end{aligned}
$$

Now $\Phi \in HS(V^*, H)$ implies that $\Phi^* \in HS(H, V^*)$ and that $(\Phi P_n)^* = P_n^* \Phi^* \in HS(H, V^*)$ where, recalling Lemma 3.1, $P_n^* = \gamma P_n \gamma^{-1}$ is the adjoint of the operator P_n considered as an element of $\mathcal{L}(V^*)$. It follows that

$$
(3.4) \qquad |P_n \Phi P_n - \Phi|_{HS(V^*, H)} \leq |P_n^* \Phi^* - \Phi^*|_{HS(H, V^*)} + |P_n \Phi - \Phi|_{HS(V^*, H)}
$$

Lemma 3.3 with $X = H$ and $Y = V^*$ together with Lemma 3.1 imply that the first term on the right hand side of the estimate (3.4) tends to zero as $n \to \infty$. Similarly, Lemma 3.3 with $X = V^*$ and $Y = H$ together with assumption (3.1) imply that the second term tends to zero as $n \to \infty$ as well. A similar argument to the one given above can be used to show that $\lim_{n\to\infty} |P_n \Phi P_n - \Phi|_{HS(H,V)} = 0$ and the lemma is proved.

The primary result of this paper is given in the following theorem.

Theorem 3.1. *Let* $\Pi \in V$ *be the unique solution to the equation (2.5) (equivalently (2.3)) and for each* $n = 1, 2, \ldots$ *let* $\Pi_n \in \mathcal{H}_n$ *be the unique solution to the approximating operator equation (3.3). Then* $\lim_{n\to\infty} \|\Pi_n - \Pi\|_V = 0$.

Proof: From the triangle inequality we obtain

$$\|\Pi_n - \Pi\|_V \le \|\Pi_n - P_n \Pi P_n\|_V + \|P_n \Pi P_n - \Pi\|_V .$$

An application of Lemma 3.4(b) yields that the second term on the right and side of the above estimate tends to zero as $n \to \infty$. As for the first term, we recall (2.4) and consider the estimate

$$
\begin{aligned}
\omega \|\Pi_n - P_n \Pi P_n\|_V^2 &\le [\mathcal{A}\{\Pi_n - P_n \Pi P_n\}, \Pi_n - P_n \Pi P_n]_{\mathcal{H}} \\
&= [\mathcal{A}\Pi_n + \mathcal{B}_n(\Pi_n) - \mathcal{A}\Pi - \mathcal{B}(\Pi), \Pi_n - P_n \Pi P_n]_{\mathcal{H}} \\
&\quad + [\mathcal{A}\Pi - \mathcal{A}P_n \Pi P_n, \Pi_n - P_n \Pi P_n]_{\mathcal{H}} \\
&\quad + [\mathcal{B}(\Pi) - \mathcal{B}(P_n \Pi P_n), \Pi_n - P_n \Pi P_n]_{\mathcal{H}} \\
&\quad + [\mathcal{B}(P_n \Pi P_n) - \mathcal{B}_n(\Pi_n), \Pi_n - P_n \Pi P_n]_{\mathcal{H}}
\end{aligned}
$$

$$
\begin{aligned}
&= [\Theta_n - \Theta, \Pi_n - P_n \Pi P_n]_{\mathcal{H}} \\
&\quad + [\mathcal{A}\{\Pi - P_n \Pi P_n\}, \Pi_n - P_n \Pi P_n]_{\mathcal{H}} \\
&\quad + [\mathcal{B}(\Pi) - \mathcal{B}(P_n \Pi P_n), \Pi_n - P_n \Pi P_n]_{\mathcal{H}} \\
&\quad - [P_n \mathcal{B}(\Pi_n) P_n - \mathcal{B}(P_n \Pi P_n), \Pi_n - P_n \Pi P_n]_{\mathcal{H}}
\end{aligned}
$$

$$
\begin{aligned}
&= [P_n \Theta P_n - \Theta, \Pi_n - P_n \Pi P_n]_{\mathcal{H}} \\
&\quad + [\mathcal{A}\{\Pi - P_n \Pi P_n\}, \Pi_n - P_n \Pi P_n]_{\mathcal{H}} \\
&\quad + [\mathcal{B}(\Pi) - \mathcal{B}(P_n \Pi P_n), \Pi_n - P_n \Pi P_n]_{\mathcal{H}} \\
&\quad - [\mathcal{B}(\Pi_n) - \mathcal{B}(P_n \Pi P_n), \Pi_n - P_n \Pi P_n]_{\mathcal{H}}
\end{aligned}
$$

$$
\begin{aligned}
&\le [P_n \Theta P_n - \Theta, \Pi_n - P_n \Pi P_n]_{\mathcal{H}} \\
&\quad + [\mathcal{A}\{\Pi - P_n \Pi P_n\}, \Pi_n - P_n \Pi P_n]_{\mathcal{H}} \\
&\quad + [\mathcal{B}(\Pi) - \mathcal{B}(P_n \Pi P_n), \Pi_n - P_n \Pi P_n]_{\mathcal{H}}
\end{aligned}
$$

where in the final estimate above we have applied assumption (2.1). Continuing we find that

$$\omega \left\| \Pi_n - P_n \Pi P_n \right\|_V^2 \leq \left\| P_n \Theta P_n - \Theta \right\|_{V^*} \left\| \Pi_n - P_n \Pi P_n \right\|_V$$
$$+ \left| \mathcal{A} \right|_{\mathcal{L}(V,V^*)} \left\| \Pi - P_n \Pi P_n \right\|_V \left\| \Pi_n - P_n \Pi P_n \right\|_V$$
$$+ \left\| \mathcal{B}(\Pi) - \mathcal{B}(P_n \Pi P_n) \right\|_{V^*} \left\| \Pi_n - P_n \Pi P_n \right\|_V ,$$

or, recalling the continuous embedding of H into V^*, that

$$\left\| \Pi_n - P_n \Pi P_n \right\|_V \leq K \left| P_n \Theta P_n - \Theta \right|_{\mathcal{H}} + \left| \mathcal{A} \right|_{\mathcal{L}(V,V^*)} \left\| \Pi - P_n \Pi P_n \right\|_V$$
$$+ K \left| \mathcal{B}(\Pi) - \mathcal{B}(P_n \Pi P_n) \right|_{\mathcal{H}} .$$

Lemma 3.4 together with the continuity assumption on $\mathcal{B}$ imply

$$\lim_{n \to \infty} \left\| \Pi_n - P_n \Pi P_n \right\|_V = 0$$

and the theorem is established.

4. An Example

In order to illustrate the application of our theory we consider an operator algebraic Riccati equation arising in the design of an optimal feedback control law for a one dimensional heat/diffusion equation. Let $H, U = L_2(0,1)$, both endowed with the usual inner product, $(\varphi, \psi) = \int_0^1 \varphi(\eta)\psi(\eta)d\eta$, and consider the linear-quadratic optimal control problem of finding $\bar{u} \in L_2(0, \infty; U)$ which minimizes the quadratic performance index

$$J(u) = \int_0^\infty (Qx(t, \cdot), x(t, \cdot)) + r(u(t, \cdot), u(t, \cdot))dt$$

subject to the linear dynamical system

$$(4.1) \qquad \frac{\partial x}{\partial t}(t, \eta) - \frac{\partial}{\partial \eta} a(\eta) \frac{\partial x}{\partial \eta}(t, \eta) = bu(t, \eta), t > 0, 0 < \eta < 1,$$

$$(4.2) \qquad x(t, 0) = 0, \; x(t, 1) = 0, \quad t > 0$$

$$(4.3) \qquad x(0, \eta) = x^0(\eta), \qquad 0 < \eta < 1,$$

where Q is a self-adjoint, nonnegative, and Hilbert-Schmidt operator from $L_2(0,1)$ into $L_2(0,1), r > 0, a \in L_\infty(0,1), a(\eta) \geq \alpha > 0$, a.e. $\eta \in (0,1), b \in \mathbb{R}$, and $x^0 \in L_2(0,1)$. If we define $\mathcal{C} = \{\Phi \in HS(L_2(0,1), L_2(0,1)), \Phi = \Phi^*, \Phi \geq 0\}$, then $Q \in \mathcal{C}$ and $(Q\varphi)(\eta) = \int_0^1 q(\eta, \zeta)\varphi(\zeta)d\zeta$ with $q \in L_2((0,1) \times (0,1)), q(\eta, \zeta) = q(\zeta, \eta), q(\eta, \zeta) \geq 0$, a.e. $(\eta, \zeta) \in (0,1) \times (0,1)$.

Define $V = H_0^1(0,1)$ endowed with the standard inner product, $\langle \varphi, \psi \rangle = \int_0^1 D\varphi(\eta)D\psi(\eta)d\eta$ and corresponding norm, $\| \cdot \|$. It follows that V is densely, continuously, and compactly embedded in H, that $V^* = H^{-1}(0,1)$, and that H is densely and continuously embedded in V^*. Define the operator $A \in \mathcal{L}(V, V^*)$ by $(A\varphi, \psi) = (aD\varphi, D\psi)$, for $\varphi, \psi \epsilon V$. It follows that $(A\varphi, \varphi) \geq \alpha \|\varphi\|^2$, $\varphi \epsilon V$, and that the restriction $-\tilde{A}$ of the operator -A to the set $Dom(\tilde{A}) = \{\varphi \epsilon V : A\varphi \epsilon H\}$ ($= H^2(0,1) \cap H_0^1(0,1)$ when a is sufficiently smooth) is densely defined in H, negative, self-adjoint, and it is the infinitesimal generator of a uniformly exponentially stable analytic semigroup, $\{T(t) : t \geq 0\}$ of bounded, self-adjoint linear opeators on H. The solution to the initial-boundary value problem (4.1)-(4.3) is given by

$$x(t) = T(t)x^0 + \int_0^t T(t-s)bu(s)ds, \quad t > 0$$

where for each $t > 0$ $x(t) = x(t, \cdot) \epsilon H = L_2(0,1)$ and $u(t) = u(t, \cdot) \epsilon U = L_2(0,1)$ for almost every $t > 0$.

The solution to the optimal control problem (see [7]) is given in closed-loop linear state feedback form by

$$\bar{u}(t) = -(b/r)\Pi x(t), \quad a.e. \ t > 0$$

where Π is the unique nonnegative self-adjoint solution to the algebraic Riccati equation

(4.4) $$A^*\Pi + \Pi A + (b^2/r)\Pi^2 = Q.$$

It is clear that the existence-uniqueness and approximation theories presented above apply with $\Theta = Q \epsilon C$ and $B(\Phi) = (b^2/r)\Phi^2$ for $\Phi \epsilon C$. It follows that there exists a unique solution $\Pi \epsilon HS(L_2(0,1), L_2(0,1))$ to the nonlinear operator equation (4.4) with $\Pi = \Pi^*, \Pi \geq 0$, and $\Pi \epsilon HS(L_2(0,1), H_0^1(0,1)) \cap HS(H^{-1}(0,1), L_2(0,1))$. Recalling that $HS(L_2(0,1), L_2(0,1))$ is isometrically isomorphic to $L_2((0,1) \times (0,1))$, $\Pi \epsilon C$ implies that there exists $\pi \epsilon L_2((0,1) \times (0,1))$ with $\pi(\eta, \zeta) = \pi(\zeta, \eta)$ and $\pi(\eta, \zeta) \geq 0$ for almost every $(\eta, \zeta) \epsilon (0,1) \times (0,1)$ for which

$$\bar{u}(t, \eta) = -(b/r) \int_0^1 \pi(\eta, \zeta)x(t, \zeta)d\zeta,$$

for almost every $\eta \epsilon (0,1)$ and $t > 0$.

We introduce linear spline based approximation. For each $n = 2, 3, \ldots$ let $H_n = $ span $\{\varphi_n^j\}_{j=1}^{n-1}$ where for $j = 1, 2, \ldots, n-1$, φ_n^j denotes the jth standard linear spline function defined on the interval [0,1] with respect to the uniform mesh $\{0, 1/n, 2/n, \ldots, 1\}$. More precisely,

$$\varphi_n^j(\eta) = \begin{cases} 0 & 0 \leq \eta \leq j^{-1}/n \\ n\eta - j + 1 & j^{-1} \leq \eta \leq j/n \\ j + 1 - n\eta & j/n \leq \eta \leq j + 1/n \\ 0 & j + 1/n \leq \eta \leq 1, \end{cases}$$

$j = 1, 2, \ldots, n-1$. Let $P_n : H \to H_n$ denote the orthogonal projection of H onto H_n with respect to the usual inner product on $H = L_2(0,1)$ and define Galerkin approximations $A_n \epsilon \mathcal{L}(H_n)$ to A in the usual way. That is let $A_n\varphi_n = \psi_n$ where for $\varphi_n \epsilon H_n, \psi_n$ is the unique element in H_n which satisfies $(A\varphi_n, \chi_n) = (\psi_n, \chi_n)$, $\chi_n \epsilon H_n$. Set $Q_n = P_n Q \epsilon \mathcal{L}(H_n)$.

Using well known approximation properties of linear interpolatory splines (see [9]) it is not difficult to argue that $\lim_{n\to\infty} \|P_n\varphi - \varphi\| = 0$, $\varphi \in H_0^1(0,1)$ and consequently that assumption (3.1) is satisfied. It follows therefore, from the theory presented in section 3 above, that there exists a unique nonnegative self-adjoint operator $\Pi_n \in \mathcal{L}(H_n)$ satisfying the algebraic Riccati equation in H_n given by

$$(4.5) \qquad\qquad A_n^*\Pi_n + \Pi_n A + (b^2/r)\Pi_n^2 = Q_n.$$

Moreover, we have the Hilbert-Schmidt norm convergence of $\Pi_n P_n$ to Π as $n \to \infty$. That is

$$(4.6) \qquad\qquad \lim_{n\to\infty} |\Pi_n P_n - \Pi|_{HS(H,H)} = 0.$$

We in fact also obtain that $\lim_{n\to\infty} |\Pi_n P_n - \Pi|_{HS(H,V)} = 0$ and that

$$\lim_{n\to\infty} |\Pi_n P_n - \Pi|_{HS(V^*,H)} = 0.$$

From a computational point of view, since the basis elements φ_n^j are not mutually orthonormal, simply replacing the operators in the finite dimensional algebraic Riccati equation (4.5) with their corresponding matrix representations will not lead to the usual symmetric matrix Riccati equation for which a variety of computational solution techniques exist. Toward this end, for a linear operator L_n with domain and/or range in H_n, we denote its matrix representation with respect to the basis $\{\varphi_n^j\}_{j=1}^{n-1}$ for H_n by L_N. Define $\phi_n : [0,1] \to \mathbb{R}^{n-1}$ by $\phi_n(\eta) = (\varphi_n^1(\eta), \varphi_n^2(\eta), \dots, \varphi_n^{n-1}(\eta))^T$ and set $M_N = (\phi_n, \phi_n^T) = \int_0^1 \phi_n(\eta)\phi_n^T(\eta)d\eta$. It then follows that $A_N = M_N^{-1}(aD\phi_n, D\phi_n^T)$ with $A_N^* = M_N^{-1}A_N^T M_N$, and $Q_N = M_N^{-1}(Q\phi_n, \phi_n^T)$. If we set $\hat{Q}_N = M_N Q_N$ and $\hat{\Pi}_N = M_N \Pi_N$, then $\hat{\Pi}_N$ is the unique nonnegative self-adjoint solution to the $(n-1) \times (n-1)$ matrix algebraic Riccati equation given by

$$(4.7) \qquad\qquad A_N^T\hat{\Pi}_N + \hat{\Pi}_N A_N + (b^2/r)\hat{\Pi}_N M_N^{-1}\hat{\Pi}_N = \hat{Q}_N.$$

The approximating optimal control laws take the form

$$\bar{u}_n(t,\eta) = -(b/r)\int_0^1 \pi_n(\eta,\zeta)x(t,\zeta)d\zeta$$

for almost every $\eta \in (0,1)$ and $t > 0$ where

$$\pi_n(\eta,\zeta) = \phi_n(\eta)^T M_N^{-1}\hat{\Pi}_N M_N^{-1}\phi_n(\zeta),$$

$(\eta,\zeta) \in [0,1] \times [0,1]$. It follows from (4.6) that $\lim_{n\to\infty} \pi_n = \pi$ in $L_2((0,1) \times (0,1))$ - that is that

$$(4.8) \qquad\qquad \lim_{n\to\infty}\int_0^1\int_0^1 |\pi_n(\eta,\zeta) - \pi(\eta,\zeta)|^2 \, d\zeta \, d\eta.$$

To illustrate we take $a(\eta) = a > 0$, a constant, and let $Q \in HS(L_2(0,1), L_2(0,1))$ be the finite rank modal projection operator given by

$$Q\varphi = \sum_{k=1}^{\nu} q_k(\varphi, \psi_k)\psi_k, \quad \varphi \in L_2(0,1)$$

where $\nu < \infty$, $\psi_k(\eta) = \sqrt{2}\sin k\pi\eta$, $\eta \in [0,1]$, $k = 1, 2, \ldots, \nu$, and $q_k \geq 0$, $k = 1, 2, \ldots, \nu$. A somewhat tedious, but rather straight forward computation yields

$$(\hat{Q}_N)_{ij} = \sum_{k=1}^{\nu} 2q_k\delta_{ki}^n\delta_{kj}^n, \quad i, j = 1, 2, \ldots, n-1,$$

where

$$\delta_{k\ell}^n = \frac{-n}{(k\pi)^2}\left\{\sin k\,\frac{\pi(\ell+1)}{n} - 2\sin k\frac{\pi\ell}{n} + \sin k\frac{\pi(\ell-1)}{n}\right\},$$

$k = 1, 2, \ldots, \nu, \ell = 1, 2, \ldots, n-1$. Setting a $= .25$, b $= 1.0$, r $= .01$, $\nu = 3$, and $q_1 = q_2 = q_3 = 1.0$, and using Schur-vector decomposition of the associated Hamiltonian matrix (see [6]) to solve the matrix Riccati equation (4.7) for various values of n we obtained the kernels, π_n, plotted in the figures below. That the convergence given in (4.8) above is achieved is immediately clear.

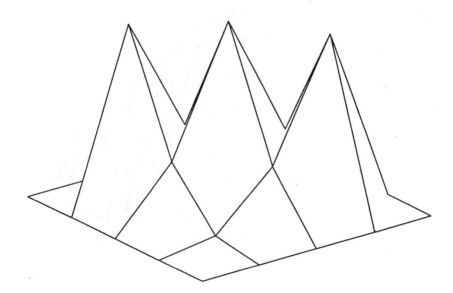

Figure 4.1a: $\pi_4(\eta, \zeta)$, $(\eta, \zeta) \in [0,1] \times [0,1]$.

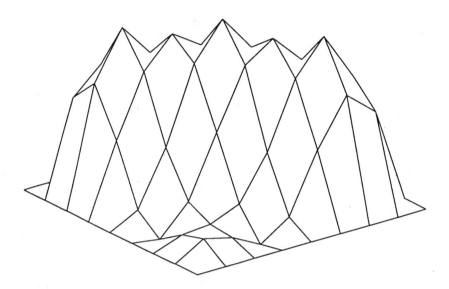

Figure 4.1b: $\pi_8(\eta,\zeta)$, $(\eta,\zeta) \in [0,1] \times [0,1]$.

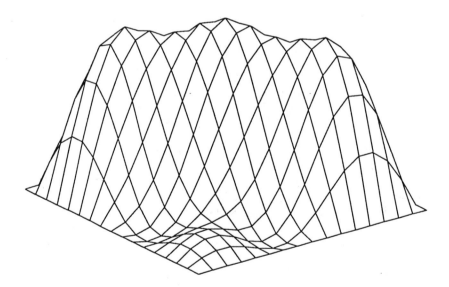

Figure 4.1c: $\pi_{16}(\eta,\zeta)$, $(\eta,\zeta) \in [0,1] \times [0,1]$.

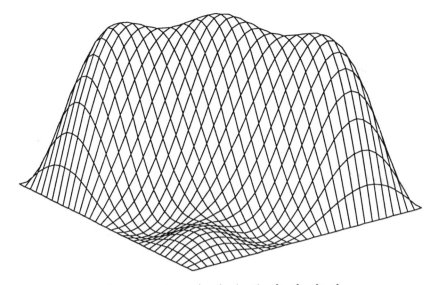

Figure 4.1d: $\pi_{32}(\eta, \zeta)$, $(\eta, \zeta) \in [0,1] \times [0,1]$.

REFERENCES

[1] V. Barbu, "Nonlinear Semigroups and Differential Equations in Banach Spaces," Noordhoff International, Leyden, The Netherlands, 1976.

[2] R.F. Curtain, and A.J. Pritchard, "Infinite Dimensional Linear System Theory," Springer Verlag, Berlin,New York, 1978.

[3] N. Dunford, and J. Schwartz, "Linear Operators, Part II, Spectral Theory, Self-Adjoint Operators in Hilbert Space," Wiley-Interscience, New York, 1963.

[4] A. Germani, L. Jetto, and M. Piccioni, *Galerkin approximation for optimal linear filtering of infinite dimensional linear systems*, SIAM J. Contr. and Opt. **26** (1988), 1287–1305.

[5] J.S., Gibson, *The Riccati integral equations for optimal control problems on Hilbert spaces*, SIAM J. Contr. and Opt. **17** (1979), 537–565.

[6] A.J. Laub, *A Schur method for solving algebraic Riccati equations*, IEEE Trans. Automatic control **AC-24** (1979), 913-921.

[7] J.L., Lions, "Optimal Control of Systems Governed by Partial Differential Equations," Springer-Verlag, New York, 1971.

[8] I.G. Rosen, *Convergence of Galerkin approximations for operator Riccati equations - a nonlinear evolution equation approach,*, submitted.

[9] B.K. Swartz and R.S. Varga, *Error bounds for spline and L-spline interpolation*, J. Approx. Theory **6** (1976), 6-49.

[10] H. Tanabe, "Equations of Evolution," Pitman, London, 1979.

[11] R. Temam, *Sur l'équation de Riccati associée à des opérateurs non bornés,en dimension infinie*, J. Functional Analysis **7** (1971), 85-115.

I.G. Rosen
Department of Mathematics
University of Southern California
Los Angeles, CA 90089
USA

International Series of
Numerical Mathematics, Vol. 91
© 1989 Birkhäuser Verlag Basel

Green's functions for chains of
Schrödinger-Helmholtz equations

Pierre C. Sabatier

Laboratoire de Physique Mathématique
U.S.T.L.

Abstract. A chain of equations is a finite set of equations linked together in $\mathbf{R}^n$ through singular surfaces by conserved quantities. Schrödinger equations reduce to Helmholtz equations if there is no potential. The Green Function of a chain of Helmholtz equations is constructed. It can be used to write more general Schrödinger chains as integral equations.

1. Introduction

Throughout the paper, we deal with a set of ordered domains Ω_i and surfaces S_i of class C^2, with $S = U_i S_i$, such that each domain is finite but Ω_{N+1}, which extends to infinity in all directions, and $i \neq j \Rightarrow S_i \cap S_j = \emptyset$. The "topological aspect" of such surfaces is that of sphere-like surfaces, each S_i enclosing a ball-like domain $\bar{D}_i = \sum_{k=0}^{k=i} \bar{\Omega}_k$. Now, we study $\Psi \in C^2(\mathbf{R}^3 \backslash S)$ that satisfies the Schrödinger equation inside each Ω_i :

$$(1) \qquad (\Delta + k^2 - W)\Psi(x) = 0 \quad (x \in \mathbf{R}^3 \backslash S)$$

and the continuity conditions

$$(2) \qquad [\Psi(x)/a(x)]_{x \in S_i^+} = \lambda_i [\Psi(x)/a(x)]_{x \in S_i^-}$$

$$(3) \qquad \left[a(x)\frac{\partial \Psi(x)}{\partial \nu_x} - b(x)\Psi(x)\right]_{x \in S_i^+} = \mu_i \left[a(x)\frac{\partial \Psi(x)}{\partial \nu_x} - b(x)\Psi(x)\right]_{x \in S_i^-}$$

where $k \in \mathbf{C}$, $W(x)$ is locally integrable, and where λ_i and μ_i are real *numbers* all different from zero, the functions $a(x) > 0$, $b(x)$ are real, belong to $C(\mathbf{R}^3 \backslash S)$ and go to finite limits as $x \to S$, but can have jumps through the surfaces S. We call the problem (1) (2) (3) a chain of Schrödinger equations, or, simply, a Schrödinger chain. If $\lambda_i = \mu_i = 1$, as we shall see later, the "current" is conserved. If, in addition $a(x)$ is continued inside domains Ω_i by a positive function $\alpha(x) \in C^2(\mathbf{R}^3 \backslash S)$, which can have jumps through the surfaces S_i, together with its normal derivative $\frac{\partial \alpha(x)}{\partial \nu(x)}$ but in such a way that everywhere $b(x) = \frac{\partial \alpha(x)}{\partial \nu(x)}$, this "conserved current" Schrödinger chain is equivalent to the problem

$$(4) \qquad [\alpha^{-2}\text{div } \alpha^2\text{grad} + k^2 - W + \alpha^{-1}\Delta\alpha]\Phi = 0 \quad (x \in \mathbf{R}^3)$$

provided $\Phi(x) = \Psi(x)/\alpha(x)$. The equation (4) appears in many physical contexts ([1],[2], [3], [4]). We call it the impedance equation, the corresponding chain is the impedance chain, and α the impedance factor. Another special case is $W = 0$. We call the corresponding chain a Schrödinger-Helmholtz chain. All the cited chains are homogeneous. A right-hand side to (1) may also be added to show inhomogeneous chains. In the following, we are interested by solutions of chains that satisfy in addition the Sommerfeld condition :

(5)
$$\begin{cases} \langle x/|x|, \quad \operatorname{grad}\Psi(x)\rangle - ik\Psi(x) = o(|x|^{-1}) \quad \text{as} \quad |x| \to \infty \\ \text{uniformly for all directions} \quad x/|x| \end{cases}$$

So as to study the general chain (1) – (2) – (3) – (5), or an inhomogeneous one, the main tool is obviously the Green's function that corresponds to the Schrödinger-Helmholtz chain (1') – (2) – (3) – (5) :

(1')
$$\Delta\Psi + k^2\Psi = 0 \qquad (x \in \mathbf{R}^3\backslash S).$$

The present paper shows a typical construction of this Green's function in the "conserved current case". The special "impedance case" would be hardly easier but is more important for applications. Other cases could be dealt analogously. The results are stated and proved in $\mathbf{R}^3$ but they can be trivially extended to $\mathbf{R}^n$ $(n \geq 3)$.

2. Basic property of homogeneous Schrödinger Helmholtz-chains

For a solution of the Helmholtz equation in Ω_{N+1}, the Sommerfeld condition (5) (see [5], p. 70) implies in particular that $\Psi(x) = O(|x|^{-1})$ as $|x| \to \infty$. It physically means that $\Psi(x)$ is a pure "outgoing" function.

The conditions (2) and (3) imply that the total current on the $(+)$ side of a surface S_i is proportional to the total current on the $(-)$ side :

(6)
$$\int_{S_i^+} (\bar{\Psi}\frac{\partial\Psi}{\partial\nu} - \Psi\frac{\overline{\partial\Psi}}{\partial\nu})ds = \lambda_i\mu_i \int_{S_i^-} (\bar{\Psi}\frac{\partial\Psi}{\partial\nu} - \Psi\frac{\overline{\partial\Psi}}{\partial\nu})ds$$

We claim that *a function $\Psi(x)$ that satisfies (1') – (2) – (3) – (5) must vanish*.

Proof: Inside each domain Ω_{i+1}, the Green's theorem ([5], p. 68) yields

(7)
$$(\int_{S_i^+} - \int_{S_{i+1}^-})(\bar{\Psi}\frac{\partial\Psi}{\partial\nu} - \Psi\frac{\partial\bar{\Psi}}{\partial\nu})ds(x) = \int_{\Omega_{i+1}} (\bar{\Psi}\Delta\Psi - \Psi\Delta\bar{\Psi})dx = 0.$$

A similar argument inside Ω_0, implies that the total current on S_0^- vanishes. Using (6) and (7) it follows that

(8)
$$\int_{S_N^+} (\bar{\Psi}\frac{\partial\Psi}{\partial\nu} - \Psi\frac{\partial\bar{\Psi}}{\partial\nu})ds = \operatorname{Im}\int_{S_N^+} \Psi\frac{\partial\bar{\Psi}}{\partial\nu}ds = 0.$$

Remember $\bar{D}_N = \sum_{k=0}^N \bar{\Omega}_k$. A known result ([5]) on the Helmholtz equation is that any function Ψ that satisfies (1') in $\mathbf{R}^3\backslash\bar{D}_N$ and the condition (5) for a given (real) k also satisfies

(9)
$$\lim_{R\to\infty}\int_{|x|=R} \{|\frac{\partial\Psi}{\partial\nu}|^2 + k^2|\Psi|^2 + 2k\operatorname{Im}(\Psi\frac{\overline{\partial\Psi}}{\partial\nu})\}ds = 0.$$

Equation (8) implies that the last term in (9) vanishes, and thus Equation (9) implies that $\lim_{R\to\infty}\int_{|x|=R}|\Psi|^2 ds = 0$. It follows from Rellich lemma ([5], p. 77) that Ψ identically vanishes in $\mathbf{R}^3\backslash\bar{D}_N$. Hence ψ and $\frac{\partial\Psi}{\partial\nu}$ vanish on S_N^+, and thanks to (2) and (3), on S_N^-, so that ψ satisfies (1') throughout $\mathbf{R}^3\backslash\bar{D}_{N-1}$ and hence vanishes therein according to Rellich lemma. Continuing the process yields $\Psi = 0$. Q.E.D.

3. A Green function for a conserved current chain

3.1. The problem.

$a(x)$, $b(x)$, the domains Ω_i and surfaces S_i being defined above, we seek $G(x,y)$ such that

$$(10) \qquad x \rightarrow G(x,y) \in C^2(\mathbf{R}^3 \backslash S, \ x \neq y)$$

$$(11) \qquad \Delta G(x,y) + k^2 G(x,y) = -\delta(x-y)$$

$$(12a) \qquad \left\{ \begin{array}{l} G(x,y)/a(x) \quad \text{and} \quad G(x,y)b(x) - a(x)\frac{\partial G(x,y)}{\partial \nu(x)} \quad \text{are} \\ \text{continuous through } S, \text{ and Sommerfeld condition.} \end{array} \right.$$

$$(12b) \qquad \left\{ \begin{array}{l} \text{If} \quad x \rightarrow S, \quad y \rightarrow G(x,y) \quad \text{and} \quad y \rightarrow \frac{\partial}{\partial \nu_x} G(x,y) \\ \qquad\qquad \text{remain locally integrable.} \end{array} \right.$$

These conditions obviously imply that for $x \neq y$, $x \rightarrow G(x,y)$ satisfies (1') and that if $\Phi(x,y)$ is the standard Helmholtz-Green function (see [5]):

$$(13) \qquad \Phi(x,y) = [4\pi|x-y|]^{-1} \exp[i|k||x-y|],$$

$x \rightarrow G(x,y) - \Phi(x,y)$ is a solution of Helmholtz Equation (1') that satisfies the Sommerfeld condition. We recall by the way that two functions $g(x)$ and $h(x)$ which satisfy the Sommerfeld condition are such that

$$(14) \qquad \lim_{R \rightarrow \infty} \int_{|x|=R} [g(x)\frac{\partial}{\partial \nu}h(x) - h(x)\frac{\partial}{\partial \nu}g(x)]ds = 0.$$

The function $G(x,y)$ is defined uniquely by (11) – (12) since the difference of two solutions would be a solution of the problem (1') – (2) – (3) – (5). If it exists, it is symmetric with respect to the exchange x, y : using the solution of (11), completed by (12), and the solution $G(x,z)$ of the same equation, with z instead of y, completed by (12), we calculate by means of Green's formula

$$
\int_{x\in\Sigma\Omega_i} [G(x,z)\Delta G(x,y) - G(x,y)\Delta G(x,z)]dx
$$

$$
= \lim_{R\rightarrow\infty} \int_{|x|=R} [G(x,z)\frac{\partial}{\partial\nu_x}G(x,y) - G(x,y)\frac{\partial}{\partial\nu_x}G(x,z)]ds_x
$$

$$(15) \qquad - \int_{S_N^+} [G(x,z)\frac{\partial}{\partial\nu_x}G(x,y) - G(x,y)\frac{\partial}{\partial\nu_x}G(x,z)]ds_x$$

$$
+ \int_{S_N^-} [G(x,z)\frac{\partial}{\partial\nu_x}G(x,y) - G(x,y)\frac{\partial}{\partial\nu_x}G(x,z)]ds_x
$$

$$
+ \dots (\int_{S_0^+} - \int_{S_0^-})[G(x,z)\frac{\partial}{\partial\nu_x}G(x,y) - G(x,y)\frac{\partial}{\partial\nu_x}G(x,z)]ds_x = 0
$$

and it follows from (11) that the left-hand side is also equal to $-G(y,z) + G(z,y)$, Q.E.D. It remains to construct the solution of (11) - (12).

3.2. Surface operators.

We first define the operators S_{ij}, K_{ij}, K'_{ij}, T_{ij}, by

$$(16) \qquad (S_{ij}f)(x) = 2 \int_{S_j} ds(z)\Phi(x,z)f(x), \quad (x \in S_i)$$

$$(17) \qquad (K_{ij}f)(x) = 2 \int_{S_j} ds(z)\frac{\partial\Phi(x,z)}{\partial\nu_z}f(x), \quad (x \in S_i)$$

$$(18) \qquad (K'_{ij}f)(x) = 2 \int_{S_j} ds(z)\frac{\partial\Phi(x,z)}{\partial\nu}f(x), \quad (x \in S_i)$$

$$(19) \qquad (T_{ij}f)(x) = 2\frac{\partial}{\partial\nu_x} \int_{S_j} ds(z)\frac{\partial\Phi(x,z)}{\partial\nu_z}f(x), \quad (x \in S_i).$$

Each of these operators maps a space of functions defined on S_j, with values in $\mathbf{C}$, says $E(S_j)$, into a space of functions defined on S_i, with values in $\mathbf{C}$, say $F(S_i)$. Stating precisely the couples E, F must be done in a different way for S, K, K', T, and may also depend whether $i = j$ or $i \neq j$. Since the surfaces are unconnected, a "non diagonal" operator is the value of a single or double layer potential, or its derivative, at a point that does not belong to the layer, and it is not diffciult to prove by means of Ascoli theorem that S_{ij}, K_{ij}, K'_{ij}, T_{ij}, are compact operators from $C(S_j)$ to $C(S_i)$. As for the "diagonal operators", they are nothing but the "one surface" operators already studied in the literature ([5]). Hence S_{ii}, K_{ii}, K'_{ii} are compact in $C(S_i)$ and $C^{0,\alpha}(S_i)$ for $0 < \alpha < 1$. But T_{ii} is an unbounded operator, defined on a set $N(S_i)$ containing $C^{1,\alpha}(S_i)$, and which is mapped into $C(S_i)$. In fact, the reader can make at this point the choice $E = C^{1,\alpha}$, and assume that not only the four operators, for any choice of i and j, but also their operator products, are mappings into the space of continuous functions on the last surface, and are compact unless a diagonal element of T is involved. In the study of integral equations to be given below, the diagonal elements of T will be eliminated, and we shall set $E(S) = F(S) = C(S)$. We have called S_{ij}, etc, "operator elements" and we have implicitly defined their "operator product", e.g. $S_{ki}T_{ij}$, by

$$(20a) \qquad (S_{ki}T_{ij}f)(x) = 4 \int_{S_i} \Phi(x,t)ds(t)\frac{\partial}{\partial\nu_t} \int_{S_j} \frac{\partial\Phi(t,y)}{\partial\nu_y}f(y)ds(y)$$
$$(\forall f \in E(S_j)), \quad (x \in S_k).$$

Now, an "operator element" like S_{ij} applies to any f defined on S_j — say $f_j(x)$. We define a function $f(x)$ on S by the equalities:

$$(21a) \qquad f(x) = f_j(x) \qquad (x \in S_j), \quad j = 0,1,\dots,N.$$

If each $f_j \in E(S_j)$, $f \in E(S)$. We define a "surface operator" S, or K, etc., as a mapping from $E(S)$ to $F(S)$ such that if $f_j(x)$ is the restriction of f to $x \in S_j$,

$$(21b) \qquad (Sf)(x) = 2\sum_j \int_{S_j} \Phi(x,t)f_j(t)ds(t) \quad (x \in S).$$

With these definitions, the operator product of two surface operators, say ST, is the operator such that for any $f \in E(S)$, $x \in S_k$, $k = 0, 1, \ldots, N$,

$$(20b) \qquad (STf)(x) = \left(\sum_i S_{ki}(Tf)(x \in S_i)\right)(x) = \left(\sum_i \sum_j S_{ki}T_{ij}f_j\right)(x).$$

Since N is finite, S, K, K' are compact mappings from $C(S)$ to $C(S)$, whereas T is not, because of its "diagonal" elements.

3.3. Relations between the surface operators.

In the standard case of one surface, there are simple relations between T and the other operators. They are more complicated in our case, but the proofs are elementary. They proceed through repeated use of Green's identity and of jump discontinuities formulas. For example (24) is obtained by summing over i the element

$$Q_{kij} = 4\frac{\partial}{\partial \nu_x} \int_{S_i} [\frac{\partial \Phi(x,t)}{\partial \nu_t} g_j(t) - \Phi(x,t)\frac{\partial g_j(t)}{\partial \nu_t}]ds(t)$$

where $x \in S_k$ and

$$g_j(t) = \int_{S_j} \frac{\partial \Phi(t,y)}{\partial \nu_y} f_j(y)ds(y)$$

is continuous unless t crosses S_j. The case $k > i > j$, given as an example, proceeds through the self explanatory formulas :

$$Q_{kij} = 4\frac{\partial}{\partial \nu_x} \int_{S_i} [\frac{\partial \Phi(x,t)}{\partial \nu_t} g_j(t) - \Phi(x,t)\frac{\partial}{\partial \nu_t} g_j(t)]ds(t)$$

$$= \cdots = 4\frac{\partial}{\partial \nu_x} \int_{S_j} [\frac{\partial \Phi(x,t)}{\partial \nu_t} g_j(t^-) - \Phi(x,t)\frac{\partial}{\partial \nu_t} g_j(t^-)]ds(t)$$

$$+ 4\frac{\partial}{\partial \nu_x} \int_{S_j} [\frac{\partial \Phi(x,t)}{\partial \nu_t} f_j(t)ds(t) = 4\frac{\partial}{\partial \nu_x} g_j(x).$$

The results are

$$(22) \qquad\qquad\qquad KK - ST = 1 + \dot{K}$$

$$(23) \qquad\qquad\qquad K'K' - TS = 1 - \dot{K'}$$

$$(24) \qquad\qquad\qquad TK - K'T = \dot{T}$$

where $\dot{M}$ is obtained from the matrix M made of operator elements M_{ij} by the ansatz

$$(25) \qquad\qquad \dot{M}_{kj} = \begin{cases} 2(k-j-1)M_{kj} & \text{for } j \le k-1 \\ 0 & \text{for } j = k \\ 2(k+j-1)M_{kj} & \text{for } j \ge k+1. \end{cases}$$

The dotted matrices involve only "non diagonal" elements. $\dot{T}$ therefore is compact on $E(S)$.

3.4. Potentials.

The right-hand side of (16), (17), (18), (19), can be continued for any $x \in \mathbf{R}^3 \backslash S$ as simple and double layer potentials and their derivatives. Well known results apply ([5]) and readily show that the single layer potential (16 cont.) and the double layer potential (17 cont.) define solutions of the Helmholtz equation in $\mathbf{R}^3 \backslash S$ that satisfy the Sommerfeld condition (5). In addition, it is well known that the single layer potential is continuous throughout $\mathbf{R}^3$ but its derivative has known discontinuities when crossing the surface where it is defined. The reverse is true for the double-layer potential. These results will be used to construct our Green Function.

3.5. Construction of the Green function.

We can choose between constructing directly a solution of (11,12), by writing down $G(x,y)$ as the sum $\Phi(x,y) + $ a single layer potential $+$ a double layer potential, or using the same ansatz to solving the inhomogeneous equation

$$(26) \qquad \Delta u(x) + k^2 u(x) = -f(x)$$

completed by the condition (12a) (on $u(x)$). $G(x,y)$ is then identified as the resolvent kernel of this problem. We use this second approach and write down

$$(27) \qquad \begin{aligned} u(x) = F(x) + \sum_{j=0}^{N} \int_{S_j} ds(z) \frac{\partial \Phi(x,z)}{\partial \nu_z} \psi_j(z) \\ + \sum_{j=0}^{N} \int_{S_j} ds(z) \Phi(x,z) \varphi_j(z) \end{aligned}$$

where

$$(28) \qquad F(x) = \int_{\mathbf{R}^3} \Phi(x,y) f(y) dy.$$

The potential representations satisfy the Helmholtz equation and the Sommerfeld condition; since $F(x)$ satisfies (26) and the Sommerfeld condition, so does also $u(x)$. Now the continuity conditions through S_i imply that for $x \in S_i$, the following equalities must be satisfied, where the index i reminds that $x \in S_i$, $i = 0, 1, 2, \ldots, N$ and the notations introduced above are used:

$$(29) \qquad \begin{aligned} {}[\psi_i(x)(a^-(x) + a^+(x))]_{x \in S_i} = \\ = \{(a^+(x) - a^-(x))[2F(x) + (\mathcal{K}\psi)(x) + (\mathcal{S}\varphi)(x)]\}_{x \in S_i} \end{aligned}$$

$$(30) \qquad \begin{aligned} {}[\varphi_i(x)(a^-(x) + a^+(x)) + \psi_i(x)(b^-(x) + b^+(x))]_{x \in S_i} = \\ = \{(a^+(x) - a^-(x))[2F'(x) + (\mathcal{T}\psi)(x) + (\mathcal{K}'\varphi)(x)] \\ - (b^+(x) - b^-(x))[2F(x) + (\mathcal{K}\psi)(x) + (\mathcal{S}\varphi)(x)]\}_{x \in S_i}. \end{aligned}$$

In these formulas, the prime used in F' stands for the normal derivative at $x(\in S_i)$, φ_i and ψ_i are respectively the restrictions of φ and ψ to $x \in S_i$. The equations (29) and (30) must be satisfied for all i's. It is convenient to rewrite them by introducing the functions on S, or "multiplicative operators" β, β', and γ, hereafter defined

$$(31a) \qquad \beta(x) = \frac{a^+(x) - a^-(x)}{a^+(x) + a^-(x)} \qquad \beta'(x) = \frac{b^+(x) - b^-(x)}{a^+(x) + a^-(x)}$$

$$(31b) \qquad \gamma(x) = \frac{b^+(x) + b^-(x)}{a^+(x) + a^-(x)}.$$

Sufficient assumptions on β, β', γ is that they belong to $C^{1,\alpha}(S)$. Hence, with these condensed notations, the continuity conditions through S yield the following equations for the functions $\varphi(x \in S)$ and $\psi(x \in S)$:

$$(32) \qquad \psi = 2\beta F + \beta S\varphi + \beta \mathcal{K}\psi$$

$$(33) \qquad \varphi + \gamma\psi = 2\beta F' - 2\beta' F + \beta T\psi - \beta' \mathcal{K}\psi + \beta \mathcal{K}'\varphi - \beta' S\varphi$$

It the system obtained from (32 - 33) by setting $F = F' = 0$ has a solution, ψ_0, φ_0, inserting it into (27) yields a solution $\Psi^0(x)$ of the homogeneous chain (1') (2) (3) (5) (with $\lambda_i = \mu_i = 1$). We know from section 2 that Ψ^0 necessarily vanishes, and using jump relations across layers shows that so do ψ^0 and φ^0. Hence, if the system (32 - 33) has a solution, it is unique.

So as to get the solution, we may try solving (32) in terms of ψ and insert the result into (33). However $(\mathcal{I} - \beta\mathcal{K})^{-1}$ exists as a bounded operator if and only if 1 is not an eigenvalue of $\beta\mathcal{K}$. Since $\mathcal{K}$ is bounded there is at least a range of values of β, small enough (say, $\|\beta\| \leq \|\mathcal{K}\|^{-1}$) where 1 cannot be an eigenvalue. On the other hand, if 1 is an eigenvalue of $\beta\mathcal{K}$, the equation (32) yields ψ if and only if $\beta(2F + S\varphi)$ is orthogonal to the n-dimensional null-space of the adjoint of $(\mathcal{I} - \beta\mathcal{K})$. Then ψ contains itself n parameters corresponding to the null-space $N(1 - \beta K)$, inserting it into (33) can give φ up to n parameters, which are eventually determined by the condition on $\beta(2F + S\varphi)$. This process works but is complicated. For the sake of simplicity, we shall not study it here and we assume in this preliminary paper that

$$(34) \qquad N(\mathcal{I} - \beta\mathcal{K}) = 0.$$

Since $\mathcal{K}$ is compact, the Fredholm alternative holds for (32), considered as an equation for ψ, $(\mathcal{I} - \beta\mathcal{K})^{-1}$ is a bounded operator and it enables us to construct ψ from φ. Reinserting the result into (34), we meet the operator

$$(35) \qquad \mathcal{U} = T(\mathcal{I} - \beta\mathcal{K})^{-1}\beta S$$

which needs a special study, because it involves the product of an unbounded operator and a compact one.

Using (23) and (24), we obtain after elementary calculations

$$(36) \quad \mathcal{U} = (\mathcal{I} - \beta \mathcal{K}')^{-1} \big\{ [\mathcal{T}, \beta] \mathcal{S} - \beta (\mathcal{I} - \mathcal{K}' \mathcal{K}' - \dot{\mathcal{K}}') + ([\mathcal{T}, \beta] \mathcal{K} + \beta \dot{\mathcal{T}}) (\mathcal{I} - \beta \mathcal{K})^{-1} \beta \mathcal{S} \big\}$$

let us show that $[\mathcal{T}, \beta]$ is a bounded operator. The only elements which need a proof are the "diagonal ones", which read for $x \in S_k$ (and applied to $f_k(x)$) :

$$
\begin{aligned}
(37) \qquad D_k(x) &= \frac{\partial}{\partial \nu_x} \int_{S_k} \frac{\partial \varphi(x,t)}{\partial \nu_t} \beta(t) f_k(t) ds(t) - \beta(x) \int_{S_k} \frac{\partial \varphi(x,t)}{\partial \nu_t} f_k(t) ds(t) \\
&= \big(\frac{\partial}{\partial \nu_x} \beta(x) \big) \int_{S_k} \frac{\partial \varphi(x,t)}{\partial \nu_t} f_k(t) ds(t) \\
&\quad + \frac{\partial}{\partial \nu_x} \int_{S_k} \frac{\partial \varphi(x,t)}{\partial \nu_t} [\beta(t) - \beta(x)] f_k(t) ds(t).
\end{aligned}
$$

The first term in the last right hand side of (37) is the product of $\frac{\partial}{\partial \nu_x} \beta(x)$ by $(K_{kk} f_k)(x)$ and thus defines a bounded operator. The boundedness of the next one follows from the lemma (2.10) of [5] (after a few easy calculations). Q.E.D.

Going back to $\mathcal{U}$ in (35), and since the product of a compact operator by a bounded operator is compact, we see that $\mathcal{U} + \beta \mathcal{I}$ is compact :

$$(38) \qquad \qquad \mathcal{U} = -\beta \mathcal{I} + \mathcal{C}$$

where $\mathcal{C}$ is complicated, but trivially written down from (36). Inserting the solution ψ of (32) into (33) and taking into account (38) yields

$$(39) \qquad \qquad (1 + \beta^2) \varphi = \mathcal{A} F + \mathcal{B} \varphi$$

where

$$(40) \quad \mathcal{A} F = 2\beta F' + [2\beta \mathcal{T} (\mathcal{I} - \beta \mathcal{K})^{-1} \beta - 2\beta' - 2\beta' \mathcal{K} (\mathcal{I} - \beta \mathcal{K})^{-1} \beta - 2\gamma (\mathcal{I} - \beta \mathcal{K})^{-1} \beta] F$$

$$(41) \qquad \mathcal{B} = -\gamma (\mathcal{I} - \beta \mathcal{K})^{-1} \beta \mathcal{S} + \beta \mathcal{C} - \beta' \mathcal{K} (\mathcal{I} - \beta \mathcal{K})^{-1} \beta \mathcal{S} + \beta \mathcal{K}' - \beta' \mathcal{S}$$

$(1 + \beta^2)^{-1} \mathcal{B}$ is compact, the Fredholm alternative holds, the homogeneous solution yields a solution of (1'- 2 – 3 – 5) and therefore vanishes. Hence φ can be constructed as

$$(42) \qquad \qquad \varphi = [(1 + \beta^2) \mathcal{I} - \mathcal{B}]^{-1} \mathcal{A} F.$$

Constructing the Green function is very much simplified in two cases:

(a) *Assumption A: $\alpha(x)$ is continuous throughout $\mathbf{R}^3$.*

Then one can use (27) with $\psi = 0$, and β vanishes identically. The remaining equation (33) reduces to

$$(43) \qquad \qquad \varphi = -\beta'(2F + \mathcal{S}\varphi).$$

With our assumptions on $a, b, \beta'S$ is compact, the homogeneous equation has only the zero solution because of section 2 results, and

(44) $$\varphi = -2(\mathcal{I} + \beta'S)^{-1}\beta'F =: -2(\mathcal{I} - \beta'\mathcal{R})\beta'F.$$

It follows from (27) and (28) that the Green's function is then :

(45)
$$G(x,y) = \Phi(x,y) - 2\sum_{k=0}^{N}\int_{S_k}ds(z)\Phi(x,z)\beta'(z)\Phi(z,y)$$
$$+ 2\sum_{k=0}^{N}\int_{S_k}ds(z)\Phi(x,z)\beta'(z)\sum_{j=0}^{N}\int_{S_j}ds(t)R(z,t)\beta'(t)\Phi(t,y)$$

where $R(z,t)$ is the kernel of $\mathcal{R}$.

(b) *Assumption B.* For each surface, the "relative discontinuity" $a^+(x)/a^-(x)$ does not depend on the position of x on the surface.

In this case, it is possible to construct, like in the one dimensional case ([2]) a piecewise constant "singular data" function say $\sigma(x)$ equal to $\sigma_{N+1} = 1$ for $x \in \Omega_{N+1}$, and

(46) $$\sigma_i = \sigma_{i+1}a(x \in S_i^-)\backslash a(x \in S_i^+)$$

in each domain Ω_i $(i = N, N-1, \ldots, 0)$. We can solve the problem made of Equation (26) completed by the Sommerfeld condition and the impedance continuity conditions (11) - (12) by setting $v = u/\sigma$, $g = f/\sigma$ and solving rather the problem

(47a) $$\Delta v(x) + k^2v(x) = -g(x)\quad x \in \mathbf{R}^3\backslash S$$

(47b)
$$\begin{cases} v(x) & \text{and} \quad \sigma^2[\frac{\partial v}{\partial \nu} - \frac{bv}{a}] \quad \text{continuous} \\ & \text{Sommerfeld condition.} \end{cases}$$

The analysis of §II applies to the homogeneous form of (47) and guarantees the uniqueness of the solution, that can be constructed by making use of (27), with $\psi = 0$, solving the Fredholm equation

(48) $$\varphi = 2\tilde{\beta}G' - 2\tilde{\beta}'G + \tilde{\beta}\mathcal{K}'\varphi - \tilde{\beta}'S\varphi$$

where $G(x) = \int_{\mathbf{R}^3}\Phi(x,y)g(y)dy$ and

(49) $$\tilde{\beta} = \frac{(\sigma^+)^2 - (\sigma^-)^2}{(\sigma^+)^2 + (\sigma^-)^2}, \quad \tilde{\beta}' = \frac{(\sigma^+)^2b^+/a^+ - (\sigma^-)^2b^-/a^-}{(\sigma^+)^2 + (\sigma^-)^2}.$$

The Fredholm alternative guarantees that $(\mathcal{I} + \tilde{\beta}'S - \tilde{\beta}\mathcal{K}')^{-1}$ always exists as a bounded operator. It yields $v(x) = \int_{\mathbf{R}^3}g(x,y)g(y)dy$. This resolvent $g(x,y)$ (which is of course not symmetric), readily yields the Green function of the original problem as the product $G(x,y) = \sigma(x)g(x,y)/\sigma(y)$. Thus, we see that our problem is much more simple than the general three-dimensional problem. Yet, this case is probably very appealing for physicists. From the mathematical point of view, it is the direct generalization of the one-dimensional case.

Remark: As it was noticed by the referee of this paper, assumption B also allows a direct treatment of the system (32) - (33) as a Fredholm equation in $C(S)$ for a vector $\begin{pmatrix} \psi \\ S\varphi \end{pmatrix}$.

4. Generalizations and applications

The generalization from $\mathbf{R}^3$ to $\mathbf{R}^n$ is easily obtained by resetting the standard Green's functions $\Phi(x, y)$ defined in (13) by $\frac{i}{4}\left(\frac{|k|}{2\pi|x-y|}\right)^{\frac{n-2}{2}} H^{(1)}_{\frac{N-2}{2}}(|k||x - y|)$, and rewriting with proper coefficient the Sommerfeld condition and the quantities associated with jump discontinuities.

The standard application of our Green's function is for transforming the chain with (1) instead of (1'), into an integral equation for $\Psi(x)$. Hence, the effects of discontinuities and those of W are studied sucessively. We use the "impedance chain Green's function" in this way to derive the generalized Lippmann-Schwinger equation for Impedance Scattering ([6]).

ACKNOWLEDGEMENTS: This work has been for us the opportunity of an interesting collaboration with A. Nachman. Parts of the work are simultaneously presented in a lecture for a meeting on Scattering Theory ([6]). More complete results will be published later in a journal.

REFERENCES

[1] Sabatier P.C., *Remark on the three-dimensional mixed impedance potential equation*, Inverse Problems **3**, **L83-L86** (1987); Corrigendum, Inverse Problems **4**, **L1** (1988).

[2] Sabatier P.C., *For an impedance scattering theory*, in "Non linear Evolutions," J. Leon Ed., World Scientific Publ., Singapore, 1988, pp. 723-745.

[3] Sabatier P.C. and Dolveck-Guilpart B., *On modeling discontinuous data, One dimensional approximations*, J. Math. Phys. 29 (1988), 861-868.

[4] Chadan C. and Sabatier P.C., "Inverse Problems in Quantum Scattering Theory," 2nd Edition revised and expanded. Springer Verlag (in press).

[5] Colton D. and Kress R., "Integral Equations Methods in Scattering Theory," . J. Wiley, New York, 1983.

[6] Sabatier P.C., *Three-dimensional Impedance Scattering Theory*, in "Proceedings of a meeting on Electromagnetic and acoustic scattering: Detection and Inverse Problem," C. Bourrely, Chiappetta and B. Torresani, Eds., World Scientific Publ., Singapore, 1989.

Pierre C. Sabatier
Laboratoire de Physique Mathématique
U.S.T.L.
Place Eugène - Bataillon
F-34060 Montpellier Cedex
France

International Series of
Numerical Mathematics, Vol. 91
© 1989 Birkhäuser Verlag Basel

Second variations for domain optimization problems

Jacques Simon

Departement de Mathématiques Appliquées
Université Blaise Pascal

Introduction

The calculation of the first variation with respect to domain variation is well known, for solutions of p.d.e. as well as for associated functionals.

Very little is known concerning the second variation, with the exception of some calculations for torsional rigidity by N. Fujii [1].

Usually the second variation is the variation of the first variation. For domain dependent functionals this is no longer true. Indeed the variation of a domain Ω is represented by a vector field u, and the variation that result from two successive variations u and v is not the sum $u + v$, that is

$$(\Omega + u) + v \neq \Omega + (u + v).$$

for not constant v.

Here we prove that, for any functional g depending on domain, the second variation g'' is related to the variation of the first variation, $(g')'$, by the following formula

$$g''(\Omega; u, u) = (g')'(\Omega; u, u) - g'(\Omega; u \cdot \nabla u).$$

Moreover g'' exists as soon as g' and $(g')'$ exist.

With this formula the second variation for solutions of p.d.e. as well as for associated functionals is calculated based on the well known formulas for first variations.

The outline is as follows

1. First variations formulas revisited
2. Second variations formulas
3. Theorem on the relation between g'' and $(g')'$
4. Second variation of the drag of a body.

In the first two sections we give general ideas and formulas; assumptions are specified in the last two sections.

The author is happy to thank the organizers of the Conference in Vorau where this paper was written, and to thank M. Vogelius for fruitful discussions.

1. First variations formulas revisited

1.1. Optimal design problem.

Let Ω be a bounded open subset of $\mathbf{R}^N$ with a boundary $\partial\Omega$. Let $y(\Omega)$ be the solution of a boundary value problem in Ω, say

$$A(y(\Omega)) = f \quad \text{in} \quad \Omega, \quad B(y(\Omega)) = h \quad \text{on} \quad \partial\Omega.$$

The cost function is the real number

$$J(\Omega) = \frac{1}{2} \int_{\Omega} |C(y(\Omega)) - d|^2 dx$$

A, B and C being partial differential operators, and f, h and d being functions given in all of $\mathbf{R}^N$.

In optimal design one seeks a domain Ω_0, which minimizes $J(\Omega)$ in some class $\mathcal{D}_{ad}$ of admissible domains.

1.2. First variation J'.

Let u be a vector field defined in all of $\mathbf{R}^N$, representing the variations of Ω; u is assumed to be in $Lip^k(\mathbf{R}^N; \mathbf{R}^N)$, $k \geq 1$, the norm of which is denoted by $\|\|_k$. The new domain is defined by

$$\Omega + u = \{x + u(x) : x \in \Omega\}.$$

We are interested in the first order expansion of $J(\Omega + u)$ with respect to u, that is

$$J(\Omega + u) = J(\Omega) + J'(\Omega; u) + o(\|u\|_k)$$

with a linear $J'(\Omega; \cdot)$.

The problem is to calculate the first variation J', and to prove that the expansion holds.

Remark. As for any optimization problem, the first variation yields a necessary optimality condition: if Ω_0 is an optimal domain then, denoting $\mathcal{D}'_{ad}$ the tangent cone to $\mathcal{D}_{ad}$ at Ω_0,

$$J'(\Omega_0; u) \geq 0, \quad \forall u \in \mathcal{D}'_{ad}.$$

Moreover, based on J' we can use a gradient method to construct a locally optimal domain Ω_0.

1.3. First local variation y'.

To find the first variation of J we need the variation of y. That is we seek an expansion of the type

$$y(\Omega + u) = y(\Omega) + y'(\Omega; u) + o(\|u\|_k).$$

However $y(\Omega + u)$ lives in the domain $\Omega + u$ which depends on u, but $y(\Omega)$ and $y'(\Omega; u)$ live on the fixed domain Ω. Therefore this expansion cannot be satisfied in all of Ω; it can be satisfied only in the intersection, for all u, of the domains $\Omega + u$.

We will say that the expansion holds locally if it holds in any $\omega \subset\subset \Omega$. This is meaningfull since then $\omega \subset \Omega + u$ for all small enough u.

The function $y'(\Omega; u)$ is defined in all of Ω by it's restrictions to any ω. It is so–called the first local variation.

1.4. Formula for first variations of the optimal design problem.

We suppose now that $y(\Omega+u)$ depends on u regularly and uniformly up to the boundary $\partial\Omega + u$.

This assumption, which will be specified later, is satisfied if, for the fixed domain Ω, $y(\Omega)$ is unique and depends regularly on the coefficients of A and B and on the functions f and h.

If A, B and C are linear, then, for any u, the first local variation $y' = y'(\Omega; u)$ is solution of the boundary value problem

$$Ay' = 0 \quad \text{in} \quad \Omega, \quad By' = -u_n \frac{\partial}{\partial n}(By - h) \quad \text{on} \quad \partial\Omega.$$

Here n is an unitary normal vector to $\partial\Omega$, $u_n = u \cdot n$ is the normal component of u and $y = y(\Omega)$.

The first variation of the cost is

$$J'(\Omega; u) = \int_\Omega (Cy - d)Cy' dx + \frac{1}{2}\int_{\partial\Omega} u_n |Cy - d|^2 ds.$$

If A, B and P are differentiable in convenient function spaces on Ω, denoting DA the derivative of A, y' satisfies more generally

$$DA(y; y') = 0 \quad \text{in} \quad \Omega, \quad DB(y; y') = -u_n \frac{\partial}{\partial n}(B(y) - h) \quad \text{on} \quad \partial\Omega.$$

And, if $J(\Omega) = \int_\Omega P(y(\Omega))dx$, J' is given by

$$J'(\Omega; u) = \int_\Omega DP(y; y')dx + \int_{\partial\Omega} u_n P(y)ds.$$

These formulas are well known in mechanics, but suitable assumptions are not so known.

1.5. Total variation y up to the boundary.

To get the boundary condition on y', and to get the variation of the integral of $C(y)$, we need uniform dependence of $y(\Omega + u)$ on u up to the boundary $\partial\Omega + u$, which will be obtained by mapping $y(\Omega + u)$ onto the fixed domain Ω.

Denoting by I the identity in $\mathbf{R}^N$, $I + u$ maps Ω onto $\Omega + u$. Thus the function $y(\Omega + u) \circ (I + u)$ lives in Ω, and the uniform dependence up to the boundary is given by

$$y(\Omega + u) \circ (I + u) = y(\Omega) + y^{\cdot}(\Omega; u) + o(\|u\|_k) \quad \text{in} \quad \Omega.$$

This condition, in a suitable function space on Ω depending on A, B and C, imply that J has a first order expansion, that locally y has a first order expansion, and that $J'(\Omega; u)$ and $y'(\Omega; u)$ satisfy the above formulas.

The reader is referred to J. Simon [2] for precise statements and for proof of these results.

The function $y^{\cdot}(\Omega; u)$ being a kind of total derivative is so-called the first total variation.

Remark: *Respective use of local and total variations.*

The existence of the total variation $y^{\cdot}$ is necessary for our proof of the existence of J'. However knowing the exact value of $y^{\cdot}$ is not required.

To calculate the value of J' we use only the value of the local variation y', which is determined by the above boundary value problem. This is the reason why we use at the same time the two different objects y' and $y^{\cdot}$.

The value of $y^{\cdot}$ may be obtained by

$$y^{\cdot}(\Omega; u) = y'(\Omega; u) + u \cdot \nabla y(\Omega).$$

1.6. Basic formulas for first local variations.

Now we will give formulas for the first variations of elementary equation, boundary condition and integral.

Let $z(\Omega + u)$ be a function defined in $\Omega + u$, which depends on u regularly and uniformly up to the boundary $\partial\Omega + u$. It's local variation $z'(\Omega; u)$ has the following properties.

(1) If, $\forall u$, $z(\Omega + u) = 0$ in $\Omega + u$ then, $\forall u$, $z'(\Omega; u) = 0$ in Ω.
(2) If, $\forall u$, $z(\Omega + u) = 0$ on $\partial\Omega + u$ then, $\forall u$, $z'(\Omega; u) + u \cdot \nabla z(\Omega) = 0$ on $\partial\Omega$.

The integral $K(\Omega + u) = \int_{\Omega+u} z(\Omega + u)dx$ has a first variation which is, $\forall u$,

(3) $$K'(\Omega; u) = \int_\Omega z'(\Omega; u) + \nabla \cdot (uz(\Omega))dx.$$

Here $\nabla = (\partial_1, \ldots, \partial_N)$ where $\partial_i = \partial/\partial x_i$, thus $u \cdot \nabla z = \Sigma_i u_i \partial_i z$ and $\nabla \cdot u = \Sigma_i \partial_i u_i$.

The regular and uniform dependence on u means that, in a convenient function space on Ω, there exists a total variation $z^{\cdot}(\Omega; u)$.

The precise statements, with assumptions in Sobolev spaces, and proofs may be found in [2], lemma 2.1, p. 657 and theorems 3.1, 3.2 and 3.3 p. 663–664. Here we will only give rough proofs.

Remark: *Explicit dependence on u_n.*

In (2), $z(\Omega) = 0$ on $\partial\Omega$, thus $\nabla z(\Omega) = n\partial z(\Omega)/\partial n$ and

$$z'(\Omega; u) = -u_n \partial z(\Omega)/\partial n \quad \text{on} \quad \partial\Omega.$$

In (3) Stokes formula yields

$$K'(\Omega; u) = \int_\Omega z'(\Omega; u)dx + \int_{\partial\Omega} u_n z(\Omega)ds.$$

In fact, for any quantity $g(\Omega)$ depending on Ω, if the first variation $g'(\Omega; u)$ exists, it depends on u only by it's normal component u_n. This is proved in [3], theorem 3.1. page III.20.

Remark. The formulas for first variations of section 1.4 follow by choosing successively z to be $A(y) - f$, $B(y) - h$, and $\frac{1}{2}|C(y) - d|^2$.

Rough proof of (1). The local expansion reduces to $z'(\Omega; u) + o(\|u\|_k) = 0$ in any fixed domain $\omega \subset\subset \Omega$. Thus $z'(\Omega; u) = 0$ in ω and therefore in Ω.

Rough proof of (2). Now $z(\Omega + u) \circ (I + u) = 0$ on the fixed boundary $\partial\Omega$, thus the first total variation satisfies $z^{\cdot}(\Omega; u) = 0$ on $\partial\Omega$. We conclude since $z^{\cdot}(\Omega; u) = z'(\Omega; u) + u \cdot \nabla z(\Omega)$.

Rough proof of (3). By the change of variable $I + u$ we obtain an integral on the fixed domain Ω:

$$K(\Omega + u) = \int_{\Omega} z(\Omega + u) \circ (I + u) Jac(I + u) dx.$$

The first variation of $Jac(I + u)$ is $\nabla \cdot u$, thus

$$K'(\Omega; u) = \int_{\Omega} z^{\cdot}(\Omega; u) + z(\Omega)\nabla \cdot u dx = \int_{\Omega} z'(\Omega; u) + u \cdot \nabla z(\Omega) + z(\Omega)\nabla \cdot u dx.$$

2. Second variations formulas

2.1. Second variation J'' and y''.

We are now interested in the second order expansion of J with respect to u, that is

$$J(\Omega + u) = J(\Omega) + J'(\Omega; u) + \frac{1}{2}J''(\Omega; u, u) + o((\|u\|_k)^2)$$

with a bilinear $J''(\Omega; \cdot, \cdot)$.

The problem is to calculate J'', and to prove that the expansion holds.

To find J'' we need the second local variation y'', that is for any $\omega \subset\subset \Omega$, an expansion

$$y(\Omega + u) = y(\Omega) + y'(\Omega; u) + \frac{1}{2}y''(\Omega; u, u) + o((\|u\|_k)^2) \quad \text{in} \quad \omega.$$

The function $y''(\Omega; u, u)$ is defined in all of Ω by it's restriction to any ω.

Remark. The second variation J'' gives a sufficient condition for local optimality: if

$$J'(\Omega_0; u) = 0 \quad \text{and} \quad J''(\Omega_0; u, u) \geq 0 \quad \forall u \in \mathcal{D}'_{ad},$$

then Ω_0 is locally optimal.

Moreover based on J'' we can improve the velocity of gradient methods. For example by choosing the $n + 1$ approximation of Ω_0 to be

$$\Omega_{n+1} = \Omega_n + t_n u_n, \quad \text{where} \quad t_n = -J'(\Omega_n; u_n)/J''(\Omega_n; u_n, u_n).$$

2.2. Formula relating J'' and $(J')'$.

Usually the second variation with respect to a parameter is the variation of the first variation. Usually means for a parameter in a linear space.

In optimal design the parameter is Ω, which is not in a linear space. The variations of Ω are represented by the parameter u which is in a linear space, however the behavior is not usual since the variation of Ω that result from two successive variations u and v is not the sum $u + v$. In fact,

$$(1) \qquad\qquad (\Omega + u) + v = \Omega + (u + v \circ (I + u)).$$

Indeed by definition $\Omega + u = (I + u)(\Omega)$, thus this follows from $(I + v) \circ (I + u) = I + u + v \circ (I + u)$.

Assume that $J(\Omega + u)$ has a first order expansion with respect to u, and that it's first variation $J'(\Omega + u; w)$ has a first order expansion with respect to u. Then $J(\Omega + u)$ has a second order expansion with respect to u, and the second variation is given by

$$(2) \qquad\qquad J''(\Omega; u, u) = (J')'(\Omega; u, u) - J'(\Omega; u \cdot \nabla u)$$

where $u \cdot \nabla u = \Sigma_i u_i \partial_i u$.

With this relation the second order expansion J'' may be obtained by using twice the formulas for first variations.

Assumptions will be specified and the whole proof will be given in section 3.

Rough proof: We define a map j by $j(u) = J(\Omega + u)$. This map is defined on a linear space of vector fields u, and the expansion of $J(\Omega + u)$ is Taylor's formula for j at 0. Thus

$$J''(\Omega; u, u) = D^2 j(0; u, u), \quad J'(\Omega; u) = Dj(0; u),$$

where $D^2 j$ and Dj are the usual derivatives:

$$D^2 j(0; u, w) = \lim_{t \to 0} \frac{1}{t}(Dj(tu; w) - Dj(0; w)), \quad Dj(tu, w) = \lim_{s \to 0} \frac{1}{s}(j(tu + sw) - j(tu)).$$

Let us calculate these quantities. At first (1) with $v = swo(I+tu)^{-1}$ yields $\Omega+(tu+sw) = (\Omega + tu) + sw \circ (I + tu)^{-1}$, thus

$$Dj(tu; w) = \lim_{s \to 0} \frac{1}{s}(J((\Omega + tu) + sw \circ (I + tu)^{-1}) - J(\Omega + tu))$$

$$= J'(\Omega + tu; w \circ (I + tu)^{-1}).$$

Thus, since $J'(\Omega + tu; \cdot)$ is linear and $w \circ (I + tu)^{-1} = w - tu \cdot \nabla w + o(t)$,

$$D^2 j(0; u, w) = \lim_{t \to 0} \frac{1}{t}(J'(\Omega + tu; w \circ (I + tu)^{-1}) - J'(\Omega; w))$$

$$= \lim_{t \to 0} \frac{1}{t}(J'(\Omega + tu; w) - J'(\Omega; w))$$

$$+ J'(\Omega + tu; \frac{1}{t}(w \circ (I + tu)^{-1} - w))$$

$$= (J')'(\Omega; u, w) + J'(\Omega; -u \cdot \nabla w).$$

Remark. This relation between variations is satisfied by any function of domain, and in particular by $y_{|\omega}$. Indeed in this calculation we have not used the particular definition of J.

2.3. Basic formulas for second local variations.

We will now deduce formulas for the second variations of an equation, of a boundary condition or of an integral, from the formulas for first variations by using the relation between g'' and $(g')'$.

Let $z(\Omega + u)$ be a function defined in $\Omega + u$, which depends on u regularly and uniformly up to the boundary $\partial\Omega + u$. The second local variation $z''(\Omega; u, u)$ has the following properties.

(1)   If, $\forall\, u,\ z(\Omega + u) = 0$ in $\Omega + u$ then, $\forall\, u,\ z''(\Omega; u, u) = 0$ in Ω.

      If, $\forall\, u,\ z(\Omega + u) = 0$ on $\partial\Omega + u$ then, $\forall\, u,$

(2) $z''(\Omega; u, u) + 2u \cdot \nabla z'(\Omega; u) + uu \cdot \nabla^2 z(\Omega) = 0$ on $\partial\Omega$.

The integral $K(\Omega + u) = \int_{\Omega+u} z(\Omega + u)dx$ has a second variation which is, $\forall\, u$,

$$(3)\ \ K''(\Omega; u, u) = \int_\Omega z''(\Omega; u, u) + 2\nabla \cdot (uz'(\Omega; u)) + \nabla \cdot (u\nabla \cdot (uz(\Omega)) - (uz(\Omega) \cdot \nabla)u)dx.$$

Here $uu \cdot \nabla^2 z = \Sigma_{ij} u_i u_j \partial_i z \partial_j z$ and $\nabla \cdot (u\nabla \cdot (uz) - (uz \cdot \nabla)u) = \Sigma_{ij} \partial_i (u_i \partial_j (u_j z) - u_j z \partial_j u_i)$.

The regular and uniform dependence on u means that $z(\Omega + u)$ has a first total variation $z^{\cdot}(\Omega; u)$ for all Ω, and therefore a first local variation $z'(\Omega; w)$, and that $z'(\Omega; u)$ has a total variation. Then locally, that is in any fixed domain $\omega \subset\subset \Omega$, $u(\Omega + u)$ has a second order expansion.

Remark: *Explicit dependence on u_n and $(u \cdot \nabla u)_n$.*

In (3) Stokes formula yields

$$(4)\qquad \begin{aligned} K''(\Omega; u, u) &= \int_\Omega z''(\Omega; u, u)dx + \int_{\partial\Omega} 2u_n z'(\Omega; u) + u_n u\nabla \cdot z(\Omega)) \\ &\quad + (u\nabla \cdot u - u \cdot \nabla u)_n z(\Omega)ds. \end{aligned}$$

The boundary condition (2) yields, since $z'(\Omega; u) + u \cdot \nabla z(\Omega)$ and $z(\Omega)$ are null on $\partial\Omega$,

(5) $z'' + 2u \cdot \nabla z' = -u_n\partial(z' + u \cdot \nabla z)/\partial n + (u \cdot \nabla u)_n\partial z/\partial n$ on $\partial\Omega$.

It may also be written in the following way

(6) $z'' = -2u_n\partial(z' + u \cdot \nabla z)/\partial n + 2(u \cdot \nabla u)_n\partial z/\partial n + uu \cdot \nabla^2 z$ on $\partial\Omega$.

Rough proof of (1): The second order expansion reduces to $z'(\Omega; u) + \frac{1}{2}z''(\Omega; u, u) + o((\|u\|_k)^2)$ in any ω. Thus $z'(\Omega; u) = z''(\Omega; u, u) = 0$ in ω and therefore in Ω.

Rough proof of (2): The boundary condition on z' given by the formula (2) in section 1.6 is satisfied for any domain, and in particular for $\Omega + u$. That is

$$z'(\Omega + u; w) + w \cdot \nabla z(\Omega + u) = 0 \quad \text{on} \quad \partial\Omega + u.$$

Again by the formula (2) of section 1.6, the variation of this new boundary condition is

$$(z'(\cdot; w) + w \cdot \nabla z)'(\Omega; u) + u \cdot \nabla(z'(\Omega; w) + w \cdot \nabla z(\Omega)) = 0 \quad \text{on} \quad \partial\Omega.$$

For $w = u$ this yields

$$(z')'(\Omega; u, u) + 2u \cdot \nabla z'(\Omega; u) + uu \cdot \nabla^2 z(\Omega) + (u \cdot \nabla u) \cdot \nabla z(\Omega) = 0 \quad \text{on} \quad \partial\Omega.$$

On the other hand the boundary condition on z' yields

$$z'(\Omega; u \cdot \nabla u) + (u \cdot \nabla u) \cdot \nabla z(\Omega) = 0 \quad \text{on} \quad \partial\Omega.$$

The desired boundary condition on $z''(\Omega; u, u) = (z')'(\Omega; u, u) - z'(\Omega; u \cdot \nabla u)$ is obtained by subtracting these two equations.

Rough proof of (3): The first variation of K given by the formula (3) of section 1.6 is satisfied for any domain, and in particular for $\Omega + u$. That is

$$K'(\Omega + u; w) = \int_{\Omega+u} z'(\Omega + u; w) + \nabla \cdot (wz(\Omega + u)) dx.$$

Again by the formuly (3) of section 1.6, the variation of this new integral is

$$(K')'(\Omega; u; w) = \int_\Omega (z'(\cdot; w) + \nabla \cdot (wz(\cdot))'(\Omega; u) + \nabla \cdot (u(z'(\Omega; w) + \nabla \cdot (wz(\Omega)))) dx.$$

Thus

$$(K')'(\Omega; u; u) = \int_\Omega (z')'(\Omega; u, u) + 2\nabla \cdot (uz'(\Omega; u) + \nabla \cdot (u\nabla \cdot (uz(\Omega))) dx.$$

The desired value of $K''(\Omega; u, u) = (K')'(\Omega; u, u) - K'(\Omega; u \cdot \nabla u)$ follows since

$$K'(\Omega; u \cdot \nabla u) = \int_\Omega z'(\Omega; u \cdot \nabla u) + \nabla \cdot ((u \cdot \nabla u)z(\Omega)) dx.$$

2.4. Formulas for second variations of the optimal design problem.

We are now in position to calculate the second variation for the optimal problem of section 1.1. The basic formulas of the preceding section yield the following results.

We assume that the solution $y(\Omega + u)$ depends on u regularly and uniformly up to the boundary $\partial\Omega + u$.

If A, B and C are *linear*, the second local variation $y'' = y''(\Omega; u, u)$ satisfies

$$Ay'' = 0 \quad \text{in} \quad \Omega, \quad By'' = -2u_n \frac{\partial}{\partial n}(z' + u \cdot \nabla z) + 2(u \cdot \nabla u)_n \frac{\partial z}{\partial n} + uu \cdot \nabla^2 z \quad \text{on} \quad \partial\Omega.$$

and the second variation of the cost is

$$J''(\Omega; u, u) = \int_{\Omega} (Cy - d)Cy'' + (Cy')^2 dx$$

$$+ \int_{\partial\Omega} 2u_n(Cy - d)Cy' + \frac{1}{2}u_n u \cdot \nabla(|Cy - d|^2) + \frac{1}{2}(u\nabla \cdot u - u \cdot \nabla u)_n |Cy - d|^2 ds.$$

The regular and uniform dependence on u means that $y(\Omega + u)$ and $y'(\Omega; u)$ have total variations in convenient function spaces depending on the operators A and B and C. It is satisfied if, for the fixed domain Ω, $y(\Omega)$ is unique and regularly depend on the coefficients of A and B and on the functions f and h.

An example of optimal design problem satisfying these properties will be given in section 4.

Remark. For nonlinear operators, similar formula are obtained by using

$$A(y)'' = DA(y; y'') + D^2 A(y; y', y'), \quad A(y)' = DA(y; y').$$

3. Theorem on the relation between g'' and $(g')'$

3.1. Preliminary estimations for composed maps.

We denote by $Lip^k(\mathbf{R}^N; \mathbf{R}^N)$, k being an integer ≥ 1, the space of bounded functions with derivatives of order $\leq k - 1$ which are uniformly Lipschitz continuous. This space coincides with the Sobolev space $W^{k,\infty}(\mathbf{R}^N; \mathbf{R}^N)$, and is provided with the norm

$$\|u\|_k = \underset{\substack{x \in \mathbf{R}^N \\ 0 \leq |\alpha| \leq k}}{Sup} |D^\alpha(x)|.$$

Lemma 3.1. Let $u \in Lip^k(\mathbf{R}^N; \mathbf{R}^N)$, $k \geq 1$, such that $\|u\|_k \leq a_k$ where $a_k = (4kN^{k+3})^{-1/2}$.

   i. The map $I + u$ invertible, $(I + u)^{-1} = I + u^*$ where $u^* \in Lip^k(\mathbf{R}^N; \mathbf{R}^N)$ and

$$\|u^*\|_k \leq c_k \|u\|_k.$$

   ii. For any $w \in Lip^k(\mathbf{R}^N; \mathbf{R}^N)$, $w \circ (I + u)$ and $w \circ (I + u)^{-1}$ are in $Lip^k(\mathbf{R}^N; \mathbf{R}^N)$, and

$$\|w \circ (I + u)\|_k \leq c_k \|w\|_k.$$
$$\|w \circ (I + u)^{-1}\|_k \leq c_k \|w\|_k.$$

   iii. If $w \in Lip^{k+1}(\mathbf{R}^N; \mathbf{R}^N)$, then

$$\|w \circ (I + u) - w\|_k \leq c_k \|w\|_{k+1} \|u\|_k$$
$$\|w \circ (I + u)^{-1} - w\|_k \leq c_k \|w\|_{k+1} \|u\|_k.$$

We denote by I the identity of $\mathbf{R}^N$ and by c_k various positive numbers depending only on k and on the space dimension N.

Proof: *Part i.* Since $\|u\|_1 \le \|u\|_k < 1$, u is a contraction, thus $I + u$ is invertible. The desired properties of u^* are given by lemma 2.4 part i page II.15 of F. Murat & J. Simon [3].

Part ii. Let $v \in Lip^k(\mathbf{R}^N; \mathbf{R}^N)$ such that $I + v$ is invertible and $(I + v)^{-1} - I \in Lip^k(\mathbf{R}^N; \mathbf{R}^N)$. Then, by lemma 2.2 part i page II.8 of [3], $w \circ (I+v) \in Lip^k(\mathbf{R}^N, \mathbf{R}^N)$ and $\|w \circ (I+v)\|_k \le \|w\|_k (1+c_k\|v\|_k)^k$. The desired properties follows by choosing successively $v = u$ and $v = u^*$.

Part iii. If $w \in Lip^{k+1}(\mathbf{R}^N; \mathbf{R}^N)$, by the lemma 2.2 part v of [3],

$$\|w \circ (I + v) - w\|_k \le \|v\|_k (1 + c_k\|w\|_{k+1}(1 + \|v\|_k)^k).$$

The desired properties follows by choosing successively $v = u$ and $v = u^*$.

3.2. Variations of a domain of $\mathbf{R}^N$.

Given $\Omega \subset \mathbf{R}^N$, and a_k as in lemma 3.1, a family of subsets of $\mathbf{R}^N$ is defined by

$$\mathcal{D} = \{\Lambda = \Omega + u : u \in Lip^k(\mathbf{R}^N; \mathbf{R}^N), \|u\|_k \le a_k\}$$

where $\Omega + u = \{x + u(x) : x \in \Omega\} = (I + u)(\Omega)$.

Remark. If Ω is open and bounded, and if it's boundary is Lip^k, then every Λ in $\mathcal{D}$ has the same properties, provided that $k \ge 2$. For $k = 1$ this is not satisfied.

We will see now that the variation of a domain Ω that result from two successive variations u and v is not $u + v$, but is $u + v \circ (I + u)$.

Lemma 3.2. *Let $\Omega \subset \mathbf{R}^N$, u and v in $Lip^k(\mathbf{R}^N; \mathbf{R}^N)$ with $\|u\|_k \le a_k$. Then*

$$(\Omega + u) + v = \Omega + (u + v \circ (I + u))$$
$$\Omega + (u + v) = (\Omega + u) + v \circ (I + u)^{-1}.$$

Proof: The first equation is given by $(I+v) \circ (I+u) = I + u + v \circ (I+u)$, and the second one is obtained by replacing v by $v \circ (I + u)^{-1}$.

Remark. The small variations preserve $\mathcal{D}$: if $\Lambda \in \mathcal{D}$ and if v is small enough, then $\Lambda + v \in \mathcal{D}$.

Indeed $\Lambda + v = \Omega + u + v \circ (I + u)$ and, by lemma 3.1, $\|u + v \circ (I + u)\|_k < a_k$ if $\|v_k\|_k < (a_k - \|u\|_k)/c_k$.

Remark. The family $\mathcal{D}$ is a neighborhood of Ω in the metric space $\mathcal{D}^k$ defined by

$$\mathcal{D}^k = \{\Lambda = \Omega + u : u \in Lip^k(\mathbf{R}^N; \mathbf{R}^N),$$
$$I + u \text{ is invertible and } (I + u)^{-1} - I \in Lip^k(\mathbf{R}^N; \mathbf{R}^N)\}$$

and provided with the metric defined by F. Murat & J. Simon [3] in section 2.5 page II.26.

3.3. The main result.

We will see now that the second variation g'' exists as soon as the first variation g' and it's variation $(g')'$ exist, and that g'' may be calculated based on g' and on $(g')'$. Now g is any function on $\mathcal{D}$, later it will be chosen to be the cost function J, or to be the restriction $y_{|\omega}$.

Theorem 3.3. Let $\Omega \subset \mathbf{R}^N$, $a_k = (4kN^{k+3})^{-1/2}$, and

$$\mathcal{D} = \{\Lambda = \Omega + u : u \in Lip^k(\mathbf{R}^N, \mathbf{R}^N), \|u\|_k < a_k\}.$$

Let g be a function on $\mathcal{D}$ with values in a Banach space E.

Assume that for any $\Lambda \in \mathcal{D}$, there exists $g'(\Lambda; \cdot) \in \mathcal{L}(Lip^k(\mathbf{R}^N; \mathbf{R}^N); E)$ satisfying, for any $u \in Lip^k(\mathbf{R}^N; \mathbf{R}^N)$ such that $\Lambda + u \in \mathcal{D}$,

$$g(\Lambda + u) = g(\Lambda) + g'(\Lambda; u) + o(\|u\|_k).$$

Assume that there exists $(g')'(\Omega; \cdot, \cdot) \in \mathcal{L}((Lip^{k+1}(\mathbf{R}^N; \mathbf{R}^N))^2; E)$ satisfying, for any u and w in $Lip^{k+1}(\mathbf{R}^N; \mathbf{R}^N)$ such that $\Omega + u \in \mathcal{D}$,

$$g'(\Omega + u; w) = g'(\Omega; w) + (g')'(\Omega; u, w) + \|w\|_{k+1} o(\|u\|_{k+1}).$$

Define $g''(\Omega; \cdot, \cdot) \in \mathcal{L}((Lip^{k+1}(\mathbf{R}^N; \mathbf{R}^N))^2; E)$ by

$$g''(\Omega; u, w) = (g')'(\Omega; u, w) - g'(\Omega; u \cdot \nabla w).$$

Then, for any $u \in Lip^{k+2}(\mathbf{R}^N; \mathbf{R}^N)$ such that $\Omega + u \in \mathcal{D}$,

$$g(\Omega + u) = g(\Omega) + g'(\Omega; u) + \frac{1}{2}g''(\Omega; u, u) + o((\|u\|_{k+2})^2).$$

We denote by $o(\|u\|_k)$ any element of E, depending on $u, \Omega, \ldots$, such that $o(\|u\|_k)/\|u\|_k \to 0$ as $\|u\|_k \to 0$.

3.4. Second order expansion in Banach spaces.

We will give now, for a function G defined on a Banach space X, conditions for the second variation G'' to exist, that is for Taylor's expansion to hold. Here $G'' = (G')'$.

The theorem 3.3 will be proved by using this result for $G(v) = g(\Omega + v)$.

Lemma 3.4. Let X be a Banach space, $x \in X$, $a > 0$ and $Y = \{y \in X : \|y - x\|_X < a\}$. Let G be a function on Y with values in a Banach space E.

Assume that for any $y \in Y$ there exists a function $G'(y; \cdot) \in \mathcal{L}(X; E)$ satisfying, for any u such that $y + u \in Y$,

$$G(y + u) = G(y) + G'(y; u) + o(\|u\|_X).$$

Assume that there exists a function $(G')'(x; \cdot, \cdot) \in \mathcal{L}(X^2; E)$ such that, for any $w \in X$ and any u such that $x + u \in Y$,

$$G'(x + u; w) = G'(x; w) + (G')'(x; u, w) + \|w\|_X o(\|u\|_X).$$

Then for any u such that $x + u \in Y$,

$$G(x + u) = G(x) + G'(x; u) + \frac{1}{2}(G')'(x; u, u) + o((\|u\|_X)^2).$$

Proof: Let u be such that $x + u \in Y$, and let $0 \le t \le 1$. By the first assumption on G, $G(x + tu)$ is differentiable with respect to t in $[0, 1]$ and it's derivative is $G'(x + tu; u)$. Thus

$$G(x + u) - G(x) - G'(x; u) = \int_0^1 (G'(x + tu; u) - G'(x; u))dt.$$

Since $(G')'(x; \cdot, u)$ is linear it follows that

$$G(x+u) - G(x) - G'(x; u) - \frac{1}{2}(G')'(x; u, u) = \int_0^1 (G'(x+tu; u) - G'(x; u) - (G')'(x; tu, u))dt.$$

Now, using the second assumption on G to bound the norm of the right hand side, we get

$$\|G(x + u) - G(x) - G'(x; u) - \frac{1}{2}(G')'(x; u, u)\|_E \le \|u\|_X o(\|u\|_X)$$

which proves the lemma.

3.5. Estimations for composed maps.

The proof of theorem 3.3 rely on the following estimates.

Lemma 3.5. *Let u and w in $Lip^{k+2}(\mathbf{R}^N; \mathbf{R}^N)$, $k \ge 1$, be such that $\|u\|_{k+1} \le a_{k+1}$. Then*

$$\|w \circ (I + u)^{-1} - w + u \cdot \nabla w\|_k \le c_k \|w\|_{k+2}(\|u\|_{k+1})^2.$$

Proof: Remind that, if $w \in W^{k+2,p}(\mathbf{R}^N; \mathbf{R}^N)$, then

$$\|w \circ (I + u)^{-1} - w + u \cdot \nabla w\|_{W^{k+1,p}(\mathbf{R}^N, \mathbf{R}^N)} = o(\|u\|_{k+2}).$$

This is proved in [3], lemma 4.4 part ii page IV-9. But it is false for $p = \infty$, by remark 4.3.

We assume for a moment that w *has a compact support*. Then $w \circ (I + u)^{-1} - w + u \cdot \nabla w$ has a compact support independent of u, thus using this equality for $p \ge N$ and Sobolev's theorem, we get

$$\|w \circ (I + u)^{-1} - w + u \cdot \nabla w\|_{W^{k,\infty}(\mathbf{R}^N, \mathbf{R}^N)} = o(\|u\|_{k+2}).$$

Now we consider s and t such that $0 \le t + s \le 1$, and we use this equality with $w \circ (I + tu)^{-1}$ and $su \circ (I + tu)^{-1}$ instead of w and u. Since $w \circ (I + tu)^{-1} \circ (I + su \circ (I + tu)^{-1})^{-1} = w \circ (I + tu + su)^{-1}$, we obtain

$$\|w \circ (I + tu + su)^{-1} - w \circ (I + tu)^{-1} + su \circ (I + tu)^{-1} \cdot \nabla(w \circ (I + tu)^{-1}\|_k$$
$$= o(\|su \circ (I + tu)^{-1}\|_{k+2}).$$

Thus the map $t \to w \circ (I + tu)^{-1}$ is differentiable from $[0, 1]$ into $Lip^k(\mathbf{R}^N; \mathbf{R}^N)$, and it's derivative at t is $u \circ (I + tu)^{-1} \cdot \nabla(w \circ (I + tu)^{-1})$. Then, using for this map the integral identity still used for G in the proof of lemma 3.4, we get

$$w \circ (I + u)^{-1} - w + u \cdot \nabla w = \int_0^1 \left((u_0(I + tu)^{-1} \cdot \nabla(w \circ (I + tu)^{-1}) - u \cdot \nabla w \right) dt.$$

Therefore

$$\|w \circ (I + u)^{-1} - w + u \cdot \nabla w\|_k \le Sup_t \|u \circ (I + tu)^{-1} \cdot \nabla(w \circ (I + tu)^{-1}) - u \cdot \nabla w\|_k.$$

Using $\|vw\|_k \le c_k \|v\|_k \|w\|_k$ and lemma 3.1, we bound

$$\|w \circ (I + u)^{-1} - w + u \cdot \nabla w\|_k$$
$$\le Sup_t \|u \circ (I + tu)^{-1} \cdot \nabla(w \circ (I + tu)^{-1} - w) + (u \circ (I + tu)^{-1} - u) \cdot \nabla w\|_k$$
$$\le Sup_t c_k \|u \circ (I + tu)^{-1}\|_k \|w \circ (I + tu)^{-1} - w\|_{k+1} + \|u \circ (I + tu)^{-1} - u)\|_k \|w\|_{k+1}$$
$$\le c_k \|w\|_{k+2} (\|u\|_{k+1})^2.$$

Now this desired inequality is proved if w has a compact support. *If the support of w is not compact*, we use a function $\varphi \in \mathcal{D}(\mathbf{R}^N)$ such that $\varphi(x) = 1$ if $|x| \le 2$, and we define, for $y \in \mathbf{R}^N$, $\varphi_y(x) = \varphi(x - y)$. Since $\varphi_y w$ has a compact support, it satisfies

$$\|(\varphi_y w) \circ (I + u)^{-1} - \varphi_y w + u \cdot \nabla(\varphi_y w)\|_k \le c_k \|\varphi_y w\|_{k+2} (\|u\|_{k+1})^2.$$

The left hand side is the norm in $W^{k,\infty}(\mathbf{R}^N; \mathbf{R}^N)$, which is greater than the norm in $W^{k,\infty}(B(y; 1); \mathbf{R}^N)$, where $B(y, 1)$ is the ball of radius 1 centered on y. In this ball $\varphi_y \circ (I + u)^{-1} \equiv \varphi_y \equiv 1$. Indeed, if $|x - y| \le 1$, then $|(I + u)^{-1}(x) - y| = |-u \circ (I + u)^{-1}(x) + x - y| \le a_k + 1 \le 2$. Therefore

$$\|w \circ (I + u)^{-1} - w + u \cdot \nabla w\|_{W^{k,\infty}(B(y;1);\mathbf{R}^N)} \le c_k \|\varphi_y w\|_{k+2} (\|u\|_{k+1})^2$$
$$\le c_k \|\varphi_y\|_{k+2} \|w\|_{k+2} (\|u\|_{k+1})^2$$

Taking the supremum for y in $\mathbf{R}^N$ we obtain the desired inequalites, since the norm of φ_y do not depend on y. Now the result is proved for all w.

3.6. Proof of theorem 3.3.

We have to obtain the second order Taylor's expansion for $G(u) = g(\Omega + u)$ with respect to u, for $u \in Lip^{k+2}$. By lemma 3.4 it is enough to obtain the first order expansion of $G(y + u)$ for any y in a neighborhood of 0, and the first order expansion of $G'(u; w)$.

Expansion of $G(y + u)$. By lemma 3.2 we have, if $\|y + u\|_k \le a_k$,

$$G(y + u) = g(\Omega + (y + u)) = g((\Omega + y) + u \circ (I + y)^{-1}).$$

The first assumption on g yields

$$g((\Omega + y) + u \circ (I + y)^{-1}) = g(\Omega + y) + g'(\Omega + y; u \circ (I + y)^{-1} + R_1$$

where $R_1 = o(\|u \circ (I + y)^{-1}\|_k)$ thus, by lemma 3.1, $R_1 = o(\|u\|_k)$.

Therefore this is the expansion of $G(y + u)$, and

$$G'(y; u) = g'(\Omega + y; u \circ (I + y)^{-1}).$$

Expansion of $G'(u; w)$. The second assumption on g yields

$$(1) \quad g'(\Omega + u; w \circ (I + u)^{-1}) = g'(\Omega; w \circ (I + u)^{-1}) + (g')'(\Omega; u, w \circ (I + u)^{-1}) + R_2$$

where $R_2 = \|w \circ (I + u)^{-1}\|_{k+1} o(\|u\|_{k+1})$. Thus by lemma 3.1

$$R_2 = \|w\|_{k+1} o(\|u\|_{k+1}).$$

We linearize now the terms in the right hand side of (1). Since $g'(\Omega, \cdot)$ is linear the first term is equal to

$$g'(\Omega; w \circ (I + u)^{-1}) = g'(\Omega; w) - g'(\Omega; u \cdot \nabla w) + R_3$$

where $R_3 = g'(\Omega; w \circ (I + u)^{-1} - w + u \cdot \nabla w)$. By lemma 3.5

$$\|R_3\|_E \leq \||g'(\Omega; \cdot)\|| \, \|w \circ (I + u)^{-1} - w + u \cdot \nabla w\|_k$$
$$\leq c_k \||g'(\Omega; \cdot)\|| \, \|w\|_{k+2}(\|u\|_{k+1})^2.$$

Since $(g')'(\Omega; u, \cdot)$ is linear, the second term in the right hand side of (1) is equal to

$$(g')'(\Omega; u, w \circ (I + u)^{-1}) = (g')'(\Omega; u; w) + R_4$$

where $R_4 = (g')'(\Omega; u; w \circ (I + u)^{-1} - w)$. By lemma 3.1

$$\|R_4\|_E \leq \||(g')'(\Omega; \cdot, \cdot)\|| \, \|u\|_{k+1} \|w \circ (I + u)^{-1} - w\|_{k+1}$$
$$\leq c_{k+1,2} \||(g')'(\Omega; \cdot, \cdot)\|| \, \|w\|_{k+2}(\|u\|_{k+1})^2.$$

Finally (1) yields

$$g'(\Omega + u; w \circ (I + u)^{-1}) = g'(\Omega; w) + (g')'(\Omega; u, w) - g'(\Omega; u \cdot \nabla w) + \|w\|_{k+2} o(\|u\|_{k+2}).$$

This is

$$G'(u; w) = G'(0; w) + (G')'(0; u, w) + \|w\|_{k+2} o(\|u\|_{k+2})$$

with

$$(G')'(0; u, w) = (g')'(\Omega; u; w) - g'(\Omega; u \cdot \nabla w).$$

Conclusion. We proved that the assumptions of lemma 3.4 are satisfied for $G(u) = g(\Omega + u)$, $x = 0$ and $X = Lip^{k+2}(\mathbf{R}^N; \mathbf{R}^N)$. Therefore the second order Taylor's expansion holds for G, and yields

$$g(\Omega + u) = g(\Omega) + g'(\Omega; u) + \frac{1}{2}((g')'(\Omega; u; u) - g'(\Omega; u \cdot \nabla u)) + o((\|u\|_{k+2})^2).$$

4. Second variation of the drag of a body

4.1. Drag of a body.

We are interested in a motionless body B in a viscous incompressible fluid moving at a uniform velocity h on the boundary of the experiment region Λ. The domain occupied by the fluid is $\Omega = \Lambda\backslash B$, and we assume $B \subset\subset \Lambda$.

The velocity $y = (y_1, y_2, y_3)$ and the pressure p satisfy

$$-\mu\Delta y + y\cdot\nabla y = -\nabla p \quad \text{in} \quad \Omega$$
$$\nabla\cdot y = 0 \quad \text{in} \quad \Omega$$
$$y = 0 \quad \text{on} \quad \partial B, \quad y = h \quad \text{on} \quad \partial\Lambda.$$

The energy dissipated by the fluid is

$$J = \frac{1}{2}\int_\Omega |Ly|^2 dx$$

where $(Ly)_{ij} = \partial_i y_j + \partial_j y_i$, $\partial_i = \partial/\partial x_i$, and the drag is $J/|h|$.

Lemma 4.1. *i. There exists $R(\Omega) > 0$ such that, if $|h|/\mu < R(\Omega)$, there exists a unique solution*

$$y \in (H^1(\Omega))^3, \quad \nabla p \in (H^{-1}(\Omega))^3.$$

ii. If $\partial\Omega$ is locally the graph of a Lip^m function, and if $|h|/\mu < R(\Omega)$, then

$$y \in (H^m(\Omega))^3, \quad p \in H^{m-1}(\Omega).$$

Proof: The existence of y is given by J.L. Lions [5] theorem 7.3 p. 102, the existence of p and the regularity of u and p follow from the theorems 1 and 2' p. 28 and 74 of O.A. Ladyzhenskaya [6], and the existence of $R(\Omega)$ is given by F. Murat & J. Simon [4], theorem 2.1 p.2.3.

Remark. Without limit on h, the uniqueness of y and therefore of J is not necessary known.

4.2. Second order expansion of the drag.

Now we will give the expansion of $J(B + u)$ with respect to a Lip^4 variation u of B.

Theorem 4.2. *Assume that*

$$\partial\Omega \text{ is locally the graph of a } Lip^4 \text{ function,}$$
$$|h|/\mu < R(\Omega).$$

Then, for any $u \in Lip^4(\mathbf{R}^3; \mathbf{R}^3)$ such that $B + u \subset\subset \Lambda$,

$$J(B + u) = J(B) + J'(B; u) + \frac{1}{2}J''(B; u, u) + o((\|u\|_4)^2)$$

where

$$J'(\mathcal{B};u) = \int_\Omega Ly \cdot Ly'\, dx = \frac{1}{2}\int_{\partial\Omega} u_n |Ly|^2\, ds$$

$y' \in (H^3(\Omega))^3$ being the unique solution, such that $p' \in H^1(\Omega)$, of

$$-\mu\Delta y' + y' \cdot \nabla y + y \cdot \nabla y' = -\Delta p' \quad \text{in} \quad \Omega$$
$$\nabla \cdot y' = 0 \quad \text{in} \quad \Omega$$
$$y' = -u_n \partial y/\partial n \quad \text{on} \quad \partial\Omega$$

and

$$J''(\mathcal{B};u,u) =$$
$$\int_\Omega Ly \cdot Ly'' + |Ly'|^2\, dx + \int_{\partial\Omega} 2u_n Ly \cdot Ly' + \frac{1}{2}u_n u \cdot \nabla(|Ly|^2) + \frac{1}{2}(u\nabla \cdot u - u \cdot \nabla u)_n |Ly|^2\, ds$$

$y'' \in (H^2(\Omega))^3$ being the unique solution, such that $p'' \in L^2(\Omega)$, of

$$-\mu\Delta y'' + y'' \cdot \nabla y + y \cdot \nabla y'' + 2y' \cdot \nabla y' = -\nabla p'' \quad \text{in} \quad \Omega$$
$$\nabla \cdot y'' = 0 \quad \text{in} \quad \Omega$$
$$y'' = -2u_n \partial(y' + u \cdot \nabla y)/\partial n + 2(u \cdot \nabla u)_n \partial y/\partial n + uu \cdot \nabla^2 y \quad \text{on} \quad \partial\Omega.$$

Remark By lemma 4.1,
$$y \in (H^4(\Omega))^3, \quad p \in H^2(\Omega).$$

4.3. Main lines of the proof.
First variation. We assume for a moment that there exists a linear map $y\dot{}(\mathcal{B};\cdot)$ such that, for u in $Lip^2(\mathbf{R}^3;\mathbf{R}^3)$,

(1) $$y(\mathcal{B}+u)\circ(I+u) = y(\mathcal{B}) + y\dot{}(\mathcal{B};u) + o(\|u\|_2) \quad \text{in} \quad (H^2(\Omega))^3.$$

Then by the lemma 2.1 p. 657 of J. Simon [2] $y(\mathcal{B}+u)$ has a first order local expansion in Ω. By the theorems 3.1 and 3.2 p. 663 and 664 of [2], the local variation y' satisfies the desired boundary value problem. The equation involving p' is obtained in the first step in a variational form as in [2], proof p. 684, and in a second step p' is obtained by the theorem 1 p. 28 of [6].

Moreover by theorem 3.3 p. 664 of [2], $J(\mathcal{B}+u)$ has a first order expansion and J has the desired value.

Second variation. Assume now that there exists a bilinear map $y'\dot{}(\mathcal{B};\cdot,\cdot)$ such that, for u and w in $Lip^3(\mathbf{R}^3;\mathbf{R}^3)$,

(2) $$y'(\mathcal{B}+u;w)\circ(I+u) = y'(\mathcal{B};w) + y'\dot{}(\mathcal{B};u,w) + o(\|u\|_3) \quad \text{in} \quad (H^2(\Omega))^3.$$

Then by our theorem 3.3 applied to the restriction $y_{|\omega}$ for $\omega \subset\subset \Omega$, $y(\mathcal{B}+u)$ has a second order local expansion in Ω with respect to u in $Lip^4(\mathbf{R}^3;\mathbf{R}^3)$. At that point the rough calculations in section 2.3 of the boundary value problem satisfied by y'' are justified by using twice the theorems 3.1 and 3.2 of [2] and our theorem 3.3 (the equation involving p'' is obtained in two steps, as we did for the equation involving p' in the first variation).

Moreover by theorem 3.3 p. 664 of [2], $J'(\mathcal{B}+u;w)$ has a first expansion and therefore, by our theorem 3.3, $J(\mathcal{B}+u)$ has a second order expansion with respect to u in $Lip^4(\mathbf{R}^3;\mathbf{R}^3)$. Now the rough calculations in section 2.3 of the J'' are justified by using twice of theorems 3.3 of [2] and our theorem 3.3. Thus J'' has the desired value.

Proof of (1): The boundary value problem satisfied by $y(\mathcal{B}+u)$ in $\Omega+u$ yields, by the map $I+u$, the boundary value problem satisfied by $y(\mathcal{B}+u)\circ(I+u)$ in Ω. This new boundary value problem has u depending coefficients, and may be written in the form,

$$F(u;y(\mathcal{B}+u)\circ(I+u))=0$$

where F maps $Lip^2(\mathbf{R}^3;\mathbf{R}^3)\times(H^2(\Omega))^3$ into $(L^2(\Omega))^3$.

The assumption (1), which is the differentiability of $u\to y(\mathcal{B}+u)\circ(I+u)$ at 0, is obtained by the theorem of differentiation of the solution of an implicit equation.

Indeed $F(0;\cdot)$ is invertible (this is the uniqueness of y), and $F(0;\cdot)$ and $F(\cdot;y(\mathcal{B}))$ are differentiable (this is easy in writing F by using $(\partial_i f)\circ(I+u)=\sum_j a_{ji}(u)\partial_j(f\circ(I+u))$, where the matrix $\{a_{ij}(u)\}$ is the inverse of $\{\delta_{ij}+\partial_i u_j\}$).

The details of a similar calculus for another optimal design problem may be found in [2], proof of theorem 6.1 pages 681-682.

Proof of (2): The outline is the same. Now $y'(\mathcal{B}+u;w)$ satisfies a boundary value problem in $\Omega+u$ with coefficients depending on $y(\mathcal{B}+u)$. By the map $I+u$, it yields the boundary value problem satisfied by $y'(\mathcal{B}+u;w)\circ(I+u)$ in Ω. The coefficients depend on u and on $y(\mathcal{B}+u)\circ(I+u)$, thus it may be written in the form

$$H(u;y(\mathcal{B}+u)\circ(I+u);y'(\mathcal{B}+u;w)\circ(I+u))=0$$

where H maps $Lip^3(\mathbf{R}^3;\mathbf{R}^3)\times(H^2(\Omega))^2\times(H^2(\Omega))^3$ into $(L^2(\Omega))^3$.

The assumption (2), which is the differentiability of $u\to y'(\mathcal{B}+u;w)\circ(I+u)$ at 0, is obtained by the theorem of differentiation of the solution of an implicit equation.

Indeed $H(0,y(\mathcal{B});\cdot)$ is invertible (this is the uniqueness of y'), and $H(0,z;\cdot)$ and $H(\cdot;y(\mathcal{B}+\cdot)\circ(I+\cdot);y'(\mathcal{B}))$ are differentiable (the last one following from the differentiability of $H(\cdot,z;y'(\mathcal{B}))$ and $H(u,\cdot;y'(\mathcal{B}))$ and from (1)).

References

[1] Fujii, N., *Second variation and it's application in domain optimization problem*, in "Control of distributed parameter systems," Proceedings of the 4th IFAC Symposium, Pergamon Press, 1987.

[2] Simon, J., *Differentiation with respect to the domain in boundary value problems*, Numerical Functional Analysis and Optimization **2** (1980), 649–687.

[3] Murat, F. and Simon, J., *Sur le contrôle par un domaine géométrique*, Research report of the Laboratoire d'Analyse Numérique, University of Paris 6 (1976), 1–222.

[4] Murat, F. and Simon, J., *Quelques résultats sur le contrôle par un domaine géométrique*, Research report of the Laboratoire d'Analyse Numérique, University of Paris 6 (1974), 1–46.

[5] Lions, J.L., "Quelques méthodes de résolution des problèmes aux limites non linéaires," Dunod, Paris, 1969.

[6] Ladyzhenskaya, O.A., "The mathematical theory of viscous incompressible flows," Gordon & Breach, 1963.

Jacques Simon
Departement de Mathématiques Appliquées
Université Blaise Pascal (Clermont-Ferrand 2)
F-63177 Aubière Cedex
France

International Series of
Numerical Mathematics, Vol. 91
© 1989 Birkhäuser Verlag Basel

Approximation for control problems with pointwise state constraints

D. Tiba and M. Tiba

Department of Mathematics
INCREST

Abstract. We address the question of the uniform estimates for the violation of the pointwise constraints in the approximation of certain control problems. This is obtained via the variational inequality method which was introduced in [3].

In the numerical examples, our procedure has good stability properties and is superior in several respects to the usual treatment by penalization.

Keywords. control problems, state constraints, variational inequalities, uniform estimates.

1980 *Mathematics subject classifications*: 49A22, 49A29, 49A36

1. Introduction

A natural question in the approximation of control problems with pointwise state constraints, is to establish pointwise estimates for their violation. This was first discussed in [3], [5] in an abstract setting and in [4], [6], [7] in the case of elliptic or parabolic state systems. The proposed method uses essentially the control theory for variational inequalities (without state constraints) and we refer to [1] for a general presentation of such problems. In particular, a delicate point is to get uniform estimates with respect to both the time and the space variables.

In this lecture we review briefly the above mentioned investigations and we give some new results related to this topic. In the last section we indicate several examples and we comment the numerical results obtained with our variational inequality approach and by the standard penalization method.

2. The abstract setting

Let V, H, U be Hilbert spaces, $V \subset H \subset V^*$ (the dual space) and $A : V \to V^*$, $B : U \to H$ be linear, continuous operators such that

$$(2.1) \qquad (Au, u) \geq \omega |u|_V^2, \quad \omega > 0, \quad u \in V,$$

$$(2.2) \qquad (Au, v) = (u, Av), \quad u, v \in V.$$

Above $| \cdot |_X$ is the norm in the space X and $(\cdot, \cdot)$ is the pairing between V and V^* or the scalar product in H.

We consider the control problem

(P_1) Minimize $\{g(y) + h(u)\}$,

(2.3) $y' + Ay = Bu + f$,   a.e. in $[0, T]$,

(2.4) $y(0) = y_0$,

(2.5) $y(t) \in C$,   in $[0, T]$

where $C \subset H$ is a closed, convex subset, $y_0 \in C$, $Ay_0 \in H$, $g : L^2(0, T; U) \to \mathbf{R}$ is convex, continuous, positive and $h : L^2(0, T; U) \to]-\infty, +\infty]$ is convex, lower semicontinuous, proper, coercive. Control constraints may be implicitly imposed in (P_1) by $u \in \mathrm{dom}(h)$.

For $f \in L^2(0, T; H)$, $u \in L^2(0, T; U)$, the state equation (2.3), (2.4) has a unique solution $y \in C(0, T; V)$, $y' \in L^2(0, T; H)$. If we assume the existence of an admissible pair, then (P_1) has at least one optimal pair $[y^*, u^*]$.

Let $\varphi : H \to]-\infty, +\infty]$ be the indicator function of C in H. We approximate the problem (P_1) by the unconstrained control problem:

$(P_{1,\epsilon})$ Minimize $\{g(y) + h(u) + \frac{1}{2}|w|^2_{L^2(0,T;V^*)}\}$,

(2.6) $y' + Ay + \epsilon w = Bu + f$, $w \in \partial\tilde{\varphi}(y)$, $\epsilon > 0$,

                                   $y(0) = y_0$,

Here $\tilde{\varphi} : V \to]-\infty, +\infty]$ is given by $\tilde{\varphi}(v) = \varphi(v)$.

For $u \in L^2(0, T; U)$, the unique solution of the variational inequality (2.6) satisfies $y \in C(0, T; H) \cap L^2(0, T; V)$, $y' \in L^2(0, T; H)$.

We denote by $[y_\epsilon, u_\epsilon]$ an optimal pair for the problem $(P_{1,\epsilon})$ and by J_1, $J_{1,\epsilon}$ the cost functionals of (P_1), $(P_{1,\epsilon})$ respectively. If y^ϵ is the solution of (2.3), (2.4) corresponding to u_ϵ, then the pair $[y^\epsilon, u_\epsilon]$ is not necessarily admissible for the problem (P_1), but we can compute $J_1(y^\epsilon, u_\epsilon)$ and we have the following suboptimality result:

Theorem 2.1. *We have*

   i) $\lim_{\epsilon \to 0} J_1(y^\epsilon, u_\epsilon) = J_1(y^*, u^*)$,

   ii) $\mathrm{dist}\,(y^\epsilon, C \cap V)_{L^\infty(0,T;H) \cap L^2(0,T;V)} \leq k \cdot \epsilon$,

where k is independent of $\epsilon > 0$.

Proof: It is quite standard to show that $u_\epsilon \to u^*$ weakly in $L^2(0, T; U)$ and $h(u_\epsilon) \to h(u^*)$. Let $z_\epsilon = y^\epsilon - y_\epsilon$. It satisfies

(2.7) $z_\epsilon' + Az_\epsilon = \epsilon w_\epsilon$, $z_\epsilon(0) = 0$,

where $w_\epsilon \in \partial\tilde{\varphi}(y_\epsilon)$ is given by (2.6) with $u = u_\epsilon$ and it is bounded in $L^2(0, T; V^*)$ when $\epsilon \to 0$.

Multiply (2.7) by z_ϵ and integrate over $[0, t]$:

$$\frac{1}{2}|z_\epsilon(t)|_H^2 + \int_0^t (Az_\epsilon, z_\epsilon)ds = \epsilon \int_0^t (w_\epsilon, z_\epsilon)ds.$$

From (2.1), by a variant of the Gronwall lemma, we get

$$|z_\epsilon|_{C(0,T;H) \cap L^2(0,T;V)} \leq k\epsilon$$

with k independent of ϵ. As $y_\epsilon(t) \in C \cap V$, a.a. $t \in [0, T]$, we infer ii). The continuity of g together with ii) and $y_\epsilon \to y^*$, $h(u_\epsilon) \to h(u^*)$ give i).

Remark 2.2: The problem $(P_{1,\epsilon})$ is a non-differentiable optimization problem and a smoothing procedure is suitable for a numerical treatment. It may be shown that the pointwise estimates are preserved by the smooth variant too, [5].

Remark 2.3: By Theorem 2.1, ii) we have pointwise estimates on $[0, T]$ in the H norm, which is an integral norm in the space variables. In order to remove this shortcoming, we study first elliptic problems.

3. The elliptic case

We focus our attention on the so called "optimal packaging problem" introduced in [8]. This is quite a challenging control problem due to several difficulties: it is governed by variational inequalities, it is an optimal design problem, it involves special state contraints.

We consider a membrane

$$\Omega(\alpha) = \{(x_1, x_2) \in \mathbf{R}^2; x_2 \in]0, 1[, 0 < x_1 < \alpha(x_2)\}$$

in possible contact with a rigid obstacle G described by a mapping φ. Here $\alpha \in U_{ad}$ is the control function defining the moving part $\Gamma(\alpha)$ of the boundary $\partial\Omega(\alpha)$:

$$\Gamma(\alpha) = \{(x_1, x_2); x_1 = \alpha(x_2), x_2]0, 1[\},$$
$$U_{ad} = \{\alpha \in W^{1,\infty}(]0, 1[); a \leq \alpha \leq b; |\alpha'| \leq c\},$$

with a, b, c positive constants.

The equilibrium position $u(\alpha)$ of the membrane $\Omega(\alpha)$ under a load f and in contact with the obstacle G is given by the unique solution of the variational inequality

$$K(\Omega(\alpha)) = \{v \in H_0^1(\Omega(\alpha)); \quad v \geq \varphi \quad \text{a.e. in} \quad \Omega(\alpha)\}$$

(3.1) $$(\text{grad}\, u(\alpha), \text{grad}\,(v - u(\alpha))_{L^2(\Omega(\alpha))} \geq (f, v - u(\alpha))_{L^2(\Omega(\alpha))},$$

$$\forall v \in K(\Omega(\alpha)), \quad u \in K(\Omega(\alpha)).$$

We assume that $f \in L^2(\hat{\Omega})$, $\hat{\Omega} =]0, b[\times]0, 1[$ $\varphi \in H^1(\hat{\Omega})$, $\varphi \leq 0$ on $\partial\hat{\Omega}$ and on $[a, b] \times [0, 1]$. We denote

$$Z(u(\alpha)) = \{x \in \Omega(\alpha) | u(\alpha)(x) = \varphi(x)\}$$

the contact (coincidence) set associated with $u(\alpha)$. The packaging problem consists in minimizing the area of $\Omega(\alpha)$ such that $Z(u(\alpha))$ contains a given subset $\Omega_0 \subset \Omega$. Therefore, we study the problem:

(P_2) Minimize $\{J_2(\alpha) = \int_0^1 \alpha(x)dx\}$

for $\alpha \in U_{ad}$ and such that $u(\alpha)$, the solution of (3.1) corresponding to α, satisfies the state constraint

$$(3.2) \qquad\qquad\qquad \Omega_0 \subset Z(u(\alpha)).$$

If there is at least one admissible pair $[u(\alpha), \alpha)]$ for the problem (P_2) then one may prove the existence of optimal pairs $[u(\alpha^*), \alpha^*]$ by the compactness of U_{ad}, [7].

In order to approximate the problem (P_2) we remark that the constraint (3.2) is equivalent with

$$(3.3) \qquad\qquad\qquad u(\alpha) \leq \varphi \quad \text{on} \quad \Omega_0$$

since the opposite inequality is always satisfied by the solution of (3.1). We relax (3.3) in the sense that we replace φ by a family $\psi_n > \varphi$, smooth approximation of the indicator function of Ω_0 plus φ. We obtain the problem

$(P_{2,n})$ Minimize $\{J_{2,n}(\alpha) = J_2(\alpha)\}$

for $\alpha \in U_{ad}$, such that

$$(3.4) \qquad\qquad\qquad u(\alpha) \leq \psi_n \quad \text{on} \quad \Omega(\alpha).$$

It is possible to give examples of subsets Ω_0 for which smooth families $\{\psi_n > \varphi\}$ may be constructed. We take $\varphi = 0$ for the sake of simplicity, in the sequel.

We remark that (3.4) is a relaxation of (3.3) in Ω_0, but it requires some additional constraints on $\Omega(\alpha) - \Omega_0$. Due to the continuity of $u(\alpha)$ in $\Omega(\alpha) \subset \mathbf{R}^2$, one may infer that any admissible pair for (P_2) is also admissible for $(P_{2,n})$ with n big enough. We denote α_n an optimal control for $(P_{2,n})$.

We apply the variational inequality technique in the problem $(P_{2,n})$. We consider the problem without state constraints:

$(P_{2,n,\epsilon})$ Minimize $\{J_2(\alpha) + \frac{1}{\epsilon}|\gamma_\epsilon(u - \psi_n)|^2_{L^2(\Omega(\alpha))}\}$

subject to

$$(3.5) \qquad\qquad\qquad -\Delta u + \beta(u) + \gamma_\epsilon(u - \psi_n) \ni f,$$

$$(3.6) \qquad\qquad\qquad u|\partial\Omega(\alpha) = 0,$$

$$(3.7) \qquad\qquad\qquad \alpha \in U_{ad}.$$

Here γ_ϵ is the Yosida approximation of the maximal monotone graph

$$\gamma(y) = \begin{cases} 0 & y < 0, \\ [0,\infty[& y = 0, \\ \Phi & y > 0 \end{cases}$$

and

$$\beta(y) = \begin{cases} 0 & y > 0, \\ ]-\infty,0] & y = 0, \\ \Phi & y < 0. \end{cases}$$

The following estimate plays a fundamental role:

Proposition 3.1. *Assume that $f \in L^\infty(\hat{\Omega})$, then*

(3.8)
$$|\gamma_\epsilon(u - \psi_n)|_{C(\Omega(\alpha))} \le |g + \Delta\psi_n|_{L^\infty(\Omega(\alpha))}$$

where u is the solution of (3.5), (3.6).

Proof: We know that $\gamma_\epsilon(z - \psi_n) \in C(\Omega(\alpha))$ by the regularity of u. Moreover as $\gamma_\epsilon(y) = 0$ for $y \le 0$, we infer

(3.9)
$$\beta(u) \cdot \gamma_\epsilon(u - \psi_n) = 0 \quad \text{in} \quad \Omega(\alpha).$$

Multiplying (3.5) by $\gamma_\epsilon^{p-1}(u - \psi_n)$, $p > 2$ even, after a short calculation we obtain

$$-\int_{\Omega(\alpha)} \Delta u \gamma_\epsilon^{p-1}(u - \psi_n)dx = \int_{\Omega(\alpha)} \text{grad}\,(u - \psi_n)\text{grad}\,\gamma_\epsilon^{p-1}(u - \psi_n)dx$$
$$-\int_{\partial\Omega(\alpha)} \frac{\partial}{\partial n}(u - \psi_n)\gamma_\epsilon^{p-1}(-\psi_n)d\sigma - \int_{\Omega(\alpha)} \Delta\psi_n\gamma_\epsilon^{p-1}(u - \psi_n)dx$$
$$\ge -\int_{\Omega(\alpha)} \Delta\psi_n\gamma_\epsilon^{p-1}(u - \psi_n)dx.$$

Then, by (3.9) and the Hölder inequality, we get

$$\|\gamma_\epsilon(u - \psi_n)\|_{L^p(\Omega(\alpha))} \le \|f + \Delta\psi_n\|_{L^p(\Omega(\alpha))}$$

and passing to the limit $p \to \infty$, it yields (3.8).

Remark 3.2: Any admissible pair for (P_2) is also admissible for $(P_{2,n,\epsilon})$ with the same cost, if n is sufficiently big. If we denote $[u(\alpha_{n,\epsilon}), \alpha_{n,\epsilon}]$ an optimal pair for $(P_{2,n,\epsilon})$, which may be shown to exist, then

$$(3.10) \qquad J_2(\alpha_{n,\epsilon}) + \frac{1}{\epsilon}|\gamma_\epsilon(u(\alpha_{n,\epsilon}) - \psi_n)\|^2_{L^2(\Omega(\alpha_{n,\epsilon}))} \leq J_2(\alpha^*),$$

for n large.

Let us now denote shortly by u^ϵ the solution of (3.1) corresponding to $\alpha_{n,\epsilon}$. By (3.20), we have

$$(3.11) \qquad |\gamma_\epsilon(u(\alpha_{n,\epsilon}) - \psi_n)|_{L^2(\Omega(\alpha_{n,\epsilon}))} \leq c\epsilon^{\frac{1}{2}}$$

with c independent of n, ϵ. We apply a result on the Lipschitzian dependence of the solution of variational inequalities on the right hand side, Brezis [2], and we obtain

$$(3.12) \qquad |u(\alpha_{n,\epsilon}) - u^\epsilon|_{L^\infty(\Omega(\alpha_{n,\epsilon}))} \leq C|\gamma_\epsilon(u(\alpha_{n,\epsilon}) - \psi_n)|_{L^2(\Omega(\alpha_{n,\epsilon}))}$$

We have the following result

Theorem 3.3. *The pair* $[\alpha_{n,\epsilon}, u^\epsilon]$ *is sub-optimal in the problem* (P_2) *in the following sense:*

    i) *it satisfies (3.1),*
    ii) $J_2(\alpha_{n,\epsilon}) \leq J_2(\alpha^*)$,
    iii) $\alpha_{n,\epsilon} \in U_{ad}$,
    iv) $0 \leq u^\epsilon|_{\Omega_0} \leq \psi_n + \epsilon|f + \Delta\psi_n|_{L^\infty(\Omega(\alpha_{n,\epsilon}))} + C\epsilon^{1/2}$.

Proof: We have only to remark that by (3.8) we infer

$$(u(\alpha_{n,\epsilon}) - \psi_n)_+ \leq \epsilon|f + \Delta\psi_n|_{L^\infty(\Omega(\alpha))}.$$

Next, (3.11) and (3.12) give iv).

Remark 3.4: We have applied the variational inequality method to a control problem governed by variational inequalities.

Remark 3.5: In this approach we obtain pointwise estimates by working only with integral norms.

4. Parabolic problems

We continue with the investigation of the question from Remark 2.3. The results of this section are weaker, due to the fact that the Lipschitz property for parabolic variational inequalities is no longer true in the form (3.12).

We consider the problem:

(P_3) Minimize $\{g(y) + h(u)\}$

(4.1)	$y_t - \Delta y = Bu + f$	in	$Q =]0, T[\times \Omega,$
(4.2)	$y(t, x) = 0$	in	$\Sigma = [0, T] \times \partial\Omega,$
(4.3)	$y(0, x) = y_0(x)$	in	Ω
(4.4)	$u \in U_{ad},$		
(4.5)	$y(t, x) \in [a, b]$	in	$Q.$

Here Ω is a bounded domain in $\mathbf{R}^N$ with regular boundary $\partial\Omega$, $f \in L^\infty(Q)$, $0 \in [a, b]$, $y_0(x) \in [a, b]$ a.e. Ω, $U_{ad} \subset U$ is a closed convex subset, $B : U \to L^2(\Omega)$ is a bounded linear operator such that BU_{ad} is bounded in $L^\infty(Q)$. For g and h we make the usual assumption, §2. If the admissibility condition is fulfilled, then the problem (P_3) has at least one optimal pair $[y^*, u^*]$.

Let β be the subdifferential of the indicator function of $[a, b]$ in $\mathbf{R} \times \mathbf{R}$ and $p > \frac{N+2}{2}$ be given. We take the approximating problem

$(P_{3,\epsilon})$ Minimize $\{g(y) + h(u) + \frac{1}{\epsilon}|\beta_\epsilon(y)|_{L^p(Q)}\}$

subject to

(4.6) $$y_t - \Delta y + \beta_\epsilon(y) = Bu + f \quad \text{in} \quad Q$$

and (4.2) - (4.4), where β_ϵ is the Yosida approximation of β.

Denote by $[y_\epsilon, u_\epsilon]$ an optimal pair for the problem $(P_{3,\epsilon})$. Obviously Bu_ϵ is bounded in $L^\infty(Q)$ since $u_\epsilon \in U_{ad}$, $\epsilon > 0$. We also remark that

(4.7) $$|\beta_\epsilon(y_\epsilon)|_{L^p(Q)} \leq C \cdot \epsilon$$

because any admissible pair for (P_3) is admissible for $(P_{3,\epsilon})$ with the same cost.

Let y^ϵ be the solution of (4.1) - (4.3) corresponding to u_ϵ and $z^\epsilon = y^\epsilon - y_\epsilon$. We have

$$z_t^\epsilon - \Delta z^\epsilon = \beta_\epsilon(y_\epsilon),$$
$$z^\epsilon|_\Sigma = 0, \quad z^\epsilon(0) = 0$$

and, by (4.7), we get $|z^\epsilon|_{W^{2,1,p}(Q)} \leq C\epsilon$.

As $p > \frac{N+2}{2}$, we obtain, [9]:

(4.8) $$|y^\epsilon(t, x) - y_\epsilon(t, x)| \leq C\epsilon, \quad (t, x) \in Q,$$

where C is a generic constant independent of $\epsilon > 0$.

Now, we multiply (4.6) by $\beta_\epsilon(y_\epsilon)^{q-1}$, q even, and integrate over $[0, t]$:

$$\int_0^t \int_\Omega \beta_\epsilon(y_\epsilon)^{q-1}(y_\epsilon)_t - \int_0^t \int_\Omega \Delta y_\epsilon \beta_\epsilon(y_\epsilon)^{q-1} + \int_0^t \int_\Omega \beta_\epsilon(y_\epsilon)^q$$
$$= \int_0^t \int_\Omega (Bu_\epsilon + f)\beta_\epsilon(y_\epsilon)^{q-1}.$$

A standard computation gives

$$\int_Q |\beta_\epsilon(y_\epsilon)|^q \leq |Bu + f|_{L^\infty(Q)} |\beta_\epsilon(y_\epsilon)|_{L^q(Q)}^{q-1}.$$

and taking $q \to \infty$ we see that

$$|\beta_\epsilon(y_\epsilon)|_{L^\infty(Q)} \leq |Bu_\epsilon + f|_{L^\infty(Q)},$$

so

(4.9) $\text{dist}(y_\epsilon(t, x), [a, b]) \leq C\epsilon, \quad (t, x) \in Q.$

Combining (4.8), (4.9), it yields:

Theorem 4.1. *The pair $[y^\epsilon, u_\epsilon]$ is suboptimal for the problem (P_3) in the sense that:*
  i) *it satisfies (4.1) - (4.4),*
  ii) $g(y^\epsilon) + h(u_\epsilon) \to g(y^*) + h(u^*)$,
  iii) $\text{dist}(y^\epsilon(t, x), [a, b]) \leq C\epsilon, \quad \epsilon > 0.$

As we have mentioned at the beginning of this section, if the state equation in (P_3) is a variational inequality we cannot proceed similarly to section 3. Another idea, applicable to a more limited range of problems, is to use an equivalence result [4]. We consider the following problem:

(P_4) Minimize $\{g(y) + h(u)\}$

(4.10) $y_t - \Delta y + \gamma(y) \ni u + f$

with the conditions (4.2), (4.3), (4.5).
   We impose the same hypotheses as in (P_3), γ is a maximal monotone graph in $\mathbf{R} \times \mathbf{R}$ and $y_0(x) \in [0, 1] \cap \text{dom}\, \gamma$ a.e. Ω.
   Let us define the problem:

(P_5) Minimize $\{g(y) + h(u - \beta(y)) + \frac{1}{2}|\beta(y)|_{L^2(Q)}^2\}$

(4.11) $y_t - \Delta y + \gamma(y) + \beta(y) \ni u + f$

with the conditions (4.2), (4.3).

Proposition 4.2. *The problems* (P_4) *and* (P_5) *are equivalent in the sense that they have the same optimal pairs and optimal value.*

Proof: If $[y^*, u^*]$ is an optimal pair for (P_4), then $[y^*, u^*]$ is admissible for (P_5) with $\beta(y^*) = 0$, that is

$$J_5(y^*, u^*) = J_4(y^*, u^*) = J_4^*$$

and $J_5^* \leq J_4^*$ (the optimal values).

If $[y, u]$ is a solution of (P_5), then $y \in [a, b]$ in Q and $[y, u - \beta(y)]$ is admissible for (P_4) with

$$J_4^* \leq J_4(y, u - \beta(y)) \leq J_5(y, u) = J_5^*.$$

Therefore $J_4^* = J_5^*$ and $\beta(y) = 0$ for all the optimal pairs of (P_5).

Remark 4.3: The significance of Proposition 4.2 is that if one obtains a minimizing sequence $\{u_n\}$ for the unconstrained problem (P_5), then $\{u_n - \beta(y_n)\}$ is a minimizing sequence with admissible states for the constrained problem (P_4). However, from a numerical point of view, it is not clear how to deal with (P_5) since usual regularizing techniques or bundle algorithms seem not applicable directly.

5. Numerical examples

We have performed several numerical experiments for a comparison between the variational inequality method and the penalization method. In the notations of §2, the penalization method consists in approximating (P_1) by the problem

(P_1^ν) Minimize $\{g(y) + h(u) + \int_0^T \psi_\nu(y) dt\}$,

subject to (2.3), (2.4), where ψ_ν is a regularization of ψ, the indicator function of C in H.

Therefore, we have computed the numerical solution of $(P_{1,\epsilon})$ and of (P_1^ν). In order to remove any other type of difficulty, we have considered the simplest case when $H = V = V^* = U = \mathbf{R}$ and (2.3) becomes an ordinary differential equation.

The example we study has the following form:

(P_6) Minimize $\{\frac{\sigma}{2} \int_0^1 (y - y_d)^2 dt + \frac{\theta}{2} \int_0^1 u^2 dt\}$,

$$y' + y = u + f \quad \text{in} \quad [0, 1]$$
$$y(0) = y_0,$$
$$u(t) \in U_{ad}, \quad y(t) \in C \quad \text{in} \quad [0, 1].$$

The subsets U_{ad} and C are real intervals and the parameters σ, θ take the values 0 or 1 according to the form of the cost functional we wish to take into account. If $\sigma = \theta = 0$ we have just the problem to find an admissible pair for a constrained system.

In order to get more significant results we fixed y_d and y_0 to be valued on the boundary of C.

In the problem $(P_{1,\epsilon})$ we also perform a smoothing of $\partial\varphi$ of parameter $\lambda > 0$ (which we don't describe here), as mentioned in Remark 2.2. We denote $\partial\varphi_\lambda$ the regularization of $\partial\varphi$ and $(P_{1,\epsilon,\lambda})$ the problem obtained by replacing $\partial\varphi$ with $\partial\varphi_\lambda$ in $(P_{1,\epsilon})$. For the solution of (P_1^ν) and $(P_{1,\epsilon,\lambda})$ we use a general gradient algorithm.

For the treatment of the control constraints $u(t) \in U_{ad}$ we use again penalization both in (P_1^ν) and in $(P_{1,\epsilon,\lambda})$, that is we add to the cost functional a term of the form

$$\int_0^1 \chi_\delta(u)dt$$

where χ_δ, $\delta > 0$, is a regularization of the indicator function of U_{ad}.

We remark that

$$\psi_\nu(y) = \frac{1}{2\nu}\text{dist}^2(y, C),$$

while $(\partial\varphi)_\lambda$, the Yosida approximation of $\partial\varphi$ (which is very close to $\partial\varphi_\lambda$, the smooth approximation of $\partial\varphi$ mentioned above), has the norm

$$|(\partial\varphi)_\lambda(y)| = |\frac{y - (1 + \lambda\partial\varphi)^{-1}(y)}{\lambda}| = \frac{1}{\lambda}\text{dist}(y, C).$$

Therefore, the cost functionals of (P_1^ν) and $(P_{1,\epsilon,\lambda})$ have a very similar structure and, as a last caution, we choose the parameters ϵ, λ, ν such that to obtain comparable values at the first iteration of the algorithm.

Another data of importance is u_0, the first iteration for u. Generally, we take a very "rough" choice of u_0, that is, such that the constraints are severely violated.

Two main types of numerical tests are associated with (P_6). In the first one we know the optimal pair, that is we choose y_d, y_0, f such that $[y_d, 0]$ is an admissible pair with cost 0, so it is optimal. In fact, this situation reduces to an exact penalty approach since $[y_d, 0]$ remains the optimal pair both for (P_1^ν), $(P_{1,\epsilon,\lambda})$. The main gain of these examples is to check the behaviour of the numerical codes and the possible errors. Both methods acted similarly and produced accurate approximations of the optimal pair in a similar number of steps.

Next, we investigate general examples. According to the above considerations, we take $C = [0, \infty[$ or $C = [0, 1]$, $U_{ad} = \mathbf{R}$ or $U_{ad} = [0, 1]$ or $U_{ad} = [1, 5]$. We fix $y_d = 0$ or

$$y_d(t) = \tilde{y}_d(t) = \begin{cases} 0 & t \leq \frac{1}{2} \\ 1 & t > \frac{1}{2}, \end{cases}$$

$y_0 = 0$ or $y_0 = 1$, $u_0 = \pm 100$, $f = 0$, $\epsilon = 0, 1$, $\lambda = 0, 01$, $\nu = 0, 002$, $\delta = 0, 01$.

Both methods behaved numerically better than predicted by the theory. In all the examples, the computed optimal pairs turned to be admissible or very close to admissible pairs for (P_6). We remark several characteristics of the variational inequality method (I) superior to the penalization method (II):

- The optimal value produced by I is generally smaller than the one produced by II,

- The number of iterations in I is smaller than in II,
- Method I reaches admissible values quickly and remains in the admissible set. So all the nonlinear terms become zero and the computations are very much shortened in fact.
- We have computed the violation of the state constraint at each iteration. The error due to method I is less than half of the error in method II.
- Method I is stable with respect to the initial guess u_0. In method II we had frequent convergence problems and we were obliged to adapt the line search rule in order to obtain convergence.

Let us denote by v_i the cost corresponding to u_0, by v_f the computed optimal cost, by N the number of iterations and by N_a the number of the iteration at which the algorithm entered the admissible set. The next table contains three examples worked by both methods:

		Example 1	Example 2	Example 3
y_0		0	0	1
u_0		100	-100	100
y_d		$\tilde{y}_d$	$\tilde{y}_d$	$\tilde{y}_d$
f		0	0	0
U_{ad}		$\mathbf{R}$	$\mathbf{R}$	$[1,5]$
C		$[0,1]$	$[0,1]$	$[0,1]$
v_i	I	$349103,28$	$359676,09$	$578458,5$
	II	$408882,91$	$427061,34$	$869901,25$
v_f	I	$0,1967274$	$0,1967292$	$0,768277$
	II	$0,1967283$	$0,2246078$	$0,7623037$
N	I	27	102	42
	II	32	71	89
N_a	I	17	11	20
	II	20	47	47

Similar numerical results were obtained for systems of ordinary differential equations and for parabolic equations. For other numerical investigations, we refer to [5], [7].

References

[1] Barbu, V., *Optimal control of variational inequalities*, in "Research Notes in Mathematics 100," Pitman, London, 1984.

[2] Brezis, H., *Problèmes unilateraux*, J. Math. Pures Appl. **51** (1972).

[3] Tiba, D., *Une approche par inéquations variationelles pour les problèmes des contrôle avec contraintes*, C.R.A.S. Paris **302**, Serie 1, no. 1 (1986).

[4] Tiba, D., Bonnans, J.F., *Equivalent control problems and applications*, in "Lecture Notes in Control and Information Sciences," Springer Verlag, 1987, pp. 154-161.

[5] Tiba, D., Neittaanmäki, P., *A variational inequality approach to constrained control problemes for parabolic equations*, Appl. Math. Optim. **17** (1988), 185-201.

[6] Tiba, D., Haslinger, J., Neittaanmäki, P., *On state constrained optimal shape design problems*, in "International Series of Numerical Mathematics 78," Birkhäuser, 1987, pp. 109-122.

[7] Tiba, D., Neittaanmäki, P., Mäkinen, R., *A variational inequality approach to the problem of the design of the optimal covering of obstacle*, Preprint **78** (1987). Univ. Jyvävkylä (Finland)

[8] Zolesio, J.P., Sokolowski, J., Benedict, B., *Shape optimization for the contact problems*, in "Lecture Notes in Control and Information Sciences 59," Springer Verlag, 1984, pp. 784-799.

[9] Ural'ceva, N.N., Solonnikov, V.A., Ladyzenskaja, O.A., *Linear and quasilinear equations of parabolic type*, Translations A.M.S. (1968). Providence, R.I.

D. Tiba and M. Tiba
Department of Mathematics
INCREST
Bd. Pacii 220
R-79622 Bucuresti
Romania

International Series of
Numerical Mathematics, Vol. 91
© 1989 Birkhäuser Verlag Basel

Uniform exponential energy decay of Euler-Bernoulli equations by suitable boundary feedback operators

R. Triggiani

Department of Applied Mathematics
University of Virginia

Abstract. We study the uniform stabilization problem for the Euler-Bernoulli equation defined on a smooth bounded domain of any dimension with feedback dissipative operators in various boundary conditions.

Keywords. Euler-Bernoulli equations; uniform stabilization.

0. Introduction

Throughout this note Ω is an open bounded domain in $\mathbf{R}^n$ with sufficiently smooth boundary $\partial\Omega = \Gamma$. In Ω we consider the Euler-Bernoulli equation with suitable boundary conditions which, once homogeneous (free dynamics), produce a *unitary* group of operators, i.e. *norm-preserving* free solutions on suitable natural function spaces. We then seek to introduce "damping" in the dynamics by virtue of expressing the non-homogeneous boundary controls as suitable feedback operators in terms of the velocity, in order to force *uniform exponential decay* of all feedback solutions. In section 1, we treat boundary controls in the Dirichlet and Neumann boundary conditions. Here, the uniform stabilization results which we present are fully consistent with recently established exact controllability and optimal regularity theories [L-6; L-7; L-T.1; L-T.2], (which, in fact, motivate the choices of spaces in the first place). In section 2, a bending moment type of condition replaces the Neumann boundary condition. Here further difficulties arise. We present results when only one boundary control is active. Thus, as expected, geometrical conditions on Ω are needed. These, however, are more restrictive than in section 1. For work in the stabilization of plates with other boundary conditions, we refer to [K-R.1], [L.1]-[L.4],[L-L.1].

1. Euler-Bernoulli equation with Dirichlet and Neumann boundary controls [B-T.1]

Consider the mixed problem

$$
\begin{array}{lll}
(1.1a) & w_{tt} + \Delta^2 w = 0 & \text{in } (0,T] \times \Omega = Q \\[4pt]
(1.1b) & w(0,\cdot) = w_0; \; w_t(0,\cdot) = w_1 & \text{in } \Omega \\[4pt]
(1.1c) & w|_\Sigma = g_1 & \text{in } (0,T] \times \Gamma = \Sigma \\[4pt]
(1.1d) & \dfrac{\partial w}{\partial \nu}\Big|_\Sigma = g_2 & \text{in } \Sigma
\end{array}
$$

in the solution $w(t, x)$ subject to suitable control functions g_1 and g_2 [L-T.1; L-T.2]. Throughout this section let A be the positive self-adjoint operator on $L^2(\Omega)$ defined by $Af = \Delta^2 f$, $\mathcal{D}(A) = H^4(\Omega) \cap H_0^2(\Omega)$. We introduce the following spaces, as in recent exact controllability studies [L-T.1; L-T.2]

$$(1.2) \qquad\qquad X = H^{-1}(\Omega) \times V'$$

$$(1.3) \qquad\qquad V = \{ f \in H^3(\Omega) : f|_\Gamma = \frac{\partial f}{\partial v}|_\Gamma = 0 \}$$

$$(1.4) \qquad\qquad Y = H_0^1(\Omega) \times H^{-1}(\Omega)$$

where we recall that

$$(1.5) \qquad\qquad \mathcal{D}(A^{1/4}) = H_0^1(\Omega); \quad \mathcal{D}(A^{3/4}) = V$$

(with equivalent norms), so that the spaces X and Y in (1.2), (1.4) can likewise be identified as

$$(1.6) \qquad\qquad X = [\mathcal{D}(A^{1/4})]' \times [\mathcal{D}(A^{3/4})]'$$

$$(1.7) \qquad\qquad Y = \mathcal{D}(A^{1/4}) \times [\mathcal{D}(A^{1/4})]'.$$

The symbol $'$ denotes duality with respect to the $L^2(\Omega)$-topology. The norms are given by

$$(1.8) \qquad\qquad \|x\|_{\mathcal{D}(A^\alpha)} = \|A^\alpha x\|_{L^2(\Omega)}; \ \|x\|_{[\mathcal{D}(A^\beta)]'} = \|A^{-\beta} x\|_{L^2(\Omega)}$$

where $\alpha, \beta \geq 0$. The problem of *exact controllability* for the dynamics (1.1) on the spaces X and Y with either one or else two controls g_1 and g_2 in suitable functions spaces was studied in [L-T.1; L-T.2]. To help motivate the spaces of uniform stabilization chosen below, we recall from these references that e.g. exact controllability of (1.1) in the space X of *optimal* regularity was obtained for an arbitrary short time $T > 0$ either with control functions

$$(1.9) \qquad\qquad g_1 \in L^2(0, T; L^2(\Gamma)); \ g_2 \equiv 0$$

under geometrical conditions for Ω; or else with controls

$$(1.10) \qquad\qquad g_1 \in L^2(0, T; L^2(\Gamma)); \ g_2 \in L^2(0, T; H^{-1}(\Gamma))$$

without geometrical conditions on Ω (except for smoothness of Γ). Thus the corresponding *uniform stabilization* problem may be stated qualitatively as follows: seek, if possible, two (linear) feedback operators $\mathcal{F}_1$ and $\mathcal{F}_2$

$$(1.11a) \qquad\qquad g_1 = \mathcal{F}_1(w_t) \in L^2(0, \infty; L^2(\Gamma))$$

$$(1.11b) \qquad\qquad g_2 = \mathcal{F}_2(w_t) \in L^2(0, \infty; H^{-1}(\Gamma))$$

based on the velocity w_t (damping) such that the corresponding closed loop problem which results from introducing (1.11a–b) into (1.1c–d), respectively, is (i) well-posed on X (in the sense of semigroup generation on X) and (ii) decays (exponentially) in the uniform operator topology of X for $t \to +\infty$. Solution to this problem is provided by the following two theorems.

Theorem 1.1. [B-T.1] (Well-posedness and uniform stabilization on X with two feedback operators.) *Consider the problem (1.1) with*

(1.12)
$$g_1 = -\frac{\partial}{\partial v}(\Delta A^{-3/2} w_t)|_\Gamma$$

(1.13)
$$g_2 = \Lambda^2[\Delta(A^{-3/2} w_t)]_\Gamma$$

where Λ denotes an isomorphism $H^s(\Gamma)$ onto $H^{s-1}(\Gamma)$, self-adjoint on $L^2(\Gamma)$ (we need only the case $s = 1$) (first order differential operator tangential to Γ with smooth coefficients: say, tangential gradient). Then, the closed loop problem obtained from inserting (1.12) and (1.13) into (1.1c) and (1.1d), respectively, possesses the following properties:

(i) *(Well-posedness) the solution map*

(1.14)
$$\{w_0, w_1\} \to \{w(t), w_t(t)\}$$

defines a strongly continuous contraction semigroup $e^{\mathcal{A}t}$ on X;

(ii) *(L_2-nature in time of feedback operators) The functions g_1, g_2 given by (1.12), (1.13) satisfy the inequality*

(1.15)
$$\int_0^\infty \{\|g_1(t)\|_{L^2(\Gamma)}^2 + \|g_2(t)\|_{H^{-1}(\Gamma)}^2\} dt \le \|\{w^0, w^1\}\|_X^2;$$

(iii) *(uniform stabilization) there exist constants M and $\delta > 0$ such that*

(1.16)
$$\left\| \begin{bmatrix} w(t) \\ w_t(t) \end{bmatrix} \right\|_X = \left\| e^{\mathcal{A}t} \begin{bmatrix} w_0 \\ w_1 \end{bmatrix} \right\|_X \le M e^{-\delta t} \left\| \begin{bmatrix} w_0 \\ w_1 \end{bmatrix} \right\|_X, \ t \ge 0 \quad \blacksquare$$

We note explicitly that no geometrical conditions are imposed on Ω in Theorem 1.1 (except for smoothness of Γ). If, however, only one feedback control action g_1 is used, then uniform stabilization is still achieved, but under geometrical conditions on Ω.

Theorem 1.2. [B-T.1] (Uniform stabilization on X with one feedback operator.) *The same conclusions of Theorem 1.1 continue to hold true if g_1 is given by (1.12) while g_2 is taken identically zero $g_2 \equiv 0$, provided Ω satisfies the following geometrical condition: There exists a vector field $h(x) \in C^2(\bar{\Omega})$ such that*

(1.17)
  (i) $h \cdot v \ge \gamma > 0$ *on* Γ; $\quad v =$ *unit outward normal;* $\quad \gamma =$ *constant*

  (ii) $\int_\Omega H(x) v(x) \cdot v(x) d\Omega \ge \rho \int_\Omega |v(x)|_{R^n}^2 d\Omega,$

(1.18)
  for some constant $\rho > 0$

and all $v(x) \in [L^2(\Omega)]^n$, where $H(x)$ is the $n \times n$ matrix with $\partial h_i(x)/\partial x_j$ as its (i,j)-th entry (H is the transpose of the Jacobian of h). We note that a sufficient checkable condition for (ii) to hold is that the symmetric matrix $H(x) + H^*(x)$ be uniformly positive definite on $\bar{\Omega}$ ∎

Next, if one whishes to study uniform stabilization of (1.1) in the space Y given by (1.4) (and corresponding to the exact controllability results in this space of [L-T.1; L-T.2]), one sees from (1.4) that we must take at the outset $g_1 \equiv 0$. A study of this case is presented in [B-T.1]. We close this section by remarking that the proof of Theorems 1.1 and 1.2 are inspired by the uniform stabilization papers [L-T.4] for the wave equation with Dirichlet feedback; [T.1] for the wave equation with Neumann feedback following the original contributions of G. Chen [C.1], [C.2] and J. Lagnese [L.5]; and [L-T.1], [L-T.2] for the corresponding exact controllability results for (1.1).

2. Euler-Bernoulli equation with boundary controls on $w|_\Sigma$ and $\Delta w|_\Sigma$ [L-T.7]

Throughout this section we consider the problem

$$(2.1a) \qquad w_{tt} + \Delta^2 w = 0 \qquad\qquad \text{in } (0,T] \times \Omega = Q$$
$$(2.1b) \qquad w(0,\cdot) = w_0; \; w_t(0,\cdot) = w_1 \quad \text{in } \Omega$$
$$(2.1c) \qquad w|_\Sigma = g_1 \qquad\qquad\qquad \text{in } (0,T] \times \Gamma = \Sigma$$
$$(2.1d) \qquad \Delta w|_\Sigma = g_2 \qquad\qquad\quad \text{in } \Sigma.$$

Regularity results (in fact, optimal) in appropriate function spaces were given in [L-T.5], while corresponding controllability results may be found in [L.5; L-T.6]. Here we study the uniform stabilization problem for (2.1) on the spaces

$$(2.2) \qquad\qquad Z = \mathcal{D}(A^{1/2}) \times L^2(\Omega)$$
$$(2.3) \qquad\qquad W = L^2(\Omega) \times [\mathcal{D}(A^{1/2})]'.$$

In (2.2) – (2.3), and throughout this section, we let $Af = \Delta^2 f$, $\mathcal{D}(A) = \{f \in H^4(\Omega) : f|_\Gamma = \Delta f|_\Gamma = 0\}$ and we have $A^{1/2}f = -\Delta f$, $\mathcal{D}(A^{1/2}) = H^2(\Omega) \cap H^1_0(\Omega)$. Hence, choice (2.2) of Z implies at the outset the condition

$$(2.4) \qquad\qquad g_1 \equiv 0 \text{ on } \Sigma$$

on problem (2.1), and we are thus left only with the control g_2 (to be suitable expressed as a feedback operator on w_t as in (2.5) below) to force *uniform stabilization* for (2.1). Not suprisingly, geometrical conditions on Ω are needed to achieve uniform stabilization. As a matter of fact, these are more restrictive, and of more difficult interpretation as well, on the class of domains Ω which are allowed. They are described in Definition 2.1 which follows the statement of Theorem 2.1.

Theorem 2.1. [L-T.7] (Well-posedness and uniform stabilization on Z.) *Consider problem (2.1) with*

$$(2.5) \qquad\qquad g_1 \equiv 0, \quad g_2 = -\frac{\partial w_t}{\partial v}\Big|_\Sigma.$$

Then:

(i) *(well-posedness) the corresponding map* $\{w_0, w_1\} \rightarrow \{w(t), w_t(t)\}$ *defines a strongly continuous contraction semigroup* e^{At} *on* Z;

(ii) *(L_2-nature of feedback operators) The feedback control g_2 in (2.5) satisfies the inequality*

$$\int_0^\infty \|g_2(t)\|_{L^2(\Gamma)}^2 dt \leq \|\{w_0, w_1\}\|_Z^2.$$

(iii) *(Uniform stabilization on Z) Let now Ω satisfy the geometrical conditions of Definition 2.1 below. Then there exist constants M and $\delta > 0$ such that*

$$(2.6) \qquad \left\| \begin{bmatrix} w(t) \\ w_t(t) \end{bmatrix} \right\|_Z = \left\| e^{At} \begin{bmatrix} w_0 \\ w_1 \end{bmatrix} \right\|_Z \leq M e^{-\delta t} \left\| \begin{bmatrix} w_0 \\ w_1 \end{bmatrix} \right\|_Z, \ t \geq 0 \quad \blacksquare$$

A similar result holds true on the space W defined by (2.3); mutatis mutandis, see [L-T.7].

The proof of Theorem 2.1 is inspired by paper [L-T.4] on the uniform stabilization of the wave equation with Dirichlet feedback; in particular, it presents a technical difficulty of the same type as one encountered in [L-T.4].

Definition 2.1. *Let Ω satisfy the following condition. There exists a vector field $h(x) \in C^2(\bar{\Omega})$ such that:*

(i) *h is parallel to v (exterior unit normal) on all of Γ; i.e. $h(\sigma) = k(\sigma)v(\sigma)$, for $k(\sigma)$ a smooth scalar function, $\sigma \in \Gamma$;*

(ii) *the following inequality holds*

$$\int_\Omega \Delta q \left(\sum_{i=1}^n \nabla h_i \cdot \nabla q_{x_i} \right) d\Omega \geq \rho \int_\Omega |\Delta q|^2 d\Omega$$

where $q(x)$ is a smooth function on Ω such that

$$q|_\Gamma \equiv 0 \ \text{ and } \ \Delta q|_\Gamma \equiv 0$$

and $\rho > 0$ is a suitable constant, possible depending on $h(x)$, Ω, and $q(x)$ $\blacksquare$

For instance, in 3-dimensions $x = x_1$, $y = x_2$, $z = x_3$ we have

$$(2.7) \qquad \sum_{i=1}^3 \nabla h_i \cdot \nabla q_{x_i} = h_{1x} q_{xx} + h_{2y} q_{yy} + h_{3z} q_{zz} + (h_{2x} + h_{1y}) q_{xy}$$

$$+ (h_{3x} + h_{1z}) q_{xz} + (h_{3z} + h_{2z}) q_{yz}.$$

Examples of domains satisfying Definition 2.1 include n-dimensional spheres with center x_0, where $h(x) = x - x_0$ and n-dimensional ellipsoids where the ratio between the axes is "sufficiently small".

3. Sketch of proof of Theorems 1.1 and 1.2

As the proof of Theorems 1.1 and 1.2 is lengthy, we can only confine ourselves here to a schematic sketch which will be concentrated on the main issue of uniform stabilization. (Well-posedness in the semigroup sense and L_2-nature of the feedback operators follow from a dissipative type of argument based on Lumer-Phillips theorem as in [L-T.4], [T.1].) We define (recall (1.2), (1.6), (1.8))

(3.1)
$$E(w,t) \equiv E(t) = \|\{w(t), w_t(t)\}\|_X^2 = \|A^{-1/4}w(t)\|_{L^2(\Omega)}^2 + \|A^{-3/4}w_t(t)\|_{L^2(\Omega)}^2 \le E(0).$$

Our main goal will be, as usual, to show that

(3.2)
$$\int_0^\infty E(t)dt \le \text{const } E(0) \quad \forall\{w_0, w_1\} \in X$$

with "const" independent of the initial data. After this, Datko's Theorem will yield the desired conclusion. As a matter of fact, it will suffice to show inequality (3.2) for initial data smooth, say in the domain of the generator of the feedback problem. To this end, we first introduce a new variable p by setting

(3.3)
$$p = A^{-3/2}w_t$$

which yields the new problem

(3.4a) $p_{tt} + \Delta^2 p = F_1 + F_2$ in $(0,\infty) \times \Omega = Q$

(3.4b) $p_0 = A^{-3/2}w_1; \ p_1 = A^{-3/2}w_{tt}(0)$ in Ω

(3.4c) $p|_\Sigma \equiv 0$ and $\dfrac{\partial p}{\partial v}|_\Sigma \equiv 0$ in $(0,\infty) \times \Gamma = \Sigma$

(3.5)
$$F_1 = -A^{-1/2}G_1 G_1^* A^{-1/2} w_{tt};$$
$$F_2 = -A^{-1/2}G_2\Lambda^2 G_2^* A^{-1/2} w_{tt}$$

(3.6) $G_1 g_1 = v \Leftrightarrow \{\Delta^2 v = 0 \text{ in } \Omega; \ v|_\Gamma = g_1; \ \dfrac{\partial v}{\partial v}|_\Gamma = 0\}$

(3.7) $G_2 g_2 = y \Leftrightarrow \{\Delta^2 y = 0 \text{ in } \Omega; \ y|_\Gamma = 0; \ \dfrac{\partial y}{\partial v}|_\Gamma = g_2\}.$

By (3.3) and the w-problem it follows that

(3.8)
$$\|A^{-3/4}w_t\|_{L^2(\Omega)} = \|A^{3/4}p\|_{L^2(\Omega)},$$
$$\text{equivalent to } \{\int_\Omega |\nabla(\Delta p)|^2 d\Omega\}^{1/2}$$

(3.9) $A^{1/4}p_t = -A^{-1/4}w + O(\|g_1\|_{L^2(\Gamma)}) + O(\|\Lambda^{-1}g_2\|_{L^2(\Gamma)})$

(3.10) $\|A^{1/4}p_t\|_{L^2(\Omega)}$ equivalent to $\{\int_\Omega |\nabla p_t|^2 d\Omega\}^{1/2}$

with g_1 and g_2 in feedback form as in (1.12), (1.13), which therefore satisfy (1.15). The equivalences in (3.8) and (3.10) have been pointed out and crucially used in the exact controllability study in [L-T.1], [L-T.2], (after Grisvard's interpolation results). Next, following an idea in the exact controllability study in [L-T.1], [L-T.2], we apply to the p-problem (3.4) two multipliers: $e^{-2\beta t}h \cdot \nabla(\Delta p)$ and $e^{-2\beta t}\Delta p\,\mathrm{div}\,h$, $\beta > 0$, with h the vector field for Ω (this is the vector field h of the statement of Theorem 1.2 if only the feedback on g_1 is used, while $g_2 \equiv 0$; instead, it may be a radial vector field $x - x_0$ in the case of both feedbacks on g_1 and g_2). After (lengthy) integrations by parts one obtains the following identity after using the boundary conditions

$$
\begin{aligned}
(3.11) \quad & \int_\Sigma e^{-2\beta t}\frac{\partial(\Delta p)}{\partial v}h \cdot \nabla(\Delta p)d\Sigma - \frac{1}{2}\int_\Sigma e^{-2\beta t}|\nabla(\Delta p)|^2 h \cdot vd\Sigma \\
& + \frac{1}{2}\int_\Sigma e^{-2\beta t}\Delta p\frac{\partial(\Delta p)}{\partial v}\mathrm{div}\,h\,d\Sigma \\
= & \int_Q e^{-2\beta t}H\nabla(\Delta p) \cdot \nabla(\Delta p)dQ + \int_Q e^{-2\beta t}H\nabla p_t \cdot \nabla p_t dQ \\
& + \frac{1}{2}\int_Q e^{-2\beta t}\Delta p\nabla(\Delta p) \cdot \nabla(\mathrm{div}\,h)dQ + \frac{1}{2}\int_Q e^{-2\beta t}p_t\nabla p_t \cdot \nabla(\mathrm{div}\,h)dQ \\
& - \beta\int_Q e^{-2\beta t}p_t\Delta p\,\mathrm{div}\,h dQ - 2\beta\int_Q e^{-2\beta t}p_t h \cdot \nabla(\Delta p)dQ \\
& + \frac{1}{2}\int_Q e^{-2\beta t}F\Delta p\,\mathrm{div}\,h\,dQ + \int_Q e^{-2\beta t}Fh \cdot \nabla(\Delta p)dQ \\
& - \frac{1}{2}\int_\Omega \nabla p_0 \cdot \Delta(p_1\,\mathrm{div}\,h)d\Omega + \int_\Omega p_1 h \cdot \nabla(\Delta p_0)d\Omega
\end{aligned}
$$

where $F = F_1 + F_2$, as the terms of integration by parts in t as $T \to \infty$ vanish (we are taking smooth initial data, as mentioned below (3.2)) [Note that $\mathrm{div}\,h \equiv 0$ if h is a radial field, the case of two feedbacks]. We have

$$
(3.12) \quad \frac{\partial(\Delta p)}{\partial v}\Big|_\Sigma = \left(\frac{\partial}{\partial v}(\Delta A^{-3/2}w_t)\right) = G_1^*AA^{-3/2}w_t = G_1^*A^{-1/2}w_t
$$

$$
(3.13) \quad -\Lambda[\Delta p]|_\Sigma = -\Lambda[\Delta(A^{-3/2}w_t)]_\Sigma = \Lambda G_2^*AA^{-3/2}w_t = \Lambda G_2^*A^{-1/2}w_t
$$

whose $L^2(0,\infty; L^2(\Gamma))$-norms are bounded by $E(0)$ by (1.15), (1.6), (3.1). Technical manipulations on the terms of (3.11) using the norm equivalences (3.8), (3.9), (3.10) for the interior terms, and (3.12), (3.13), (1.15) for the boundary terms yield the following estimates for the left hand side (L.H.S.) and right hand side (R.H.S.) of identity (3.11):

$$
(3.14) \quad C_h E(0) \geq C_h\{\int_0^\infty [\|g_1\|_{L^2(\Gamma)}^2 + \|g_2\|_{H^{-1}(\Gamma)}^2]dt \geq \text{L.H.S. of (3.11)}
$$

$$
(3.15) \quad \text{R.H.S. of (3.11)} \geq c_h^2\int_0^\infty e^{-\beta t}E(t)dt - K_h^2 E(0)
$$

in the case of radial vector field (Theorem 1.1), where the positive constants c_h^2 and K_h^2 do not depend on β. Combining (3.14), (3.15) and letting $\beta \downarrow 0$ yields (3.2) as desired. In the case of general vector field h the right hand side contains also an additional term: an integral over $\int_0^\infty$ in time with *lower order* terms in p (or w). These then can be "obsorbed" through a theorem like Theorem 2 in [L.5] or Theorem 1.2 in [T.1] (our proof follows the operator proof in [T.1]) ∎

4. Sketch of proof of Theorem 2.1

With reference to problem (2.1), (2.5), we now introduce a new variable p by setting $p = A^{-1/2}w_t$, where now A realizes Δ^2 with homogeneous boundary conditions $f|_\Gamma = \Delta f|_\Gamma = 0$. We thus obtain a corresponding problem in p:

(4.1a) $$p_{tt} + \Delta^2 p = F_1$$
(4.1b) $$p|_\Sigma = \Delta p|_\Sigma = 0$$
(4.1c) $$p_0, p_1$$

(4.2) $$F_1 = -G_1 G_1^* A^{1/2} w_{tt}$$

counterpart of (3.4), (3.5). In (4.2) we have that G_1 is the Dirichlet map D in the notation of [L-T.4]: $G_1 = D$, i.e.

(4.3) $$G_1 g_1 = v \Leftrightarrow \Delta v = 0 \text{ in } \Omega; \; v|_\Gamma = g_1.$$

Now however we apply the multipliers $e^{-2\beta t}h \cdot \nabla p$ and $e^{-2\beta t}p \, \text{div} \, h$ to the problem (4.1), integrate by parts (extensively), use the boundary conditions (4.1b), and finally obtain the following identity, counterpart of identity (3.11)

(4.4)
$$-\int_\Sigma e^{-2\beta t}\frac{\partial \Delta p}{\partial v}\frac{\partial p}{\partial v}h \cdot vd\Sigma = 2\int_Q e^{-2\beta t}\Delta p\left(\sum_{i=1}^n \nabla h_i \cdot \nabla p_{x_i}\right)dQ$$
$$+\frac{1}{2}\int_Q e^{-2\beta t}p\Delta p\Delta(\text{div } h)dQ + \int_Q e^{-2\beta t}\Delta p\nabla p \cdot \nabla(\text{div } h)dQ$$
$$+\int_Q e^{-2\beta t}\Delta p[\Delta h_1,\ldots,\Delta h_n] \cdot \nabla pdQ + \beta\int_Q e^{-2\beta t}pp_t\text{div } hdQ$$
$$+2\beta\int_Q e^{-2\beta t}p_t h \cdot \nabla pdQ - \int_Q e^{-2\beta t}F_1 h \cdot \nabla pdQ$$
$$-\frac{1}{2}\int_Q e^{-2\beta t}F_1 p \, \text{div } h \, dQ - \frac{1}{2}(p_t(0), p(0) \, \text{div } h)_\Omega - (p_t(0), h \cdot \nabla p(0))_\Omega$$

as the terms of integration by parts in t vanish as $T \to +\infty$ (we are taking smooth initial data). Next we have

(4.5) $$\frac{\partial(\Delta p)}{\partial v} = G_1^* Ap = G_1^* A^{1/2}w_t = \frac{\partial w_t}{\partial v}\Big|_\Sigma$$

which is the boundary feedback whose $L_2(0, \infty; L^2(\Gamma))$-norm is bounded by the initial "energy" $\|A^{1/2}w_0\|_\Omega^2 + \|w_1\|_\Omega^2 = E(0)$ by the dissipativity argument. Note that the change of variable implies

$$(4.6) \qquad \|w_t\|_{L^2(\Omega)}^2 = \|A^{1/2}p\|_{L^2(\Omega)}^2 = \int_\Omega (\Delta p)^2 \, d\Omega$$

$$(4.7) \qquad p_t = -A^{1/2}w + O(\|G_1^* A^{1/2} w_t)\|_{L^2(\Gamma)})$$

where now the "energy" (norm) of interest is

$$(4.8) \qquad E(t) = \|A^{1/2}w(t)\|_{L^2(\Omega)}^2 + \|w_t(t)\|_{L^2(\Omega)}^2.$$

It is now the term in identity (4.4) due to $F_1 h \cdot \nabla p$ that gives rise to a technical difficulty of the same type as the one encountered in the wave equation with Dirichlet feedback in [L-T.4; Lemma 3.3], as one sees by comparing the term $F_1 h \cdot \nabla p$ with F_1 as in (4.2) where $G_1 = D$ and the corresponding "right hand side" term (3.24) of problem (3.23) in [L-T.4]. This forces the condition that h be parallel to v on Γ. On the other hand the first term on the right hand side of (4.4) requires the inequality of definition (2.1). The remaining argument proceeds along the lines of section 3 below Eq (3.11), mutatis mutandis, using (4.5) - (4.8). Details are given in [L-T.7].

REFERENCES

[B-T.1] J. Bartolomeo and R. Triggiani, *Uniform stabilization of the Euler-Bernoulli equation with Dirichlet and Neumann boundary feedback*, Report, Dept. of Applied Mathematics, University of Virginia, 1988 (to appear).

[C.1] G. Chen, *Energy decay estimates and exact controllability of the wave equation in a bounded domain*, J. Math. Pures at Appl. 58(9) (1979), 249–274.

[C.2] G. Chen, *A note on the boundary stabilization of the wave equation*, SIAM J. Control & Optimization 19 (1981), 106–113.

[F-L-T.1] F. Flandoli, I. Lasiecka and R. Triggiani, *Algebraic Riccati equations with non-smoothing observation arising in hyperbolic and Euler-Bernoulii equations*, Annali di Matematica Pura e Applicata. (to appear February 1989)

[K-R.1] J.V. Kim and Y. Renardy, *Boundary control of the Timoshenko beam*, SIAM J. Control & Optimization (1988).

[L.1] J. Lagnese, *Asymptotic energy estimates for Kirchhoff plates subject to weak viscoelastic damping*, in these proceedings.

[L.2] J. Lagnese, *Boundary stabilization of thin elastic plates*, SIAM Studies in Applied Mathematics (to appear).

[L.3] J. Lagnese, *Uniform boundary stabilization of homogeneous, isotropic plates*, in "Lecture Notes in Control and Information Sciences Vol. 102," Springer-Verlag, 1987, pp. 204–215. Proceedings of the 1986 Vorau Conference on Distributed Parameter Systems

[L.4] J. Lagnese, *A note on boundary stabilization of wave equations*, SIAM J. Control & Optimization 26 (1988), 1250–1256.

[L.5] J. Lagnese, *Decay of solutions of wave equations in a bounded region with boundary dissipation*, J. Diff. Eqs. 50(2) (1983), 163–182.

[L.6] J.L. Lions, *Exact controllability, stabilization and perturbations*, SIAM Review, March 1988, to appear in extended version by Masson 1988.

[L.7] J.L. Lions, *Un resultat de regularite*, (paper dedicated to S. Mizohata), in "Current Topics on Partial Differential Equations," Kinokuniya Company, Tokyo, 1986.

[L-L.1] J. Lagnese and J.L. Lions, "Modeling, analysis and control of thin plates," Masson, 1988.

[L-T.1] I. Lasiecka and R. Triggiani, *Exact controllability of the Euler-Bernoulli equation with $L^2(\Sigma)$-control only in the Dirichlet boundary conditions*, Atti della Accademia Nazionale dei Lincei, Rendiconti Classe di Scienze fisiche, matematiche e naturali **LXXXI**. Roma, August 1987

[L-T.2] I. Lasiecka and R. Triggiani, *Exact controllability of the Euler-Bernoulli equation with controls in the Dirichlet and Neumann boundary conditions: a non-conservative case*, SIAM J. Control & Optimization **27(2)** (1989), 303–373.

[L-T.3] I. Lasiecka and R. Triggiani, *A direct approach to exact controllability for the wave equation with Neumann boundary control and to an Euler-Bernoulli equation*, in "Proceedings 26th IEEE Conference," Los Angeles, December 1987, pp. 529–534.

[L-T.4] I. Lasiecka and R. Triggiani, *Uniform exponential energy decay of the wave equation in a bounded region with $L_2(0, \infty; L_2(\Gamma))$-feedback control in the Dirichlet boundary conditions*, J. Diff. Eqs. **66** (1987), 340–390.

[L-T.5] I. Lasiecka and R. Triggiani, *Regularity theory for a class of non-homogeneous Euler-Bernoulli equations: a cosine operator approach*, Bollettino Unione Matematica Italianan **(7)2-B** (December 1988).

[L-T.6] I. Lasiecka and R. Triggiani, *Exact controllability of the Euler-Bernoulli equation with boundary controls for displacement and moments*, J. Mathem. Analysis & Applic. (to appear).

[L-T.7] I. Lasiecka and R. Triggiani, *Uniform exponential energy decay of the Euler-Bernoulli equation on a bounded region with boundary feedback acting on the bending moment.* (to appear).

[T.1] R. Triggiani, *Wave equation on a bounded domain with boundary dissipation: an operator approach*, J. Mathem. Anal. & Applic. **137** (1989), 438–461. in "Lectures Notes in Pure and Applied Mathematics, Vol. 108, Operator Methods for Optimal Control Problems," (S. Lee Ed.), Marcel Dekker, 1987, pp. 283–310; also in Recent Advances in Communication and Control Theory, honoring the sixtieth anniversary of A.V. Balakrishnan (R.E. Kalman and G.I. Marchuk, Eds.), Optimization Software, New York, 1987, pp. 262–286.

R. Triggiani
Department of Applied Mathematics
University of Virginia
Charlottesville, VA 22902
USA

International Series of
Numerical Mathematics, Vol. 91
© 1989 Birkhäuser Verlag Basel

The representation of regular
linear systems on Hilbert spaces

George Weiss

Department of Theoretical Mathematics
The Weizmann Institute

Abstract. We introduce the concept of an abstract linear system, which is the Hilbert space analogue of finite dimensional systems described by $\dot{x}(t) = Ax(t) + Bu(t)$, $y(t) = Cx(t) + Du(t)$. Essentially, the definition of an abstract linear system requires that on any finite time interval, the final state and the output function should depend continuously on the initial state and the input function. We prove that if a certain regularity assumption is satisfied, which is satisfied in all cases of practical interest, then the equations describing an abstract linear system are the same as those describing a finite dimensional linear system. The operators A, B and C appearing in these equations may, however, be unbounded.

Keywords. Unbounded control and observation operators, admissibility, well posedness.

1980 *Mathematics subject classifications*: 93C25

1. Abstract linear systems

An imprecise statement of the main result of this paper would be the following.

Main Result. *Let Σ be an infinite dimensional linear system. If the output functions of Σ corresponding to constant input functions on $[0, \infty)$ and zero initial conditions do not oscillate too wildly at $t = 0$, then Σ is described by equations of the form*

$$(1.1) \qquad \begin{cases} \dot{x}(t) &= A\,x(t) + B\,u(t), \\ y(t) &= C\,x(t) + D\,u(t), \end{cases}$$

where A, B and C are (possibly) unbounded operators and D is a bounded operator.

By "do not oscillate too wildly" we actually mean that these functions should have a Lebesgue point at 0. Systems satisfying this condition will be called *regular*. To make the above statement precise, we need some preliminaries. The rigorous version of the main result appears in Section 4 (see Theorem 4.6).

In this section we want to motivate and rigorously define the concept of an abstract linear system. Such a system should be the (possibly) infinite dimensional analogue of a finite dimensional linear system described by the equations (1.1), where A, B, C and D are (constant) matrices of appropriate dimensions, $u(\cdot)$ is the input function, with values in the input space U, $x(t)$ is the state of the system at time t and is an element of the state space X, and $y(\cdot)$ is the output function, with values in the output space Y.

We want to describe those features of the finite dimensional system described by (1.1) which the infinite dimensional version should also have. We consider the input and output functions as elements of $L^2_{loc}([0, \infty), U)$ and $L^2_{loc}([0, \infty), Y)$, respectively, and we denote by P_τ the projection operator defined by truncation to $[0, \tau)$, i.e.,

(1.2) $$(\mathrm{P}_\tau u)(t) \;=\; \begin{cases} u(t), & \text{for } t \in [0, \tau), \\ 0, & \text{for } t \geq \tau, \end{cases}$$

and similarly for y.

Then on any finite time interval $[0, \tau]$, the final state $x(\tau)$ and the truncated output function $\mathrm{P}_\tau y$ depend linearly and continuously on the initial state $x(0)$ and the truncated input function $\mathrm{P}_\tau u$. That means that for any $\tau \geq 0$ there are four operators $\mathbb{T}_\tau$, Φ_τ, $\mathbb{L}_\tau$ and $\mathbb{F}_\tau$ such that

(1.3) $$\begin{pmatrix} x(\tau) \\ \mathrm{P}_\tau y \end{pmatrix} \;=\; \begin{pmatrix} \mathbb{T}_\tau & \Phi_\tau \\ \mathbb{L}_\tau & \mathbb{F}_\tau \end{pmatrix} \cdot \begin{pmatrix} x(0) \\ \mathrm{P}_\tau u \end{pmatrix}.$$

A trivial computation gives

$$\mathbb{T}_\tau \;=\; e^{A\tau},$$

$$\Phi_\tau u \;=\; \int_0^\tau e^{A(\tau - \sigma)} \, B \, u(\sigma) \, d\sigma,$$

$$(\mathbb{L}_\tau x)(t) \;=\; C \, e^{At} x, \qquad \text{for } t \in [0, \tau),$$

$$(\mathbb{F}_\tau u)(t) \;=\; C \int_0^t e^{A(t - \sigma)} \, B \, u(\sigma) \, d\sigma + D \, u(t), \qquad \text{for } t \in [0, \tau).$$

For $t \geq \tau$, $(\mathbb{L}_\tau x)(t) = 0$ and $(\mathbb{F}_\tau u)(t) = 0$.

The families of operators $\mathbb{T}$, Φ, $\mathbb{L}$ and $\mathbb{F}$ satisfy four functional equations, which reflect natural systems theoretical axioms in the spirit of Kalman *et al* [5]. The first functional equation is $\mathbb{T}_{\tau + t} = \mathbb{T}_t \mathbb{T}_\tau$, the other three are listed below as (1.4), (1.5) and (1.6). These functional equations should remain valid for infinite dimensional linear systems. Actually, we shall define below an abstract linear system to be a quadruple of families of operators which satisfy the four functional equations mentioned above, plus a natural continuity assumption: $\mathbb{T}_t x \to x$ as $t \to 0$, for any $x \in X$. The spaces U, X and Y will be assumed to be Hilbert spaces.

Since for $u \in L^2_{loc}([0, \infty), U)$ and $\tau \geq 0$, $\mathrm{P}_\tau u \in L^2([0, \infty), U)$, we shall consider for simplicity Φ_τ and $\mathbb{F}_\tau$ to be defined on $L^2([0, \infty), U)$, but their extension to L^2_{loc} is obvious. An analogous remark can be made for the range of $\mathbb{L}_\tau$ and $\mathbb{F}_\tau$.

We need the the notion of concatenation on $L^2_{loc}([0, \infty), W)$, where W is a Hilbert space. Let $u, v \in L^2_{loc}([0, \infty), W)$ and let $\tau \geq 0$. Then the τ-*concatenation of u and v*, denoted $u \underset{\tau}{\Diamond} v$, is an element of $L^2_{loc}([0, \infty), W)$ given by

$$(u \underset{\tau}{\Diamond} v)(t) \;=\; \begin{cases} u(t), & \text{for } t \in [0, \tau), \\ v(t - \tau), & \text{for } t \geq \tau. \end{cases}$$

Now we give the formal definition of an abstract linear system in the Hilbert space context (the definition has an obvious generalization for Banach spaces and inputs and outputs of class L^p_{loc}, see Weiss [14]).

Definition 1.1. Let U, X and Y be Hilbert spaces, $\Omega = L^2([0,\infty), U)$ and $\Gamma = L^2([0,\infty), Y)$.

An *abstract linear system* on Ω, X and Γ is a quadruple $\Sigma = (\mathbb{T}, \Phi, \mathbb{L}, \mathbb{F})$, where

($\imath$) $\mathbb{T} = (\mathbb{T}_t)_{t \geq 0}$ is a strongly continuous semigroup of bounded linear operators on X,

($\imath\imath$) $\Phi = (\Phi_t)_{t \geq 0}$ is a family of bounded linear operators from Ω to X such that

$$(1.4) \qquad \Phi_{\tau+t} \left(u \underset{\tau}{\diamond} v \right) = \mathbb{T}_t \Phi_\tau u + \Phi_t v \,,$$

for any u, $v \in \Omega$ and any τ, $t \geq 0$,

($\imath\imath\imath$) $\mathbb{L} = (\mathbb{L}_t)_{t \geq 0}$ is a family of bounded linear operators from X to Γ such that

$$(1.5) \qquad \mathbb{L}_{\tau+t} x = \mathbb{L}_\tau x \underset{\tau}{\diamond} \mathbb{L}_t \mathbb{T}_\tau x \,,$$

for any $x \in X$ and any τ, $t \geq 0$, and $\mathbb{L}_0 = 0$,

($\imath v$) $\mathbb{F} = (\mathbb{F}_t)_{t \geq 0}$ is a family of bounded linear operators from Ω to Γ such that

$$(1.6) \qquad \mathbb{F}_{\tau+t} \left(u \underset{\tau}{\diamond} v \right) = \mathbb{F}_\tau u \underset{\tau}{\diamond} \left(\mathbb{L}_t \Phi_\tau u + \mathbb{F}_t v \right) ,$$

for any u, $v \in \Omega$ and any τ, $t \geq 0$, and $\mathbb{F}_0 = 0$.

U is the *input space* of Σ, X is the *state space* of Σ and Y is the *output space* of Σ. The operators Φ_τ are called *input maps*. The operators $\mathbb{L}_\tau$ are called *output maps*. The operators $\mathbb{F}_\tau$ are called *input/output maps*.

A definition equivalent to the above (but formulated differently) was given by Salamon in [10], and a definition related to 1.1 appeared in Yamamoto [15]. In Weiss [12], an *abstract linear control system* was defined as a pair $(\mathbb{T}, \Phi)$ satisfying conditions ($\imath$) and ($\imath\imath$) above. In Weiss [13], an *abstract linear observation system* was defined as a pair $(\mathbb{L}, \mathbb{T})$ satisfying conditions ($\imath$) and ($\imath\imath\imath$) above. (The last two mentioned papers were written in the Banach space and L^p context.)

From the definition of an abstract linear system given above, one can derive the following two formulae expressing *causality*:

$$(1.7) \qquad \Phi_\tau \mathrm{P}_\tau = \Phi_\tau \,, \qquad \mathbb{F}_\tau \mathrm{P}_\tau = \mathbb{F}_\tau \,,$$

for any $\tau \geq 0$. We also have that for $0 \leq \tau \leq T$

$$(1.8) \qquad \mathrm{P}_\tau \mathbb{L}_T = \mathbb{L}_\tau \,, \qquad \mathrm{P}_\tau \mathbb{F}_T = \mathbb{F}_\tau \,.$$

The proofs of (1.7) and (1.8) are easy and we omit them. The formulae concerning Φ and $\mathbb{L}$ can be found also in Weiss [12] and [13].

We give a very simple example to illustrate the concept introduced in Definition 1.1. We shall return to this example in the following sections.

Example 1.2. We model a delay line as an abstract linear system. Let $X = L^2[-h, 0]$, where $h > 0$, and let $\mathbb{T}$ be the left shift semigroup on X with zero entering from the right, i.e., for any $\tau \geq 0$ and $\zeta \in [-h, 0]$

$$(\mathbb{T}_\tau x)(\zeta) = \begin{cases} x(\zeta + \tau), & \text{for } \zeta + \tau \leq 0, \\ 0, & \text{for } \zeta + \tau > 0. \end{cases}$$

Let $U = \mathbb{C}$ and for any $\tau \geq 0$ and $\zeta \in [-h, 0]$

$$(\Phi_\tau u)(\zeta) = \begin{cases} u(\zeta + \tau), & \text{for } \zeta + \tau \geq 0, \\ 0, & \text{for } \zeta + \tau < 0. \end{cases}$$

Let $Y = \mathbb{C}$ and for any $\tau \geq 0$ and $t \in [0, \tau)$

$$(\mathbb{L}_\tau x)(t) = \begin{cases} x(t - h), & \text{for } t - h \leq 0, \\ 0, & \text{for } t - h > 0. \end{cases}$$

For $t \geq \tau$ we put $(\mathbb{L}_\tau x)(t) = 0$. Finally, let for any $\tau \geq 0$ and $t \in [0, \tau)$

$$(\mathbb{F}_\tau u)(t) = \begin{cases} u(t - h), & \text{for } t - h \geq 0, \\ 0, & \text{for } t - h < 0. \end{cases}$$

For $t \geq \tau$ we put $(\mathbb{F}_\tau u)(t) = 0$. Then $\Sigma = (\mathbb{T}, \Phi, \mathbb{L}, \mathbb{F})$ is an abstract linear system.

2. Extensions to infinite time

We shall use the notation introduced in Definition 1.1. We consider $\tilde{\Omega} = L^2_{loc}([0, \infty), U)$ as a Fréchet space with the family of seminorms $p_n(u) = \|P_n u\|_{L^2}$, $n \in \mathbb{N}$. The operators Φ_τ and $\mathbb{F}_\tau$, which are defined on Ω, can be extended to $\tilde{\Omega}$ by continuity, which is the same as using formulae (1.7) to define the extensions. This enables us to apply these operators to constant input functions, for example. The extensions of Φ_τ and $\mathbb{F}_\tau$ to $\tilde{\Omega}$ will be denoted by the same symbols.

The space $\tilde{\Gamma} = L^2_{loc}([0, \infty), Y)$ is a Fréchet space similar to $\tilde{\Omega}$. The following limits exist in $\tilde{\Gamma}$, for any $x \in X$ and any $u \in \tilde{\Omega}$:

$$\mathbb{L}_\infty x = \lim_{\tau \to \infty} \mathbb{L}_\tau x,$$

$$\mathbb{F}_\infty u = \lim_{\tau \to \infty} \mathbb{F}_\tau u.$$

The operators $\mathbb{L}_\infty : X \to \tilde{\Gamma}$ and $\mathbb{F}_\infty : \tilde{\Omega} \to \tilde{\Gamma}$ are continuous. Formulae (1.8) extend to

$$(2.1) \qquad\qquad P_\tau \mathbb{L}_\infty = \mathbb{L}_\tau, \qquad P_\tau \mathbb{F}_\infty = \mathbb{F}_\tau,$$

for any $\tau \geq 0$. Formula (1.5) extends to

$$(2.2) \qquad \mathbb{L}_\infty \, x \; = \; \mathbb{L}_\infty \, x \underset{\tau}{\Diamond} \mathbb{L}_\infty \, \mathbb{T}_\tau \, x \, ,$$

for any $x \in X$ and any $\tau \geq 0$. Formula (1.6) extends to

$$(2.3) \qquad \mathbb{F}_\infty \, (u \underset{\tau}{\Diamond} v) \; = \; \mathbb{F}_\infty \, u \underset{\tau}{\Diamond} (\mathbb{L}_\infty \, \mathbf{\Phi}_\tau \, u + \mathbb{F}_\infty \, v) \, ,$$

for any u, $v \in \tilde{\Omega}$ and any $\tau \geq 0$. The operator $\mathbb{F}_\infty$ can be extended further to certain function spaces on $(-\infty, \infty)$, see Salamon [10].

It is obvious that $\|\mathbb{F}_t\|$ is nondecreasing with respect to t. In the next proposition we investigate the growth rate of $\|\mathbb{F}_t\|$ for large t. We shall use this proposition later, but not in the proof of the representation theorem of Section 4, so the reader interested only in the main result may skip the rest of this section.

Proposition 2.1. *Let* $\Sigma = (\mathbb{T}, \mathbf{\Phi}, \mathbb{L}, \mathbb{F})$ *be an abstract linear system on* Ω, X *and* Γ. *Let* $M \geq 1$ *and* $\omega \in \mathbb{R}$ *be such that*

$$(2.4) \qquad \|\mathbb{T}_t\| \; \leq \; M \, e^{\omega t} \, , \qquad \forall \, t \geq 0 \, .$$

If $\omega > 0$, *then there is an* $L \geq 0$ *such that*

$$\|\mathbb{F}_t\| \; \leq \; L \, e^{\omega t} \, , \qquad \forall \, t \geq 0 \, .$$

If $\omega = 0$, *then there is an* $L \geq 0$ *such that*

$$\|\mathbb{F}_t\| \; \leq \; L \, (1 + t) \, , \qquad \forall \, t \geq 0 \, .$$

If $\omega < 0$, *then* $\|\mathbb{F}_t\|$ *is bounded.*

Proof. Let $n \in \mathbb{N}$. Let $u \in \Omega$ be supported on $[0, n]$ (i.e. $u(t) = 0$ for $t > n$) and let $y = \mathbb{F}_n u$, so y is supported on $[0, n]$. We decompose both u and y into n components supported on $[0, 1]$:

$$u \; = \; (\cdots (u_1 \underset{1}{\Diamond} u_2) \underset{2}{\Diamond} \cdots \underset{n-1}{\Diamond} u_n) \, , \qquad y \; = \; (\cdots (y_1 \underset{1}{\Diamond} y_2) \underset{2}{\Diamond} \cdots \underset{n-1}{\Diamond} y_n) \, .$$

Clearly

$$(2.5) \qquad \|u\|^2 \; = \; \sum_{k=1}^{n} \|u_k\|^2 \, , \qquad \|y\|^2 \; = \; \sum_{k=1}^{n} \|y_k\|^2 \, .$$

One can show by induction, using (1.4), (1.5) and (1.6), that for $1 \leq k \leq n$

$$y_k \; = \; \mathbb{F}_1 \, u_k + \sum_{j=1}^{k-1} \mathbb{L}_1 \, \mathbb{T}_{k-j-1} \, \mathbf{\Phi}_1 \, u_j \, .$$

Denoting

$$m_l = \begin{cases} \| \mathbb{F}_1 \|, & \text{for } l = 0, \\ \| \mathbb{L}_1 \mathbb{T}_{l-1} \Phi_1 \|, & \text{for } 1 \leq l \leq n - 1, \end{cases}$$

we have that

$$\| y_k \| \leq \sum_{j=1}^{k} m_{k-j} \| u_j \| \, .$$

By a well known estimate concerning convolutions of sequences we get, using (2.5), that

$$\| y \| \leq \left(\sum_{l=0}^{n-1} m_l \right) \cdot \| u \| \, ,$$

whence by (2.4)

$$\| \mathbb{F}_n \| \leq \| \mathbb{F}_1 \| + \| \mathbb{L}_1 \| \cdot \| \Phi_1 \| \cdot M \sum_{l=1}^{n-1} e^{\omega(l-1)} \, .$$

From here it is very elementary to complete the proof, using the fact that $\| \mathbb{F}_t \|$ is nondecreasing. ∎

For the growth rate of $\| \Phi_t \|$ and $\| \mathbb{L}_t \|$, similar results have been obtained in Weiss [12] and [13].

Remark 2.2. It follows from the previous proposition that if $\mathbb{T}$ is *exponentially stable* (i.e. $\omega < 0$ in (2.4)), then the restriction of $\mathbb{F}_\infty$ to Ω has its range in Γ and

$$\mathbb{F}_\infty \in \mathcal{L}(\Omega, \Gamma) \, .$$

Remark 2.3. For W a Hilbert space and $w \in L^1_{loc}([0, \infty), W)$, $\hat{w}$ will denote the *Laplace transform* of w, i.e. the function

$$\hat{w}(s) = \int_0^\infty e^{-st} w(t) dt \, ,$$

defined for those $s \in \mathbb{C}$ for which the above integral converges absolutely. The set of such s may be empty, and we say that $\hat{w}$ exists if it is nonempty. If $\hat{w}$ exists then $\hat{w}(s)$ is defined for all $s \in \mathbb{C}$ with Re s sufficiently large. Proposition 2.1 implies that if $u \in \Omega$ and $y = \mathbb{F}_\infty u$, then $\hat{y}$ exists. (This observation will be useful in Section 4.)

3. A short review of known results

We shall frequently need the spaces X_1 and X_{-1}, as introduced in Salomon [9], Nagel *et al* [7], or Weiss [12]. They are defined as follows. Let $\mathbb{T}$ be a semigroup on the Hilbert space X, with generator A. X_1 is $D(A)$ with the norm $\|x\|_1 = \|(\beta I - A)x\|$, where $\beta \in \rho(A)$ is fixed. This norm is equivalent to the graph norm. X_{-1} is defined as the completion of X with respect to the norm $\|x\|_{-1} = \|(\beta I - A)^{-1} x\|$, where β is as above. The restriction of $\mathbb{T}$ to X_1 is a semigroup on X_1, which is isomorphic to the original one, and $\mathbb{T}$ has an extension to a semigroup on X_{-1}, which is also isomorphic to the original semigroup. We shall denote all these semigroups by the same symbol. The generator of $\mathbb{T}$ on X_{-1} is an extension of A to X. Denoting this extension also by A, we have $A \in \mathcal{L}(X_1, X)$ and $A \in \mathcal{L}(X, X_{-1})$.

Let $\Sigma = (\mathbb{T}, \Phi, \mathbb{L}, \mathbb{F})$ be an abstract linear system, with input space U, state space X and output space Y. The generator of $\mathbb{T}$ will be denoted A. We know from Salamon [10] or Weiss [12] that there is a (unique) operator $B \in \mathcal{L}(U, X_{-1})$, called the *control operator* of the system, such that

$$(3.1) \qquad \Phi_\tau u \;=\; \int_0^\tau \mathbb{T}_{\tau-\sigma}\, B\, u(\sigma)\, d\sigma \;.$$

The integration is carried out in X_{-1} but the result is in X. The operator B is called *bounded* if it is in $\mathcal{L}(U, X)$ and *unbounded* otherwise. For any $x_0 \in X$ and any $u \in \tilde{\Omega}$, the *state trajectory* of the system, i.e. the function of $t \geq 0$ given by

$$(3.2) \qquad x(t) \;=\; \mathbb{T}_t\, x_0 + \Phi_t\, u \;,$$

is the (unique) *continuous state strong solution* of the differential equation

$$(3.3) \qquad \dot{x}(t) \;=\; A\, x(t) + B\, u(t)$$

with $x(0) = x_0$. By a continuous state strong solution of (3.3) we mean a continuous function $x : [0, \infty) \to X$ such that for any $t \geq 0$

$$x(t) - x(0) \;=\; \int_0^t [\, A x(\sigma) + B u(\sigma)\,]\, d\sigma \;.$$

That implies that, as an X_{-1}-valued function, $x(\cdot)$ is absolutely continuous, almost everywhere differentiable and (3.3) holds for a.e. $t \geq 0$. For details on this, we refer to Weiss [12].

We know from Salamon [10] or Weiss [13] that there is a (unique) operator $C \in \mathcal{L}(X_1, Y)$, called the *observation operator* of the system, such that for $x \in D(A)$ and $t \geq 0$

$$(3.4) \qquad (\mathbb{L}_\infty x)(t) \;=\; C\,\mathbb{T}_t\, x \;.$$

This fully determines $\mathbb{L}_\infty$, because of the density of $D(A)$ in X. (The operators $\mathbb{L}_\tau$ follow by (2.1).) C is called *bounded* if it can be extended to an operator in $\mathcal{L}(X, Y)$ and *unbounded* otherwise. For systems having unbounded control and/or observation operators see e.g. Curtain and Pritchard [1], Fuhrmann [3], Helton [4], Lasiecka, Lions and Triggiani [6] or Salamon [9], [11].

It is also possible to give a formula for $\mathbb{L}_\infty x$ valid for any $x \in X$, by using the *Lebesgue extension* of C, as introduced in Weiss [13]. The Lebesgue extension of an operator $C \in \mathcal{L}(X_1, Y)$ is defined by

$$(3.5) \qquad C_L x \;=\; \lim_{\tau \to 0} C\, \frac{1}{\tau} \int_0^\tau \mathbb{T}_\sigma\, x\, d\sigma \;,$$

with domain

$$D(C_L) = \{\, x \in X \mid \text{the limit in (3.5) exists}\,\}\,.$$

The space $D(C_L)$ with the norm

$$\|x\|_{D(C_L)} = \|x\| + \sup_{\tau \in (0,1]} \left\| C \frac{1}{\tau} \int_0^\tau \mathbb{T}_\sigma\, x\, d\sigma \right\|$$

becomes a Banach space, we have

$$X_1 \subset D(C_L) \subset X$$

with continuous embeddings, and $C_L \in \mathcal{L}(D(C_L), Y)$. For any $x \in X$ and any $t \geq 0$, we have that $\mathbb{T}_t\, x \in D(C_L)$ if and only if $\mathbb{L}_\infty x$ has a Lebesgue point in t. That implies

$$(3.6) \qquad\qquad \mathbb{T}_t\, x \in D(C_L), \qquad \text{for a.e. } t \geq 0\,.$$

Further, $\mathbb{L}_\infty$ has the following representation:

$$(3.7) \qquad\qquad (\mathbb{L}_\infty x)(t) = C_L \mathbb{T}_t\, x\,, \qquad \text{for a.e. } t \geq 0\,.$$

That implies that we can rewrite (3.5) in the form

$$(3.8) \qquad\qquad C_L x = \lim_{\tau \to 0} \frac{1}{\tau} \int_0^\tau (\mathbb{L}_\infty x)(\sigma)\, d\sigma\,,$$

and $x \in D(C_L)$ if and only if the limit exists. For details on this, see Weiss [13].

We know from Salamon [10] that for sufficiently smooth input functions u, the function $\mathbb{F}_\infty u$ is given by

$$(3.9) \qquad (\mathbb{F}_\infty u)(t) = C\left[\int_0^t \mathbb{T}_{t-\sigma} B\, u(\sigma)\, d\sigma - (\beta I - A)^{-1} B\, u(t) \right] + \mathbf{H}(\beta)\, u(t)\,,$$

for any $t \geq 0$. In the above formula, one has to assume that u belongs to the space $W^{1,2}_{0,loc}([0,\infty), U)$, which consists of those functions $u : [0,\infty) \to U$ which admit the representation

$$u(t) = \int_0^t \gamma(\sigma)\,d\sigma\,,$$

for some $\gamma \in L^2_{loc}([0,\infty), u)$ (in particular, $u(0) = 0$). In (3.9), $\mathbf{H}$ denotes an analytic $\mathcal{L}(U, Y)$-valued function on $\rho(A)$ called the *transfer function* of Σ and β is a fixed element of $\rho(A)$. Formula (3.9) completely determines $\mathbb{F}_\infty$, because $W^{1,2}_{0,loc}([0,\infty), U)$ is dense in $\tilde{\Omega} = L^2_{loc}([0,\infty), U)$. The operators $\mathbb{F}_\tau$ follow by (2.1). (Salamon wrote (3.9) in a different, but equivalent way.)

In the paper Weiss [14], it is proved that formula (3.9) remains valid for any $u \in L^2_{loc}([0,\infty), U)$ and for a.e. $t \geq 0$, if C is replaced by its Lebesgue extension C_L. Thus, we get a representation for the operator $\mathbb{F}_\infty$ valid on all of its domain. (The paper [14] is written in the Banach space and L^p context.)

The two terms inside the big paranthesis on the right-hand side of (3.9) do not have to belong to $D(C_L)$, only their difference. However, in most cases of practical interest the regularity assumption (presented in more detail in the next section) holds, implying that the operator $C_L(\beta I - A)^{-1}B$ makes sense and is an element of $\mathcal{L}(U,Y)$. In this case, denoting

$$D = \mathbf{H}(\beta) - C_L(\beta I - A)^{-1}B\,,$$

it is easy to check that D is independent of β and we get from (3.9) (the version with C_L instead of C) that

$$(3.10) \qquad (\mathbb{F}_\infty u)(t) = C_L \int_0^t \mathbb{T}_{t-\sigma}\, B\, u(\sigma)\, d\sigma + D\, u(t)\,,$$

for any $u \in \tilde{\Omega}$ and a.e. $t \geq 0$, just as for finite dimensional systems.

Our aim in the next section is to give a direct proof of (3.10). By that we mean that we do not derive (3.10) as a special case of (3.9) (the version with C_L instead of C), but prove it directly using the regularity assumption. Thus the proof becomes considerably simpler.

Example 1.2 *(continued)*. For the semigroup of this system, X_1 consists of the absolutely continuous functions x on $[-h,0]$ with $x(0) = 0$ and $x' \in L^2[-h,0]$. X_{-1} is the completion of $L^2[-h,0]$ with respect to the norm

$$\|x\|_{-1} = \left\| \int_0^\zeta x(\sigma)d\sigma \right\|_{L^2[-h,0]}\,.$$

Since U is one dimensional, the control operator of Σ can be identified with an element $b \in X_{-1}$, namely with the "delta function" at 0. More precisely, b is the limit in X_{-1} of the sequence (b_n), where $b_n \in X$ is defined by

$$b_n(\zeta) = \begin{cases} n, & \text{for } \zeta \geq -\frac{1}{n}, \\ 0, & \text{for } \zeta < -\frac{1}{n}. \end{cases}$$

The observation operator of Σ is an element $c \in X_1^*$, defined by

$$cx = x(-h)\,.$$

The Lebesgue extension of c, c_L is defined for those $x \in L^2[-h,0]$ which have a Lebesgue point at $-h$, and for such x

$$c_L x = \lim_{\tau \to 0} \frac{1}{\tau} \int_0^\tau x(-h+\zeta)d\zeta\,.$$

For details on how to get the control operator and the observation operator from $\mathbb{T}$, Φ and $\mathbb{L}$ in general, see Weiss [12] and [13].

4. The representation theorem

Let U, X, Y be Hilbert spaces, let the function spaces Ω, $\tilde{\Omega}$, Γ and $\tilde{\Gamma}$ be defined as in the previous section and let $\Sigma = (\mathbb{T}, \Phi, \mathbb{L}, \mathbb{F})$ be an abstract linear system on Ω, X and Γ. For any $v \in U$, χ_v will denote the constant function on $[0, \infty)$ equal to v everywhere, so $\chi_v \in \tilde{\Omega}$. The function

$$(4.1) \qquad\qquad\qquad y_v = \mathbb{F}_\infty \chi_v$$

will be called the *step response* of Σ corresponding to v. By the definition of $\mathbb{F}_\infty$, $y_v \in \tilde{\Gamma}$. For example, if Σ is finite dimensional and described by equations (1.1), then

$$y_v(t) = C \int_0^t e^{A\sigma} B v \, d\sigma + D v \ .$$

In the finite dimensional case, y_v is continuous (actually analytic) in t. (One might be tempted to think of y_v as a function on $(-\infty, \infty)$ and to object that y_v has a jump at $t = 0$. But y_v is defined only on $[0, \infty)$ and we identify functions equal a.e., so the jump disappears.) In the infinite dimensional case, y_v does not have to be continuous (think of the step response of the delay line of Example 1.2).

Definition 4.1. Let Σ be an abstract linear system, with input space U and output space Y. We say that Σ is *regular* if for any $v \in U$, the corresponding step response y_v (given by (4.1)) has a Lebesgue point at 0, i.e., the following limit exists (in Y):

$$(4.2) \qquad\qquad\qquad Dv = \lim_{\tau \to 0} \frac{1}{\tau} \int_0^\tau y_v(\sigma) \, d\sigma \ .$$

In that case, the operator D defined by (4.2) is called the *feedthrough operator* of the system Σ.

Most systems arising in practice (for example from PDE's) actually satisfy a much stronger condition than (4.2), namely, the limit $\lim_{\tau \to 0} y_v(\tau)$ exists for any $v \in U$.

Remark 4.2. For any $\tau > 0$, the operator D_τ given by

$$D_\tau v = \frac{1}{\tau} \int_0^\tau y_v(\sigma) \, d\sigma$$

is in $\mathcal{L}(U, Y)$. If Σ is regular, then it follows by the uniform boundedness principle that the feedthrough operator is bounded:

$$D \in \mathcal{L}(U, Y) \ .$$

The next proposition is not needed in the proof of the representation theorem, but it gives some insight into the concept of regularity.

Proposition 4.3. *Let* $\Sigma = (\mathbb{T}, \Phi, \mathbb{L}, \mathbb{F})$ *be an abstract linear system, with input space* U, *state space* X *and output space* Y. *Let* A *be the generator of* $\mathbb{T}$, *let* B *be the control* *operator of Σ and let C be the observation operator of Σ. We denote by C_L the Lebesgue extension of C. The following conditions are equivalent:*

   (ı) Σ is regular,

   (ıı) for some $s \in \rho(A)$ and any $v \in U$, $(sI - A)^{-1} Bv \in D(C_L)$,

   (ııı) $C_L(sI - A)^{-1}B$ is an analytic $\mathcal{L}(U, Y)$-valued function of s on $\rho(A)$.

Proof. Let us prove that $(ı)$ and $(ıı)$ are equivalent. From (2.3) we have that for any $v \in U$, any $\tau \geq 0$ and a.e. $t \geq 0$

$$y_v(\tau + t) = (\mathbb{L}_\infty \, \Phi_\tau \chi_v)(t) + y_v(t),$$

where y_v is as in (4.1). By (3.7) this becomes

(4.3) $$y_v(\tau + t) = C_L \, \mathbb{T}_t \, \Phi_\tau \chi_v + y_v(t) .$$

For any $v \in U$, y_v has a Laplace transform $\hat{y}_v$ (see Remark 2.3) and from Proposition 2.1 we can deduce that if $\omega \geq 0$ is such that (2.4) holds, then the Laplace integral defining $\hat{y}_v(s)$ is absolutely convergent for any complex s with Re $s > \omega$. Using some basic facts about semigroups (see e.g. Pazy [8]), we get that for s as described above and any $v \in U$

$$e^{s\tau}\hat{y}_v(s) - e^{s\tau} \int_0^\tau e^{-s\sigma} y_v(\sigma)d\sigma = C(sI - A)^{-1} \Phi_\tau \chi_v + \hat{y}_v(s) .$$

(To see that C_L can be taken out of the Laplace integral, consider the Laplace transform of $C \mathbb{T}_t x$ for $x \in D(A)$ and then extend by continuity.) For $\tau > 0$ we get, by rearranging the last formula and using (3.1)

(4.4) $$\frac{e^{s\tau} - 1}{\tau} \hat{y}_v(s) = C\frac{1}{\tau} \int_0^\tau \mathbb{T}_\sigma (sI - A)^{-1} Bv d\sigma + e^{s\tau}\frac{1}{\tau} \int_0^\tau e^{-s\sigma} y_v(\sigma)d\sigma .$$

The left-hand side of (4.4) tends to $s\hat{y}_v(s)$, as $\tau \to 0$. Therefore, if one of the two terms on the right-hand side has a limit for $\tau \to 0$, the other one has a limit too. By the definition of C_L (see (3.5)), the first term on the right hand-side of (4.4) has a limit for $\tau \to 0$ if and only if $(sI - A)^{-1}Bv \in D(C_L)$. The second term on the right-hand side of (4.4) has a limit for $\tau \to 0$ if and only if y_v has a Lebesgue point at 0, as is easy to show. Therefore $(ı)$ and $(ıı)$ are equivalent.

    Let us prove that $(ıı)$ and $(ııı)$ are equivalent. It is obvious that $(ııı)$ implies $(ıı)$. Conversely, suppose $(ıı)$ holds. Taking limits in (4.4) we get, using (4.2), that for $s \in \mathbb{C}$ with large real part

(4.5) $$s\hat{y}_v(s) = C_L(sI - A)^{-1} Bv + Dv .$$

Since $\hat{y}_v(s)$ and Dv both depend continuously on v, it follows that for s with large real part

(4.6) $$C_L(sI - A)^{-1}B \in \mathcal{L}(U, Y)$$

(this could have been obtained also with the closed graph theorem, using that C_L is bounded from $D(C_L)$ to Y).

Let $s_0 \in \mathbb{C}$ be such that (4.6) holds with s_0 instead of s and let $s \in \rho(A)$. Then clearly

$$C_L(s_0 I - A)^{-1} B + C(s_0 - s)(sI - A)^{-1}(s_0 I - A)^{-1} B \in \mathcal{L}(U, Y) .$$

Factoring out C_L and applying the resolvent identity, we get that $(sI - A)^{-1} B$ has its range in $D(C_L)$ and

$$C_L(sI - A)^{-1} B = C_L(s_0 I - A)^{-1} B + C(s_0 - s)(sI - A)^{-1}(s_0 I - A)^{-1} B .$$

Using once again the resolvent identity, the last formula can be rewritten as

$$C_L(sI - A)^{-1} B = C_L(s_0 I - A)^{-1} B + (s_0 - s)C(s_0 I - A)^{-2} B$$
$$+ (s_0 - s)^2 C(s_0 I - A)^{-1}(sI - A)^{-1}(s_0 I - A)^{-1} B .$$

Since $(s_0 I - A)^{-1} B \in \mathcal{L}(U, X)$ and $C(s_0 I - A)^{-1} \in \mathcal{L}(X, Y)$, all terms on the right-hand side are analytic $\mathcal{L}(U, Y)$-valued functions of s on $\rho(A)$. Hence ($\imath\imath\imath$) holds. ∎

Lemma 4.4. *Let U be a Hilbert space and $u \in L^2_{loc}([0, \infty), U)$. Then for almost every $t \geq 0$*

$$(4.7) \qquad \lim_{\tau \to 0} \frac{1}{\tau} \int_0^\tau \|u(t + \sigma) - u(t)\|^2 \, d\sigma = 0 .$$

Obviously it is sufficient to prove the statement for $u \in L^2([0, 1], U)$. For $u \in L^1([0, 1], U)$ and exponent 1 instead of 2 under the integral, it is proved in Diestel and Uhl [2, p. 49]. For our case, the proof is very similar and we omit the details.

We now state the representation theorem for $\mathbb{F}$, already announced as (3.10).

Theorem 4.5. *With the notation of Proposition 4.3, suppose Σ is regular, let D be its feedthrough operator and let $u \in L^2_{loc}([0, \infty), U)$. Then for almost every $t \geq 0$*

$$(4.8) \qquad \int_0^t \mathbb{T}_{t-\sigma} B u(\sigma) \, d\sigma \in D(C_L)$$

and

$$(4.9) \qquad (\mathbb{F}_\infty u)(t) = C_L \int_0^t \mathbb{T}_{t-\sigma} B u(\sigma) \, d\sigma + D u(t) .$$

Proof. Let $y = \mathbb{F}_\infty u$, so $y \in L^2_{loc}([0, \infty), Y)$. Both u and y are equivalence classes of functions modulo equality almost everywhere. We choose one representative for u and one for y, and for the rest of this proof we consider u and y to be well defined in every point $t \geq 0$.

Let $\mathcal{T}$ be the set of points $t \in [0, \infty)$ where the following two conditions are satisfied:

($\imath$) y has a Lebesgue point at t and

$$y(t) = \lim_{\tau \to 0} \frac{1}{\tau} \int_0^\tau y(t + \sigma) d\sigma ,$$

($\imath\imath$) the equality (4.7) holds.

By a well known theorem on Lebesgue points (see Diestel and Uhl [2, p. 49]) and by Lemma 4.4, almost every $t \geq 0$ belongs to $\mathcal{T}$. We will show that (4.8) and (4.9) hold for any $t \in \mathcal{T}$.

Let $t \in \mathcal{T}$. Let $\epsilon \in L^2_{loc}([0, \infty), U)$ be defined by

$$\epsilon(\sigma) = u(t + \sigma) - u(t) ,$$

so with notation from (4.1)

$$u = u \underset{t}{\diamondsuit} (\chi_{u(t)} + \epsilon) .$$

The functional equation (2.3) implies that for a.e. $\sigma \geq 0$

$$
\begin{aligned}
(4.10) \qquad y(t + \sigma) &= \left[\mathbb{L}_\infty \Phi_t u + \mathbb{F}_\infty (\chi_{u(t)} + \epsilon) \right] (\sigma) \\
&= (\mathbb{L}_\infty \Phi_t u)(\sigma) + y_{u(t)}(\sigma) + (\mathbb{F}_\infty \epsilon)(\sigma) ,
\end{aligned}
$$

where $y_{u(t)}$ is the step response corresponding to $u(t)$, as in (4.1).

Let us show that

$$(4.11) \qquad \lim_{\tau \to 0} \frac{1}{\tau} \int_0^\tau (\mathbb{F}_\infty \epsilon)(\sigma) d\sigma = 0 .$$

For any $\tau \in (0, 1]$ we have that

$$
\begin{aligned}
\left\| \frac{1}{\tau} \int_0^\tau (\mathbb{F}_\infty \epsilon)(\sigma) d\sigma \right\| &\leq \frac{1}{\tau} \int_0^\tau \|(\mathbb{F}_1 \epsilon)(\sigma)\| d\sigma \\
&\leq \left(\frac{1}{\tau} \int_0^\tau \|(\mathbb{F}_1 \epsilon)(\sigma)\|^2 d\sigma \right)^{\frac{1}{2}} \\
&\leq \|\mathbb{F}_1\| \cdot \left(\frac{1}{\tau} \int_0^\tau \|\epsilon(\sigma)\|^2 d\sigma \right)^{\frac{1}{2}} .
\end{aligned}
$$

By the definition of ϵ and by condition ($\imath\imath$) above, the last expression tends to 0 as $\tau \to 0$. Thus we have proved (4.11).

Let us integrate both sides of (4.10) from 0 to τ, divide everything by τ and take limits for $\tau \to 0$. By condition ($\imath$) above, by the regularity assumption (4.2) and by (4.11) we get that $\mathbb{L}_\infty \Phi_t u$ has a Lebesgue point in 0 and

$$y(t) = \lim_{\tau \to 0} \frac{1}{\tau} \int_0^\tau (\mathbb{L}_\infty \Phi_t u)(\sigma) d\sigma + Du(t) .$$

By (3.8) we get that $\Phi_t u \in D(C_L)$ and

$$y(t) = C_L \Phi_t u + D u(t) .$$

Because of the representation (3.1), we have obtained (4.8) and (4.9). ∎

We are now able to give the rigorous version of the main result stated at the beginning of the paper.

Theorem 4.6. *Let $\Sigma = (\mathbb{T}, \Phi, \mathbb{L}, \mathbb{F})$ be an abstract linear system, with input space U, state space X and output space Y. Let A be the generator of $\mathbb{T}$, let B be the control operator of Σ and let C be the observation operator of Σ. We denote by C_L the Lebesgue extension of C. Suppose Σ is regular and let D be given by (4.2).*
Then for any $x_0 \in X$ and any $u \in L^2_{loc}([0,\infty), U)$, the functions $x : [0,\infty) \to X$ and $y \in L^2_{loc}([0,\infty), Y)$ defined by

$$\text{(4.12)} \qquad \begin{cases} x(t) &= \mathbb{T}_t x_0 + \Phi_t u , \\ y &= \mathbb{L}_\infty x_0 + \mathbb{F}_\infty u , \end{cases}$$

satisfy the equations

$$\text{(4.13)} \qquad \begin{cases} \dot{x}(t) &= A x(t) + B u(t) , \\ y(t) &= C_L x(t) + D u(t) , \end{cases}$$

in the following sense. The function x is a continuous state strong solution of the first equation in (4.13) (see Section 3, after (3.2)), in particular, x satisfies this equation for a.e. $t \geq 0$. We have $x(t) \in D(C_L)$ for a.e. $t \geq 0$ and the second equation in (4.13) is also satisfied for a.e. $t \geq 0$.
The pair (x, y) given by (4.12) is the unique solution of (4.13) (in the above sense) which satisfies the initial condition $x(0) = x_0$.

Proof. The fact that x is the unique solution of the first equation in (4.13) with $x(0) = x_0$ has been already discussed in Section 3.
From (3.6), (4.8) and (3.1) we get that $x(t) \in D(C_L)$, for a.e. $t \geq 0$. For those $t \geq 0$ where both (3.7) and (4.9) hold (so for a.e. $t \geq 0$) we have, taking (3.1) into account,

$$\begin{aligned} y(t) &= C_L \mathbb{T}_t x_0 + C_L \int_0^t \mathbb{T}_{t-\sigma} B u(\sigma) \, d\sigma + D u(t) \\ &= C_L \left(\mathbb{T}_t x_0 + \Phi_t u \right) + D u(t) \\ &= C_L x(t) + D u(t) . \end{aligned}$$

We have obtained the second equation in (4.13). It is obvious that this equation determines y uniquely in terms of x and u. ∎

Corresponding to the time domain description of Σ given in Theorem 4.6, there is a description of Σ in terms of Laplace transforms, as follows.

Proposition 4.7. *With the notation of Theorem 4.6, if $u \in L^2([0,\infty), U)$ then $\hat{y}$, the Laplace transform of y, exists and for $s \in \mathbb{C}$ with $\mathrm{Re}\ s$ sufficiently large*

$$(4.14) \qquad \hat{y}(s) = C(sI - A)^{-1} x_0 + \mathbf{H}(s)\,\hat{u}(s)\,,$$

where $\mathbf{H}$ *is given by*

$$\mathbf{H}(s) = C_L(sI - A)^{-1} B + D\,.$$

This is a consequence of (4.13) and Remark 2.3, but in the proof one has to be careful to show that C_L can be taken out of the Laplace integral. For details see Weiss [14]. It is easy to see that the function $\mathbf{H}$ appearing above is the transfer function of Σ, as defined in Section 3.

It can be shown using Weiss [13, Proposition 4.7], that for any $v \in U$

$$(4.15) \qquad \lim_{\lambda \to \infty} \mathbf{H}(\lambda)v = Dv\,,$$

where λ is real. If Σ is not regular then (4.14) still holds but $\mathbf{H}$ is more complicated to express. The limit in (4.15) may not exist. For details see again Weiss [14].

Remark 4.8. Everything in this paper remains valid, with very minor modifications, if we replace Hilbert spaces with Banach spaces and L^2 with L^p, where $p \in [1, \infty)$.

Acknowledgments. The author is grateful to Dietmar Salamon for enlightening discussions and to Ruth Curtain for helpful suggestions.

REFERENCES

[1] R.F. Curtain and A.J. Pritchard, "Infinite Dimensional Linear Systems Theory," Lecture Notes in Control and Information Sciences, Vol. 8, Springer-Verlag, Berlin, 1978.

[2] J. Diestel and J.J. Uhl, "Vector Measures," Amer. Math. Soc. Surveys, Vol. 15, Amer. Math. Soc., Providence, Rhode Island, 1977.

[3] P.A. Fuhrmann, "Linear Systems and Operators in Hilbert Space," McGraw-Hill, New York, 1981.

[4] J.W. Helton, *Systems with Infinite-Dimensional State Space: The Hilbert Space Approach*, Proceedings of the IEEE 64 (1976), 145–160.

[5] R.E. Kalman, P.L. Falb, M.A. Arbib, "Topics in Mathematical Systems Theory," McGraw-Hill, New York, 1969.

[6] I. Lasiecka, J.L. Lions, R. Triggiani, *Non homogeneous boundary value problems for second order hyperbolic operators*, J. Math. pures et appl. 65 (1986), 149–192.

[7] R. Nagel (editor), *One-parameter Semigroups of Positive Operators*, in "Lecture Notes in Mathematics vol. 1184," Springer-Verlag, New York, 1986.

[8] A. Pazy, "Semigroups of Linear Operators and Applications to P.D.E.'s," Applied Mathematical Sciences Vol. 44, Springer-Verlag, New York, 1983.

[9] D. Salamon, "Control and Observation of Neutral Systems," Research Notes in Math. Vol. 91, Pitman, London, 1984.

[10] D. Salamon, *Realization theory in Hilbert space*, University of Wisconsin-Madison, Technical Summary Report **2835** (1985). Revised version submitted in 1988

[11] D. Salamon, *Infinite dimensional systems with unbounded control and observation: A functional analytic approach*, Transactions of the A.M.S. 300 (1987), 383-431.

[12] G. Weiss, *Admissibility of unbounded control operators*, SIAM J. Control & Optim. 27 (1989).

[13] G. Weiss, *Admissible observation operators for linear semigroups*, Israel J. Math. **64** (1989).

[14] G. Weiss, *The representation of abstract linear systems on Banach spaces*, in preparation.

[15] Y. Yamamoto, *Realization theory of infinite-dimensional linear systems, Part I*, Math. Systems Theory **15** (1981), 55–77.

George Weiss
Department of Theoretical Mathematics
The Weizmann Institute
Rehovot 76100
Israel

International Series of
Numerical Mathematics, Vol. 91
© 1989 Birkhäuser Verlag Basel

Equivalence of external and internal stability
for a class of infinite-dimensional systems

Yutaka Yamamoto

Department of Applied Systems Science
Kyoto University

Abstract. It is known that the notion of L^2-input/output stability does *not* necessarily imply
exponential stability of the internal realization. Similarly, for distributed parameter systems,
stability is not necessarily determined by the location of spectrum. On the other hand, there
are a number of systems for which stability *is* determined by the location of spectrum. Delay-
differential systems provide an example. However, not much has been known on the precise
characterization of such a class in terms of the external behavior, e.g., transfer functions. This
paper provides a class of impulse response matrices in which 1) internal stability agrees with
external L^2–input/output stability, and 2) internal (exponential) stability is determined by
the location of spectrum. This also implies that for a certain class of systems, the so-called
small-gain theorem yields internal exponential stability.

Keywords. L^2-input/output stability, internal exponential stability, H^∞ space, transfer functions, small-gain theorem.

1. Introduction

It is well known ([17]) that the location of spectrum does *not* necessarily determine
stability of infinite-dimensional systems. The difficulty here is that the multiplicity of
eigenvalues is not globally bounded, thereby allowing a response of type $t^{n_i} e^{\lambda_i t}$ where n_i
may tend to ∞ as $i \to \infty$. For similar reasons, external stability (for example, the notion
of L^2-input/output stability [3]) need not imply internal stability. For example, consider
the impulse response function $A(t) = e^{-t} \sin(e^{2t})$. The Laplace transform of the function
belongs to $H^\infty(\mathbf{C}^+)$, so that this impulse response is L^2-input/output stable in the sense
of [3]. However, the decay rate e^{-t} has no control over the high-frequency behavior, and
one can easily construct a sequence of outputs which converge on any bounded interval
(therefore, in any practical sense, they must be regarded convergent), but their growth
order approach e^t. Apparently such behavior would not occur for finite-dimensional systems
where the condition $\hat{A}(s) \in H^\infty$ simultaneously guarantees mild behavior of the high-
frequency response.

This gap attracts the recent research interest, and there are now several attempts to
establish equivalence between the two notions of stability. For example, Callier and Winkin
[2] gave a subclass of the ring of stable impulse responses $\mathcal{A}$ in which this equivalence is
valid. The recent work of Jacobson and Nett [5] gave the equivalence under the condition
of stabilizability and detectability.

This paper presents a different type of approach, which is a continuation of [15]. We restrict our attention to a class of infinite-dimensional systems which admits a special type of
fractional representation. This class is introduced in [10], [13], and is called *pseudo-rational*.
While restricting the class, we do not impose the condition of stabilizability/detectability.
This has the advantage that we need not place a condition on the internal realization

when we deal with the external behavior. Therefore, the class of systems in which the aforementioned equivalence holds is characterized solely in terms of external behavior.

The paper is organized as follows: Some preliminary results on pseudo-rational impulse responses are given in Section 2. Then a subclass, called $\mathcal{R}$ of pseudo-rational impulse responses is introduced in Section 3, and our main results on the equivalence of internal and external stability are given. We prove

1. *in the class $\mathcal{R}$, the canonical realization of an impulse response matrix is exponentially stable iff the spectrum of the system belongs strictly in the open left-half complex plane;*

2. *this realization is exponentially stable iff the transfer function matrix belongs to $H^\infty(\mathbf{C}^+)$;*

3. *therefore, the well known small-gain theorem ([3]) assures internal stability, not just L^2 input/output stability.*

Notation and convention.

Throughout we assume that the reader is familiar with materials on Schwartz distributions as can be found in the standard textbooks [8] or [9].

$\mathcal{D}'_+$ denotes the space of distributions on $\mathbf{R}$ with support bounded on the left. $\mathcal{E}'(\mathbf{R}^-)$ denotes the space of distributions on $\mathbf{R}$ with *compact* support contained in $(-\infty, 0]$). They constitute a convolution algebra with identity δ (the Dirac delta distribution at the origin). The delta distribution at point a will be denoted by δ_a, and its derivative will be denoted by δ'_a. Convolution will be denoted by $*$, as usual.

By Γ we denote the space $L^2_{loc}[0, \infty)$ of locally square integrable functions on $[0, \infty)$. This is a Fréchet space with generating seminorms

$$\|\psi\|_{[0,a]} := \{\int_0^a |\psi(t)|^2 dt\}^{1/2}, \ a > 0.$$

This space is equipped with the left-shift semigroup:

$$(1) \qquad\qquad (\sigma_t \gamma)(\tau) := \gamma(\tau + t).$$

As usual, $H^\infty(\mathbf{C}^+)$ (or simply, H^∞) denotes the space of functions analytic on the open right-half complex plane and having nontangential limit almost everywhere on the imaginary axis such that this limit function is essentially bounded. For $\varphi \in H^\infty$, $\|\varphi\|_\infty$ denotes its H^∞-norm.

We say that a square matrix Q over $\mathcal{E}'(\mathbf{R}^-)$ is of *normal type* if i) $\det Q$ is invertible with respect to convolution over $\mathcal{D}'_+$, ii) supp $Q^{-1} \subset [0, \infty)$, and iii) ord $(\det Q)^{-1} = -$ord $(\det Q)$, where ord α denotes the *order* of a distribution α ([8]).

2. Pseudo-rational impulse responses

Let A be a $p \times m$ impulse response matrix. It can be a measure (i.e., distributions of order 0), but we do not allow any higher irregularity as that of differentiation.

Let us give a definition of pseudo-rationality:

Definition 2.1. *Let A be a p×m impulse response matrix. It is said to be pseudo-rational if it can be written in the form*

$$(2) \qquad\qquad A = Q^{-1} * P$$

for some matrices Q and P over $\mathcal{E}'(\mathbf{R}^-)$ where the $p \times p$ matrix Q is of normal type.

When the impulse response matrix A is pseudo-rational, we can always associate with it a Fuhrmann-type realization $\Sigma^{Q,P}$ which is topologically observable in bounded time ([10]). This realization $\Sigma^{Q,P}$ is given as follows:

- State Space : $X^Q := \{x(\tau) \in L^2_{loc}[0,\infty) : \pi(Q * x) = 0\}$,
  where $\pi\varphi := \varphi|_{[0,\infty)}$.
- State Transition:

$$(3) \qquad\qquad \frac{d}{dt}x_t(\cdot) = Fx_t(\cdot) + A(\cdot)u(t)$$

$$(4) \qquad\qquad y(t) = x_t(0)$$

$$(5) \qquad\qquad Fx(\tau) := \frac{dx}{d\tau}, \qquad D(F) = W^1_{2,loc}[0,\infty) \cap X^Q.$$

The space X^Q is closed under the left shifts (1), and this is the semigroup generated by the operator F. Therefore, the free state transition of $\Sigma^{Q,P}$ is governed by σ_t. It is known that the class of pseudo-rational impulse responses contains those of all delay-differential systems ([10], [12]). Furthermore, the above realization coincides with the well-known M_2-model for such systems ([12]) which has been employed also for neutral systems ([1]).

The following facts are known for this realization:

Facts 2.2. ([10], [11], [13])

1. *The space X^Q is isomorphic to a Hilbert space. The norm of this space is given by the L^2-norm $\| \cdot \|_{[0,T]}$ for some bounded interval $[0,T]$.*
2. *The system $\Sigma^{Q,P}$ is always observable.*
3. *$\Sigma^{Q,P}$ is approximately reachable if and only if the pair (Q,P) is approximately left coprime, i.e.,*

$$(6) \qquad\qquad Q * X_n + P * Y_n \to \delta I$$

   in the space $\mathcal{E}'(\mathbf{R}^-)$. When this condition holds, the system $\Sigma^{Q,P}$ gives a canonical realization of A.
4. *The spectrum of F is given by*

$$(7) \qquad\qquad \sigma(F) = \{\lambda \in \mathbf{C} : \hat{Q}(\lambda) = 0\},$$

   and they are all eigenvalues. Furthermore, the generalized eigenspace corresponding to each λ is finite-dimensional, and it is spanned by elements of type

$$\{e^{\lambda t}, \dots, t^m e^{\lambda t}\}.$$

When the above condition 3) is not satisfied, the realization $\Sigma^{Q,P}$ is not canonical. In this case, the canonical realization is given by changing the state space to

(8) $$X_A := \overline{\operatorname{im} A*} = \overline{\{A * \omega : \omega \in L^2[-a, 0]\}}$$

for some $a > 0$. This realization, given by replacing X^Q by X_A, is denoted by Σ_A.

The space X_A is a closed subspace of X^Q, so that the above property 1) also holds for this realization. In any case, in view of property 1), we can unambiguously speak about the exponential stability of such systems.

Definition 2.3. *The system $\Sigma^{Q,P}$ is exponentially stable if its transition semigroup σ_t is exponentially stable, i.e.,*

$$\|\sigma_t\| \leq M e^{-\beta t}, \ \forall t \geq 0.$$

The same definition applies to the canonical realization Σ_A of A, even when (Q, P) is not approximately left coprime. In a more concrete term, the canonical realization Σ_A is exponentially stable iff

(9) $$\|\sigma_t(A * \omega)\|_{[0,T]} \leq M e^{-\beta t} \|A * \omega\|_{[0,T]}, \ \forall t \geq 0$$

for some $T, M, \beta > 0$.

Concerning the stability of $\Sigma^{Q,P}$, the following result is given in [15]:

Theorem 2.4.

1. *If $\Sigma^{Q,P}$ is exponentially stable, then $\hat{Q}^{-1}, \hat{A} \in H^\infty$.*
2. *Conversely, if $s^r / \det \hat{Q}(s) \in H^\infty$ then $\Sigma^{Q,P}$ is exponentially stable.*

The difficulties here are

1. the higher-order condition $(s^r / \det \hat{Q}(s) \in H^\infty)$ is required;
2. since an (approximately) coprime factorization is not known to exist, one cannot directly discuss the stability of the canonical realization Σ_A.

In view of these, we now pose the question: *Is there a subclass in which stability is determined either by the location of spectrum or by the condition $\hat{A} \in H^\infty$?*

3. Class $\mathcal{R}$

We first confine ourselves to the single-input/single-output case.

Definition 3.1. *An impulse response A belongs to the class $\mathcal{R}$ if and only if there exists $q, p \in \mathcal{E}'(\mathbf{R}^-)$, with q of normal type such that*

1. $A = q^{-1} * p$;
2. $\operatorname{ord} q^{-1}|_{(T,\infty)} < \operatorname{ord} q^{-1}$ *for some $T > 0$.*

This means that the regularity of q^{-1} becomes higher after $T > 0$. It is easy to see that impulse responses of retarded systems satisfy this condition. (Consider, for example, the impulse response $A(t) = \delta/(\delta'_{-1} - \delta)$.)

Our target is the following theorem:

Theorem 3.2. *Suppose that $A \in \mathcal{R}$. Then the canonical realization of A is exponentially stable iff either one of the following conditions holds:*

1. *The poles of $\hat{A}(s)$ belong to the strict left-half plane $\{s \in \mathbf{C} : \Re\, s < -c\}$ for some $c > 0$.*
2. *$\hat{A}$ belongs to H^∞.*

As a corollary, we can give the small-gain theorem with internal stability as follows:

Corollary 3.3. *(Small-Gain Theorem with Internal Stability) Suppose that $A \in \mathcal{R}$ and that each entry of A is a function in $L^2_{loc}[0, \infty)$. Suppose also that $A \in H^\infty$ and*

$$(10) \qquad\qquad\qquad \|\hat{A}\|_\infty < 1.$$

Then the closed-loop system given by

$$(11) \qquad\qquad\qquad y = A * e,$$
$$(12) \qquad\qquad\qquad e = u - y.$$

is also exponentially stable.

Remark 3.4. The condition $A \in L^2_{loc}[0, \infty)$ is necessary to assure that the closed-loop impulse response $(I + A)^{-1} * A$ (the inverse being taken in the sense of convolution) belongs to the class $\mathcal{R}$. Otherwise, $A = \delta_1 \in \mathcal{R}$ (the Dirac distribution placed at point 1) gives a counterexample to $(I + A)^{-1} * A \in \mathcal{R}$. The rest of the arguments is similar to the usual small-gain theorem ([3]) involving the condition (10).

4. Proof of Theorem 3.2

Let us first give an outline of the proof.

Outline of Proof.

1. Show that the shift-semigroup σ_t is a **compact** operator for any $t \geq T$.
2. The spectrum of $\sigma_t = e^{\sigma(F)t} \cup \{0\}$ for such t.
3. Show that when $\sigma(F) \subseteq \{\lambda : \Re\, \lambda < -c, c > 0\}$, $\|\sigma_t\| < 1$, and $(\sigma_t)^n \to 0$ as $n \to \infty$.

The above outline is motivated by the proof of the stability theorem in ([4], Corollary 7.4.1); what is new here is that we have given an *externally characterized* class $\mathcal{R}$ in which σ_t becomes a compact operator for some $t > 0$.

Let us first complete the easier part 2), 3).

Proof of 2). Since σ_t is compact, $\sigma(\sigma_t)$ consists only of point spectrum and $\{0\}$ ([7]). For the correspondence of point spectrum,

$$P_\sigma(\sigma_t) = e^{\sigma(F)t} [\cup\{0\}]$$

is well known ([16]). Thus 2) follows. ■

Proof of 3). If $\sigma(F) \subseteq \{\lambda : \Re \lambda < -c, c > 0\}$, then by 2) above, the spectral radius $r(\sigma_t) < 1$. Hence $\limsup_{n \to \infty} \|(\sigma_t)^n\|^{1/n} < 1$. This, along with the semigroup property ([4], Lemma 7.4.2), implies

$$\|\sigma_t\| \leq Me^{-\beta t}. \blacksquare$$

Proof of 1). Fix any $t > T$, and consider the state transition σ_t. It is easier to work with the adjoint $\sigma'_t : (X^q)' \to (X^q)'$.

Facts 4.1. ([10],[13]) *There exists $a > 0$ such that*

1. $(X^q)' \cong L^2[-a, 0] / \overline{(q * L^2[-a, 0] \cap L^2[-a, 0])}$
2. $\sigma'_t x = q * \pi(q^{-1} * x(\cdot + t))$ *for* $x \in L^2[-a, 0] / \overline{(q * L^2[-a, 0] \cap L^2[-a, 0])}$.

From 2) above, we see that $\sigma'_t x = q * \pi(q^{-1}(\cdot + t) * x)$. Observe now that

$$\text{ord } \pi(q^{-1}(\cdot + t) * x) < \text{ord } \pi(q^{-1} * x)$$

by $A \in \mathcal{R}$ and $t > T$. This implies that the regularity of $\sigma'_t x$ is at least one higher than x. Therefore,

$$\sigma'_t(L^2[-a, 0]) \subseteq W^1_2[-a, 0].$$

This implies, along with Rellich's theorem ([16]) that σ'_t is a compact operator. Therefore, its adjoint σ_t is also compact.

Thus, we have proved $\sigma_t : X^q \to X^q$ is compact. When the pair (q, p) is not approximately coprime, we must take $X_A = \overline{\{A * \omega : \omega \in L^2[-a, 0]\}}$ as the state space. The property $\sigma_t = $ compact is, however, preserved under this change. This completes the proof of 3). $\blacksquare$

To complete the proof of Theorem 3.2, we need the following lemmas:

Lemma 4.2. *The spectrum $\sigma(F)$ (considered in $X_A = \overline{\text{im } A*}$) consists precisely of poles of $\hat{A}(s)$.*

For an indication of proof, see Appendix. $\blacksquare$

Clearly, this lemma, along with property 3) in the outline, completes the proof for 1) in Theorem 3.2. Furthermore, if $\hat{A}(s)$ belongs to H^∞, then the poles of $\hat{A}(s)$ should lie in the strict left-half plane, so that $\hat{A}(s) \in H^\infty$ implies $\sigma(F) \subseteq \{\lambda : \Re \lambda < -c, c > 0\}$, and hence exponential stability. The converse is obvious from 1) of Theorem 2.4. This establishes Theorem 3.2 for the single-input/single-output case. To generalize the result to the multivariable case, we need the following lemma.

Lemma 4.3. *Let $A = (a_{i_j})$ be a $p \times m$ matrix with $a_{i_j} \in \mathcal{R}$. Then the canonical realization of A is internally stable iff the same is true of each a_{i_j}.*

Proof: Omitted. $\blacksquare$

As a final remark, we here emphasize the point that the stability theorem obtained here does *not* depend on the existence of a coprime fractional representation. This is a great advantage over some other stability results, say the one for delay systems, in which stability is discussed in terms of a particular realization. In such a case, when there are infinitely many pole-zero cancellations, it is not trivial to deduce the stability of the canonical realization, while this problem does not arise here, even when there exists no coprime fractional representation.

Acknowledgment.

The author wishes to thank Foundation for C& C Promotion, Japan for the travel support. He also wishes to acknowledge that the work here has been supported in part by the Ministry of Education under Grant-in-Aid # 62302034 for Co-operative Research.

Appendix

Sketch of Proof of Lemma 4.2.

Recall that, by Fact 4.2, the spectrum of the system $\Sigma^{q,p}$ consists precisely of zeros of $\hat{q}(s)$. Therefore, if $A = q^{-1} * p$ and if there is no pole-zero cancellation between $\hat{q}(s)$ and $\hat{p}(s)$, then the conclusion follows.

So let us consider the case where there are (possibly infinitely many) pole-zero cancellations between $\hat{q}(s)$ and $\hat{p}(s)$. The complete proof is rather lengthy to be given here, so we only give an outline. First observe that q can be written as $q = \delta_{-a} * q_1$ $(a \geq 0)$ such that

$$\ell(q_1) := \{t \in \text{supp } q_1\} = 0.$$

According to Theorem 3.2 of [14], the space X^{q_1} is eigenfunction complete, and this space, denoted M, agrees with the space of eigenfunctions of X^q. It is not difficult to prove that

$$X^q = X^{\delta_{-a}} \oplus X^{q_1}.$$

Since $X^{\delta_{-a}}$ is irrelevant to the spectral properties, we assume without loss of generality that $a = 0$, i.e., $q = q_1$, so that X^q is eigenfunction complete.

The idea of the proof is to show that a complex number λ is a spectrum of F in X_A depending on whether or not the corresponding eigenvector $e^{\lambda t}$ remains to be in X_A, and this is governed by the condition if it remains to be a pole of $\hat{A}(s)$, i.e., if it is not canceled by the numerator $\hat{p}(s)$.

Take any $\lambda \in \mathbf{C}$. If $\hat{q}(\lambda) \neq 0$, then λ belongs to the resolvent set in $\Sigma^{q,p}$ (Facts 2.2). This clearly induces a resolvent operator in X_A, so that λ belongs also to the resolvent set in Σ_A.

So, suppose λ belongs to the spectrum of $\Sigma^{q,p}$ (but not necessarily so for Σ_A). It is known that the eigenspace M_λ in $\Sigma^{q,p}$ corresponding to λ is finite-dimensional, and consists of elements of form $\{e^{\lambda t}, \ldots, t^m e^{\lambda t}\}$ (Facts 2.2). Either λ is a pole of $\hat{A}(s)$ or not. When λ is a pole, we claim that $e^{\lambda t}$ belongs to X_A. Otherwise, the whole generalized eigenspace M_λ is not contained in X_A. Since the realization $X^{q,p}$ is always observable, this means that M_λ is not reachable. Therefore, there must be a pole-zero cancellation at λ. This means that we can cancel the common factor $s - \lambda$ both from $\hat{q}(s)$ and $\hat{p}(s)$. Repeating this argument, we would obtain a factorization $\hat{q}_1^{-1}\hat{p}_1$ in which $\hat{q}_1(\lambda) \neq 0$. But this contradicts our assumption that λ is a pole, so that $e^{\lambda t} \in X_A$. Since the eigenvector $e^{\lambda t}$ belongs to X_A, it is easy to show that λ is also an eigenvalue when considered in X_A.

It remains to show that if λ is not a pole, then it belongs to the resolvent set. This can be done by computing the resolvent operator as in [11]. ∎

REFERENCES

[1] J. A. Burns, T. L. Herdman and H. W. Stech, *Linear functional differential equations as semigroups on product spaces*, SIAM J. Math. Anal. **14** (1983), 98–116.

[2] F. M. Callier and J. Winkin, *Distributed system transfer functions of exponential order*, Int. J. Control **43** (1986), 1353–1373.

[3] C. A. Desoer and M. Vidyasagar, "Feedback Systems: Input/Output Properties," Academic Press, 1975.

[4] J. K. Hale, "Theory of Functional Differential Equations," Springer, 1977.

[5] C. A. Jacobson and C. N. Nett, *Linear state-space systems in infinite-dimensional space: the role and characterization of joint stabilizability/detectability*, IEEE Trans. Autom. Control **33** (1988), 541–549.

[6] A. Manitius, *Necessary and sufficient conditions of approximate controllability for general linear retarded systems*, SIAM J. Control & Optimiz. **19** (1981).

[7] W. Rudin, "Functional Analysis," McGraw-Hill, 1973.

[8] L. Schwartz, "Théorie des Distributions," 2me Edition, Hermann, 1966.

[9] F. Treves, "Topological Vector Spaces, Distributions and Kernels," Academic Press, 1967.

[10] Y. Yamamoto, *A note on linear input/output maps of bounded type*, IEEE Trans. Autom. Control **AC-29** (1984), 733–734.

[11] Y. Yamamoto, *Realization of pseudo-rational input/output maps and its spectral properties*, Mem. Fac. Kyoto Univ. **47** (1985), 221–239.

[12] Y. Yamamoto and S. Ueshima, *A new model for neutral delay-differential systems*, Int. J. Control **43** (1986), 465–472.

[13] Y. Yamamoto, *Pseudo-rational input/output maps and their realizations: a fractional representation approach to infinite-dimensional systems*, SIAM J. Control & Optimiz. **26** (1988).

[14] Y. Yamamoto, *Reachability of a class of infinite-dimensional linear systems: an external approach with applications to general neutral systems*, SIAM J. Control & Optimiz. (to appear).

[15] Y. Yamamoto and S. Hara, *Relationships between internal and external stability with applications to a servo problem*, IEEE Trans. Autom. Control **AC-33, No. 11** (1988).

[16] K. Yosida, "Functional Analysis," Springer, 1980.

[17] J. Zabczyk, *A note on C_0-semigroups*, Bull l'Acad. Pol. de Sc. Serie Math. **23** (1975), 895–898.

Yutaka Yamamoto
Department of Applied Systems Science
Faculty of Engineering
Kyoto University
Kyoto 606
Japan

International Series of
Numerical Mathematics, Vol. 91

425

Some remarks on open and closed loop stabilizability for infinite dimensional systems

Hans Zwart

Department of Applied Mathematics
University of Twente

Abstract. In this paper we study the stabilizability of the infinite dimensional system $\dot{x} = Ax + bu$, with a one-dimensional input operator. One of the main results that we shall give states that if this system is open-loop stabilizable i.e. $x(\cdot)$ decays exponentially to zero for some $u(\cdot)$, then the unstable part of the spectrum of A consists of only point spectrum with finite multiplicity. Furthermore we shall show that for a large class of systems there is no difference between open-loop and closed-loop stabilizability.

Keywords. Stabilizability, infinite dimensional systems.

1980 *Mathematics subject classifications*: 93C25, 93D99

1. Introduction

In this paper we shall study the stabilizability of a single input, infinite dimensional, linear and time-invariant system, which we represent as

(1.1)
$$\dot{x}(t) = Ax(t) + bu(t)$$
$$x(0) = x_0$$

where x_0 is an element of the Banach space $\mathcal{X}$, A is an infinitesimal generator of the semigroup $T(t)$ on $\mathcal{X}$, b is an element of $\mathcal{X}$, $b \neq 0$, and $u(\cdot)$ is locally square integrable function with values in $\mathbf{R}$, $(\mathcal{L}^2_{\text{loc}}([0, \infty); \mathbf{R}))$. By the solution of (1.1) we mean the mild solution, see Curtain and Pritchard [4]

(1.2)
$$x(t) = T(t)x_0 + \int_0^t T(t - s)bu(s)ds$$

and we denote such a system (1.1) and (1.2) by (A, b).

The stability and stabilizability of this system is one of the main topics in infinite-dimensional system theory, see e.g. [13], [16] and the references therein.

In the second section we shall recapulate recent results on closed loop stabilizability, meaning under what conditions on (A, b) does there exists a $F \in \mathcal{L}(\mathcal{X}, \mathbf{R})$ such that the solution of the closed loop system

(1.3)
$$\dot{x}(t) = (A + bF)x(t)$$
$$x(0) = x_0$$

converges to zero exponentially, for every $x_0 \in \mathcal{X}$.

We shall see that the closed loop stabilizability implies that in the unstable part of the spectrum of A there are *finitely* many eigenvalues with *finite* multiplicity.

In the third section we shall introduce a weaker concept of stabilizability, namely the concept of open-loop stabilizability, which states exponentially decay to zero of $x(t)$ for some input $u(t)$, depending on x_0. As in the closed loop case the solvability of this problem implies that the unstable part of the spectrum of A consists of only eigenvalues with finite multiplicity. In the last section we shall investigate the relation between open and closed loop stabilizability.

We remark that throughout this paper we only consider exponential decay of $x(\cdot)$, for other definitions of stability, see e.g. [1] and [13].

2. Closed-loop stabilizability

We shall start by giving a formal definitions of the concept of closed loop stabilizability.

Definition 2.1. *Closed loop stabilizability.*[1] *The system (A, b) is closed loop stabilizable if there exists a bounded operator F, such that the semigroup generated by $A + bF$, $T_F(t)$ is exponentially stable, i.e $\|T_F(t)\| \leq Me^{\beta t}; \beta < 0$.*

Theorem 2.2. *For the system (A, b) the following assertions are equivalent:*

  i) *The system is closed loop stabilizable.*
  ii) *There exists a $\alpha < 0$, such that the state space $\mathcal{X}$ can be decomposed as $\mathcal{X} = \mathcal{X}_u \oplus \mathcal{X}_s$, which are both $T(t)$-invariant, $\mathcal{X}_u$ is finite dimensional, $\sigma(A|_{\mathcal{X}_u}) \subset \mathbf{C}_{\alpha,+}$, $\|T(t)|_{\mathcal{X}_s}\| \leq Me^{\alpha t}$, and the system restricted to $\mathcal{X}_u$ is controllable.*

Remark 2.3: This theorem also holds if the input space is not one, but finite dimensional.

Remark 2.4: Theorem 2.2 reflects one of the most important results in the theory of infinite dimensional systems. The implication ii) to i) is known to hold for a long time, see Triggiani [17]. Of the other implication, it was known that the state space could be decomposed as $\mathcal{X} = \mathcal{X}_u \oplus \mathcal{X}_s$, where the spaces $\mathcal{X}_u$ and $\mathcal{X}_s$ are both $T(t)$-invariant, and the spectrum of A restricted to $\mathcal{X}_u$ consisted only of point spectrum, Kato [11, pp. 244-250]. However, the finite dimensionality of $\mathcal{X}_u$ was unknown and so was the decay of $T(t)|_{\mathcal{X}_s}$. The implication i) $\rightarrow$ ii) was proved independently by Desch and Schappacher [6], Jacobson and Nett [10] and Nefedov and Sholokhovich [12]. Desch and Schappacher were the first to publish their result, and their result is the most general one. It treats the Banach space case and a class of unbounded input operators. In [3], Curtain extended the result of [12] to a class of unbounded operators. Although [3] and [6] treat different classes of unbounded input operators, yet most examples will be in both the classes.

3. Open loop stabilizability

In the previous section we have seen that if our system operator A has infinitely many eigenvalues in the closed right halft-plane, then it is not stabilizable by a bounded feedback law. So we may ask the question if this system may be stabilized by another type of feedback. Another reason for introducing a weaker definition of stabilizability lies in the

---

[1] This notion is also called exponentially stabilizability of just stabilizability

following observation. Suppose that we can stabilize our system by some (crazy) control law. Then the control input that is applied to the system will no longer be a linear, time-invariant function of the state. To describe this weak form of stabilizability we need an "open loop" concept. We shall start with defining this concept of stabilizability.

Let $\mathcal{L}^2_{\alpha,-}([0,\infty); \mathcal{Z})$ denote all functions $[0,\infty)$ to a Banach space $\mathcal{Z}$, that satisfy $\int_0^\infty e^{-\alpha t} \|f(t)\|^2 dt < \infty$.

Definition 3.1. *Open loop stabilizability. The system (A, b) is open loop stabilizable if there exists a $\alpha < 0$ such that for every $x_0 \in \mathcal{X}$ there exists an input $u(\cdot) \in \mathcal{L}^2_{loc}([0,\infty); \mathbf{R})$ for which the mild solution (1.2) of (1.1) satisfies $x(\cdot) \in \mathcal{L}^2_{\alpha,-}([0,\infty); \mathcal{X})$.*

In this section we are interested in the properties of open-loop stabilizable systems (A, b). An natural question that one may pose is whether all systems are open-loop stabilizable. In theorem 3.5 we will answer this question negatively. The proof of this theroem will occupy the rest of this section.

In order to formulate the main theorem of this section we shall need some notation

$$(3.1) \quad \begin{cases} \mathbf{C}_{\alpha,+} := \{z \in \mathbf{C} | \mathcal{R}e\, z > \alpha\} \text{ for } \alpha \in \mathbf{R} \\ \mathbf{C}_{\alpha,-} := \{z \in \mathbf{C} | \mathcal{R}e\, z \leq \alpha\} \text{ for } \alpha \in \mathbf{R} \\ \sigma(A) \text{ is the spectrum of } A \\ \sigma_{\alpha,+}(A) = \sigma(A) \cap \mathbf{C}_{\alpha,+} \\ \sigma_{\alpha,-}(A) = \sigma(A) \cap \mathbf{C}_{\alpha,-} \\ \mathcal{H}(\mathbf{C}_{\beta,+}; \mathcal{Z}) \text{ denotes the set of all holomorphic functions on } \mathbf{C}_{\alpha,+} \\ \text{with values in a Banach space } \mathcal{Z}. \\ \mathcal{H}_-(\mathcal{Z}) \text{ will denote the set of all functions } f(\cdot) \text{ such that} \\ f \in \mathcal{H}(\mathbf{C}_{\beta,+}; \mathcal{Z}) \text{ for some } \beta < 0. \end{cases}$$

Before we can state our theorem we need to define two operators.

Lemma 3.2. *Let Γ_λ be a simpled closed curve in $\mathbf{C}_{\beta,+}$ with only one point of $\sigma(A)$, λ, in its interior and no points of $\sigma(A)$ on Γ_λ. By Ω_λ we shall denote all functions which are holomorphic in the interior of Γ_λ and continuous on Γ_λ. We now have the following standard result.*

If we impose on Ω_λ the following norm $\|\cdot\|_{\Omega_\lambda}$; $\|\omega(\cdot)\|_{\Omega_\lambda} = \sup_{s \in int \Gamma_\lambda} \|\omega(s)\| = \sup_{s \in \Gamma_\lambda} \|\omega(s)\|$, then Ω_λ is a Banach algebra.

Definition 3.3. *$G_\lambda(\cdot)$. Let Ω_λ be the Banach space as defined in lemma 3.2. Then $G_\lambda(\cdot)$ will denote the following operator form Ω_λ to $\mathcal{X}$.*

$$G_\lambda(\omega) = \frac{1}{2\pi i} \int_{\Gamma_\lambda} (s - A)^{-1} b\omega(s) ds.$$

Definition 3.4. *P_λ. Let Γ_λ be a subset of $\mathbf{C}$ as defined in lemma 3.2. Then P_λ will denote the projection operator on the eigenspace of the, by Γ_λ enclosed, spectral value λ, i.e.*

$$P_\lambda x = \frac{1}{2\pi i} \int_{\Gamma_\lambda} (s - A)^{-1} x ds.$$

Using the above made definitions and notation we can formulate our main result precisely.

Theorem 3.5. *Assume that the system (A, b) is open loop stabilizable and let α be the negative number from definition 3.1, then*

> i) *for every x_0 in $\mathcal{X}$ there exists $\xi(\cdot) \in \mathcal{H}(\mathbf{C}_{\alpha,+}; D(A))$ and $\omega(\cdot) \in \mathcal{H}(\mathbf{C}_{\alpha,+}; \mathbf{C})$ such that on $\mathbf{C}_{\alpha,+}$ the following representation holds*

$$(3.2) \qquad\qquad x_0 = (s - A)\xi(s) - b\omega(s).$$

> ii) *$\sigma_{\beta,+}(A)$ contains no finite accumulation point for all $\beta > \alpha$.*
> iii) *In the half-plane $\mathbf{C}_{\alpha,+}$ the spectrum of the operator A consists of only point spectrum with finite multiplicity.*
> iv) *The representation (3.2) satisfies*

$$P_\lambda x_0 = -G_\lambda(\omega) \text{ for all } \lambda \text{ in } \sigma_{\alpha,+}(A).$$

> v) *If $(\xi_1, \omega_1) \in \mathcal{H}(\mathbf{C}_{\gamma,+}, D(A)) \oplus \mathcal{H}((\mathbf{C}_{\gamma,+}, \mathbf{C})$ also satisfies the relation (3.2) on $\mathbf{C}_{\gamma,+}$ for some γ, then $G_\lambda(\omega_1) = G_\lambda(\omega)$ for all $\lambda \in \sigma_{\alpha,+}(A) \cap \sigma_{\gamma,+}(A)$.*

Remark: The representation (3.2) is called a (ξ, ω)- representation of x_0, and was introduced for the finite dimensional case by Hautus in [9].

Under stronger assumptions a similar theorem was found by Fattorini [8] and Voigt [18].

Before we shall prove this theorem we shall try to explain what this result tell us. So from ii) and iii) we have that in every compact subset of $\mathbf{C}_{\alpha,+}$ there are only finitely many points of the spectrum of A, counted with multiplicity.

Furthermore properties ii) and iii) give that even with arbitrary inputs one cannot open loop stabilize a system which has an accumulation point in the unstable part of its spectrum and one cannot remove the residual part, nor the continuous part of the spectrum nor eigenvalues with infinite multiplicity. So if A is the identity operator on an infinite dimensional Banach space, then it is not open-loop stabilizable. Finally iv) and v) gives an uniqueness result, i.e. if a $\omega(s)$ stabilizes the initial condition x_0, then this ω is uniquely determined at the unstable spectral values.

The proof of theorem 3.5 is based on the concept of (ξ, ω)- representation, see (3.2). This representation is essentially the Laplace transform of (1.2) and it has as advantage that instead of working with functions from $\mathcal{L}^2_{\alpha,-}([0, \infty); \mathcal{X})$ or $\mathcal{L}^2_{\text{loc}}([0, \infty); \mathbf{R})$ one works with functions which are holomorphic on some right half plane. Since a lot is known about holomorphic functions and they have very nice properties we can prove strong results using this (ξ, ω) representation, (see also Zwart [19] and [20]). Before we can prove theorem 3.5 we need some technical lemmas.

Lemma 3.6. *Suppose that $h(\cdot)$ satisfies the following convolution equation on $[0, \infty)$.*

$$f(t) = \int_0^t k(t - s)h(s)ds,$$

where $k(t)$ is continuous and has a continuous derivative on $[0, \infty)$, $k(0) = 1$, and $|e^{\gamma \cdot} k(\cdot)|$, $|e^{\gamma \cdot} k'(\cdot)|$, $|e^{\gamma \cdot} f(\cdot)|$ are uniformly bounded for some $\gamma \in \mathbf{R}$. Then h is Laplace transformable (in the sense of distributions), the Laplace transform of h, $L(h)$ is an element of $\mathcal{H}(\mathbf{C}_{\beta,+}; \mathbf{R})$ and $L(f) = L(k) * L(h)$ on $\mathbf{C}_{\beta;+}$, for some β in $\mathbf{R}$.

Proof: See Doetsch [7, Ch. 40].

Lemma 3.7. Let the system (A, b) be open loop stabilizable, and let α be the number defined in definition 3.1. Then for every x_0 in $\mathcal{X}$ there exist a $\xi(\cdot) \in \mathcal{H}(\mathbf{C}_{\alpha,+}; D(A))$ and a $\omega(\cdot) \in \mathcal{H}(\mathbf{C}_{\alpha,+}; \mathbf{C})$ such that

$$(3.3) \qquad\qquad x_0 = (s - A)\xi(s) - b\omega(s); \quad s \in \mathbf{C}_{\alpha,+}.$$

Proof: By $\mathcal{X}'$ we shall denote the dual space of $\mathcal{X}$ and A' will denote the dual operator of A with domain $D(A')$.

Let $y_b \in D(A') \subset \mathcal{X}'$ be such that $\langle y_b, b \rangle = 1$. Operating y_b on $x(t)$ in (1.2) we obtain

$$(3.4) \qquad\qquad \langle y_b, x(t) \rangle = \langle y_b, T(t)x_0 \rangle + \int_0^t \langle y_b, T(t-s)b\rangle u(s)ds.$$

By the properties of y_b and the assumptions on $x(t)$ we have from lemma 3.6 that $u(\cdot)$ is Laplace transformable and $\omega(\cdot) := L(u)$ is holomorphic in a right half-plane. So we can take the Laplace transform of equation (1.2). If we do this, then we obtain

$$(3.5) \qquad\qquad \xi(s) = (s - A)^{-1}x_0 + (s - A)^{-1}b\omega(s),$$

where $\xi(\cdot)$ is the Laplace transform of $x(t)$ and (3.5) is valid in some right half-plane. Rewriting (3.5) gives

$$(3.6) \qquad\qquad\qquad x_0 = (s - A)\xi(s) - b\omega(s).$$

Since $\int_0^\infty e^{-\alpha t}\|x(t)\|^2 dt < \infty$, we have that $\xi(\cdot)$ is holomorphic on $\mathbf{C}_{\alpha,+}$, [11] and

$$(3.7) \quad \langle y_b, x_0 \rangle = \langle y_b, (s - A)\xi(s) \rangle - \langle y_b, b\rangle\omega(s) = \langle y_b, s\xi(s)\rangle - \langle A'y_b, \xi(s)\rangle - \omega(s)$$

implies that $\omega(\cdot)$ has a holomorphic continuation on $\mathbf{C}_{\alpha,+}$. Since this continuation is unique we have proved the lemma. ∎

Before we can prove our main result we need some further properties of the operators P_λ and G_λ as defined by definitions 3.3 and 3.4; we have the following result.

Lemma 3.8. The operator G_λ in definition 3.3 has the following properties:
  i) G_λ is compact as an operator from Ω_λ to $\mathcal{X}$, and
  ii) the range of G_λ is contained in the range of P_λ.

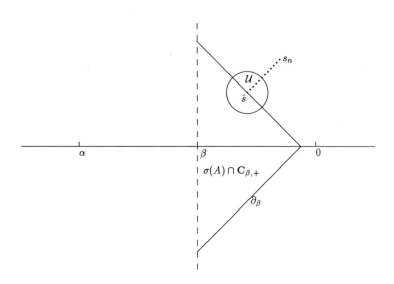

Figure 1.

Proof: The proof of part i) is a simple adaptation of the proof of Proposition 3.2 of Curtain and Pritchard [4] and will be omitted here.

The proof of part ii) can be found basically in Rudin [15]. ∎

With these definitions and technical lemmas we can now prove theorem 3.5.

Proof of Theorem 3.5: Part i) has already been proved in lemma 3.7. We shall now prove assertion ii).

Let β be an arbitrary real number larger than α, where α is the strictly negative number defined in definition 3.1. We shall begin by showing that the spectrum of A contains no finite accumulation point in $\mathbf{C}_{\beta,+} := \{s \in \mathbf{C} | \mathcal{R}e\, s > \beta\}$.

With $x_0 = b$ in equation (3.2) we have on $\mathbf{C}_{\alpha,+}$ and hence $\mathbf{C}_{\beta,+}$

$$(3.8) \qquad\qquad b = (s - A)\xi_b(s) - b\omega_b(s).$$

We show that if s is an element of the spectrum of A with real part larger than β, then $\omega_b(s) = -1$. We prove this first for elements on the boundary.

Let ∂_β denote the boundary of $\sigma(A) \cap \mathbf{C}_{\beta,+}$, in $\mathbf{C}_{\beta,+}$, and let $\hat{s}$ be an element of ∂_β. Then there exists a sequence $\{s_n\}$, $\{s_n\} \subset \rho(A)$, such that $\mathcal{R}e\, s_n > \beta$ and s_n converges to $\hat{s}$.

As mentioned above we claim that $\omega_b(\hat{s}) = -1$. For if this were not so, then by the analyticity of $\omega_b(\cdot)$ on $\mathbf{C}_{\alpha,+}$, there would exist a neighbourhood, $\mathcal{U}$, of $\hat{s}$ that is contained in $\mathbf{C}_{\alpha,+}$ such that $\omega_b(s) \neq -1$, for all s in $\mathcal{U}$. Thus pictorially we have the following situation.

Since for $s \in \mathcal{U}$, $\omega_b(s)$ is unequal to minus one, from (3.8) we have on $\mathcal{U}$ that

$$(3.9) \qquad b = (s - A)\frac{\xi_b(s)}{1 + \omega_b(s)}.$$

Let x_0 be an arbitrary element of $\mathcal{X}$, then from lemma 3.7 we have that x_0 has the following decomposition on $\mathbf{C}_{\alpha,+}$

$$x_0 = (s - A)\xi(s) - b\omega(s); \qquad x \in \mathbf{C}_{\alpha,+}$$

with $\xi(\cdot) \in \mathcal{H}(\mathbf{C}_{\alpha,+}; D(A))$ and $\omega(\cdot) \in \mathcal{H}(\mathbf{C}_{\alpha,+}; \mathbf{C})$. Combining this result with (3.9) gives for $s \in \mathcal{U} \cap \mathbf{C}_{\alpha,+}$ the representation

$$(3.10) \qquad x_0 = (s - A)\left\{\xi(s) - \frac{\xi_b(s)}{1 + \omega_b(s)}\omega(s)\right\}.$$

Since $\{s_n\}$ is contained in the resolvent set of A and in $\mathbf{C}_{\alpha,+}$, we have that

$$(3.11) \qquad (s_n - A)^{-1}x_0 = \xi(s_n) - \frac{\xi_b(s_n)}{1 + \omega_b(s_n)}\omega(s_n).$$

Since s_n converges to $\hat{s}$ we have by the analyticity of $\xi(\cdot)$, $\xi_b(\cdot)$, $\omega(\cdot)$ and $\omega_b(\cdot)$ on $\mathbf{C}_{\alpha,+}$ the right hand side of (3.11) converges. Thus for every x_0 in $\mathcal{X}$, the limit of $(s_n - A)^{-1}x_0$ exists as $n \to \infty$ and this implies that $\hat{s} \in \rho(A)$, Kato [11; p. 174], providing the contradiction.

Thus we have proved that if $\hat{s}$ is an element of ∂_β, then $\omega_b(\hat{s}) = -1$. Suppose that the set ∂_β has a finite accumulation point which is contained in the open set $\mathbf{C}_{\beta,+}$, then by the analyticity of $\omega_b(\cdot)$ on $\mathbf{C}_{\alpha,+}$. However this is not the Laplace transform of a $\mathcal{L}^2_{\text{loc}}[0, \infty)$ function, and thus the boundary of $\mathbf{C}_{\beta,+} \cap \sigma(A)$ cannot contain a finite accumulation point in $\mathbf{C}_{\beta,+}$. By standard topology arguments (e.g. see Armstrong [2]) we can conclude that ∂_β is equal to $\mathbf{C}_{\beta,+} \cap \sigma(A)$. Since β is an arbitrary real number larger than α, we have proved the second assertion of theorem 3.5, i.e. for all $\beta > \alpha$, $\mathbf{C}_{\beta,+}$ contains no finite accumulation point.

Now we shall prove the fourth assertion of theorem 3.5. So let Γ_λ be a simple closed curve in $\mathbf{C}_{\alpha,+}$ with only one point, λ, of $\sigma_{\alpha,+}(A)$ in its iterior and no points of $\sigma_{\alpha,+}(A)$ on Γ_λ, and $G_\lambda(\cdot)$ the operator from all bounded holomorphic functions in Γ_λ to $\mathcal{X}$ defined by

$$(3.12) \qquad G_\lambda(\omega) = \frac{1}{2\pi i}\int_{\Gamma_\lambda}(s - A)^{-1}b\omega(s)ds.$$

Furthermore P_λ is the following projection operator on $\mathcal{X}$.

$$(3.13) \qquad P_\lambda x = \frac{1}{2\pi i}\int_{\Gamma_\lambda}(s - A)^{-1}xds.$$

From lemma 3.8 we have that the range of G_λ is contained in $P_\lambda\mathcal{X}$. Here we shall show that the ranges of both operators are equal if the system (A, b) is open loop stabilizable.

Let x be an element of $P_\lambda \mathcal{X}$, then with lemma 3.7 we have that

$$
x = P_\lambda x = \frac{1}{2\pi i} \int_{\Gamma_\lambda} (s - A)^{-1} x ds
$$

(3.14)
$$
= \frac{1}{2\pi i} \int_{\Gamma_\lambda} (s - A)^{-1} \left\{ (s - A)\xi(s) - b\omega(s) \right\} ds
$$

$$
= \frac{1}{2\pi i} \int_{\Gamma_\lambda} \left\{ \xi(s) - (s - A)^{-1} b\omega(s) \right\} ds
$$

$$
= -G_\lambda(\omega), \text{ since } \xi(\cdot) \in \mathcal{H}(\mathbf{C}_{\alpha,+}; D(A)).
$$

So we have proved the fourth assertion.

Now the third assertion will follow easily. As a consequence of assertion four we have the range of G_λ is $P_\lambda \mathcal{X}$. Since P_λ is a projection we have that the range of this operator is closed. Thus we have that G_λ is a compact opertor (lemma 3.8) with closed range, and this implies that its range is finite dimensional (see Kato [11]). So $P_\lambda \mathcal{X}$ is finite dimensional and this implies that λ is an eigenvalues of A with finite multiplicity. So we have proved that for alle $\beta > \alpha$, $\sigma_{\beta,+}(A)$ is contained in the point of spectrum of A and the multiplicity of these poles is finite.

It remains to prove the last assertion, but this is easily proved by using equation (3.14). ■

4. The relation between open-loop and closed loop stabilizability

In this section we shall investigate the relation between open-loop and closed-loop stabilizability. If the state space is finite dimensinal, then it is well known that open-loop stabilizability is equivalent to closed-loop stabilizability. A natural question to pose is whether the same holds for infinite dimensional systems. The next theorem will affirm this question for a large class of systems on a Hilbert space.

Theorem 4.1. *Let $\mathcal{X}$ be a Hilbert space, and assume that the system (A, b) is open-loop stabilizable. Assume furthermore that for some $\beta < 0$ and some $\rho > 0$ the following supremum is finite, $\sup\{\|(s - A)^{-1}\|$, where $s \in \mathbf{C}_{\beta,+} \cap \{s : |s| > \rho\}\}$. Then open-loop stabilizability is equivalent to closed-loop stabilizability.*

Proof: Assume that the system (A, b) is open-loop stabilizable. Let $\gamma < 0$ be larger than β and α (see definition 3.1). Since (A, b) is open-loop stabilizable, and since the half circle $\mathbf{C}_{\beta,+} \cap \{s : |s| > \rho\}$ is contained in the resolvent set of A, we have from theorem 3.5 that there are only finitely many eigenvalues with finite multiplicity in the half-plane $\mathbf{C}_{\gamma,+}$. So we may decompose that state space $\mathcal{X}$ as $\mathcal{X} = \mathcal{X}_u \oplus \mathcal{X}_s$, where $\mathcal{X}_u$ is the span of all (generalized) eigenvectors with corresponding eigenvalue in $\mathbf{C}_{\gamma,+}$. So $\mathcal{X}_u$ is finite dimensional and $(\lambda - A)|_{\mathcal{X}_s}$ is invertable for any $\lambda \in \mathbf{C}_{\gamma,+}$. This together with the assumption from the theorem implies that $\sup_{s \in \mathbf{C}_{\gamma,+}} \|(s - A)^{-1}|_{\mathcal{X}_s}\|$ is finite, and by Prüss [14] we obtain that $\|T(t)|_{\mathcal{X}_s}\| \leq M e^{\gamma t}$. The controllability of the system restricted to $\mathcal{X}_u$ follows easily from the (ξ, ω)-representation, (3.2). So by theorem 2.2 we have that the system (A, b) is closed loop stabilizable. ■

Systems which satisfies the conditions in theorem 4.1 are e.g. systems with a bounded generator, systems where A is a generator of an analytic semigroups, a large class of delay systems. These are all systems of practical importance. The most importance class of systems which do not satisfy the conditions in theorem 4.1 are systems described by hyperbolic partial differential equations, e.g. undamped beams. The relation between open-loop stabilizability and the existence of a (unbounded) feedback law for general systems is the subject of Zwart [21].

The results as stated in this paper are related to the following well-known result of Datko [5].

Fact 4.2. Let $\mathcal{X}$ be a Hilbert space, then closed loop stabilizability is equivalent with open loop stabilizability with $L^2([0,\infty))$ as the class of input functions.

So the class of input functions is important, however, one can prove that if we extend the class of input functions in the definition of open-loop stabilizability further, say to distributions, then non of the results as stated in this paper changes, Zwart [22]. Due to space limitations we can not present the proof of this result.

REFERENCES

[1] Arendt, W. and C.J.K. Batty, *Tauberian Theorems and Stability of One-Parameter Semigroups*, Trans Am. Math. Soc. **306** (1988).

[2] Armstrong, M.A., *Basic Toplogy*, in "Undergraduate Text in Mathematics," Springer Verlag, 1983.

[3] Curtain, R.F., *Equivalence of Input-Output Stability and Exponential Stability for Infinite Dimensional Systems*, Math. Systems Theory **21** (1988), 19–48.

[4] Curtain, R.F. and A.J. Pritchard, *Infinite Dimensional Linear System Theory*, in "Lecture Notes in Control and Information Sciences," Springer Verlag, 1978.

[5] Datko, R., *A Linear Control Problem in an Abstract Hilbert Space*, J. Diff. Eq. **9** (1971), 346–359.

[6] Desch, W. and W. Schappacher, *Spectral Properties of Finite-Dimensional Perturbed Linear Semigroups*, J. Diff. Eq. **59** (1985), 80–102.

[7] Doetsch, G., "Introduction to the Theory and Application to the Laplace Transform," Springer Verlag, 1974.

[8] Fattorini, H.O., *Exact Controllability of Linear Systems in Infinite Dimensional Spaces*, in "Partial and Differential Equations and Related Topics," (Program, Tulane Univ., New Orleans, L.A., 1974), Lecture Notes in Math., Vol. 446, Springer, Berlin, 1975, pp. 166–183.

[9] Hautus, M.L.J., *(A, B)-invariant and stabilizability subspace, a frequency domain description*, Automatica **16** (1980), 703–707.

[10] Jacobson, C.A. and C.N. Nett, *Linear State-Space Systems in Infinite Dimensional Space: The Role and Characterization of joint Stabilizability/Detectability*, IEEE Trans. on AC **33** (1988), 541–549.

[11] Kato, T., *Perturbation Theory for Linear Operators*, in "Grundlehren der Mathematischen Wissenschaften 132," Springer Verlag, 1984.

[12] Nefedov, S.A. and F.A. Sholokhovisch, *A Criterium for the Stabilizability of Dynamical Systems with Finite-Dimensional Input*, Differentsial'nye Uraveneniya **22** (1986), 163–166. New York, Plenum

[13] Pritchard, A.J. and J. Zabczyk, *Stability and Stabilizability of Infinite Dimensional Systems*, SIAM Review **23** (1971), 25–52.

[14] Prüss, J., *On the Spectrum of C_0-semigroups*, Trans. Am. Math. Soc. **284** (1984).

[15] Rudin, W., "Functional Analysis," McGraw-Hill, New York.

[16] Russell, D.L., *Controllability and Stabilizability Theory for Linear Partial Differential Equations: recent progress and open problems*, SIAM Review **20** (1978), 639–739.

[17] Triggiani, R., *On the Stabilizability Problem in Banach Space*, J. Math. An. and Appl. **52** (1975), 383–403.

[18] Voigt, J., *A Perturbation Theorem for the Essential Spectral Radius of Strongly Continuous Semigroups*, Monatshefte für Mathematik **90** (1980), 153–161.

[19] Zwart, H.J., *Equivalence between Open loop and Closed loop invariance for Infinite Dimensional Systems: A Frequency Domain Approach*, SIAM Journal on Control and Optimization **26** (1988), 1175–1199.

[20] Zwart, H.J., *Geometric Theory for Infinite Dimensional Systems*, in "Lecture Notes in Control and Information Science," Vol 105, 1988.

[21] Zwart, H.J., *Open Loop Stabilizability, a research note*, Proceedings of the IFAC conference on Distributed Parameter systems, Perpignan, France, 26/29 June 1989, 505-509.

[22] Zwart, H.J., *On stability of Infinite Dimensional Systems*, in preparation, 1989.

Hans Zwart
Department of Applied Mathematics
University of Twente
P.O. Box 217
NL-7500 AE Enschede
The Netherlands